Gottfried Wilhelm Leibniz (1646-1716) in front of the lecture hall (*left*) of the IAMP Congress. Leibniz was a student at the University of Leipzig. The university's main building may be seen in the background.

K. Schmüdgen (Ed.)

Mathematical Physics X

Proceedings of the Xth Congress
on Mathematical Physics
Held at Leipzig, Germany, 30 July - 9 August 1991

M∩Φ

Springer-Verlag
Berlin Heidelberg New York
London Paris Tokyo
Hong Kong Barcelona
Budapest

Prof. Dr. Konrad Schmüdgen
Fachbereich Mathematik
Universität Leipzig
Augustusplatz 10
O-7010 Leipzig, FRG

ISBN-13: 978-3-642-77305-1 e-ISBN-13: 978-3-642-77303-7
DOI: 10.1007/978-3-642-77303-7

Softcover reprint of the hardcover 1st edition 1992

Typesetting: Camera ready by author
55/3140-543210 - Printed on acid-free paper

Preface

This volume contains the proceedings of the Xth Congress of the International Association of Mathematical Physics, held at the University of Leipzig from 30 July until 9 August 1991. There were more than 400 participants, from 29 countries, making it a truly international gathering.

The congress had the support of the Deutsche Forschungsgemeinschaft, the European Economic Community, the International Association of Mathematical Physics, the International Mathematical Union and the International Union of Pure and Applied Physics. There were also sponsors from industry and commerce: ATC Mann, Deutsche Bank AG, Miele & Cie GmbH, NEC Deutschland GmbH, Rank Xerox, Siemens AG and Stiftungsfonds IBM Deutschland. On behalf of the congress participants and the members of the International Association of Mathematical Physics, I would like to thank all these organisations for their very generous support.

The congress took place under the auspices of the Ministerpräsident des Freistaates Sachsen, K. Biedenkopf.

The conference began with an address by A. Uhlmann, Chairman of the Local Organizing Committee. This was followed by speeches of welcome from F. Magirius, City President of Leipzig; C. Weiss, Rector of the University of Leipzig; and A. Jaffe, President of the International Association of Mathematical Physics.

The main part of the scientific program consisted of 18 plenary lectures and 13 special sessions. The plenary speakers were invited by the scientific Organizing Committee, chaired by W. Thirring (Vienna), to present reviews of their respective fields, accessible to all participants. The committee also chose the session organizers. The special sessions comprised both invited lectures in topical sessions and shorter contributions, invited and submitted, in poster sessions. The session organizers were responsible for the program and are also responsible for the choice of contributions to the proceedings from their sessions.

Two seminars were organized during the conference, one on "Geometric phase" by A. Bohm (Austin) and one on "Energy levels and moments of the Schrödinger equation and relativistic generalizations" by A. Martin (Geneva). There was also a particularly lively round table discussion "On the foundations of quantum mechanics" directed by A. Wightman (Princeton), with contributions from S. Goldstein (Rutgers), R. Haag (Hamburg), R. Omnés (Paris), H. Primas (Zurich) and W. Zurek (Los Alamos).

Additionally, a free poster session was held on Sunday. A list of these posters and their authors is included in these proceedings.

As at all such congresses, a great deal of work was done by many people. In particular, the congress could not have taken place without the dedicated work of members of the Mathematical Physics groups at the Department of Mathematics and the Department of Physics. Their work began more than one year ago, and culminated during the congress itself. Special thanks must also go to the staff of the Naturwissenschaftlich-Theoretisches Zentrum and to K. Scharnhorst, Secretary INTSEM.

Finally, I would like to express my personal thanks to R. Becker, J. Dittmann and R. Schildt for their assistance in preparing these proceedings, in particular, for their help in formatting, proofreading and even retyping manuscripts.

Leipzig,
November 1991

K. Schmüdgen

Contents

Part II Session Lectures

Section 1: Analysis on Manifolds and Classical Mechanics

Section 2: Quantum Groups and Non-Commutative Differential Geometry

Section 3: Chaotic Quantum Systems

Section 4: Equilibrium Statistical Mechanics

Section 5: Classical Dynamical Systems and Random Perturbations

Section 6: Disordered Systems

Section 7: Nonequilibrium Statistical Mechanics

Section 8: General Theory of Quantized Fields

Section 9: Nonrelativistic Quantum Mechanics (Stationary)

Section 10: Special Systems in Quantum Field Theory

Section 11: Scattering Theory, Inverse Problems

Section 12: Conformal and Topological Field Theory

Section 13: General Relativity and Classical Field Theory

Poster Session and Program

Committees

Scientific Organizing Committee

W. Thirring (Vienna, Chairman), D. Buchholz (Hamburg), R.L. Dobrushin (Moscow), J.R. Klauder (Gainesville), B. Simon (Pasadena), Ya.G. Sinai (Moscow), A. Trautman (Warsaw), A.S. Wightman (Princeton)
G. Lassner, A. Uhlmann (Local members)

Scientific Advisory Committee

A. Ashtekar (Syracuse), R.J. Baxter (Canberra), J. Bellisard (Marseille), R. Benguria (Santiago), Ph. Blanchard (Bielefeld), J. Bricmont (Louvain-la-Neuve), D. Brydges (Charlottesville), J. Chayes (Los Angeles), P. Collet (Palaiseau), J. Cuntz (Heidelberg), I. Daubechies (Murray Hill), A. Degasperis (Rome), R. de la Llave (Austin), V. Enss (Aachen), J. Ford (Atlanta), A. Galindo (Madrid), F. Gesztesy (Columbia), L.P. Horwitz (Tel Aviv), W. Hunziker (Zurich), A. Martin (Geneva), A.A. Migdal (Moscow), J. Niederle (Prague), G. Roepstorff (Aachen), S. Ruijsenaars (Amsterdam), H. Siedentop (Birmingham), F. Strocchi (Pisa), A.A. Slavnov (Moscow), H. Spohn (Munich), D. Szász (Budapest), S.L. Woronowicz (Warsaw), Y. Ingvason (Reykjavik), D. Zwanziger (New York)

For the Local Organizing Committee:

B. Geyer, A. Uhlmann (Chairman)

List of Participants

Alberti, Peter M.	Universität Leipzig, Germany
Alicki, Robert	University of Gdansk, Poland
Anderson, Malcolm	Australian National Univ., Canberra, Australia
Aneva, Boyka	Bulgarian Academy of Sciences, Sofia, Bulgaria
Antoine, Jean-Pierre	Univ. Catholique Louvain-la-Neuve, Belgium
Apfeldorf, Karyn	University of California, Berkeley, USA
Araki, Huzihiro	RIMS, Kyoto University, Japan
Arlt, Dietmar	Universität Bonn, Germany
Avron, Joseph	Technion-Israel Inst.of Techn., Haifa, Israel
Bach, Volker	ETH Zürich-Hönggerberg, Switzerland
Balasin, Herbert	TU Wien, Austria
Barashenkov, Igor	JINR Dubna, Moscow, Soviet Union
Barth, Joachim	Universität Freiburg i. Br., Germany
Barthel, Werner	TU Chemnitz, Germany
Baumann, Klaus	Universität Göttingen, Germany
Baumgartner, Bernhard	Universität Wien, Austria
Becker, Ralf	Universität Leipzig, Germany
Beckers, Jules	University of Liege, Belgium
Beggs, Edwin	University College Swansea, United Kingdom
Behn, Ulrich	Universität Leipzig, Germany
Beiglböck, Wolf	Springer - Verlag, Heidelberg, Germany
Benatti, Fabio	University of Trieste, Italia
Benguria, Rafael D.	Cath. University of Chile, Santiago, Chile
Berndl, Karin	Universität München, Germany
Blanchard, Philip	Universität Bielefeld, Germany
Blank, Michail	USSR Academy of Sciences, Moscow, Soviet Union
Bleher, Pavel	Tel Aviv University, Ramat Aviv, Israel
Boenkost, Wolfram	Universität Frankfurt a.M., Germany
Böttger, Wulf	RWTH Aachen, Germany
Bogolubsky, Igor L.	JINR Dubna, Moscow, Soviet Union
Bohm, Arno	University of Texas, Austin, USA
Boldrighini, Carlo	University of Camerino, Italia
Bongaarts, Peter J.M.	Lorentz Institute, Leiden, Netherlands
Borchers, Hans-Jürgen	Universität Göttingen, Germany
Bordag, Michael	Universität Leipzig, Germany
Bordemann, Martin	Universität Freiburg i. Br., Germany
Borgs, Christian	FU Berlin, Germany
Borowiec, Andrej	University of Wroclaw, Poland
Bose, Subir Kumar	University of Gorakhpur, India
Boseck, Helmut	EMAU Greifswald, Germany
Boutet de Monvel, Anne	Universite Paris VII, France
Boya, Luis J.	University of Zaragoza, Spain
Bramson, Maury	University of Wisconsin, Madison, USA
Breitenecker, Manfred	Universität Wien, Austria
Bros, Jacques	Centre d'Etudes de Saclay, Gif-sur-Yvette, France
Brüning, Erwin	Univ. of Cape Town, Rondebusch, South Africa
Brunetti, Romeo	University "Federico II", Napoli, Italia
Buchholz, Detlev	Universität Hamburg, Germany
Bunimovich, Leonid A.	USSR Academy of Sciences, Moscow, Soviet Union
Busch, Paul	Universität Köln, Germany

Bux, Kai-Uwe	J.W.Goethe-Univ. Frankfurt a.M., Germany
Campanino, Massimo	University of Bologna, Italia
Cangemi, Daniel	University of Lausanne, Switzerland
Cantoni, Vittorio	University of Milano, Italia
Carfora, Mauro	University of Pavia, Italia
Casati, Paolo	University of Torino, Italia
Cheremshantsev, Sergei	Steklov Math. Institute, Leningrad, Soviet Union
Chirikov, Boris V.	Inst. of Nucl. Physics, Novosibirsk, Soviet Union
Chrusciel, P.T.	Australian National Univ., Canberra, Australia
Combescure, Monique	University of Paris-Sud, Orsay, France
Conlon, Joseph	University of Michigan, Ann Arbor, USA
Connes, Alain	IHES Bures-sur-Yvette, France
Contucci, Pierluigi	School for Advanced Studies, Trieste, Italia
Cook, Joanna	Forschungszentrum Jülich, Germany
Corbett, John	Macquarie University, North Ryde, Australia
Crell, Bernd	Universität Leipzig, Germany
Cuntz, Joachim	Universität Heidelberg, Germany
Dabrowski, Ludwik	School of Advanced Studies, Trieste, Italia
Dankel, Thad	University of North Carolina, Wilmington, USA
Davies, Edward Brian	King's College London, United Kingdom
Debergh, Nathalie	University of Liege, Belgium
Demkin, Paul	Vilnius University, Soviet Union
Demuth, Michael	Karl-Weierstrass-Institut, Berlin, Germany
Deuschel, Jean-Dominique	ETH Zürich, Switzerland
Devchand, Chandrashekar	JINR Dubna, Moscow, Soviet Union
Dimock, Jonathan	State University of New York, Buffalo, USA
Dittmann, Jochen	Universität Leipzig, Germany
Dittrich, Peter	Karl-Weierstrass-Institut, Berlin, Germany
Dittrich, Jaroslav	Czechoslovak Acad. of Sci., Rez, Czechoslovakia
Dobrev, Vladimir K.	Bulgarian Acad. of Sciences, Sofia, Bulgaria
Dobrushin, R.L.	USSR Acad. of Sciences, Moscow, Soviet Union
Doebner, H.-D.	Universität Clausthal-Zellerfeld, Germany
Doliwa, Adam	University of Warsaw, Poland
Dostoglou, Stamatis	Univ. of Warwick, Coventry, United Kingdom
Dubin, Daniel A.	The Open Univ., Milton Keynes, United Kingdom
Dürr, Detlef	Universität München, Germany
Eckmann, J.-P.	University of Geneva, Switzerland
Eizenberg, Alex	The Hebrew University, Jerusalem, Israel
Elworthy, Kenneth David	Univ. of Warwick, Coventry, United Kingdom
Emch, Gerard	University of Florida, Gainesville, USA
Emmrich, Claudio	Universität Freiburg i. Br., Germany
Englisch, Harald	Universität Leipzig, Germany
Enss, Volker	RWTH Aachen, Germany
Exner, Pavel	Czechoslovak Acad. of Sci., Rez, Czechoslovakia
Ezawa, Hiroshi	Gakushuin University of Tokyo, Japan
Faddeev, Ludwig	Steklov Math. Institute, Leningrad, Soviet Union
Fannes, Mark	KU Leuven, Belgium
Fernandez, Roberto	ETH Zürich-Hönggerberg, Switzerland
Figotin, Aleksandr	University of North Carolina, Charlotte, USA
Filk, Thomas	Universität Freiburg i. Br., Germany
Fontes, Luiz	Courant Institute, New York, USA
Fournier, Jean-Daniel	Observatory of Cote d'Azur, Nice, France
Frank, Michael	Universität Leipzig, Germany
Frank, Karin	Universität Leipzig, Germany

Fredenhagen, Klaus	Universität Hamburg, Germany
Freudenberg, Wolfgang	FSU Jena, Germany
Friedrich, Jürgen	Universität Leipzig, Germany
Fritz, Jozsef	Hungarian Acad. of Sci., Budapest, Hungary
Frochaux, Etienne	Ecole Polytechnique Federale, Lausanne, Switzerland
Fröhlich, Jürg	ETH Zürich-Hönggerberg, Switzerland
Gärtner, Jürgen	TU Berlin, Germany
Galley, Thierry	University of Geneva, Switzerland
Gavrilik, Alexander	Institute of Theor. Physics, Kiev, Soviet Union
Gawedzki, Krzystof	IHES Bures-sur-Yvette, France
Gentile, Guido	Univ. of Roma "La Sapienza", Italia
Georges, Antoine	Princeton University, USA
Georgescu, Vladimir	Bucharest University, Romania
Geyer, Bodo	Universität Leipzig, Germany
Ghikas, Demetris P.K.	University of Patras, Greece
Gielerak, Roman	University of Wroclaw, Poland
Ginibre, Jean	University of Paris-Sud, Orsay, France
Goldin, Gerald A.	Rutgers University, Piscataway, USA
Goldsheid, Ilya	Ruhr Universität Bochum, Germany
Goldstein, Sheldon	Rutgers University, New Brunswick, USA
Gole, Christophe	ETH-Zentrum, Zürich, Switzerland
Golodets, Valentin	Ukrainian Acad. of Sci., Kharkov, Soviet Union
Gonzaga de Sa, Paula	Univ. Catholique de Louvain-la-Neuve, Belgium
Gorini, Vittorio	University of Bari, Italia
Gough, John	University College Dublin, Ireland
Grabowski, Marian	N.Copernicus University Torun, Poland
Graf, Gian Michele	California Inst.of Technology, Pasadena, USA
Grigore, Dan Radu	CERN, Geneva, Switzerland
Grosse, Harald	Universität Wien, Austria
Haag, Rudolf	Univ. Hamburg, Germany
Hagedorn, George	Virginia TECH Blacksburg, USA
Hamm, Andreas	Universität-GH-Essen, Germany
Hartsleben, Gerhard	University of Geneva, Switzerland
Hempel, Rainer	Universität München, Germany
Hennings, Mark	University of Cambridge, United Kingdom
Herbst, Ira	National Science Foundation, Washington, USA
Herdegen, Andrzej	Jagellonian University Krakow, Poland
Hjorth, Poul G.	Technical Univ. of Denmark, Lyngby, Danmark
Hoffmann-Ostenhof, Thomas	Universität Wien, Austria
Hoffmann-Ostenhof, Maria	Universität Wien, Austria
Hofmann, Gerald	Universität Leipzig, Germany
Hollevoet, Jan	KU Leuven, Belgium
Holthaus, Martin	Universität Bonn, Germany
Horowski, Maciej	University of Warsaw, Bialystok, Poland
Huber, Albrecht	Christ.-Albrechts-Univ. Kiel, Germany
Hudetz, Thomas	Universität Wien, Austria
Hurd, Thomas	McMaster Univ. Ontario, Hamilton, Canada
Hurst, C. Angas	University of Adelaide, Australia
Iagolnitzer, Daniel	Centre d'Etudes Nucl. Saclay, Gif-sur-Yvette, France
Ingarden, Roman Stanisław	N.Copernicus University Torun, Poland
Ionescu, Lucian	Bucharest University, Romania
Jackiw, Roman	Massachusetts Inst.of Techn., Cambridge, USA

Jaffe, Arthur	Harvard University, Cambridge, USA
Jamiołkowski, Andrzej	N.Copernicus University Torun, Poland
Jauslin, Hans Rudolf	Rutgers University, New Brunswick, USA
Jensen, Arne	Aalborg University, Danmark
Jezierski, Jacek	University of Warsaw, Poland
Joye, Alain	Ecole Polytechnique Federale, Lausanne, Switzerland
Jüttner, Hans-Ulrich	Christ.-Albrechts-Univ. Kiel, Germany
Jurco, Branislav	Palacky University Olomouc, Czechoslovakia
Kaiser, Rainer	Universität Leipzig, Germany
Kanatchikov, Igor	Lebedev Physical Institute, Moscow, Soviet Union
Kandilarov, Boris	Bulgarian Academy of Sciences, Sofia, Bulgaria
Karner, Gunther	Universität Bielefeld, Germany
Kasparek, Andreas	Universität Leipzig, Germany
Kastler, Daniel	Bandol, France
Kaushal, R.S.	University of Delhi, India
Kay, Bernard S.	University of Cambridge, United Kingdom
Keating, Jon	University of Bristol, United Kingdom
Keller, Georg	Max-Planck-Institut für Physik, München, Germany
Khan, Iqbal Ahmad	B.N.College of Engineering, Pusad, India
Khoruzhenko, Boris	Inst. for Low Temperature, Kharkov, Soviet Union
Khuri, Nicola	Rockefeller University, New York, USA
Kifer, Yuri	Hebrew University of Jerusalem, Israel
Kijowski, Jerzy	Polish Academy of Sciences, Warsaw, Poland
Kim, Hoil	Max-Planck-Institut, Bonn, Germany
Kirschner, Roland	Universität Leipzig, Germany
Klainerman, Sergiu	Princeton University, USA
Klauder, John R.	University of Florida, Gainesville, USA
Kleespies, Frank	J.W.Goethe-Univ. Frankfurt a.M., Germany
Klein, Abel	University of California, Irvine, USA
Knauf, Andreas	TU Berlin, Germany
Koch, Martin	Universität Freiburg i. Br., Germany
Köbler, Kirstin	J.W.Goethe-Univ. Frankfurt a.M., Germany
Königs, Robert Paul	Deutsche Forschungsgemeinschaft, Bonn, Germany
Kopec, Sławomir	Jagellonian University Krakow, Poland
Kopper, Christoph	Ecole Polytechnique Palaiseau, France
Kotecky, Roman	Charles University Prague, Czechoslovakia
Koukiou, Flora	Ecole Polytechnique Palaiseau, France
Kowalski-Glikman, Jerzy	Kluwer Academic Publ., Dordrecht, Netherlands
Kozlov, Valerij Pavlovich	Byelorussian State University, Minsk, Soviet Union
Kraus, Elisabeth	DESY, Hamburg, Germany
Krause, Jorge	University of Chile, Santiago, Chile
Kröger, Pawel	King's College London, United Kingdom
Krüger, Tyll	Universität Bielefeld, Germany
Kürsten, Klaus-Detlef	Universität Leipzig, Germany
Kupiainen, Antti	Rutgers University, New Brunswick, USA
Kupsch, Joachim	Universität Kaiserslautern, Germany
Laartz, Jürgen	Universität Freiburg i. Br., Germany
Lam, Mai Mai	Portsmouth Polytechnic, United Kingdom
Landsman, Nicolaas	University of Cambridge, United Kingdom
Lassner, Gerd	Universität Leipzig, Germany
Lassner, Gisela A.	Universität Leipzig, Germany
Lazar, Aldo	Tel Aviv University Aviv, Israel
Le, Hung Son	Polytechnical University Hanoi, Vietnam

Leboeuf, Patrick	Institute of Nuclear Physics, Orsay, France
Lenard, Andrew	Indiana University, Bloomington, USA
Lewis, Roger	University of Alabama, Birmingham, USA
Lisiecki, Wojciech	University of Warsaw, Bialystok, Poland
Loeffel, Jean-Jacques	University of Lausanne, Switzerland
Löffelholz, Jürgen	Universität Leipzig, Germany
Löffler, Frank	Universität Leipzig, Germany
Lopuszanski, Jan	University of Wroclaw, Poland
Lukierski, Jerzy	University of Wroclaw, Poland
Maes, Christian	KU Leuven, Belgium
Magnon, Anne	University Blaise Pascal, Aubiere, France
Majewski, Władysław A.	University of Gdansk, Poland
Mania, Alessandro	University of Milano, Italia
Manin, Yuri	Steklov Math. Institute, Moscow, Soviet Union
Martin, Andre	CERN, Geneva, Switzerland
Martinelli, Fabio	University of Roma, Italia
Marzuoli, Annalisa	University of Pavia, Italia
Matsui, Taku	Tokyo Metropolitan University, Japan
Matthes, Rainer	Universität Leipzig, Germany
Mayer, Meinhard E.	University of California, Irvine, USA
Mazel, Alexander	Inst.of Earthquake Prediction, Moscow, Soviet Union
McLenaghan, Raymond	University of Waterloo, Canada
Meinrenken, Eckhard	Universität Freiburg i. Br., Germany
Menotti, Pietro	University of Pisa, Italia
Messer, Joachim	Universität Göttingen, Germany
Miekisz, Jacek	Univ. de Louvain-la-Neuve, Belgium
Mirlin, Alexander D.	Leningrad Nucl.Phys.Institute, Gatchina, Soviet Union
Mitter, Werner	Böhler Ges.m.b.H., Kapfenberg, Austria
Möbius, Peter	TU Dresden, Germany
Mrugala, Ryszard	N.Copernicus University Torun, Poland
Müger, Michael	TH Darmstadt, Germany
Nachman, Adrian	University of Rochester, USA
Nachtergaele, Bruno	KU Leuven, Belgium
Nakagami, Yoshiomi	Yokohama City University, Japan
Nakamura, Touru	Sundai Preparatory School, Tokyo, Japan
Napiórkowski, Kazimierz	Warsaw University, Poland
Narnhofer, Heide	Universität Wien, Austria
Naudts, Jan	University of Antwerpen, Belgium
Nazin, Georgy	Tyumen University, Soviet Union
Negi, Om Prakash Sing	Kumaun University Campus, Almora, India
Neidhardt, Hagen	Karl-Weierstrass-Institut, Berlin, Germany
Nesterenko, Vladimir	JINR Dubna, Moscow, Soviet Union
Neugebauer, Gernot	FSU Jena, Germany
Newman, Charles	Courant Institute, New York, USA
Nikolova, Lora	Bulgarian Academy Sciences, Sofia, Bulgaria
Nill, Florian	FU Berlin, Germany
Oberaigner, Eduard R.	Montanuniversität Leoben, Austria
Omnes, Roland	University of Paris Sud, Orsay, France
Osipov, Edward P.	Inst. for Mathematics, Novosibirsk, Soviet Union
Pachucki, Krzysztof	Polish Academy of Sciences, Warsaw, Poland
Panfil, Semen	Inst. for Mathematics, Novosibirsk, Soviet Union
Pang, Michael Man Ho	King's College London, United Kingdom

Pastur, Leonid	Ukrainian Acad. of Sciences, Kharkov, Soviet Union
Patrick, Anatoly	School of Theor. Physics, Dublin, Ireland
Petrina, Dmitry	Ukrainian Acad. of Sciences, Kiev, Soviet Union
Petritis, Dimitri	University of Rennes Italia, Rennes, France
Pflaum, Markus	Universität München, Germany
Pi, So-Young	Boston University, USA
Pittner, Ludwig	Karl-Franzens-Universität Graz, Austria
Pizzocchero, Livio	University of Milano, Italia
Pohle, H.-J.	Universität Leipzig, Germany
Pokorny, Martin	University of Arizona, Tucson, USA
Popov, Valentin	USSR Acad. of Sciences, Tomsk, Soviet Union
Pordt, Andreas	University of Lausanne, Switzerland
Präve, Alexandra	J.W.Goethe-Univ. Frankfurt a.M., Germany
Priezzhev, Vjatcheslav	JINR Dubna, Moscow, Soviet Union
Primas, Hans	ETH Zürich-Zentrum, Switzerland
Proksch, Martin	Universität Würzburg, Germany
Pule, Joseph V.	University College Dublin, Ireland
Pusz, Wiesław	University of Warsaw, Poland
Quadt, Ralf	Universität Köln, Germany
Radzikowski, Marek	Princeton University, USA
Ramadas, Ramakrishnan T.	Tata Inst. of Fund. Research, Bombay, India
Raptis, Sotiris	University of Patras, Greece
Rebenko, Alexei	Ukrainian Acad. of Sciences, Kiev, Soviet Union
Redig, Frank	University of Antwerpen, Belgium
Reeh, Helmut	Universität Göttingen, Germany
Reentz, Georg	Universität Würzburg, Germany
Rehren, Karl-Henning	Universität Hamburg, Germany
Rendall, Alan	Max-Planck-Institut, München, Germany
Richter, Olaf	Universität Leipzig, Germany
Riela, Giovanni	University of Palermo, Italia
Rivero, Alejandro	University of Zaragoza, Spain
Robaschik, Dieter	Universität Leipzig, Germany
Robert, Didier	University of Nantes, France
Roberts, John	Universität Osnabrück, Germany
Roepstorff, Gert	RWTH Aachen, Germany
Rogers, Alice	King's College London, United Kingdom
Romano, Silvano	University of Pavia, Italia
Roos, Hansjörg	Universität Göttingen, Germany
Rudnicki, Sławomir	University of Gdansk, Poland
Rudolph, Gerd	Universität Leipzig, Germany
Ruelle, David	IHES Bures-sur-Yvette, France
Ruskai, Mary Beth	University of Lowell, USA
Rybko, A.	Inst.of Earthquake Prediction, Moscow, Soviet Union
Salmhofer, Manfred	Max-Planck-Institut, München, Germany
Samoilenko, Juri Stefan	Ukrainian Acad. of Sciences, Kiev, Soviet Union
Sanchez-Valenzuela, O.Adolfo	Centro de Investigacion, Guanajuato, Mexico
Sauder, Thorsten	RWTH Aachen, Germany
Sauer, Hans Martin	Univ. des Saarlandes, Saarbrücken, Germany
Schäper, Ulrich	Universität Freiburg i. Br., Germany
Scharnhorst, Klaus	Universität Leipzig, Germany
Scheffold, Bernhard	Universität Freiburg i. Br., Germany
Schimming, Rainer	EMAU Greifswald, Germany
Schlichenmaier, Martin	Universität Mannheim, Germany

Schlieker, Michael	Universität München, Germany
Schmidt, Bernd	Max-Planck-Institut, München, Germany
Schmidt, Christian	RWTH Aachen, Germany
Schmitt, Andreas	TH Darmstadt, Germany
Schmüdgen, Konrad	Universität Leipzig, Germany
Scholl, Marcus	RWTH Aachen, Germany
Scholz, Erasmus	HfV Dresden, Germany
Schomerus, Volker	Universität Hamburg, Germany
Schroeck, Franklin	Florida Atlantik University, Boca Raton, USA
Schröder, Ulrich E.	Universität Frankfurt a.M., Germany
Schroers, Bernd	University of Cambridge, United Kingdom
Seba, Petr	Ruhr Universität Bochum, Germany
Seco, Luis	California Inst. of Technology, Pasadena, USA
Seiler, Erhard	Max-Planck-Institut, München, Germany
Seiler, Ruedi	TU Berlin, Germany
Sen, R.N.	Ben Gurion University, Beer Sheva, Israel
Sewell, Geoffrey	Queen Mary College London, United Kingdom
Shcherbina, Maria	Ukrainian Acad. of Sci., Kharkov, Soviet Union
Shima, Kazunari	Saitama Inst. of Technology, Japan
Shlosman, Senya	USSR Academy of Sci., Moscow, Soviet Union
Sibold, Klaus	Max-Planck-Universität, München, Germany
Siedentop, Heinz	University of Alabama, Birmingham, USA
Sirugue-Collin, Madeleine	Universite de Provence, Marseille, France
Smirnov, Vladimir	Max-Planck-Institut, München, Germany
Sobczyk, Jan	University of Wroclaw, Poland
Soloveitchik, Mikhail	Universität Heidelberg, Germany
Solovej, Jan Philip	Princeton University, USA
Sommer, Gustav	Universität Bielefeld, Germany
Spohn, Herbert	Universität München, Germany
Stanilov, Grosio	University of Sofia, Bulgaria
Stefanov, Plamen	Bulgarian Academy of Sciences, Sofia, Bulgaria
Steiner, Frank	Universität Hamburg, Germany
Steinmann, Othmar	Universität Bielefeld, Germany
Stichel, Peter C.	Universität Bielefeld, Germany
Stöltzner, Michael	Universität Wien, Austria
Stolz, Günter	Universität Frankfurt a.M., Germany
Stolz, Sonja	J.W.Goethe-Univ. Frankfurt a.M., Germany
Stovicek, Pavel	Technical University Prague, Czechoslovakia
Stroffolini, Roberto	University "Federico II", Napoli, Italia
Stubbe, Joachim	University of Geneva, Switzerland
Summers, Stephen	University of Florida, Gainesville, USA
Sznitman, Alain-Sol	ETH Zürich, Switzerland
Thaller, Bernd	Karl-Franzens-Universität Graz, Austria
Thirring, Walter	Universität Wien, Austria
Timmermann, Werner	TU Dresden, Germany
Tirozzi, Brunello	University of Roma, Italia
Todorov, Ivan	Bulgarian Academy of Sciences, Sofia, Bulgaria
Trofimov, V.V.	Moscow State University, Soviet Union
Truman, Aubrey	University College Swansea, United Kingdom
Uhlmann, Armin	Universität Leipzig, Germany
Vainermann, Leonid	Kiev State University, Soviet Union
Vaninsky, Kirill	Courant Institute, New York, USA
Varadhan, Srinivasa	Courant Institute, New York, USA
Venables, Herwin	Portsmouth Polytechnics, United Kingdom

Verlinde, Erik	Institute for Advanced Studies, Princeton, USA
Villani, Matteo	University of Bari, Italia
Vogt, Andrew	Georgetown University, Washington, USA
Voiculescu, Dan	University of California, Berkeley, USA
Vojta, Günter	Universität Leipzig, Germany
Wang, Wei-Min	Ecole Polytechnique, Palaiseau, France
Wayne, Clarence	Pennsylvania State Univ., University Park, USA
Wehr, Jan	Institute for Advanced Studies, Princeton, USA
Wehrl, Alfred	Universität Wien, Austria
Weich, Wolfgang	Universität München, Germany
Weikard, Rudi	University of Alabama, Birmingham, USA
Weixler, Ralf	Universität München, Germany
Weller, Wolfgang	Universität Leipzig, Germany
Werner, Reinhard	Universität Osnabrück, Germany
Whiting, Bernard	University of Florida, Gainesville, USA
Wieczerkowski, Christian	ETH Zürich-Hönggerberg, Switzerland
Wiesbrock, Hans-Werner	FU Berlin, Germany
Wightman, Arthur	Princeton University, USA
Winnink, Marinus	University of Groningen, Netherlands
Wipf, Andreas	ETH Zürich-Hönggerberg, Switzerland
Woronowicz, Stanisław L.	University of Warsaw, Poland
Wyss, Walter	University of Colorado, Boulder, USA
Wystup, Uwe	Universität Frankfurt a.M., Germany
Yamagishi, Kengo	University of California, Livermore, USA
Yau, Horng-Tzer	Courant Institute, New York, USA
Yelkhovsky, Alexander S.	USSR Acad. of Sci., Novosibirsk, Soviet Union
Yngvason, Jakob	University of Iceland, Reykjavik, Iceland
Young, Lai-Sang	University of California, Los Angeles, USA
Zagrebnov, Valentine	JINR Dubna, Moscow, Soviet Union
Zahradnik, Milos	Charles University Prague, Czechoslovakia
Zakrzewski, Stanisław	University of Warsaw, Poland
Zartl, Alexander	Universität Wien, Austria
Zavialov, Oleg I.	Steklov Math. Institut, Moscow, Soviet Union
Zegarlinski, Boguslaw	Ruhr Universität Bochum, Germany
Zeidler, Eberhard	Universität Leipzig, Germany
Zhislin, Gregory	Radiophysical Research Inst., N.Novgorod, Soviet Union
Ziep, Otto	HU Berlin, Germany
Ziermann, Andre	Charles University Prague, Czechoslovakia
Zumino, Bruno	University of California, Berkeley, USA
Zurek, Wojciech	Los Alamos National Laboratory, USA
Zylka, Christian	Universität Leipzig, Germany
da Veiga, Paulo A.	Harvard University, Cambridge, USA
de Groote, Hans F.	J.W.Goethe-Univ. Frankfurt a.M., Germany
de Jager, Eduard	University of Amsterdam, Netherlands
de la Llave, Rafael	University of Texas, Austin, USA
den Hollander, Frank	Delft University of Technology, Netherlands
d'Antoni, Claudio	Univ. of Roma "La Sapienza", Italia
van Enter, Aernout	University of Groningen, Netherlands

Part I

Plenary Lectures

Recent Progress in Classical Mechanics

R. de la Llave

Department of Mathematics, The University of Texas at Austin,
Austin, Texas 78712 USA

The goal of this lecture is to review several developments in classical mechanics that have taken place in the last years, that will fit in the time of the talk and that I have become aware of. Unfortunately, the latter is a constraint more severe than what I would like and I apologize to the authors and the audience for many things that have been left out. In particular, I have left out topics such as "twist mappings", "geometric quantization" or "geometric phases" and "quantum chaos" that are generating a great deal of activity in the literature. I would also like to recommend the excellent set of reviews [AANS] for the many developments that took place up to the time the books were published.

Geometric Theory of Integrable Systems

One of the central problems of mechanics has been to integrate Hamilton's equations of motion, or at least decide if such an integration is impossible.

The most geometrically natural notion of integrability is that the system should have as many conserved quantities with vanishing Poisson brackets as degrees of freedom.

It has long been known that the fact that a system is integrable severely restricts the topology of the phase space and of the energy surface. For example, if the system admits action-angle variables — which is strictly stronger than being integrable in the above sense — the phase space should be $\mathbb{R}^n \times \mathbb{T}^n$ so that the topology is determined.

The first foothold in the geometric theory of integrable systems is the Liouville/Arnol'd theorem that says that the system can be decomposed in pieces $\mathbb{R}^{n_i} \times \mathbb{T}^{m_i}$, but the n_i, m_i can change. For example in the Kepler system we have bounded trajectories along ellipses and unbounded ones along hyperbolas. These two sets get "glued" in the intermediate set of parabolic trajectories.

An important realization was that the gluing of the different pieces has to be done in very precise ways so as to preserve the very rigid structure imposed by integrability. Hence, the phase space of an integrable system can be considered as pieces of $\mathbb{R}^{n_i} \times \mathbb{T}^{m_i}$ glued in very precise ways. This turns out to impose severe restrictions on the phase spaces and energy surfaces of integrable systems. (Notice that having standard pieces that get glued in well defined ways according to the singular submanifolds of a vector field is somewhat reminiscent of Morse Theory.)

This beautiful theory, which one may start to learn in the paper [Fo] written by the indefatigable leader of the method, has many spin-offs.

Since one of the best known methods to generate integrable systems has been to create systems in manifolds with lots of symmetries — e.g., groups or quotients of groups — it is possible to obtain many results about topological consequences of algebraic structures. The machinery of cutting and pasting manifolds while preserving a rich geometric structure can be used to generate examples and counterexamples in low dimensional topology. One of the methods of choice to attempt the classification of low dimensional manifolds has been to show that on them one can define canonical geometric structures that produce invariants, and, one hopes, provide with methods to show they are equivalent. Once one has an integrable system on a manifold, it is very easy to generate other types of structures. Hence, these methods have tested the boundaries of what geometric structures determine.

Analytically Integrable Systems

On the other side of the spectrum one can try to decide which systems can be integrated using only algebraic or rational transformations.

For example, if one takes particles on a line which are very repulsive, the resulting system will be integrable [Hub] [Gu]— essentially, the integration is achieved by the scattering operator — nevertheless, the Calogero potential is special because it can be integrated by algebraic methods.

There are several criteria to show that a system cannot be integrated by algebraic methods. The most time honored is the one used by S. Kowaleska to narrow down the search of algebraically integrable cases for the rigid body to a few particular ones. (Basically, we observe that an integrated system has no singularities and that an algebraic integration can only introduce algebraic singularities. The possible singularities of the system under study can be ascertained by solving the differential equation.)

A more recent method for complex integrability has been introduced by Ziglin, based on the observation that if a system is integrable, the periodic orbits can be deformed, along complex paths. This imposes contraints on the algebraic properties of the variational equations.

To my knowledge, there are no definitive results about converses of these negative criteria. One would like to know if an algebraic system satisfying

Kowaleska criterion, and maybe some other global condition, is algebraically integrable.

A much more severe obstruction to the possibility of integrability is the existence of homoclinic points. The verification of this hypothesis usually is done by perturbation theory. We mention that in several cases, this is quite difficult since the perturbation expansions vanishe to all orders.

A very easy to read review of the developments in the physical literature is [RGB]. See also the corresponding chapters in [AANS], [Koz], [Mor].

One could wonder why one should worry about the difference between algebraic integrability versus, say C^{50}. Let me mention two reasons. One is that there are several structures that only manifest themselves when we consider complex extensions. (We will discuss one of those when we discuss perturbations beyond all orders.) Another is that there are many natural systems that are algebraic, for example, one step of algorithms acting on matrices can be considered as an algebraic dynamical system. It has been known for a long time that there were important similarities between the Jacobi algorithm to diagonalize matrices and the scattering of particles on a line. This has been used to analyze in quite detailed fashion the Jacobi algorithm and in turn this has lead to implementations that soon will improve in many respects the now standard algorithms [DLT]. From the mathematical point of view, the theory of algebraically integrable systems has important connections with algebraic geometry. For example, Hilbert's 21st problem is related to issues raised by Ziglin's method.

Asymptotics Beyond All Orders

There are several interesting quantities in classical mechanics that, do not vanish even if a perturbation expansion yields identically zero.

It has been known for a long time to use these perturbation expansions (physicists called them "divergent" even if they are the sum of zero term) to get upper bounds. In a typical situation, by truncating the series to order N we obtain $|Q| \leq \varepsilon^N N^N$. Then, if we take $N = (\varepsilon e)^{-1}$ we obtain $|Q| \leq e^{-1/\varepsilon e}$.

Such upper bounds appear very frequently. See for example [Nek] for transport, [Ne2], [FS] for splitting of separatrices. Getting lower bounds seems much harder. Nevertheless, there is a trick that has yielded results in several cases. If we consider complex extensions of radii $O(1/\varepsilon)$ for the objects under consideration, perturbation theory will produce results $Q(x) = A(x)\varepsilon + O(\varepsilon^2)$. Then, using Cauchy estimates, it is possible to prove lower bounds of the right order of magnitude.

Needless to say, this broad sketch does not do justice to the difficulty of concrete situations. Some recent examples are the treatment of the Landau-Zener formula for the corrections to the adiabatic limit [Ha], [JaSe] or the crossing of separatrices in two dimensional twist mappings, [An], [ACKR] a problem that has plagued the literature for several years.

Let us point out that the upper bounds in [Nek] do not have a corresponding set of lower bounds. This is the famous problem of Arnol'd diffusion and it involves, besides analytical lower bounds, a good understanding of several geometric structures. It is known that this phenomenon occurs in examples [Ar1] and there are now more systematic constructions [Do] that show it is generic. Nevertheless, this is a far cry from being able to decide whether a concrete system presents it or not or calculate its magnitude.

Let us point out that in this problem of Nekhorosev bounds/Arnol'd diffusion there has been significant recent progress. For upper bounds, the paper [Nek] was significantly clarified in [BGG] and a radically different proof has appeared [Lo]. For Arnol'd diffusion, there has been progress in the computation of one of the ingredients, the whiskered tori. There are two different perturbative calculations of whiskered tori that yield disjoint sets of tori [LW], [Tr].

Non-Collision Singularities in the N-Body Problem

The problem of N-bodies moving classically under their mutual gravitational attraction is perhaps the oldest in mathematical physics.

The first question to ask is whether solutions are defined for all time. A first glance to the differential equations reveals that they become singular if two bodies collide. Nevertheless, a more detailed analysis reveal that the singularity when only two bodies collide is only apparent. If we modify the conditions so that the bodies miss by a small amount, we get a well defined limit as this modification goes to zero (a back of the envelope argument: the conservation of momentum and energy and the symmetry of the situation determines what happens). Unfortunately, for triple collisions, this argument does not work and indeed, it is possible to show that if the bodies miss by small amounts many things can happen and one does not get a well defined limit (exercise: show that the same happens with billiard balls). Very detailed information is nevertheless available (see [SM], Ch.I for the classical work of Sundmann, [De]) for the three body collision and glimpses for more bodies.

One natural question to ask is how pervasive these singularities are and whether there are any others. Important results were obtained at the beginning of the century and completed in the early 70's (See [SM]) that showed that if a solution cannot be continued, either there is a triple collision or the moment of inertia of the whole system has to become unbounded.

In [MMcG] it was shown that, indeed there are singularities different from collisions. Their example consists of 4 particles in a line. Two of them are oscillating and a third one is far apart. A "messenger particle" collides with the oscillating pair and, since this is almost a triple collision, leaves at an enormous speed obtained from the potential energy of the pair. It catches the other particle, gives momentum to it and bounces back, just in time to catch the other two in an almost triple collision, in precisely the conditions that will make it bounce back even more violently and so on. The net result is that the fourth particle escapes to infinity in finite time.

Unfortunately, it was quite difficult to generalize their method to higher dimension so as to avoid collisions completely. (This was included in the list of problems in mathematical physics by B. Simon.)

Recently, however, there have been two remarkable papers [Xi], [Ge] in which examples with collisionless singularities are produced. Both are based on having a messenger particle going back and forth between others managing to always arive near a triple collision (but not colliding). The configuration in [Xi] has two pairs rotating on parallel planes directly above each other and the messenger particle moving perpendicularly. The configuration of [Ge] consists of pairs orbiting around — roughly the vertices of an equilateral polygon and the several messenger particles going around them.

These two papers entail quite involved estimates and are somewhat difficult to read (I have not checked them in detail myself) but it is clear that they are important.

Let me point out that it is not known whether such singularities occupy a set of positive measure.

The most obvious quantum version of the problem of showing that the dynamics is well defined for all times for almost all trajectories is to show that the quantum Hamiltonian is self-adjoint, which is much easier than the classical one. (See, nevertheless, [RS] for some more detailed physical discussion of the analogy.) It is amusing to note that the effect of a messenger particle oscillating wildly between two channels is, however, the enemy to beat in the proofs of quantum asymptotic completeness.

Examples of Classical Ergodic Systems

The main argument to prove that a system is ergodic originated in the work of [Hed], [Ho] (the latter published in Leipzig!) which established ergodicity of geodesic flows.

Basically the main ingredients are a geometric study of the trajectories, which establishes that trajectories that converge in the future or the past form a manifold and that by moving alternatively along these stable/unstable manifolds we can go from every point to every point. Secondly, an abstract theorem due to Birkhoff that shows that for almost all trajectories, the statistical behaviour in the future is the same as that in the past. (Unfortunately, due to the "almost all" rather than all in Birkhoff theorem we need an extra technical condition on the foliations.)

The argument was generalized and streamlined in [A]. Further generalizations were due to [Si] who introduced allowing singularities and [Pe] who allowed for non-uniformity of the approaches in the future and the past but constructed the stable and unstable manifolds and concluded that, if they are sufficiently long, the system is indeed ergodic, if they are not, there are counterexamples [Pe2], [W].

Nevertheless, one can say that, in spite of its beauty, the theory was too abstract and the the only concrete examples with physical appeal were the dispersing billiards [Si], joined later by the celebrated stadium [Bu], a non-dispersing ergodic billiard.

Recently, the situation has changed, manageable conditions to verify that concrete systems satisfy the abstract hypothesis of [Pe] were introduced in [W]. This unified the examples of dispersing and non-dispersing billiards and examples of ergodic geodesic flows and flows on scattering potentials were constructed [Don], [DoL].

More or less at the same time, there appeared another elaboration of the basic strategy that can produce ergodicity even in the case that the manifolds are short [SCh] or that there are singularities The basic idea is perhaps the concept of local ergodicity. These methods were applied to show that certain sets of hard balls in two or three dimensions are ergodic. The methods of this paper have been extended in [KSS] to systems of more balls in higher dimensions.

Let us mention that all the above verifications of ergodicity, almost automatically yield ergodic properties such as K-property, positive entropy, Bernoulli. It is by now well known how to prove that uniformly hyperbolic systems are Bernoulli and satisfy the central limit theorem. The method is to use Markov partions that reduce the sytem to an spin system in one dimension. Unfortunatelly, in the non-uniformly hyperbolic case, the partitions have infinitely many components and the statistical mechanics porblem involves systems with unbounded spins.

Maybe I should mention that presumably proving ergodicity is not quite the only problem one wants to tackle for physical applications. For Hamiltonian systems, it is very easy to create little islands that, even if they will be negligible for practical applications, would destroy ergodicity. A very interesting problem that I learned from T. Spencer is to prove that the well known system $T_\varepsilon(A,\theta) = (A + \varepsilon \sin\theta, \theta + A + \varepsilon\sin\theta)$ has an ergodic component of positive measure for some large ε's. Let me mention that if one takes in place of sin a piecewise linear version

$$f(x) = \begin{cases} x & \text{if } x \in [0, \pi/2] \\ \pi - x & \text{if } x \in [\pi/2,\, 3\pi/2] \\ x - 2\pi & \text{if } x \in [3\pi/2,\, 2\pi] \end{cases}$$

it is reasonably easy to verify hypothesis that imply that the piecewise linear version of T is ergodic for $\varepsilon > 100$. Can one make a prove the same result for some values of ε when we round the corners of the piecewise linear version of T?.

Rigidity of Dynamical Systems

An object is called rigid if, whenever there is another object equivalent to it in a certain sense, it is also equivalent in another stronger sense. For example, a triangle is rigid because any polygon equivalent to it in the sense of having sides of the same length is equivalent in the much stronger sense of being isometric. A somewhat weaker version of rigidity is rigidity under deformations. We say that a system is rigid under deformations if all the deformations that preserve some structure are trivial. Typically when one starts working with equivalences one tries to attach invariants and it is usual to also call a rigidity theorem a result that shows that some set of invariants determines the object up to trivial changes.

One of the reasons why rigidity theorems are amusing to study is that they cross category lines. We investigate measure theoretic consequences of a Riemannian hypothesis and so on.

One set of objects for which many rigidity theorems are known is negatively curved Riemannian manifolds. A very interesting line of development was started in the papers [GK1][GK2] which showed that, for negatively curved manifolds (either two dimensional or satisfying pinching conditions), isospectral deformations of the metric are isometries (this is a version of the famous "can you hear the shape of a drum?" question). The proof of the theorem consists of two steps. One, showing that if we deform metrics of negative curvature keeping the spectrum of the laplacian constant the lengths of closed geodesics are kept constant, and then showing that this implies that the deformations are isometric. The later is purely a problem in Riemannian geometry.

Another investigation of the consequences of keeping the lengths constant was undertaken in [CEG] who showed that, for surfaces of constant negative curvature, any Hamiltonian deformations that kept some action invariants had to be smooth canonical transformations. This result was generalized in [LMM] who showed the same result is true for Hamiltonian Anosov flows in any dimension. The methods of this paper were extended in [L1] [LM] [L2] to show that, for two dimensional Anosov diffeomorphisms or three dimensional Anosov flows, the eigenvalues at periodic orbits — which are obviously invariants of C^1equivalence — form a complete set of invariants for C^∞ or C^ω equivalence. A different proof using the theory of SRB measures was constructed in [Po] and generalized in [L3] to the case of partially hyperbolic systems. So it seems that the fact that there is complete set of local invariants of peridic orbits for smooth conjugacy of Anosov systems is false in higher dimensions. Nevertheless,it is true that in some cases the eigenvalues at periodic orbits are complete sets of invariants for $C^{\tilde{k}}$ conjugacy and that C^{k^*} conjugacies are C^∞ or C^ω. Unfortunately, $\tilde{k} < k^*$ so that there is a range of regularities without counterexamples or theorems.

Related to the geodesic flows, the horocycle flows — which have no periodic orbits — have been shown to be extremely rigid. In that case, measure theoretic equivalence can be bootstrapped to differentiable equivalence and, by the previous results to smooth equivalence. See for example [FO] and the references there to previous work of a more algebraic nature.

Another set of problems which I believe is strongly related to the previous ones is to show that Cantor sets generated by dynamical processes are differentiably equivalent if they are topologically equivalent. For the Feigenbaum Cantor set, there are several results already [Ra] [Pa][Su] that show that the topology of the Cantor sets and the fact that they are dynamically generated forces them to be C^1 equivalent.

Other problems with a similar flavor are the study of diffeomorphisms that only commute only with their powers –i.e., diffeomorphisms without any differentiable symmetry.– (See [PY1],[PY2] for theorems that show diffeomorphisms with this property and [PM] for examples of germs of diffeomorphisms with non trivial centralizer) or the study of hyperbolic diffeomorphisms with differentiable stable and unstable foliations. (See [HK],[Ghy].) It is sometimes possible to characterize the systems for which inequalities among dynamical quantities or characteristic classes are saturated [HK],[KKW].

An interesting generalization of dynamical systems are actions of groups on manifolds. On the one hand, they can be considered as non-linear generalizations of representations and on the other hand as dynamical systems with a more complicated time — a point of view emphasized by the thermodynamic formalism approach to dynamical systems. There has been very active program (e.g. [Z] and references there) based on the first point of view to show that actions of *"large"* groups are rigid in the sense that, by a smooth change of variables they can be reduced to a canonical one. The dynamical systems point of view has been exploited in [Hu] to obtain results of rigidity of $SL(n, \mathbb{Z})$ on $\mathbb{T}^n$.

Coming back to the original problems posed by the papers [GK1],[GK2] there has been considerable progress made in two papers that are infuriatingly devoid of heavy machinery and depend only on stark cleverness. In [Sn] there is an extremely simple machinery to produce manifolds which are isospectral but not isometric. In [O] there is a very clever proof that knowledge of the lengths of closed geodesics and their homotopy classes determines the manifold up to isometry. For all that we know [Sn] provides the only mechanism to produce isospectral but not isometric manifolds and for a while it was discussed whether it should be possible to show that isospectral two dimensional manifolds are related as in [Sn].

I would be embarrassed not to mention the new developments in symplectic geometry that are related to symplectic capacities. Symplectic capacities are invariants of symplectic maps. The most transparent constructions of these invariants use the Floer cohomology. Nevertheless, since Floer cohomology is defined through the gradient flows of a variational principle, one can also give a variational definition. There are also rather simple axiomatic characterizations.

Symplectic capacities have many applications. A tentative one that jumps into mind is that, for the study of pseudodifferential operators via the uncertainty principle, one has to pack some regions of phase space with symplectic images of rectangles. The theory of symplectic capacities places severe constraints on what these images could be. Can this be used to prove bounds for pseudodifferential operators?.

More firmly, the theory of symplectic capacities has produced a proof of the celebrated result that the group of C^∞ symplectic diffeomorphisms is C^0 closed in the group of diffeomorphisms. Very readable surveys with references are the papers by Viterbo and Hofer in [Mi].

K. A. M. Theory

Use of Frequencies as Parameters

The first theorem of K. A. M. theory had two very different proofs. The first one, proposed by Kolmogorov (see e.g [Ba] for a modern exposition of the method) consisted in selecting one frequency and producing a torus though several canonical transformations chosen to exhibit that tori of this frequency indeed existed. The other strategy, proposed and carried out in great detail in [Ar2] consisted in studying the resonance regions and performing the transformations suggested by first order perturbation theory in the regions where this transformations were well defined.

The first method of proof is simpler and became the dominant method in the mathematical literature. Especially since it was discovered that it allowed smoothing schemes that were much sharper than the simple truncating of Fourier series required by Arnol'd's method and hence yielded superior results with respect to the differentiability assumptions required or the differentiability of the conclusions. The method could also be abstracted into implicit function theorems, which could be applied in a variety of situations.

One of the consequences of this point of view is that, since the formulation of implicit function theorems requires one to keep the frequency fixed and each step of the perturbation changes the frequencies, it becomes necessary to have extra parameters to adjust so that the frequency is restored. Hence, it became customary, when trying to apply a theorem to try to count free parameters to see if one had enough to generate counterterms that allowed the application of an implicit function theorem.

In the recent mathematical physics literature, the situation was a little bit more divided due, in good part to the very nice exposition of Arnol'd's method [CG]. Also, there were theorems such as [FSW] in which the role of adjusting extra parameters had to be played by performing probability estimates in a measure space. An important motivation for overcoming the constraints imposed by parameter counting is applications to infinite dimensions, that we will discuss in the next paragraph.

A very interesting intermediate point of view is the method of Rüssmann [Rü] that keeps the frequency fixed by adding extra artificial parameters to keeps the frequency fixed and, at the end uses the geometry to show the cancel. A very direct attack on a problem that parameter counting said was impossible was [El], who studied the problem of the preservation of lower dimensional tori. The method used was to try to change the frequency considered at every step of the iteration. Soon afterwards it was remarked that similar results could be understood as applying an extended version of Rüssmann method or, inspired by that, as a modification of Arnold's method [DL]. Then, there has been a flurry of activity using this type of ideas. Many problems that had been put on the backburner because the parameter counting said they were impossible became accessible. One application that I particularly like is the existence of two dimensional invariant tori in three dimensional volume preserving flows. Flows such as those to be considered in the theorem can be produced in fluid mechanics experiments and the existence of the tori has important physical meaning such as the lack of mixing. Moreover the tori can be seen by injecting dye! Such experiments have been performed in the Center for Nonlinear Dynamics at U. Texas. Before that, there were numerical experiments [Su] confirmed and extended in [FKP]. Proofs of the theorem were finally obtained in [ChSu], [DL]. Other theorems in which the frequencies play the role of extra parameters is [JS]. I certainly expect that there will be many more as Rüssmann or Arnol'd's methods of proof becomes better known and are as thoroughly explored as the implicit function methods have.

Incidentally, another advantage of using Arnol'd's method is that the non-degeneracy conditions one obtains are much better than those required using an implicit function theorem type of proof. Unfortunately, besides making more difficult to obtain sharp results with respect to differentiability it also makes it more difficult to obtain quantitative versions of the result or to use it to validate numerical computations.

K.A.M. Theory for Infinite Systems

For several years, there has been an active program to study the thermodynamic limit of K.A.M. Theorem. One aspect of this program was to show that a system of N particles, each of them with d degrees of freedom and interacting though short range forces has invariant tori of positive measure for values of the perturbation that decrease like $O(N^{-\alpha})$ [W1], [V]. Moreover in [W3], [W4], there is an analogue of Nekhoroshev upper bound when the number of degrees of freedom increases.

In this program, one has to use the short range of the interaction to show that the system is effectively almost finite dimensional. This gives a precise meaning to the physicists intuition that degrees of freedom are frozen and that the interactions do not count.

Perhaps the frist result for infinitely many degrees of freedom is [FSW]. Nevertheless, the system is such that the amplitude of the oscillations in the degrees of freedom decreases very fast with the distance to a center. In effect, the oscillations are localized – in the sense of the word in solid state physics. An abstract version of this result is in [Pö2].

It is interesting to realize the technical similarities of this with the series of papers [BS] [PS] in which the opposite is attempted, namely, to develop a hyperbolic perturbation theory for hyperbolic systems each of which is hyperbolic but which are coupled through a short range interaction.

Another interesting development is the application of methods of K.A.M. theory to partial differential equations. This is not just a case of taking the thermodynamic limit. If one did that, one would obtain tori of as many dimensions as the number of degrees of freedom i.e., infinite. What does it mean to have a function with infinitely many frequencies? It is not a very interesting result. There are two interesting papers. One is the paper [W2] in which the method used is very similar to Arnol'd's method applied by defining a measure in the set of potentials entering in the definition of the problem. In another set of papers, [K], [K2] performs an iteration in which the frequency changes at every stage, but one obtains the control by direct methods.

Quantitative Versions of the K.A.M. Theorem

From the beginning of K.A.M. theory it has been a point of friction with the physicists that the values for which the theory applies are quite hard to ascertain rigorously [Mo]. While the mathematicians were using as the main figure of merit the differentiability properties, physicists were emphasizing numbers. In some particular examples, it was possible to obtain reasonable values [L4] [He] — incidentally, if one chases though the constants in [Br], one gets better numbers than those in [L4] as was pointed out to me by M. Herman. Nevertheless, it is clear that the methods are quite specific.

One different method was afforded by the use of computer assisted proofs. The method consists in getting computers to perform calculations that verify the conditions of a theorem. The calculations should be performed taking care to prevent the computers from using approximations as they do since they were abandoned by the mathematicians to the engineers. (See [La] for a review.)

There are two basic strategies. One [Ce], [CC1], [CC2], [CCF] considers perturbation expansions using methods similar to the Linstedt series and estimates the remainders and applies K.A.M. theory to them. This has several advantages. First, it produces results which are valid in open ranges. Second, the analysis is somewhat easier since the analytical theorem to be proved are proved for perturbations of the identity.

The other method, [R], [LR] consists in proving an implicit function theorem that shows that, if one has an approximate solution that verifies some extra hypothesis, then, there is a true one nearby. Secondly, one has to verify the hypothesis, again with a computer. This method has the advantage that it requires less calculations and is insensitive to singularities in the complex domain that, even if they do not affect the true results, could affect the convergence of expansions. Indeed, in several cases, there are proofs that these methods — up to the limits of numerical analysis — will not miss any solution.

Perhaps the main inconvenience of the method as applied so far is that the programs developed require severe modifications in order to be adapted to other problems. (Nevertheless, the programs in [R] have been used in four different projects.) It is to be hoped that, with the new developments in software tools, soon there will be packages that not only meet a new version of Salam's criterion for renormalization theories but which become routinely used.

The method of computer assisted proofs can also be used to produce upper bounds of values for which the theorem cannot be true. The first such paper is [MP]. [Ju] produces a very clever method which is not only easy to implement but also gets very close to the presumed value. In [Mu] [MMS] there is a generalization of the method of [MP] to higher dimensions.

The results of K.A.M. theory are nicely counterpointed with the mathematical results of Aubry-Mather theory and with the very delicate numerical work based on renormalization group ideas. Both these subjects would merit more detailed discussion.

Other mathematical papers that construct systematically examples which fail to exhibit the conclusions of K.A.M. theorems are [He2] – whose methods were used in [NMS] -, [Yo1],[Yo2].

References

[ACKR] Amik, C., King, E.C.S., Kadanoff, L.P., Rom-Kedar, V.: Beyond all orders: singular perturbations in a mapping.

[An] Angenent, S.: Lecture given at IMA conference on twist mappings. To appear

[A] Anosov, D.V.: Geodesic flows on closed riemannian manifolds of negative curvature. Proc. Stekelov Inst. **90** (1967)

[AANS] Anosov, D.V., Arnol'd, V.I., Novikov, S.P., Sinai, V.G. (eds.): *Dynamical systems I-IV* (Encyclopedia of Mathematical Sciences). Springer-Verlag, 1988

[Ar1] Arnol'd, V.I.: Instability of dynamical systems with several degrees of freedom. Soviet Math.Dokl. **5**, 581–585 (1964)

[Ar2] Arnol'd, V.I.: Proof of a theorm of A.N. Kolmogorov on the invariance of quasi-periodic motions under small perturbations of the hamiltonian. Russian Math.Surveys **18**, 9–36 (1963)

[BGG] Benettin, G., Galgani, L., Giorgilli, A.: A proof of Nelshoroshev's theorem for the stability times in nearly integrable systems.

[BS] Bunimovich L.A., Sinai, Y.G.: Space time chaos in coupled map lattices. Nonlinearity **1**, 491-516 (1988)

[Bar] Barrar, R.: Convergence of the VonZeipel procedure. Celestial Mech. **2**, 491-504 (1970)

[Br] Brjno, A.D.: Analytical form of differential equations. I. Trans.Moscow Math.Soc. **25**, 131–288 (1971) II. Ibid. **26**, 199–239 (1972)

[Bu] Bunimovitch S.: On the ergodic porperties of nowhere dispersing billiards. Comm.Math.Phys. **65**, 295–312 (1979)

[CC1] Celletti, A.,Chierchia, L.: Construction of analytic KAM surfaces and effective stability bounds Comm.Math.Phys. **118**, 119–161 (1988)

[CC2] Celletti, A.,Chierchia, L.: Invariant curves for area-preserving twist maps far from integrable. J.Statist.Phys. to appear.

[Ce] Celletti, A.: Analysis of resonances in the spin orbit problem in celestial mechanics. Part I., ZAMP. **41**, 174–204 (1990); Part II., Ibid. 453–479

[ChSu] Cheng, Chong-qüing, Sun, Yi-Sui: Existence of invariant tori in three-dimesional measure preserving maps. Cel.Mech.Dyn.Ast. **47**, 275–292 (1990)

[CG] Chierchia, L., Gallavotti, G.: Smooth prime integrals for quasi-integrable hamiltonian systems. Nuovo Cimento **B67**, 277–295 (1982)

[CEG] Collet, P., Epstein, H., Gallavotti, C.: Perturbations of geodesic flows on surfaces of constant negative curvature and their mixing properties. Comm.Math.Phys. **95**, 61–112 (1984)

[DLT] Delft, P., Li, L.-C., Tomei, C.: The Bidiagonal Singular Value Decomposition and Hamiltonian Mechanics.

[DL] Delshams, A., de la Llave, R.: Existence of quasi-periodic orbits and absence of transport for volume preserving transformation and flows. Preprint

[De] Devaney, R.: Singularities in classical mechanical systems. In: Ergodic Theory and Dynamical Systems, I. Birkhauser, Boston, 1981

[Don] Donnay, V.: Geodesic flow on the two sphere, Part I, positive topological entropy. Ergodic Theory Dynamical Systems **8**, 531–553 (1988)

[DoL] Donnay, V., Liverani, C.: Potentials on the two torus for which the Hamiltonian flow is ergodic. Comm.Math.Phys. **135**, 367–303 (1991)

[Do] Douady, R.: Stabilité ou Instabilité des points fixes elliptiques. Ann.Sci. ENS **21**, 1–46 (1988)

[El] Eliasson, L.H.: Perturbations of stable invariant tori for Hamiltonian systems. Publ.Sc.Norm.Piser **15**, 115–147 (1988)

[FKP] Feingold, M., Kadanoff, L.P., Piro, O.: Passive scalars, three dimensional volume preserving maps and chaos. J.Statist.Phys. **50**, 529–565 (1988)

[FO] Feldman, J., Ornstein, D.: Semi-rigidity of horocycle flows over compact surfaces of variable negative curvature. Ergodic Theory Dynamical Systems **7**, 49–72 (1987)

[Fo] Fomenko, A.: The symplectic topology of completely integrable Hamiltonian systems. Russian Math.Surveys **44** (1989)

[FS] Fontich, E., Simó, C.: The splitting of separatices for analytic diffeomorphisms. Ergodic Theory Dynamical Systems **10**, 295–318 (1990)

[FSW] Frölich, J., Spencer, T., Wayne, C.E.: Localization in disordered nonlinear dynamical systems. J.Statist.Phys. **42**, 247–274 (1986)

[Ge] Gever, J.L.: The existence of pseudocollisions in the plane. To appear in J.Differential Equations

[Gh] Ghys, E.: Flots d'Anosov dont les feulletages stables sont différentiables. Ann.Sci.ENS **20**, 251–271 (1987)

[GK1] Guillemin, V., Kazhdan, D.: Some inverse spectral results for negatively curved 2-manifolds. Topology **19**, 301–312 (1980)

[GK2] Guillemin, V., Kazhdan, D.: Some inverse spectral results for negatively curved n manifolds. AMS Proc.Sympos.Pure Math. **35**, 153–180 (1982)

[Gu] Gutkin, E.: Asymptotic velocities in classical mechanics. J.Math.Phys. **30**, 1245–1249 (1989)

[Ha] Hagedorn, G.A.: Proof of the Landau-Zenner formula in an adiabatic limit with small eigenvalue gaps. Comm.Math.Phys. **136**, 433–451 (1991)

[He] Hedlund, G.: The dynamics of geodesic flows. Bull.Amer.Math.Soc. **45**, 241–260 (1969)

[He] Herman, M.R.: Sur les courbes invariantes par les diffeomorphismes de l'anneau. I. Asterisque 103–104 (1983); II. Ibid. 144 (1986)

[He2] Herman, M.R.: Inégalités a priori pour des tores Lagrangunes invariants par des diffeomorphisms symplectiques. Publ.Math.IHES **63**, 47–101 (1990)

[Ho] Hopf, E.: Statistik der geodätischen Linien in Mannigfaltigkesten negativer Krümmung. Akad.Wiss.Leipzig **91**, 261–304 (1939)

[Hu] Hubacher, A.: Classical scattering theory in 1 dimension. Preprint

[Hu] Hurder, S.: Rigiditity for Anosov actions of higher rank lattices. Preprint

[HK] Hurder, S., Katok, A.: Differentiability, rigidity and Godbillon-Vey classes for Anosov flows. To appear in Publ.Math.IHES

[JS] Jakšić, V., Segert, J.: Exponential approach to the limit and the Landau-Zenner formula. Preprint

[JS] Jorba, A., Simó, C.: On the reducibility of linear differential equations with quasi-periodic coefficients. Preprint

[Ju] Jungreis, I.: Ergodic Theory Dynamical Systems **11**, 79–84 (1991)

[KKW] Katok, A., Knieper, G., Weiss, H.: Formulas for the derivative and critical points of topological entropy for Anosov and geodesic flows. Comm.Math.Phys. **130**, 19–33 (1991)

[KSS] Krámli, A., Simányi, N., Szász, D.: A "transversal" fundamental theorem for semi dispersing billiards. Comm.Math.Phys. **129**, 535–560 (1989); Erratum Comm.Math.Phys. **138**, 207 (1991)

[Koz] Kozlov, V.V.: Integrability nad non-integrability in Hamiltonian mechanics. Russian Math.Surveys **38**, 1-76 (1983)

[Ku] Kuksin, S.B.: Perturbation theory for quasiperiodic solutions of infinite dimensional hamiltonian systems and its applications to the Korteweg-deVries equation. Math.USSR-Sb. **64**, 397–413 (1989)

[Ku2] Kuksin, S.B.: Perturbation theory for quasi-periodic solutions of infinite dimensional Hamiltonian systems, Parts I, II, III. Max Plank preprints

[La] Lanford, O.E.: Computer assisted proofs in Analysis. Physica A **124**, 465–470 (1984)

[Lo] Lochak, P.: Stabilité en temps exponentiels des systèmes hamiltoniens proches de systèmes integrables: Resonances et orbites fermées. ENS preprint

[L1] de la Llave, R.: Invariants for smooth conjugacy of hyperbolic dynamical systems II. Comm.Math.Phys. **109**, 369–378 (1987)

[L2] de la Llave, R.: Analytic regularity of solutions of Livsic's cohomology equation and application to smooth conjugacy of hyperbolic dynamical systems. To appear in Ergodic Theory Dynamical Systems

[L3] de la Llave, R.: Smooth conjugacy and SRB measures for uniformly and non-uniformly hyperbolic systems. To appear in Comm.Math.Phys.

[L4] de la Llave, R.: A simple proof of a particular case of C Siegel's center theorem. J.Math.Phys. **24**, 2118–2121 (1983)

[LMM] de la Llave, R., Marco, J.M., Moriyón, R.: Canonical perturbation theory for Anosov systems and regularity properties of Livsic's cohomology equation. Ann.of Math. **123**, 537–611 (1986)

[LM] de la Llave, R., Moriyón, R.: Invariants for smooth conjugacy of hyperbolic dynamical systems IV. Comm.Math.Phys. **116**, 185–192 (1988)

[LR] de la Llave, R., Rana D.: Accurate strategies for K.A.M. bounds and their implementation. In: Computer aided Proofs in Analysis. Meyer, K., Schmidt, D. (eds), Springer Verlag (1991)

[LW] de la Llave, R., Wayne, C.E.: Whiskered and low dimensional tori in nearly integrable dynamical systems. Preprint

[MMcG] Mather, J., McGehee, R.G.: Solutions of the four body problem which become unbounded in finite time. Lecture Notes in Phys. **38** (1975)

[MMS] McKay R.S., Meiss J.D., Stark, J.: Converse KAM theory fro symplectic twist maps. Nonlinearity **2**, 555-570 (1989)

[MP] McKay, R.S., Percival, I.C.: Converse K.A.M. theory and practice. Comm.Math.Phys. **98**, 569–512 (1985)

[Mi] Milojevic, P.S. (ed): *Nonlinear functional analysis.* Marcel Decker, New York, 1990

[Mo] Moser, J.: Is the solar system stable? Math.Intelligencer **1**, 65–71 (1978)

[Mor] Morales, J.J.: Técnicas algebraicas para el estudio de la integrabilidad algebraica de sistemas hamiltonianos. Tesis U. Barcelona (1989)

[Mu] Muldoon, M.: Gohsts of order on the frontier of chaos. PhD thesis, Caltech, 1990

[Nek] Nehoroshev, A.N.: An exponential estimate for the time of stability of nearly integrable Hamiltonian systems. Russian Math.Surveys **32**, no.6, 1–65 (1977)

[Ne2] Neishdtadt, A.: The separation of motions in systems with a rapidly rotating phase. P.M.M. USSR **40**, 133–139 (1984)

[Ni] Nikolenko, N.V.: Invariant asymptotically stable tori of the perturbed Korteweg-deVries equation. Russian Math.Surveys **35**, 139–207 (1980)

[O] Otal, J.P.: Le spectre marqué des longeurs des surfaces à courbure negative. Ann.of Math. **131**, 151–162 (1990)

[PY1] Palis, J., Yoccoz, J-C.: Rigidity of centralizers of diffeomorphisms. Ann.Sci.ENS **22**, 81–99 (1989)

[PY2] Palis, J., Yoccoz, J-C.: Centralizers of Anosov diffeomorphisms on tori. Ann.Sci.ENS **22**, 99–109 (1989)

[PM] Perez Marco, R.: Centralizers I: Uncountable centralizers for non linearizable germs of (C,0) IHES preprint 91/23

[PS] Pesin, Y., Sinai, Y.G. Space time chaos in the system of weakly interactive hyperbolic systems. Preprint

[Pa] Paluba, W.: The Lipschitz condition for the conjugacies of Feigenbaum-like mappings. Fund.Math. **132**, 227–257 (1989)

[Pe] Pesin, Y.: Characteristic Lyapunov exponents and smooth ergodic theory. Russian Math.Surveys **32**, 55-114 (1976)

[Pe2] Pesin, Y.: An example of a non-ergodic flow with non-zero characteristic exponents. Functional Anal.Appl. **8**, 263–264 (1974)

[Po] Pollicott, M.: C^k rigidity for hyperbolic flows. To appear in Israel J.Math.

[Pö] Pöschel, J.: On elliptic lower dimensional tori in hamiltonian systems. Math.Z. **202**, 559–608 (1989)

[Pö] Pöschel, J.: Small divisors with spatial structure in infinite dimensional dynamical systems. Comm.Math.Phys. **127**, 351-393 (1990)

[R] Rana, D. Proof of accurate upper and lower bounds to stability domains in small denominator problems: Princeton thesis (1987)

[RGB] Ramani, A., Gramatikos, B., Bountis, T.: The Panilevé Property and singularity analysis of integrable and non-integrable systems. Phys.Rep. **180** (1989)

[RS] Radin, C., Simon, B.: Invariant domains for the time- dependent Schrödinger equation. J.Differential Equations **29**, 289-296 (1978)

[Ra] Rand, D.: Global phase space universality, smooth conjugacies and renormalisations. The $C^{1+\alpha}$ case. Nonlinearity **1**, 181–202 (1988)

[Rü] Rüssmann, H.: On a new proof of Moser's twist mapping theorem. Celestial Mech. **14**, 19–31 (1976)

[SM] Siegel, C.L., Moser, J.: *Lectures on Celestial Mechanics.* Springer-Verlag, 1971

[SCh] Sinai, Y.G., Chernoff, S.: Ergodic porperties of certain systems of two dimensional discs and three dimensional balls. Russian Math.Surveys **42**, 181-207 (1987)

[Si] Sinai, Ya.G.: Dynamical systems with elastic reflections. Ergodic properties of dispersing billiards. Russian Math.Surveys **25** no.2, 137–189 (1970)

[Su] Sullivan, D.: Bounds, quadratic differentials and renormalization conjectures. Preprint

[Sn] Sunada, T.: Riemannian coverings and isospectral manifolds. Ann.of Math. **121**, 169–186 (1985)

[Sun] Sun, Y.S.: On the measure preserving mappings with odd dimension. Celestial Mech. **30**, 7–19 (1983)

[Tr] Treshev, D.V.: The mechanisms of destruction of resonance tori of hamiltonian systems. Math.USSR-Sb. **68**, 181–203 (1991)

[V] Vitott, M. Théorie classique des perturbations en grand nombre de degres de liberté. Univ. de Marseille, Thesis (1985)

[W] Wojtkowski, M.: Principles for the design of Billiards with Non-vanishing Lyapunov exponents. Comm.Math.Phys. **105**, 391–414 (1986)

[W1] Wayne, C.E.: The K.A.M. theory of systems with short range interactions. Part I, Comm.Math.Phys. **96**, 311-329 (1984); Part II ibid. 331–344

[W2] Wayne, C.E.: Periodic and quasi-periodic solutions of nonlinear wave equations via K.A.M. Theory. Comm.Math.Phys. **127**, 479–528 (1990)

[W3] Wayne, C. E.: On the elimination of non-resonant harmonics. Comm. Math.Phys. **103**, 351-386 (1986)

[W3] Wayne, C. E.: Uniform estimates on trajectories in hamiltonian systems with many degrees of freedom. Comm.Math.Phys. **104**, 21-36 (1986)

[Xi] Xia, Z.: The existence of the non-collision singularities in Newtonian systems. PhD thesis, Northwestern University, 1988

[Y1] Yoccoz, J-C.: Théorème de Siegel, polynômes quadratiques et nombres de Brjuno. Preprint

[Y2] Yoccoz, J-C.: Linéarization des germes de diffeomorphismes holomorphes de (C,0). C.R.Acad.Sci.Paris **306**, 555-558 (1988)

[Z] Zimmer, W.: Infinitesimal rigidity for smooth actions of discrete subgroups of Lie groups.

Introduction to the Differential Geometry of Quantum Groups

Bruno Zumino

Department of Physics, University of California, and Theoretical Physics Group, Lawrence Berkeley Laboratory, Berkeley, California 94720

Abstract: An introduction to the noncommutative differential calculus on quantum groups. The invariant group average is also discussed.

1 Introduction

In this lecture I shall try to describe those aspects of the mathematics of quantum groups which I believe may be relevant for the formulation of new physical theories based on noncommutative calculus and noncommutative geometry. As a consequence, there is no attempt here to cover the standard applications of the theory to integrable systems, knot theory and conformal field theory in two dimensions.

A very helpful review of the more conventional mathematical approach can be found in [1], which contains numerous references. This standard approach is based on the concept of quasitriangular Hopf algebras. It provides a satisfactory all-embracing theory of quantum groups but it ignores the existence of the quantum exterior differential calculus of Woronowicz on quantum groups and therefore the connection with noncommutative differential geometry [2]. Here I shall underplay the importance of Hopf algebras and concentrate instead on the calculus aspect. Comultiplication is barely mentioned, in (2.17) and (3.4). Instead I emphasize equations such as (3.5),(4.2) and (4.15) from which the comultiplication rule can be derived. The word antipode will never appear here, although plenty of inverse quantum matrices will play a role. I shall emphasize the computational aspect which I believe will be important in the "new physics". The quantum groups $SL_q(2)$ and $SU_q(2)$ provide useful simple examples in the following. This restricted choice of material can also be justified by space limitations. Many details are contained in the papers quoted here, where numerous additional references can also be found.

2 Quantum Groups and Lie Algebras

This section is a review of some very well known facts about quantum groups which are needed in the following. We follow mostly the approach of Faddeev, Reshetikin and Takhtajan [3]. A quantum group can be defined in terms of the defining representation of the corresponding Lie group. For instance, $GL_q(2)$ can be defined in terms of the two by two matrix

$$T = \begin{pmatrix} \alpha & \beta \\ \gamma & \delta \end{pmatrix}. \tag{2.1}$$

The space of the group parameters α, β, γ and δ can be quantized by turning the parameters into non commuting quantities satisfying the commutation relations

$$\begin{array}{rl} \alpha\beta = q\beta\alpha & \alpha\gamma = q\gamma\alpha \\ \beta\delta = q\delta\beta & \gamma\delta = q\delta\gamma \\ \alpha\delta - \delta\alpha = \lambda\beta\gamma & \beta\gamma = \gamma\beta, \end{array} \tag{2.2}$$

where q is a generic complex number and we shall consistently use the abbreviation

$$\lambda = q - q^{-1}. \tag{2.3}$$

Using these commutation relations it is easy to check that the "quantum determinant"

$$\det_q T = \alpha\delta - q\beta\gamma = \delta\alpha - q^{-1}\beta\gamma \tag{2.4}$$

is central, i.e. it commutes with α, β, γ and δ. We shall call a matrix like (2.1) whose elements satisfy (2.2) a q-matrix.

The commutation relations (2.2) have the following remarkable property. Let $\alpha', \beta', \gamma', \delta'$ commute with $\alpha, \beta, \gamma, \delta$ and let them satisfy the same commutation relations (2.2). So

$$\alpha'\beta' = q\beta'\alpha' \quad etc..., \tag{2.5}$$

which means that

$$T' = \begin{pmatrix} \alpha' & \beta' \\ \gamma' & \delta' \end{pmatrix} \tag{2.6}$$

is a q-matrix with elements commuting with those of T. Then the matrix

$$T'' = TT' = \begin{pmatrix} \alpha'' & \beta'' \\ \gamma'' & \delta'' \end{pmatrix} = \begin{pmatrix} \alpha\alpha' + \beta\gamma' & \alpha\beta' + \beta\delta' \\ \gamma\alpha' + \delta\gamma' & \gamma\beta' + \delta\delta' \end{pmatrix} \tag{2.7}$$

obtained from T and T' by matrix multiplication, rows by columns, is also a q-matrix. We refer to this fact as to the quantum group property of (2.2).

The commutation relations (2.2) can be written compactly in terms of the R-matrix of the quantum group [3]. Consider the tensor product of two vector spaces and let the matrix T operate on it as

$$T_1 = T \otimes 1, \tag{2.8}$$

i.e. T on the first space and the identity on the second, or explicitly

$$(T_1)^{ik}_{j\ell} = T^i{}_j \delta^k{}_\ell. \tag{2.9}$$

Similarly define

$$T_2 = 1 \otimes T, \tag{2.10}$$

i.e.

$$(T_2)^{ik}_{j\ell} = \delta^i{}_j T^k{}_\ell. \tag{2.11}$$

If the matrix elements of T were commuting quantities, T_1 and T_2 would commute. Because of (2.2) their products in opposite order are not equal but they are related by conjugation in the tensor product space

$$R_{12} T_1 T_2 = T_2 T_1 R_{12}. \tag{2.12}$$

The matrix R_{12} operates in the tensor product space and is given for $GL_q(2)$ by

$$(R_{12})^{ik}_{j\ell} = \begin{pmatrix} q & 0 & 0 & 0 \\ 0 & 1 & 0 & 0 \\ 0 & \lambda & 1 & 0 \\ 0 & 0 & 0 & q \end{pmatrix} \tag{2.13}$$

where the rows (and columns) are numbered as 11,12,21,22. We shall call (2.12) the "RTT equation". Using it, it is very easy to prove the quantum group property, since T' also satisfies an equation like (2.12) and T_1' and T_2 commute.

The R-matrix satisfies the "Yang-Baxter equation" in the triple tensor product space

$$R_{12} R_{13} R_{23} = R_{23} R_{13} R_{12} \tag{2.14}$$

which ensures consistency of the RTT equation (2.12). In this approach classifying all possible quantum groups is equivalent to finding all solutions of the Yang-Baxter equation[3].

Since the determinant (2.4) is central, one can set it equal to unity:

$$\alpha\delta - q\beta\gamma = 1. \tag{2.15}$$

The quantum group $GL_q(2)$ is then restricted to the "subgroup" $SL_q(2)$. However beware: quantum groups are not groups.

For quantum groups other than $GL_q(n)$ or $SL_q(n)$ in addition to giving the relevant R matrix one must impose further restrictions compatible with

(2.12) and (2.14), such as orthogonality or other conditions for the quantum matrices. See reference [3] where the cases of quantum orthogonal and symplectic groups are described in detail.

Jimbo [4] and Drinfeld [5] have provided us with a consistent quantum deformation of any simple Lie algebra. For the case of $SL(2)$ one has the generators H, X_+ and X_- and their commutation relations are deformed to

$$\begin{aligned}[][H, X_\pm] &= \pm 2X_\pm \\ [X_+, X_-] &= \frac{q^H - q^{-H}}{q - q^{-1}}.\end{aligned} \tag{2.16}$$

The comultiplication is given by

$$\begin{aligned}\Delta(H) &= H \otimes 1 + 1 \otimes H \\ \Delta(X_\pm) &= X_\pm \otimes q^{H/2} + q^{-H/2} \otimes X_\pm.\end{aligned} \tag{2.17}$$

It is easy to check that (2.16) and (2.17) are consistent with each other. For $q \to 1$ they become the standard relations for $SL(2)$ (or $SU(2)$ if appropriate reality conditions are imposed).

3 Relation Between the Quantum Group and the Quantum Lie Algebra

In this section we follow again the approach of reference [3] with some minor modification. We continue to use the example of $SL(2)$.

Define the upper diagonal matrix

$$L^+ = \begin{pmatrix} q^{-H/2} & \lambda X_+ \\ 0 & q^{H/2} \end{pmatrix} \tag{3.1}$$

and the lower diagonal matrix

$$L^- = \begin{pmatrix} q^{H/2} & 0 \\ -\lambda X_- & -q^{-H/2} \end{pmatrix}. \tag{3.2}$$

It is easy to see that the algebra (2.16) can be written compactly as

$$\begin{aligned}R_{12} L_2^+ L_1^+ &= L_1^+ L_2^+ R_{12} \\ R_{12} L_2^- L_1^- &= L_1^- L_2^- R_{12} \\ R_{12} L_2^+ L_1^- &= L_1^- L_2^+ R_{12}.\end{aligned} \tag{3.3}$$

The first two equations (3.3) state that L^+ and L^- are q^{-1}-matrices. From their definition (3.1) they have quantum determinant equal to one. The comultiplication rule (2.17) is equivalent to

$$\Delta((L^+)^i{}_j) = (L^+)^i{}_k \otimes (L^+)^k{}_j$$
$$\Delta((L^-)^i{}_j) = (L^-)^i{}_k \otimes (L^-)^k{}_j \tag{3.4}$$

(sum over repeated indices).

The matrix elements of L^+ and L^- can be identified with quantum differential operators on group space. Their action on the group variables is given by the equations

$$L_1^+ T_2 = T_2 \mathcal{R}_{21} L_1^+$$
$$L_1^- T_2 = T_2 \mathcal{R}_{12}^{-1} L_1^-. \tag{3.5}$$

Here $\mathcal{R}$ is a suitably normalized R-matrix. For $SL(n)$

$$\mathcal{R} = q^{-1/n} R. \tag{3.6}$$

The normalization factor is determined by the property that L^+ and L^- have q-determinant equal to one. It is easy to check, using (2.14) that (3.5) are consistent with (2.12), i.e. the action on the group preserves the quantum structure of the group. The algebra (3.3) and the comultiplication law (3.4) can actually be derived from (3.5), which can be taken as the basic relations.

Equations (3.3-5) are general. Consistent relations among the matrix elements of L^+ and among those of L^- (or an appropriate ansatz generalizing (3.1)) must be given for different groups so that the number of independent generators agrees with that of the classical Lie algebra (see [3], [6]).

Let us define a right "vacuum"$>$ and a left "vacuum" $<$ such that

$$L^+ >= L^- >= I > \tag{3.7}$$

and

$$< T =< I \tag{3.8}$$

where I is the unit matrix. Using (3.5) one can compute vacuum values of products. For instance

$$< L_0^+ T_1 >=< T_1 \mathcal{R}_{10} L_0^+ >= \mathcal{R}_{10}. \tag{3.9}$$

More generally

$$< L_0^+ T_1 T_2 ... T_n > = \mathcal{R}_{10} \mathcal{R}_{20} ... \mathcal{R}_{n0}$$
$$< L_0^- T_1 T_2 ... T_n > = \mathcal{R}_{01}^{-1} \mathcal{R}_{02}^{-1} ... \mathcal{R}_{0n}^{-1}. \tag{3.10}$$

The knowledge of all vacuum values is equivalent to the basic relations (3.5). We see that the enveloping algebra of the quantum Lie algebra is dual to the algebra of functions on the quantum group and consists of "regular" linear functionals of these functions.

4 Bicovariant Calculus

The bicovariant calculus on quantum groups is due to Woronowicz [7]. Here we follow the approach of Jurčo [9], which provides a direct connection between the calculus of Woronowicz and the work of Faddeev, Reshetikin and Takhtajan [3].

Define the matrix

$$Y = L^+(L^-)^{-1} \tag{4.1}$$

which is neither upper nor lower triangular. It is not hard to see that (3.5) imply (see the explicit example (6.8) below)

$$Y_1 T_2 = T_2 \mathcal{R}_{21} Y_1 \mathcal{R}_{12} \tag{4.2}$$

and that the algebra relations (3.3) imply

$$\mathcal{R}_{21} Y_1 \mathcal{R}_{12} Y_2 = Y_2 \mathcal{R}_{21} Y_1 \mathcal{R}_{12}. \tag{4.3}$$

These equations have the remarkable property that they are covariant under the transformation

$$T \to T'T, \quad Y \to Y, \tag{4.4}$$

as well as under

$$T \to TT', \quad Y \to (T')^{-1} Y T'. \tag{4.5}$$

The matrix elements of T' are taken to commute with those of T as well as with those of Y. Furthermore, the matrix T' satisfies

$$R_{12} T'_1 T'_2 = T'_2 T'_1 R_{12}.$$

We shall say that (4.2) and (4.3) are "left-invariant" (because of (4.4)) and "right-covariant" (because of (4.5)). The matrix elements of Y are the differential operators of a "bicovariant" calculus on the quantum group.

Take the case of $SL(2)$ and write

$$Y = \begin{pmatrix} y_1 & y_+ \\ y_- & y_2 \end{pmatrix}. \tag{4.6}$$

Using (4.3) one can verify that

$$D = y_1 y_2 - q^2 y_+ y_- \tag{4.7}$$

commutes with y_1, y_2, y_+ and y_- and that

$$Y^{-1} = \frac{1}{D} \begin{pmatrix} y_2 & -q^2 y_+ \\ -q^2 y_- & q^2 y_1 + (1-q^2) y_2 \end{pmatrix}. \tag{4.8}$$

Furthermore, when Y is given by (4.1), it follows from (3.3) that

$$D = (\det_{q^{-1}} L^+)(\det_{q^{-1}} L^-)^{-1}. \tag{4.9}$$

The special matrices L^+ and L^- given in (3.1) and (3.2) have q^{-1}-determinant equal one, therefore

$$D = 1. \tag{4.10}$$

We could have used Y^{-1} instead of Y, but we see from (4.8) that the corresponding operators are linear combinations of the elements of Y. The calculus of Y^{-1} is completely equivalent to that of Y. An alternative choice is given by $(L^+)^{-1}L^-$ or its reciprocal. This gives operators which belong to the same enveloping algebra but are not linearly related to those of Y and Y^{-1}. The resulting calculus is right invariant and left covariant.

The "determinant" D, given by (4.7), commutes not only with the elements of Y but also with the matrix elements of T, as one can check using (4.2). Therefore (4.10) is a consistent condition. In general there are other quantities which commute with the elements of Y but not with those of T. These correspond to the classical Casimir operators, which are central in the Lie algebra, but have a nontrivial action on the group. For $SL(2)$ the Casimir operator is given, by

$$\mathcal{C} = y_1 + q^{-2}y_2. \tag{4.11}$$

A general formula for Casimir operators [3] can be given in terms of the invariant quantum trace discussed at the beginning of the next section.It is

$$\mathcal{C}_k = Tr(\mathcal{D}^{-1}Y^k) \tag{4.12}$$

where $k = 1, 2, ...r$ and r is the rank of the group.

As $q \to 1$, the matrices L^+, L^- and Y tend to the unit matrix I. To establish a connection with the classical Lie algebra, define the matrix X by

$$Y = I - \lambda X. \tag{4.13}$$

The matrix elements of

$$X = \begin{pmatrix} \chi_1 & \chi_+ \\ \chi_- & \chi_2 \end{pmatrix} \tag{4.14}$$

correspond to the generators of the classical Lie algebra (differential operators on the group). From (4.2) and (4.3), one can easily obtain the analogous relations for X. They are

$$X_1T_2 = T_2\mathcal{R}_{21}X_1\mathcal{R}_{12} - \frac{1}{\lambda}T_2(\mathcal{R}_{21}\mathcal{R}_{12} - I_{12}) \tag{4.15}$$

and

$$\begin{aligned} &\mathcal{R}_{21}X_1\mathcal{R}_{12}X_2 - X_2\mathcal{R}_{21}X_1\mathcal{R}_{12} \\ &= \frac{1}{\lambda}(\mathcal{R}_{21}\mathcal{R}_{12}X_2 - X_2\mathcal{R}_{21}\mathcal{R}_{12}). \end{aligned} \tag{4.16}$$

The condition (4.10) becomes, in terms of X,

$$\chi_1 + \chi_2 - \lambda\chi_1\chi_2 + q^2\lambda\chi_+\chi_- = 0. \tag{4.17}$$

Using this equation one can eliminate, for instance, χ_1. The Casimir operator (4.11) becomes then

$$\mathcal{C} = 1 + q^{-2} + q^{-2}\lambda^2(1-\lambda\chi_2)^{-1}[q\chi_2 + \chi_2^2 + q^4\chi_+\chi_-]. \tag{4.18}$$

As $q \to 1$ the expression in square brackets tends to the well known classical expression for the $SL(2)$ Casimir operator.

Although we started from he matrices L^+ and L^- and defined Y by (4.1) in terms of them, we can now formulate the bicovariant calculus directly in terms of Y and the eqs. (4.2), (4.3) (4.7) and (4.10) which it satisfies. Then X will still be defined from (4.13). The matrices L^+ and L^- are left-invariant but (because of the triangularity conditions they satisfy) they transform in a very complicated way under the right multiplication (4.5). Nevertheless, as shown in [6], one can always decompose a matrix like Y into a product like (4.1) where L^+ is upper and L^- lower triangular. It seems that the triangularity properties of L^+ and L^- should be considered as special choices of gauge.

5 Quantum Differential Forms

Just as for ordinary Lie groups, one can introduce exterior differential forms [7]. We shall derive their properties from those of the differential operators on the group. We first notice that, if a matrix M transforms as

$$M \to (T')^{-1}MT' = M' \tag{5.1}$$

where T' is a q-matrix whose elements commutes with those of M, then the quantum trace

$$Tr(\mathcal{D}^{-1}M) = Tr(\mathcal{D}^{-1}M') \tag{5.2}$$

is invariant, where $\mathcal{D}$ is a suitable matrix and Tr denotes the usual trace. In general, $\mathcal{D}$ must satisfy, for any q-matrix T,

$$\mathcal{D}^t(T^{-1})^t(\mathcal{D}^t)^{-1} = (T^t)^{-1} \tag{5.3}$$

where t denotes the ordinary transposed of a matrix. It turns out that $\mathcal{D}$ can be chosen diagonal. For $SL(n)$,

$$\mathcal{D} = diag\left(1, q^2, q^4, ...q^{2(n-1)}\right). \tag{5.4}$$

Let us now introduce a matrix of differential one-forms

$$\Omega = \begin{pmatrix} \omega^1 & \omega^- \\ \omega^+ & \omega^2 \end{pmatrix}. \tag{5.5}$$

The exterior differential

$$\begin{aligned} d &= Tr(\mathcal{D}^{-1}\Omega X) \\ &= \omega^1\chi_1 + \omega^-\chi_- + q^{-2}\omega^+\chi_+ + q^{-2}\omega^2\chi_2 \end{aligned} \tag{5.6}$$

is invariant if one transforms

$$X \to (T')^{-1}XT' \tag{5.7}$$

(see (4.5) and (4.13)) and

$$\Omega \to (T')^{-1}\Omega T'. \tag{5.8}$$

At the same time the quantum trace of Ω

$$\xi = Tr(\mathcal{D}^{-1}\Omega) = \omega^1 + q^{-2}\omega^2 \tag{5.9}$$

is invariant.

The exterior differential is required to satisfy the standard underformed relations

$$d^2 = 0, \quad d(\text{constant}) = 0 \tag{5.10}$$

and

$$d(fg) = dfg + (-1)^{p(f)}fdg, \tag{5.11}$$

where $p(f)$ is the parity of f. The properties of the differential one-forms can be derived from those of the differential operators by using (5.10) and (5.11). For instance, it must be

$$\begin{aligned} 0 = d^2 = & \, d\omega^1\chi_1 + d\omega^-\chi_- + q^{-2}d\omega^+\chi_+ + q^{-2}d\omega^2\chi_2 \\ & - \omega^1 d\chi_1 - \omega^- d\chi_- - q^{-2}\omega^+ d\chi_+ - q^{-2}\omega^2 d\chi_2. \end{aligned} \tag{5.12}$$

Substitute (5.6) in the last four terms of this equation. We know the commutation relations among the χ's from (4.16). With a little work one obtains the commutation relations for the one-forms

$$\begin{aligned} \omega^+\omega^- + \omega^-\omega^+ &= 0 \\ \omega^1\omega^+ + \omega^+\omega^1 &= 0 \\ \omega^1\omega^- + \omega^-\omega^1 &= 0 \\ \omega^2\omega^+ + q^2\omega^+\omega^2 &= q\lambda\omega^+\omega^1 \\ \omega^2\omega^- + q^{-2}\omega^-\omega^2 &= -q^{-1}\lambda\omega^-\omega^1 \\ \omega^1\omega^2 + \omega^2\omega^1 &= -q^{-1}\lambda\omega^+\omega^- \\ (\omega^1)^2 = (\omega^+)^2 &= (\omega^-)^2 = 0 \\ (\omega^2)^2 &= q\lambda\omega^+\omega^- \end{aligned} \tag{5.13}$$

and the quantum Maurer-Cartan equations

$$
\begin{aligned}
d\omega^1 &= -q^{-3}\omega^+\omega^- \\
d\omega^2 &= q^{-1}\omega^+\omega^- \\
d\omega^+ &= q^{-1}\omega^+(\omega^1-\omega^2) \\
d\omega^- &= q^{-1}(\omega^1-\omega^2)\omega^- .
\end{aligned} \tag{5.14}
$$

These are the bicovariant relations of Woronowicz [7] for the case of $SL(2)$. They have a very interesting property as pointed out by Woronowicz in general. Although the group $SL(2)$ has only three parameters, there are four one-forms, one too many, it would seem at first. However, the invariant form ξ of (5.9) can be easily seen to satisfy

$$\xi^2 = 0, \quad d\xi = 0 \tag{5.15}$$

and, for all $\omega^\alpha, \alpha = 1, -, +, 2$

$$\xi\omega^\alpha + \omega^\alpha\xi = \lambda d\omega^\alpha . \tag{5.16}$$

As $q \to 1, \lambda \to 0$ and ξ decouples. What happens to $\lambda^{-1}\xi$ in the limit? We leave this as an interesting exercise for the reader.

Finally, it is clear that (4.15) and (5.6) allow us to derive the commutation relations between the one-forms and the matrix elements of the matrix T. One finds, in general

$$\Omega_1 T_2 = T_2 \mathcal{R}_{12}^{-1}\Omega_1\mathcal{R}_{21}^{-1} . \tag{5.17}$$

The derivation of this equation requires the identity, satisfied by the $\mathcal{R}$ matrix,

$$\left((\mathcal{R}_{12}^{-1})^{t_1}\right)^{-1} = \mathcal{D}_1^{-1}\mathcal{R}_{12}^{t_1}\mathcal{D}_1 , \tag{5.18}$$

where t_1 denotes transposition in the space 1 of the tensor product. For $SL(2)$ one obtains, from (5.17),

$$
\begin{aligned}
\alpha\omega^1 &= q\omega^1\alpha \\
\alpha\omega^- &= \omega^-\alpha + \lambda\omega^1\beta \\
\alpha\omega^+ &= \omega^+\alpha \\
\alpha\omega^2 &= q^{-1}\omega^2\alpha + q^{-1}\lambda\omega^+\beta
\end{aligned} \tag{5.19}
$$

and

$$
\begin{aligned}
\beta\omega^1 &= q^{-1}\omega^1\beta \\
\beta\omega^- &= \omega^-\beta \\
\beta\omega^+ &= \omega^+\beta + q^2\lambda\omega^1\alpha \\
\beta\omega^2 &= q\omega^2\beta + q\lambda^2\omega^1\beta + q\lambda\omega^-\alpha .
\end{aligned} \tag{5.20}
$$

The same equations are valid with α replaced by γ and β replaced by δ. For functions on the group the invariant form ξ plays a similar role as in (5.16) for forms: if f is a function of α, β, γ and δ, one has

$$\xi f - f\xi = \lambda df. \tag{5.21}$$

Notice that, from (5.19)and (5.20), all one-forms ω^α, for $\alpha = 1, -, +, 2$, commute with the quantum determinant (2.4) of T. Therefore the invariant form ξ also commute with it and, by (5.21), it is identically

$$d(\det_q T) \equiv 0. \tag{5.22}$$

As a consequence one can impose (2.15).

There is an alternative way to introduce differential forms on a quantum group, which is perhaps closer to that which one often follows in the classical case. Define the new matrix of forms

$$\widetilde{\Omega} = T^{-1} dT, \tag{5.23}$$

which satisfies

$$d\widetilde{\Omega} = -\widetilde{\Omega}^2. \tag{5.24}$$

Clearly $\widetilde{\Omega}$ is left-invariant (under $T \to T'T$) and right-covariant, i.e. under $T \to TT'$ it transforms as

$$\widetilde{\Omega} \to (T')^{-1} \widetilde{\Omega} T'. \tag{5.25}$$

How are Ω and $\widetilde{\Omega}$ related? Since they transform in the same way, we are led to the identification

$$\widetilde{\Omega} = c_1 \Omega + c_2 \xi I \tag{5.26}$$

where c_1 and c_2 are constants. For $SL(n)$ one has

$$c_1 = -q^{2/n-2n+1}, \; c_2 = q \left[\frac{1}{n} \right], \tag{5.27}$$

where in general

$$[x] \equiv \frac{1 - q^{2x}}{1 - q^2}. \tag{5.28}$$

6 Invariant Measure

The general properties of a left and right invariant Haar measure for compact quantum groups were discussed by Woronowicz [8]. Here we consider briefly the case of $SU_q(2)$ and, using a different technique, we compute explicitly the invariant measure.

For $SU_q(2)$ the unitarity condition

$$T^\dagger = T^{-1} \tag{6.1}$$

for the matrix (2.1) gives

$$\begin{pmatrix} \overline{\alpha} & \overline{\gamma} \\ \overline{\beta} & \overline{\delta} \end{pmatrix} = \begin{pmatrix} \delta & -q^{-1}\beta \\ -q\gamma & \alpha \end{pmatrix}, \tag{6.2}$$

where the bar indicates an involution of the algebra of functions on the group which changes the order of factors in a product. For consistency it must be

$$\overline{q} = q. \tag{6.3}$$

In the following we shall keep in mind that

$$\delta = \overline{\alpha}, \quad \beta = -q\overline{\gamma} \tag{6.4}$$

but we shall continue to use the letters β and δ for those matrix elements. Ordered monomials in α, β, γ and δ can be taken as a basis for functions on the group or at least for polynomials. Using the determinant condition

$$\alpha\delta - q\beta\gamma = 1 = \delta\alpha - q^{-1}\beta\gamma \tag{6.5}$$

and the commutation relations (2.2) one can transform any monomial to the form $\alpha^k\beta^\ell\gamma^m$ or to the form $\delta^k\beta^\ell\gamma^m$, where the exponents take all integer values $0, 1, 2, \dots$. We shall take these monomials as a complete basis.

We wish to associate to a function $\langle f\rangle$ on the group a real number, its invariant group average, which we shall denote as $\langle f\rangle$. There are different ways to state the invariance of the group average. For algebraic manipulations a convenient way is to require that

$$\langle \chi_\alpha f\rangle = 0, \tag{6.6}$$

where χ_α are the differential operators of a bicovariant calculus, the matrix elements of the matrix X in (4.13). For $q \neq 1$, we can use (4.13) and rewrite (6.6) in the very convenient form

$$\langle Y^i{}_j f\rangle = \delta^i{}_j. \tag{6.7}$$

This condition actually allows us to compute the average for all basic monomials, by means of the commutation relations (4.2).

We write explicitly the commutation relations (4.2) for the example of $SU_q(2)$. They are

$$\begin{aligned} y_+\alpha &= \alpha y_+ + q^{-1}\lambda\beta y_2 \\ y_-\alpha &= \alpha y_- \\ y_2\alpha &= q^{-1}\alpha y_2 \\ y_+\beta &= \beta y_+ \\ y_-\beta &= \beta y_- + q^{-1}\lambda\alpha y_2 \\ y_2\beta &= q\beta y_2, \end{aligned} \tag{6.8}$$

together with the equations obtained by replacing α with γ and β with δ. We have not written the relations involving y_1, which we consider as defined by (4.7), (4.10). Using (6.8) one can easily compute

$$(y_2\alpha^k\beta^\ell\gamma^m) = q^{-k+\ell-m}\alpha^k\beta^\ell\gamma^m. \tag{6.9}$$

Together with (6.7), which is now

$$\langle y_2\alpha^k\beta^\ell\gamma^m\rangle = 1 \tag{6.10}$$

this implies that $\langle\alpha^k\beta^\ell\gamma^m\rangle$ must vanish unless $k+m-\ell=0$. Computing $(y_-\alpha^k\beta^\ell\gamma^m)$ one finds, since (6.7) gives

$$\langle y_-\alpha^k\beta^\ell\gamma^m\rangle = 0, \tag{6.11}$$

that

$$\langle\alpha^k\beta^\ell\gamma^m\rangle = 0 \text{ unless both } k=0 \text{ and } \ell=m. \tag{6.12}$$

A similar argument gives the result

$$\langle\delta^k\beta^\ell\gamma^m\rangle = 0 \text{ unless } k=0 \text{ and } \ell=m. \tag{6.13}$$

To obtain the remaining nonvanishing average values $\langle\beta^k\gamma^k\rangle$, compute

$$\begin{aligned}&(y_+\alpha\gamma^k\beta^{k-1})\\ &\quad= \lambda q^{-1}[k]\alpha\delta\gamma^{k-1}\beta^{k-1} + q^{-2}\lambda\gamma^k\beta^k,\end{aligned} \tag{6.14}$$

where we have used the notation of (5.28). The average of the left hand side of (6.14) vanishes. Using (6.5) we obtain the recursion relation

$$\langle\beta^k\gamma^k\rangle = -q\frac{[k]}{[k+1]}\langle\beta^{k-1}\gamma^{k-1}\rangle. \tag{6.15}$$

Choosing the normalization

$$\langle 1\rangle = 1, \tag{6.16}$$

we obtain finally

$$\langle\beta^k\gamma^k\rangle = \frac{(-q)^k}{[k+1]}. \tag{6.17}$$

With the unitarity conditions (6.3) and (6.4) the above results agree with those obtained by Woronowicz [8] with a different method. In particular (6.18) becomes

$$\langle\overline{\gamma}^k\gamma^k\rangle = \frac{1}{[k+1]}. \tag{6.18}$$

It is remarkable that the results (6.12), (6.13) and (6.17) make sense even without the unitarity conditions, although one loses the positivity property of (6.18). So, it seems possible to define a left and right invariant average for polynomials on $SL_q(2)$. As $q\to 1$ this average still makes sense and is invariant, but apparently cannot be defined in terms of an integral over the group, even though a left and right invariant volume element exists on $SL(2)$.

7 Conclusion

The methods and results described above are examples of differential and intergal calculus on non commutative spaces. Other related examples can be found in [10] to [13].

The way seems to be open for the construction of consistent deformations of quantum mechanics and quantum field theory. Perhaps these deformations will provide a form of realistic regularization (too many as yet inconclusive papers to cite on this). In any case, it is remarkable that our present physical laws seem to allow consistent deformations, which are not required or suggested by experiment.

I am very grateful to Julius Wess, Chryss Chryssomalakos, Peter Schupp and Paul Watts for helping me to understand the subjects discussed here.

Refrences

1. Majid, S.: Internat.J.Modern Phys. A **5**, 1 (1990)
2. Connes, A.: *Géométrie non commutative.* InterEditions, Paris (1990)
3. Reshetikin, N.Yu., Takhtajan, L.A., Faddeev, L.D.: Leningrad Math.J. **1**, 193 (1990) (translated from the Russian Algebra and Analysis **1** (1989)); Takhtajan, L.A.: Advanced Studies in Pure Math. **19**, (1989)
4. Jimbo, M.: Internat.J.Modern Phys. A **4**, 3759 (1989)
5. Drinfeld, V.G.: Proc.Int.Congr.Math., Berkeley, 798 (1986)
6. Burroughs, N.: Comm.Math.Phys. **127**, 109 (1990)
7. Woronowicz, S.L.: Comm.Math.Phys. **122**, 125 (1989)
8. Woronowicz, S.L.: Comm.Math.Phys. **111**, 613 (1987)
9. Jurčo, B.: Lett.Math.Phys. **22**, 177 (1991)
10. Pusz, W., Woronowicz, S.L.: Rep.Math.Phys. **27**, 231 (1989)
11. Pusz, W.: Rep.Math.Phys. **27**, 349 (1989)
12. Wess, J., Zumino, B.: Nuclear Phys. B (Proc.Suppl.) **18**, 302 (1990)
13. Zumino, B.: Modern Phys.Lett. A **6**, 1225 (1991)

Chaotic Quantum Systems

Boris V. Chirikov

Institute of Nuclear Physics
630090 Novosibirsk, USSR

Abstract: New ideas concerning the peculiar phenomenon of quantum chaos are presented with special emphasis on a number of unsolved problems and current apparent contradictions.

1 Introduction

This lecture[1] is primarily addressed to mathematicians with the main purpose to explain new physical ideas in the so-called *quantum chaos* which since recently attracts ever growing interest of many researchers [1-5],[10].

The breakthrough in understanding of this phenomenon has been achieved, particularly, due to a new philosophy accepted, explicitly or more often implicitly, in most studies of quantum chaos. Namely, the whole physical problem of quantum dynamics is separated into two different parts: (*i*) the proper quantum motion described by a specific dynamical variable $\psi(t)$ which obeys, e.g., the Shrödinger equation, and (*ii*) the quantum measurement including ψ collapse which, as yet, has no dynamical description. In this way one can single out the vague problem of the fundamental randomness in quantum mechanics which is related to the second part only, and which in a sense is foreign to the proper quantum system. The remaining first part is then perfectly fits the general theory of dynamical systems.

The importance of quantum chaos is not only in that it represents a new unexplored field of nonintegrable quantum dynamics with many applications but also, and this is most interesting for fundamental science, in reconciling the two seemingly different dynamical mechanisms for the statistical laws in physics.

Historically, the first mechanism is related to the *thermodynamic limit* $N \to \infty$ in which the completely integrable system becomes chaotic [6]. A

[1] Here is the abridged version, for the full text see [33].

natural question, what happens for large but finite number of freedoms N, has still no rigorous answer but the new phenomenon of quantum chaos, at least, presents an insight into this problem too. We call this mechanism, which is equally applicable in both classical and quantum mechanics, the *traditional statistical mechanics* (TSM).

The second (new) mechanism is based upon the strong (exponential) local instability of motion characterized by positive Lyapunov's exponent $\Lambda > 0$ [6],[7]. It is not at all restricted to large N, and is possible, e.g., for $N > 1$ in a Hamiltonian system. However, this mechanism has been operative, until recently, in the classical mechanics only. We term this the *dynamical chaos* as it does not require any random parameters or any noise in the equations of motion.

The quantum system bounded in phase space has a discrete energy (frequency) spectrum and, in this sense, is always completely integrable similar to the finite-N TSM. Yet, the fundamental correspondence principle requires the transition to the classical mechanics, including the dynamical chaos, in the *classical limit* $q \to \infty$, where q is some quasi-classical parameter, e.g., the quantum number n (the action variable, $\hbar = 1$). Again, a natural physical conjecture is that for finite but large q there must be some chaos similar to finite-N TSM. Yet, in a chaotic quantum system the number of freedoms N does not need to be large as well as in the classical chaos. The quantum counterpart of N is q, both quantities determining the number of frequencies which control the motion. Thus, mathematically, the problem of quantum chaos is the same as that of the finite-N TSM.

The main difficulty in both problems (especially for mathematicians) is in that they suggest some chaos in discrete spectrum which is completely contrary to the existing theory of dynamical systems and the ergodic theory where the discrete spectrum corresponds to the opposite limit of regular motion.

The ultimate origin of the quantum integrability is discreteness of the phase-space (but not, as yet, of the space-time!) or, in modern mathematical language, the noncommutative geometry of the former.

As an illustration I will make use of the simple model described classically by the *standard map* (SM) [7],[8]:

$$\bar{n} = n + k \cdot \sin\theta; \qquad \bar{\theta} = \theta + T \cdot \bar{n}\,, \tag{1.1}$$

with action-angle variables n, θ, and pertubation parameters k, T. The quantized standard map (QSM) is given by [9],[10]

$$\bar{\psi} = \exp(-ik \cdot \cos\hat{\theta}) \cdot \exp\left(-i\frac{T}{2}\hat{n}^2\right)\psi\,, \tag{1.2}$$

where momentum operator $\hat{n} = -i\partial/\partial\theta$. To provide the boundedness of motion we consider SM on a torus of circumference (in n)

$$L = \frac{2\pi m}{T} \tag{1.3}$$

with integer m to avoid discontinuities. The quasi-classical transition corresponds to quantum parameters $k \to \infty$, $T \to 0$, $L \to \infty$ while classical parameters $K = kT = \text{const}$, and $m = LT/2\pi = \text{const}$.

QSM models the *energy shell* of a conserved system which is the quantum counterpart of the classical energy surface.

In the studies of dynamical systems, both classical and quantal, most problems unreachable for rigorous mathematical analysis are treated "numerically" using computer as a universal model. With all obvious drawbacks and limitations such "numerical experiments" have very important advantage as compared to the laboratory experiments, namely, they provide the complete information about the system under study. In quantum mechanics this advantage becomes crucial as in laboratory one cannot observe (measure) the quantum system without a radical change of its dynamics.

2 Definition of Quantum Chaos

The common definition of the classical chaos in physical literature is the *strongly unstable motion*, that is one with positive Lyapunov's exponents $\Lambda > 0$. The Alekseev — Brudno theorem then implies that almost all trajectories of such a motion are unpredictable, or random (see [11]). A similar definition of quantum chaos fails because, for the bounded systems, the set of such motions is empty due to the discreteness of the phase space and, hence, of the spectrum.

The common definition of quantum chaos is *quantum dynamics of classically chaotic systems* whatever it could happen to be. Logically, this is most simple and clear definition. Yet, in my opinion, it is completely inadequate from the physical viewpoint just because such a chaos may turn out to be a perfectly regular motion as, for example, in case of the *perturbative localization* [12]. In QSM this corresponds to $k \leq 1$ when all quantum transitions are suppressed independent of classical parameter K which controls the chaos.

I would like to define the quantum chaos in such a way to include some essential part of the classical chaos. The best definition I have managed to invent so far reads: *the quantum chaos is statistical relaxation in discrete spectrum* .This definition is certainly in contradiction with the existing ergodic theory as the relaxation (particularly, correlation decay) requires the mixing, hence, a continuous spectrum. In what follows I will try to explain a new, modified, concept of mixing which is necessary to describe the peculiar phenomen of quantum chaos.

3 The Time Scales of Quantum Dynamics

Already the first numerical experiments with QSM revealed the quantum diffusion in n close to the classical one under conditions $K \geq 1$ (classical stability border) and $k \geq 1$ (quantum stability border) [9]. Further studies confirmed this conclusion and showed that the former followed the latter in all details but on a *finite time interval* only [10],[13]. The latter fact was the clue to understanding the dynamical mechanism of the diffusion, which is apparently an aperiodic process, in discrete spectrum. Indeed, the fundamental uncertainty principle implies that the discreteness of the spectrum is not resolved for sufficiently short time interval. Whence, the estimate for the *diffussion (relaxation) time scale* :

$$t_R \sim \varrho_0 \leq \varrho \,. \tag{3.1}$$

Here ϱ is the density of (quasi)energy levels, and ϱ_0 is the same for the *operative eigenstates* which are actually present in the initial quantum state $\psi(0)$. In QSM the quasi-energies are determined mod $2\pi/T$, hence, $\varrho = LT/2\pi = m$ is a classical parameter (1.3). As to ϱ_0, it depends on the dynamics and is given by the estimate [10],[13]:

$$\frac{\varrho_0}{T} \sim \frac{t_R}{T} \equiv \tau_R \sim D \equiv \frac{\langle(\Delta n)^2\rangle}{\tau} \leq \frac{m}{T} \,. \tag{3.2}$$

Here τ is discrete map's time (the number of iterations), and D is the classical diffusion rate. This remarkable expression relates an essentially quantum characteristic (τ_R) to the classical one (D). The latter inequality in Eq. (3.2) follows from that in Eq. (3.1), and it is explained by the boundedness of QSM on a torus.

In the quasi-classical region $\tau_R \sim k^2 \to \infty$ (see Eq.(1.1)) in accordance with the correspondence principle. Yet, the transition to the classical limit is (conceptually) difficult to understand (and still more to accept) as it involves two limits ($k \to \infty$ and $t \to \infty$) which do not commute. The second limit is related to the existing ergodic theory which is asymptotic in t. Meanwhile the new phenomenon of the quantum chaos requires the modification of the theory to a finite time which is a difficult mathematical problem still to be solved.

Besides relatively long time scale (3.2) there is another one given by the estimate [14],[10]

$$t_r \sim \frac{\ln q}{\Lambda} \to \frac{T|\ln T|}{\ln(K/2)} \,, \tag{3.3}$$

where q is some (large) quasi-classical parameter, and where the latter expression holds for QSM. It is called some time the *random time scale* since here the quantum motion of a narrow wave packet is as random as classical trajectories according to the Ehrenfest theorem. This was well confirmed in a number of numerical experiments [15]. The physical meaning of t_r is in fast spreading of a wave packet due to the strong instability of classical motion.

Even though the random time scale t_r is very short it grows indefinitely in the quasi-classical region ($q \to \infty$, $T \to 0$), again in agreement with the correspondence principle.

Big ratio t_R/t_r implies another peculiarity of quantum diffusion: It is dynamically stable as was demonstrated in striking numerical experiments [16].

4 The Quantum Steady State

As a result of quantum diffusion and relaxation some steady state is formed whose nature depends on the *ergodicity parameter*

$$\lambda = \frac{l_s}{L} \approx \frac{D}{L} \,. \tag{4.1}$$

If $\lambda \gg 1$ the quantum steady state is close (at average) to the classical statistical equilibrium which is described by ergodic phase density $g_{cl}(n)$ =const where n is continuous variable. In quantum mechanics n is integer, and the quantum phase density $g_q(n,\tau)$ in the steady state fluctuates according to the Gauss law approximately [17],[5], the ergodicity meaning

$$g_q(n) = \overline{|\psi_s(n,\tau)|^2} = \frac{1}{L} \,, \tag{4.2}$$

where the bar denotes time averaging.

According to numerical experiments the ergodicity does not depend on initial state which implies that all eigenfunctions $\phi_m(n)$ are also ergodic at average with Gaussian fluctuations [17],[5]:

$$\langle|\phi_m(n)|^2\rangle = \frac{1}{L} \,. \tag{4.3}$$

This is always the case sufficiently far in the quasi-classical region as $\lambda \sim$ $\sim k^2/L \sim Kk/m \to \infty$ with $k \to \infty$ ($K = kT$ and $m = LT/2\pi$ remain constant) in accordance with old Shnirelman's theorem [18].

Finite fluctuations (4.2) show that a single chaotic quantum system, described by $\psi_s(n,\tau)$, represents, in a sense, finite statistical ensemble of $M \sim L$ "particles". The fluctuations can result in partial recurrences toward the initial state but the recurrence time is much longer as compared to the relaxation time scale τ_R and sharply depends on the recurrence domain.

If $\lambda \ll 1$ the quantum steady state is qualitatively different from the classical one. Namely, it is localized in n within the region of size l_s around the initial state if the size of the latter $l_0 \ll l_s$. Numerical experiments show that the phase space density, or the *quantum statistical measure* , is approximately exponential [10],[13]

$$g_s(n) \approx \frac{1}{l_s} \exp\left(-\frac{2|n|}{l_s}\right); \qquad l_s \approx D \tag{4.4}$$

for initial $g(n,0) = \delta(n)$. The quantum ensemble is now characterized by $M \sim l_s \sim k^2$ "particles".

The relaxation to this steady state is called *diffusion localization* , and it is described approximately by the diffusion equation [19],[33]

$$\frac{\partial g}{\partial \tau'} = \frac{1}{2}\frac{\partial}{\partial n} D \frac{\partial g}{\partial n} \pm \frac{\partial g}{\partial n} \tag{4.5}$$

for initial $g(n,0) = \delta(n)$ where the signs correspond to $n \gtrless 1$, and where new time

$$\tau' = \tau_R \ln\left(1 + \frac{\tau}{\tau_R}\right) \tag{4.6}$$

accounts for the discrete motion spectrum [20]. The last term in Eq. (4.5) describes "backscattering" of ψ wave propagating in n which eventually results in the diffusion localization. The fitting parameter $\tau_R \approx 2D$ was derived from the best numerical data available (see Ref. [21] where a different theory of diffusion localization was also developed).

5 Intermediate Statistics

Statistical properties of ergodic states are well described by the random matrix theory (RMT) [22]. Particularly, the statistics of nearest-neighbour level spacings s is given by the Wigner — Dyson distribution

$$p(s) = A s^\beta e^{-Bs^2}, \tag{5.1}$$

where constants A and B are found from normalization and condition $\langle s \rangle = 1$, and where level repulsion parameter $\beta = 1$; 2 or 4 depending on system's symmetry. This is to compare with the Poisson distribution for integrable systems

$$p(s) = e^{-s} \tag{5.2}$$

with $\beta = 0$ (independent levels). Extension of these statistics onto the quasi-energy levels was made in [23],[5].

The impact of diffusion localization on the level statistics is described by the *Izrailev distribution* [24]

$$p(s) \approx A s^\beta \exp\left(-\frac{\pi^2}{16}\beta s^2 - \left(B - \frac{\pi\beta}{4}\right) s\right), \tag{5.3}$$

which is also called *intermediate* (between (5.1) and (5.2)) *statistics* . Repulsion parameter β is now continuous and may take any value in the whole interval (0, 4). The *limiting statistics* ($\lambda = l_s/L \to \infty$) corresponds, for a given l_s, to small L which in RMT is matrix's size. This shows that RMT is a local theory which holds true within the localization length l_s only.

Repulsion parameter in (5.3)

$$\beta \approx \beta_H = \exp(\langle H \rangle - H_e) \approx \frac{2l_H}{L} , \tag{5.4}$$

where $\langle H \rangle$ is the mean entropy of eigenstates

$$H = -\sum_n |\phi(n)|^2 \ln |\phi(n)|^2 , \tag{5.5}$$

$l_H = \exp(H)$ is another localization length and $H_e \approx \ln(L/2)$ is the entropy of ergodic state. No explanation of a simple relation (5.4) has been given as yet, nor for the important scaling $\beta_H(\lambda)$ obtained numerically [5].

The discovery of intermediate statistics stimulated the development of a new RMT which makes use of band random matrices (BRM) [25] with the scaling parameter

$$\lambda_r = \frac{b^2}{L} , \tag{5.6}$$

where $2b$ is the band width. In the limit $L \to \infty$ this scaling has been proved recently in Ref. [26].

6 Concluding Remarks

In conclusion I would like to briefly mention a few important results for unbounded quantum motion. In SM it corresponds to $L \to \infty$. First, there is an interesting analogy between localization in momentum space and the celebrated Anderson localization in disordered solids. It was discovered in [27] and essentially developed in [28]. The analogy is based upon (and restricted by) the equations for eigenfunctions. The most striking (and less known) difference is in the absence of diffusion regime in $1D$ solids [29]. This is because the energy level density of the operative eigenfunctions in solids

$$\rho_0 \sim \frac{l dp}{dE} \sim \frac{l}{u} \sim t_R , \tag{6.1}$$

which is the localization (relaxation) time scale, is always of the order of the time interval for a free spreading of the initial wave packet at caracteristic velocity u.

Another similarity between the two problems is in that the Bloch extended states in periodic potential correspond to a pecular quantum resonance in QSM for rational $T/4\pi$ [9],[10]. An interesting open question is the dynamics for irrational Liouville's (transcendental) $T/4\pi$. As was proved in [30] the motion can be unbounded in this case unlike a typical irrational value. In [33] the conjecture is put forward, supported by some semiqualitative considerations, that depending on a particular Liouville's number the broad range of motions is possible, from purely resonant one ($|n| \sim \tau$) down to complete localization ($|n| \leq l$).

If quantum motion is not only unbounded but its rate in unbounded variables is exponential, the "true" chaos (not restricted to a finite time scale) can occur. A few exotic examples together with considerations from different viewpoints can be found in [10],[31]. However, such chaos does not seem to be a typical quantum dynamics.

The final remark is that the quantum chaos, as defined in Section 2, comprises not only quantum systems but also any linear, particularly classical, waves [32]. So it is essentially the *linear wave chaos* . Moreover, a similar mechanism works also in completely integrable nonlinear systems like Toda lattice, for example. From mathematical point of view all these new ideas require reconsideration of the existing ergodic theory. Perhaps, better to say, that a new ergodic theory is needed which, instead of benefiting from the asymptotic approximation ($|t| \to \infty$ or $N \to \infty$), could analyze the finite-time statistical properties of dynamical systems. This is the most important conclusion from first attempts to comprehend the quantum chaos.

References

1. Proc. Les Houches Summer School on Chaos and Quantum Physics, Elsevier, (1990)
2. Haake, F.: *Quantum Signatures of Chaos.* Springer (1991)
3. Eckhardt, B.: Phys.Rep. **163**, 205 (1988)
4. Casati, G., Molinari, L.: Prog.Theoret.Phys.Suppl. **98**, 287 (1989)
5. Izrailev, F.M.: Phys.Rep. **196**, 299 (1990)
6. Kornfeld, I., Fomin, S., Sinai, Ya.: *Ergodic Theory.* Springer (1982)
7. Lichtenberg, A., Lieberman, M.: *Regular and Stochastic Motion.* Springer (1983)
   Zaslavsky, G.M.: *Chaos in Dynamic Systems.* Harwood (1985)
8. Chirikov, B.V.: Phys.Rep. **52**, 263 (1979)
9. Casati, G. et al.: Lecture Notes in Physics **93**, 334 (1979)
10. Chirikov, B.V., Izrailev, F.M., Shepelyansky, D.L.: Soviet Sci.Rev. C **2**, 209 (1981);
    Chirikov, B.V., Izrailev, F.M., Shepelyansky, D.L.: Physica D **33**, 77 (1988)
11. Alekseev, V.M., Yakobson, M.V.: Phys.Rep. **75**, 287 (1981)
12. Shuryak, E.V.: Zh.Eksper.Teoret.Fiz. **71**, 2039 (1976)
13. Chirikov, B.V., Shepelyansky, D.L.: Radiofizika **29**, 1041 (1986)
14. Berman, G.P., Zaslavsky, G.M.: Physica A **91**, 450 (1978)
15. Toda, M., Ikeda, K.: Phys.Lett. A **124**, 165 (1987);
    Bishop, A. et al.: Phys.Rev. B **39**, 12423 (1989)

16. Shepelyansky, D.L.: Physica D **8**, 208 (1983);
Casati, G. et al.: Phys.Rev.Lett. **56**, 2437 (1986)
17. Izrailev, F.M.: Phys.Lett. A **125**, 250 (1987)
18. Shnirelman, A.I.: Uspekhi Mat.Nauk **29**, No 6, 181 (1974);
Shnirelman, A.I.: On the Asymptotic Properties of Eigenfunctions in the Regions of Chaotic Motion. Addendum in: V.F. Lazutkin: The KAM Theory and Asymptotics of Spectrum of Elliptic Operators. Springer (1991)
19. Chirikov, B.V.: Chaos **1**, 95 (1991)
20. Chirikov, B.V.: Uspekhi Fiz.Nauk **139**, 360 (1983);
Berman, G.P., Izrailev, F.M.: Oper.Theory: Adv.Appl. **46**, 301 (1990)
21. Cohen, D.: Quantum Chaos, Dynamical Correlations and the Effect of Noise on Localization. (1991) (unpublished)
22. Brody, T. et al.: Rev.Modern Phys. **53**, 385 (1981)
23. Izrailev, F.M.: Phys.Rev.Lett. **56**, 541 (1986)
24. Izrailev, F.M.: Phys.Lett. A **134**, 13 (1988);
Izrailev, F.M.: J.Phys. A **22**, 865 (1989)
25. Casati, G. et al.: Phys.Rev.Lett. **64**, 1851 (1990)
26. Fyodorov, YA., Mirlin, A.D.: Scaling Properties of Localization in Random Band Matrices: A σ-Model Approach. Phys.Rev.Lett. (To appear)
27. Fishman, S. et al.: Phys.Rev. A **29**, 1639 (1984)
28. Shepelyansky, D.L.: Physica D **28**, 103 (1987)
29. Nakhmedov, E.P. et al.: Zh.Eksper.Teoret.Fiz. **92**, 2133 (1987)
30. Casati, G., Guarneri, I.: Comm.Math.Phys. **95**, 121 (1984)
31. Weigert, S.: Z.Phys. B **80**, 3 (1990);
Berry, M.: True Quantum Chaos? An Instructive Example, Proc. Yukawa Symposium (1990);
Benatti, F. et al.: Lett.Math.Phys. **21**, 157 (1991)
32. Chirikov, B.V.: Linear Chaos. Preprint INP 90-116, Novosibirsk (1990)
33. Chirikov, B.V.: Chaotic Quantum Systems. Preprint INP 91-83, Novosibirsk (1991)

Dynamical Zeta Functions: Where Do They Come from and What Are They Good for ?

David Ruelle

Institut des Hautes Etudes Scientifiques, F-91440 Bures-sur-Yvette, France

Abstract: The properties and usefulness of dynamical zeta functions associated with maps and flows are discussed, and they are compared with the more traditional number-theoretic zeta functions.

0 Dynamical Zeta Functions

Let (M, f) be a dynamical system, i.e., M is a space and $f : M \to M$ a map. Let also $g : M \mapsto \mathbb{C}$ be a function and consider the formal power series

$$\zeta(z) = \exp \sum_{m=1}^{\infty} \frac{z^m}{m} \sum_{x \in \operatorname{Fix} f^m} \prod_{k=0}^{m-1} g(f^k x)$$

where $\operatorname{Fix} f^m$ is the set of fixed points of $f^m = f \circ \cdots \circ f$. The *dynamical zeta function* $\zeta(z)$ is a natural object from a combinatorial point of view, as is the corresponding zeta function for a flow (f^t).*) Under suitable conditions, the sum over $\operatorname{Fix} f^m$ in the definition of ζ converges, and $\zeta(z)$ is a meromorphic function in a certain domain. In this talk we shall discuss dynamical zeta functions in a relatively informal way. We shall try to explain where they come from and what they are good for. We shall not try to show how their properties are proved, but note that some properties of zeta functions are notoriously hard to prove ! In a sense the study of dynamical zeta functions is part of the theory of dynamical systems, but we shall see that it is also intimately related to statistical mechanics.

*) At this level of generality, these objects were first introduced in Ruelle [21], [22], [23].

1 Where Do They Come from ?

1.1 Riemann Zeta Function

At the beginning there is the Riemann zeta function:

$$\begin{aligned}\zeta(s) &= \sum_{n=1}^{\infty} n^{-s} \\ &= \prod_{p \text{ prime}} \left(1 - p^{-s}\right)^{-1} .\end{aligned}$$

The representation as a product over primes reflects the unique factorization of integers into primes, and is due to Euler. Of course Euler is earlier than Riemann, but the function is associated with Riemann rather than Euler because of Riemann's detailed work on the analytic properties of ζ. In particular, writing

$$\xi(s) = \pi^{-s/2}\, \Gamma(s/2)\zeta(s)$$

we have the *functional equation* $\xi(s) = \xi(1-s)$. The Riemann zeta function has been used by Hadamard and de la Vallée-Poussin to prove the *prime number theorem* (if $\pi(x)$ is the number of primes $\leq x$, then $\pi(x) \sim x/\log x$). In view of other arithmetic applications, various other number-theoretic zeta functions have been introduced after that of Riemann. For a survey of this fascinating field we refer to the article *Zeta functions* in the Encyclopedic Dictionary of Mathematics [15]. This brings us somehow to the next item.

1.2 Weil Conjectures

Let V be a nonsingular projective algebraic variety of dimension n over a finite field k with q elements. The variety V is thus defined by homogenous polynomial equations with coefficients in k for $n+1$ variables $x_0, \ldots, x_n$. These variables are in the algebraic closure $\bar{k}$ of k, and constitute the homogeneous coordinates of a point of V. The variety V is invariant under the *Frobenius map* $F : (x_0, \ldots, x_n) \mapsto (x_0^q, \ldots, x_n^q)$. Arithmetic considerations lead one to introduce a zeta function which counts the points of V with coordinates in the different finite extensions of the field k, or equivalently points of V which are fixed under F^m for some $m \geq 1$:

$$Z(z, V) = \exp \sum_{m=1}^{\infty} \frac{z^m}{m} \operatorname{card} \operatorname{Fix} F^m .$$

Conjectures made by Weil [34] about this function led to a lot of work by many people including A. Weil himself, B. Dwork, A. Grothendieck and finally P. Deligne [5] who concluded the proof of the conjectures. Here are (roughly) the results.

The function Z is rational:

$$Z = \prod_{\ell=0}^{2n} P_\ell(z)^{(-1)^{\ell+1}} .$$

The zeros of the polynomial P_ℓ have absolute value $q^{-\ell/2}$. The P_ℓ have a cohomological interpretation: P_ℓ is roughly the characteristic polynomial associated with the action of F on ℓ-dimensional cohomology.

1.3 Selberg Zeta Function

Given a compact surface of curvature -1, the Selberg zeta function (see [30]) is related to

$$\zeta(s) = \prod_{\gamma\in P} \left(1 - e^{-s\,T(\gamma)}\right)^{-1} \tag{1}$$

where P is the set of closed geodesics and $T(\gamma)$ is the length of γ. Preferably we interpret P as the set of (minimal) periodic orbits γ for the geodesic flow, and $T(\gamma)$ is the period of γ.

In fact the Selberg zeta function is $Z(s) = \left[\prod_{n=0}^{\infty} \zeta(s+n)\right]^{-1}$. It is an entire analytic function, and satisfies a functional equation. Furthermore its zeros can be analysed, and the "nontrivial zeros" are related to the eigenvalues of the Laplace-Beltrami operator on the compact surface of curvature -1 from which we started. (This yields a relation between the "classical mechanics" of the geodesic flow and the "quantum mechanics" of the Laplace operator). The localization of the "nontrivial zeros" corresponds to the *Riemann hypothesis* which asserts that the nontrivial zeros of the Riemann zeta function are located on the line $\mathrm{Re}s = \frac{1}{2}$ (this, of course, remains unproved).

1.4 Dynamical Zeta Functions

Inspired by the zeta function of an algebraic variety over a finite field, Artin and Mazur [1] defined

$$\zeta(z) = \exp \sum_{m=1}^{\infty} \frac{z^m}{m} \operatorname{card} \operatorname{Fix} f^m . \tag{2}$$

for a C^1 diffeomorphism f of a compact manifold, and proved that, for a dense set of such diffeomorphisms, ζ has nonvanishing radius of convergence.

The Artin-Mazur zeta function counts periodic points for a diffeomorphism, and the same definition was adopted later by Milnor and Thurston to count periodic points for a piecewise monotone map of the interval.

Instead of the discrete dynamical system generated by a map f, let us consider a flow (f^t). Denoting by $T(\gamma)$ the minimal period of a periodic orbit γ, it is natural to define a zeta function by (1). This form corresponds in fact closely to the Euler product formula for the Riemann zeta function.

Following Artin-Mazur and Selberg we have thus obtained natural definitions (2) and (1) for dynamical zeta functions associated with maps or flows. Knowledge of equilibrium statistical mechanics induces one however to make more general definitions, as follows.

A. Map Case

Given a map $f : M \to M$ and a function $g : M \to$ matrices, we let

$$\zeta(z) = \exp \sum_{m=1}^{\infty} \frac{z^m}{m} \sum_{x \in \mathrm{Fix}\, f^m} \mathrm{Tr} \prod_{k=0}^{m-1} g\left(f^k x\right) .$$

In the particular case where $g = e^A$, $A : M \to \mathbb{C}$, we have

$$\zeta(z) = \exp \sum_{m=1}^{\infty} \frac{z^m}{m} \sum_{x \in \mathrm{Fix}\, f^m} \exp \sum_{k=0}^{m-1} A(f^k x) .$$

We have, in these formulae, replaced simple counting by counting with weights, which is the essence of equilibrium statistical mechanics.

B. Flow Case

Given a flow $(f^t) : M \to M$ and a function $B : M \to \mathbb{R}$ or $\mathbb{C}$, we let

$$\zeta(s) = \prod_{\gamma \in P} \left[1 - \exp\left(-\int_0^{T(\gamma)} \left(s - B(f^t x_\gamma)\right) dt\right)\right]^{-1} . \tag{3}$$

There are extensions of this formula, which we shall not consider here.

The zeta functions for maps and flows look different, but the zeta function for a *special flow* (suspension) reduces to the form given for a map. In fact, starting from (1) one arrives not at (2) but at the more general form given above, which makes such a generalization more or less unavoidable.

The above generalizations were introduced in [21], [22], based on the ideas of statistical mechanics. Let me indicate another natural way of obtaining the general expression for the dynamical zeta function of a map.

Let S be the set of functions $X : M \to \mathbb{Z}_+$ such that $X = X \circ f$ and $|X| = \sum_{x \in M} X(x) < \infty$ (i.e., X is f-invariant and vanishes except at finitely many points). Each $X \in S$ has a unique decomposition

$$X = \sum_{\gamma} n_\gamma X_\gamma$$

where X_γ is the characteristic function of the f-periodic orbit γ, and the n_γ are integers ≥ 0 (only a finite number being > 0). This unique (additive) decomposition is analogous to the unique (multiplicative) decomposition of an integer into primes, and the analogy with the Riemann zeta function suggests to define

$$\zeta(z) = \sum_{X \in S} z^{|X|} = \prod_{\gamma \in P} \left(1 - z^{|\gamma|}\right)^{-1}$$

(where $-\log z$ replaces s, and $e^{|X|}$ replaces n), which is precisely the definition (2). But number theory suggest to consider more generally zeta functions of the form

$$\zeta(z) = \sum_{X \in S} g(X)\, z^{|X|}$$

where $g(X+Y) = g(X) \cdot g(Y)$, at least when X and Y are "relatively prime" (i.e., $X \to g(X)$ is a "multiplicative function"). This allows to represent ζ as a Euler product over prime periodic orbits. Given $g : M \to$ matrices, define

$$g(X) = \prod_{\gamma} \operatorname{Tr} \prod_{k} g\left(f^k x_\gamma\right)$$

where $X = \sum_{\gamma} n_\gamma X_\gamma$, x_γ is an arbitrary point of γ and the product over k extends from 0 to $n_\gamma |X_\gamma| - 1$; then $X \to g(X)$ is indeed multiplicative. In particular, if $g = e^A$, $A : M \to \mathbb{C}$, we have

$$\sum_{X \in S} g(X)\, z^{|X|} = \prod_{\gamma \in P} \left(1 - g(X_\gamma)\, z^{|\gamma|}\right)^{-1} .$$

2 What Are They Good for?

As the case of the Selberg zeta function shows, there are some dynamical zeta functions which have number-theoretic interest. This is an active area of research, which we can only mention in passing, see Sarnak [28], [29], Fried [6], Series [31], Pollicott [19], Cartier and Voros [4], Mayer [12], [13], [14], etc). In general, however, the dynamical zeta functions are different from those having number-theoretic interest, and are for instance not known to satisfy a functional equation. This is compensated by the fact that dynamical zeta functions have relations with statistical mechanics (entropy, pressure, Gibbs states, equilibrium states) while such relations are not known for number-theoretic zeta functions.

I shall now discuss some selected examples, the selection corresponding of course to my own interests. As a background reference to much of the material below, see the excellent monograph of Parry and Pollicott [17].

2.1 Expanding/Contracting Maps (See Tangerman [33], Haydn [9], Ruelle [25], [26])

Let $f : m \to M$ be of class C^r (M a compact Riemann manifold, $r > 0$ not necessarily integer), and assume that f is expanding by a factor $> \theta^{-1}$ (for some $\theta \in (0,1)$). Consider the *transfer operator* $\mathcal{L}_0$ such that

$$\mathcal{L}_0 \Phi(x) = \sum_{y: fy=x} g(y)\Phi(y)$$

with g of class C^r. Then one can define a *Fredholm determinant* $\det(1 - z\mathcal{L}_0)$ analytic for $|z| < \theta^{-r} \exp(-P(log|g|))$, where the *pressure* $P(A)$ is the max of $h(\rho) + \rho(A)$ taken over ergodic measures ρ, $h(\rho)$ being the entropy. There is a similar Fredholm determinant associated with transfer operators $\mathcal{L}_k$ acting on k-forms. The zeta function (2) is of the form

$$\zeta(z) = \prod_{k \geq 0} \left(\det(1 - z\mathcal{L}_k)\right)^{(-1)^{k+1}}$$

and meromorphic for $|z| < \theta^{-r} \exp(-P(\log|g|))$. The Hölder version of this result is due to Haydn, the C^k version (in somewhat weaker form) to Tangerman. These results can be extended in various ways, in particular to the definition of Fredholm determinants for a *linear* class of operators [26]. This extension of the theory of Fredholm (and Grothendieck [7]) to kernels with δ-singularities may be expected to have applications in physics. For a different type of applications see [2] and references given there.

2.2 Hyperbolic Maps

We discuss here an Axiom A diffeomorphism f in the sense of Smale [32], restricted to a basic set K. We assume that the contraction (expansion) in the contracting (expanding) direction is by a factor $< \theta (> \theta^{-1})$ with $0 < \theta < 1$. The "counting" zeta function

$$\zeta(z) = \exp \sum_{m=1}^{\infty} \frac{z^m}{m} \text{card Fix} f^m$$

is rational, as conjectured by Smale [32], and proved by Guckenheimer [8] and Manning [11], but no general cohomological interpretation is known.

If A is a real α-Hölder continuous function, the inverse

$$\frac{1}{\zeta(z)} = \exp - \sum_{m=1}^{\infty} \frac{z^m}{m} \sum_{x \in \text{Fix} f^m} \exp \sum_{k=0}^{m-1} A(f^k x)$$

has a zero at $e^{-P(A)}$ and converges for

$$|z| \leq e^{-P(A)}\theta^{-\alpha/2} .$$

This extends earlier results of Ruelle [24] and Pollicott [18], and can be proved by a method due to Haydn [9].

2.3 Hyperbolic Flows

We consider an Axiom A flow in the sense of Smale [32], restrict it to a basic set, and assume mixing. (An example is the geodesic flow on a compact Riemann manifold with sectional curvature < 0 everywhere, non constant in general). Given a real α-Hölder function B, we have defined the zeta function $\zeta(s)$ by (3). One can show that this function is analytic for $Re\ s > P(B)$, where $P(B)$ is the "pressure of the function B with respect to the flow", $\zeta(s)$ has a simple pole at $P(B)$, no other pole or zero with $Re\ s = P(B)$, and is meromorphic for $Re\ s > P(B) - \frac{1}{2}\log\ \theta$. (This is an extension of earlier results of Ruelle, and Pollicott [20]).

Parry and Pollicott [16], considering the case $B = 0$, realized that they could apply to the study of the distribution of the periods $T(\gamma)$ the same method that is used with the Riemann zeta function to prove the prime number theorem. (This method is based on the Wiener-Ikehara Tauberian theorem, see for instance [10]). The result is

$$\begin{aligned} \pi(x) &= (\#\,\text{closed orbits of minimal period } \leq x) \\ &\sim \frac{e^{hx}}{hx} \end{aligned}$$

where h is the topological entropy $P(0)$ of the flow. In the special case of the geodesic flow on a manifold of negative curvature one recovers an earlier result of Margulis on the distribution of the lengths of closed geodesics.

2.4 Piecewise Monotone Maps of the Interval

Baladi and Keller [3] have obtained for the zeta function of these maps a result in the same spirit as for hyperbolic maps.

2.5 Open Problems

One also expects analyticity results for zeta functions associated with rational maps of the Riemann sphere, see [27].

A fascinating possibility, which is probably a red hering, is that the Lee-Yang circle theorem may be somehow connected with zeta functions (see [23], the idea has later also been hit upon by B. Julia).

An unexplored topic is braids. Let $N(n)$ be the number of inequivalent braids (on k pieces of string) for which the minimal number of crossings is n. It is easily seen that

$$H = \lim_{n \to \infty} \frac{1}{n} \log N(n)$$

exists, and is > 0 if $k \geq 3$. This suggests doing statistical mechanics with braids (the ideas of quantum groups provide interactions) and one can also study zeta functions. I do for instance not know if one can replace $N(n)$ by a "number of crossings with periodic boundary conditions" and retain the limit H.

References

1. Artin, M., Mazur, B.: On periodic points. Ann.of Math. (2) **81**, pp. 82-99 (1965)
2. Artuso, R., Aurell, E., Cvitanović, P.: Recycling of strange sets: I. Cycle expansions, II. Applications. Nonlinearity **3**, pp. 325-359, 361-386 (1990)
3. Baladi, V. and Keller, G.: Zeta functions and transfer operators for piecewise monotone transformations. Comm.Math.Phys. **127**, pp. 459-477 (1990)
4. Cartier, P. and Voros, A.: Une nouvelle interprétation de la formule des traces de Selberg. C.R. Acad.Sci.Paris, **307**, Ser. I, pp. 143-148 (1988)
5. Deligne, P.: La conjecture de Weyl. Publ.math. IHES, **43**, pp. 273-307 (1974)
6. Fried, D.: The zeta functions of Ruelle and Selberg I. Ann.Sci. E.N.S. **19**, pp. 491-517 (1986)
7. Grothendick, A.: La théorie de Fredholm. Bull.Soc.Math.France **84**, pp. 319-384 (1956)
8. Guckenheimer, J.: Axiom A + no cycles $\Rightarrow \zeta_f(t)$ rational. Bull.AMS **76**, pp. 592-594 (1970)
9. Haydn, N.: Meromorphic extension of the zeta function for Axiom A flows. Ergodic Therory Dynamical Systems **10**, pp. 347-360 (1990)
10. Korevaar, J.: On Newman's quick way to the prime number theorem. Math. Intelligencer **4**, pp.108-115 (1982)
11. Manning, A.: Axiom A diffeomorphisms have rational zeta functions. Bull. London Math.Soc. **3**, pp. 215-220 (1971)
12. Mayer, D.: On a zeta function related to the continued fraction transformation. Bull.Soc.Math. France **104**, pp. 195-203 (1976)
13. Mayer, D.: On the thermodynamic formalism for the Gauss map. Comm. Math.Phys. **130**, pp. 311-333 (1990)
14. Mayer, D.: The thermodynamic formalism approach to Selberg's zeta function for PSL $(2,\mathbb{Z})$. Bull.Amer.Math.Soc. **25**, pp. 55-60 (1991)
15. Nihon Sugakkai, ed.: *Encyclopedic dictionary of mathematics.* MIT Press, Cambridge, Mass., (1977)
16. Parry, W., Pollicott, M.: An analogue of the prime number theorem for closed orbits of Axiom A flows. Ann.Math. **118**, pp. 573-591 (1983)
17. Parry, W., Pollicott, M.: Zeta functions and the periodic orbit structure of hyperbolic dynamics. Astérisque **187-188**, pp. 1-268 (1990)
18. Pollicott, M.: Meromorphic extensions of generalised zeta functions. Invent.Math. **85**, pp. 147-164 (1986)

19. Pollicott, M.: Distribution of closed geodesics on the modular surface and quadratic irrationals. Bull.Soc.Math. France **114**, pp. 431-446 (1986)
20. Pollicott, M.: Margulis distributions for Anosov flows. Comm.Math.Phys. **113**, pp. 137-154 (1987)
21. Ruelle, D.: Generalized zeta-functions for axiom *A* basic sets. Bull.Amer. Math.Soc. **82**, pp. 153-156 (1976)
22. Ruelle, D.: Zeta functions for expanding maps and Anosov flows. Invent. Math. **34**, pp. 231-242 (1976)
23. Ruelle, D.: Zeta functions and statistical mechanics. Astérisque **40**, pp. 167-176 (1976)
24. Ruelle, D.: *Thermodynamic Formalism. Encyclopedia of Math. and its Appl.* Vol. **5**, Addison-Wesley, Reading, Mass. (1978)
25. Ruelle, D.: The thermodynamic formalism for expanding maps. Comm. Math.Phys. **125**, pp. 239-262 (1989)
26. Ruelle, D.: An extension of the theory of Fredholm determinants. Publ. Math. IHES, **72**, pp. 175-193 (1990)
27. Ruelle, D.: Spectral properties of a class of operators associated with conformal maps in two dimensions (preprint)
28. Sarnak, P.: Class numbers of indefinite binary quadratic forms. J.Number Theory **15**, pp. 229-247 (1982)
29. Sarnak, P.: Determinants of Laplacians. Comm.Math.Phys. **110**, pp. 113-120 (1987)
30. Selberg, A.: Harmonic analysis and discontinuous subgroups in weakly symmetric Riemannian spaces with applications to Dirichlet series. J.Indian Math.Soc. **20**, pp. 47-87 (1956)
31. Series, C.: The modular surface and continued fractions. J.London Math. Soc. (2) **31**, pp. 69-80 (1985)
32. Smale, S.: Differentiable dynamical systems. Bull.Amer.Math.Soc.**73**, pp. 747–817 (1967)
33. Tangerman, F.: Meromorphic continuation of Ruelle zeta functions. Boston University thesis (1986) unpublished
34. Weil, A.: Numbers of solutions of equations in finite fields. Bull.Amer.Math. Soc. **55**, pp. 497–508 (1949)

Billiard-Type Systems with Chaotic Behaviour and Space-Time Chaos

Leonid A. Bunimovich

Institute of Oceanology Ac. Sci. USSR, Moscow 117218, USSR
Forschungszentrum BiBoS, Universität Bielefeld,
W-4800 Bielefeld 1, Deutschland

0 Introduction

In the last two decades the subject of the branch of mathematics that is called Dynamical Systems is almost identified with the study of chaotic motion of such systems. However until recently there were very few examples of in a sense realistic systems that have been proven to be chaotic. The first found models of that type were geodesic flows on manifolds of negative curvature (see [Ha], [He], [H1]). The general ideas of these papers were developed by D. V. Anosov, Ya. G. Sinai and S. Smale (see [An], [AS], [S2], [Sm]) and lead to the concept of hyperbolicity as the basic mechanism of chaos in classical dynamical systems. This concept together with the closely related notion of Lyapunov exponents (see [O], [P]) serve as the foundation of the modern theory of nonuniformly hyperbolic dynamical systems.

The soviet physicists N. S. Krylov was first to mention [K] the analogy between the dynamics of hard spheres gas and geodesic flows on surfaces of negative curvature. The mathematical theory of chaotic billiards was initiated and developed by Ya. G. Sinai who proved in particular that dispersing (Sinai) billiards are hyperbolic and chaotic [S1], [S4].

The mechanism of hyperbolicity in Sinai billiards and in the closely related semi-dispersing billiards is the dispersion of nearby trajectories by the boundary. The mechanism of dispersing of trajectories by a negative curvature in geodesic flows is quite similar.

For many years the general opinion was that dispersing is the only mechanism of chaos (hyperbolicity) in Hamiltonian systems. Examples of chaotic nowhere dispersing billiards ([B3], [B4]) lead to the understanding that there exists another mechanism of chaos (of hyperbolicity) in Hamiltonian systems which is natural to call the mechanism of defocusing (see [B5], [B7], [B8]).

The discovery of this mechanism together with the recently developed techniques for the study of nonuniformly hyperbolic systems with singularities allowed to construct some new realistic examples of dynamical systems with chaotic behaviour. We shall give a brief review of these results (following in some places to [S10]) and also discuss the first models of extended dynamical systems which were proven to demonstrate turbulent behavior, i. e. space-time chaos.

1 Definition of Billiards

Billiards are dynamical systems corresponding to the motion by inertia of a point particle inside a connected compact domain on a Riemannian manifold with a piecewise smooth boundary. Upon reaching the boundary the point reflects from it in accordance to the rule: "the angle of incidence is equal to the angle of reflection". Thus billiards are geodesic flows on manifolds with boundaries and with elastic reflections from it.

The modern theory of billiards refers mainly to the situation when a configurational space billiard table Q is a closed connected subset of the Euclidean space $\mathbb{R}^d$ or of the d-dimensional torus Tor^d with the Euclidean metric. We suppose that the boundary ∂Q is piecewise C^3-smooth. Since the absolute value of the velocity of the particle is a first integral of motion the phase space M of a billiard can be identified with the unit tangent bundle over Q, i. e. $M = Q \times S^{d-1}$ where S^{d-1} is the unit $(d-1)$-dimensional sphere. Thus every phase point has the form $x = (q, v)$, where $x \in M, q \in Q, v \in T_q(Q), \| v \| = 1$ and $T_q(Q)$ is the tangent space to Q at q. The natural projection $\pi : M \to Q$ is defined by $\pi(q, v) = q$. The boundary ∂Q is the union $\partial Q = \cup_{i=1}^r \partial Q_i$, where ∂Q_i are the smooth components of the boundary. The points of a component ∂Q_i which do not belong to other components $\partial Q_j, j \neq i$, are called regular points of the boundary. At regular points $q \in \partial Q$ the unit normal vector $n(q)$ and the operator $K(q)$ of the second fundamental form of the boundary are defined via the relations $n(q + dq) = n(q) + K(q)dq, dq \in T_q(\partial Q)$. The phase space M is also a Riemannian manifold with piecewise smooth boundary $\partial M = \cup_{i=1}^r \partial M_i, \partial M_i = \pi^{-1}(\partial Q_i)$. a point $x \in \partial M$ is regular if its projection $\pi(x) \in \partial Q$ is regular.

Consider a subset $M' \subset M$ that consists of such x whose trajectories pass only through regular points of the boundary. Let $d\mu =$ const $dq d\omega_q(v)$, where $d\omega_q(v)$ is the uniform measure on the unit sphere $S_q^{d-1} \subset T_q(Q)$ and dq is the normalized Liouville measure on Q. It is easy to show that $\mu(M') = 1$ (see [CFS]). Thus in M' the flow $\{S^t\}$ is defined that preserves the measure μ. The dynamical system $(M, \{S^t\}, \mu)$ is neither everywhere defined nor smooth. It is also useful to consider the induced map T that is defined on the set $\partial M^+ = \{x = (q, v) \in \partial M \cap M'$ and $(v, n, (q)) \geq 0\}$. The automorphism T preserves the measure $d\mu_1 = \text{const}(n(q), v) d\sigma(q) dw_q(v)$ where $d\sigma(q)$ is the Riemannian volume on ∂Q.

2 Hyperbolicity and Lyapunov Exponents

The necessary condition for classical dynamical systems to be chaotic is the strong local instability of their motion. The latter one means an exponential divergence of nearby orbits in phase space and is called hyperbolicity. Let $x = (q, v) \in M'$ and $\tilde{\gamma} \subset Q$ be a $(d-1)$-dimensional open C^3manifold that is homeomorphic to a $(d-1)$-dimensional disc and orthogonal to v at q. Denote by γ its continuous framing by unit normal vectors such that $x \in \gamma$.

Definition 1 A submanifold γ is called a local stable manifold (LSM) of x if for any $y \in \gamma$ positive semitrajectories $\{S^t x\}$ and $\{S^t y\}$ have reflections from the same regular components of the boundary ∂M for all $t > 0$ and $\exists C(x), \lambda(x)$ such that $\forall t > 0$

$$\text{dist}(\pi(S^t x), \pi(S^t y)) \leq C(x) \exp\{-\lambda(x) t\} \quad .$$

Taking $t < 0$ we shall get the definition of a local unstable manifold (LUM) of x. LSM and LUM will be denoted by $\gamma^{(s)}(x)$ and $\gamma^{(u)}(x)$ respectively.

Definition 2 A global stable (unstable) manifold $\Gamma^{(s)}(x)(\Gamma^{(u)}(x))$ consists of all y for which one can find $\tau \geq 0 (\tau \leq 0)$ such that $S^\tau y$ belongs to $\gamma^{(s)}(S^\tau x)(\gamma^{(u)}(S^\tau x))$.

GSM $\Gamma^{(s)}(x)$ and GUM $\Gamma^{(u)}(x)$ in billiards can have singularities which arise from trajectories tangent to the boundary ∂Q and/or hitting singular points of ∂M. The same definitions can be introduced for the induced automorphism T.

Definition 3 The billiard in a domain Q is called hyperbolic if μ-almost every point $x \in M$ has GSM and GUM and $\dim \gamma^{(s)}(x) = \dim \gamma^{(u)}(x) = d - 1$.

Let $\mathcal{M}$ be a piecewise C^3 compact, connected Riemannian manifold and ν a measure on $\mathcal{M}$ that is equivalent to the Riemannian volume. Suppose that f^t is a measurable flow on $\mathcal{M}$ generated by a vector field X, that preserves ν. Oseledec's multiplicative ergodic theorem (see [O]) implies that for ν-a.e. $z \in \mathcal{M}$ there exist Lyapunov exponents for $\xi \in T_z\mathcal{M}, \xi \neq 0$

$$\lambda^+(z, \xi) = \lim_{t \to +\infty} \frac{1}{t} \log \| df^t \xi \|$$

and

$$\lambda^-(z,\xi) = \lim_{t\to-\infty} \frac{1}{t} \log \| df^t\xi \|$$

if

$$\int_{\mathcal{M}} \log^+ \sup_{t\in[0,1]} \| df^t(x) \| \, d\nu(x) < \infty.$$

If the dynamical system (M,f^t,ν) is hyperbolic then ν-a.e. $\lambda^\pm(z,\xi)$ are non-zero except for the vector $\xi = \alpha X(z)$ and vice versa. If all Lyapunov exponents are non-zero almost everywhere then ergodic components of f have positive measure (see [P], [KS], [S9]). In case $\lambda^\pm(z,\xi) \neq 0$ for some but not for all $\xi \neq \alpha X(z)$ the system is called partially hyperbolic. Pesin's theory (see [P]) ensures the positivity of Kolmogorov-Sinai entropy if at least one Lyapunov exponent is positive on a subset of positive measure. Thus hyperbolicity implies a chaotic behavior of a dynamical system. Therefore a mechanism of local instability of motion which implies hyperbolicity it is natural to call mechanism of chaos.

3 Sinai (dispersing) and Semi-dispersing Billiards

A rich class of hyperbolic billiards was introduced and investigated by Sinai in [S 4].

Definition A billiard is called Sinai billiard if at all regular points of the boundary ∂Q the operator $K(q)$ is strictly positive.

In [S4] such billiards were called dispersing ones because at any regular point ∂Q is strictly convex inwards of the "billiard table".

Theorem 1([S1], [S4]) *Every Sinai billiard is hyperbolic.*

The mechanism of local exponential instability in Sinai billiards is analogous to geodesic flows on manifolds of negative curvature. The dispersing role of negative curvature in these billiards is played by dispersing boundary. Thus this mechanism of chaos in Hamiltonian systems is natural to call the mechanism of dispersing. A more general class is formed by semi-dispersing billiards that are analogous to geodesic flows on manifolds of non-positive curvature.

Definition A billiard is called semi-dispersing if at all regular points of the boundary $K(q) \geq 0$.

The most famous example of a semi-dispersing billiard is the gas of N hard spheres or disks (see [CFS], [S7], [S9]). In this model the boundary ∂Q

is the union of $N(N-1)/2$ cylinders that correspond to collisions of some pair of spheres. In case $N = 2$ we get a Sinai billiard but for $N \geq 3$ there are flat directions in the boundary ∂Q that correspond to different partitions of spheres into noninteracting groups. It is easy to show that in this case the system is partially hyperbolic (see [S1], [S4], [S7]) but it is a much more difficult problem to prove its (complete) hyperbolicity. The approach to the proof of hyperbolicity of semi-dispersing billiards was proposed in [SC1]. The general sufficient conditions were obtained in [C1] (see also [KS]).

4 Nowhere Dispersing Billiards and Two Mechanisms of Chaos in Hamiltonian Systems

Contrary to geodesic flows on manifolds of negative curvature there were known for a long time some classical examples of the integrable geodesic flows on manifolds of positive (or sectionally positive) curvature such as surfaces of revolution or ellipsoids in $\mathbb{R}^3$. Geodesics on manifolds of positive curvature do not diverge localy but even have a tendency to converge and intersect in so called conjugate points. The simplest example of such geodesics are large circles on a two-dimensional sphere S^2. Billiards which correspond to the motion of such geodesic flows reflect in the focusing part of the boundary ∂Q which produces the same effect. Transparent examples are integrable billiards in circles and ellipses.

E. Hopf [H2] has mentioned that if trajectories of a geodesic flow spend almost all the time in a part of a manifold with negative curvature then this system could be still chaotic, i. e. it is possible to have small pieces with positive curvature but it is necessary to have (much) more area with negative curvature.

The first examples of that type were constructed in [B1], [B2] as pertubations of Sinai billiards. Focusing components of the boundary ∂Q in [B1], [B2] were arcs of circles and the only condition was that complements (up to the whole circles) of these arcs have to be inside the domain Q. From the first sight the results of [B1], [B2] just prove Hopf's conjecture but the true situation turns out to be quite different. The point is that there is another mechanism of chaos different from the mechanism of dispersing. Consider a billiard in a domain Q which is a perturbation of the boundary of a Sinai billiard. It was shown in [B3], [B4] that in some cases the billiard will still be completely hyperbolic if we remove the dispersing part of the boundary, leaving only the perturbed part of the boundary.

That means that there exist chaotic billiards in domains with nowhere dispersing boundary ∂Q and even in some convex domains in Euclidean plane.

These results have demonstrated evidently that there is the other mechanism of chaos in billiards (see [B3], [B4], [B6], [B7], [B10]). To understand the essence of this mechanism consider the reflections of a local (narrow) beam of trajectories (rays) in the focusing part of the boundary ∂Q. Suppose that after being focused the beam will be defocused before the next reflection. Then if the periods of defocusing are longer then focusing ones in the average along the trajectory then the neighboring trajectories will diverge in the phase space. Thus it is naturally to call this mechanism of chaos mechanism of defocusing.

In [B6], [B7], [W2], [M1], [M2], [D3], [B12], [B13] it was shown that there are hyperbolic billiards with focusing components of more general shapes than the arcs of circles. It is again the mechanism of defocusing that is responsible for their hyperbolicity.

In conclusion of this section we want to mention that the wide spreaded idea is wrong that negative curvature (dispersing part of the boundary) serves always to produce chaos in geodesic flows (in billiards). A transparent example was considered in [B10] where a mixing billiard in a "stadium" (see [B2], [B3]) gets an island of stability by cutting out from it some convex subset. Thus dispersing and defocusing are really different mechanisms of chaos. They can work both in the same dynamical system (see [B1], [B6], [W2], [B7], [B9], [B10]) but only if they are consistent.

5 Fundamental Theorem in the Theory of Hyperbolic Systems of Billiard-Type

Now we shall discuss the problems on global statistical properties of billiard-type systems like ergodicity, mixing, K-property, etc. (see eg. [CFS], [S9]). The first step is still local and has the aim to show that ergodic components of T and $\{S^t\}$ are open. It is easy to show ([S2]) that hyperbolicity of the flow $\{S^t\}$ implies that the induced map T is also hyperbolic. The dimensions of LSM and LUM are equal to $d-1$ while the dimension of ∂M is $2(d-1)$. The basic idea is due to Hopf [H1] and says that the ergodic component that contains (almost every) point x contains as well its GSM and GUM. One has first to show that ergodic components of T and $\{S^t\}$ have positive measure. This fact is based on the property of absolute continuity of foliations generated by GSM and GUM. The complete proof of this property for hyperbolic systems with singularities is contained in [KS].

A much more difficult problem is to prove the uniqueness of the ergodic component of $\{S^t\}$ (or T). The reason is that because of singularities the sizes of LSM and LUM can be arbitrarily small. Hence it is not clear whether two generic points can be connected by a Hopf chain of LSM and LUM (see [H1]).

The general approach how to overcome that difficulty was proposed by Sinai (see [S2]). In [S2] he gave the proof of this fact for two-dimensional dispersing (Sinai) billiards. The idea is to show that in a neighborhood of a typical point for a given LUM $\gamma^{(u)}$ the probability to find $\gamma^{(s)}(y), y \in \gamma^{(u)}$, whose length is greater than $Cl(\gamma^{(u)})$, where l is the length, C is an arbitrary constant, tends to 1 as the radius of the neighborhood tends to zero. Hence (see [S2]) a small neighborhood of this point lies within one ergodic component. It is easy to show in case dim $Q = 2$ that the set of typical points is connected. The proofs of this theorem that contain some modifications and improvements see also in [BS], [S8], [G], [Ku]. In [B1] (see also [B2]) this theorem was proved for some classes of two-dimensional billiards with dispersing components and focusing components which are arcs of circles. Because this theorem can be applied for all classes of hyperbolic billiards it is natural to call it fundamental theorem in the theory of such billiards.

The very essential contribution to the field was made by Sinai and Chernov [SC2]. They proposed the new approach to the proof of this theorem for dispersing and semi-dispersing billiards in any (finite) dimension. Their main condition (which is called now Sinai-Chernov ansatz) is formulated in the following form. Let $R \subset \partial M$ be the union of the singular points of ∂M and the points of discontinuity of T. Then for a. e. point of R the image under dynamics of a plane framed by unit normals locally extends in all directions. Under this condition Sinai and Chernov proved that typical points have neighborhoods belonging to one ergodic component (this property is called local ergodicity). This result implies immediately (see [SC2]) the following

Theorem *Every Sinai billiard is ergodic.*

As well in [SC2] there was shown the ergodicity of some systems of 2D discs and 3D balls.

Then Kramli, Simanyi and Szász made an important observation that one of the conditions imposed on the systems of hard balls in [SC2] can be weakened. They used also the old idea by Sinai (see [S2]) that for the establishing ergodicity it is not necessary to have two invariant (stable and unstable) foliations but instead one and to consider in addition some transversal field to it.

Using these improvements they proved the following

Theorem [KSS3] *Consider the system of three balls on $Tor^d, d \geq 2$. Then it is ergodic on the subspace of the phase space where the total momentum is zero and the centre of mass is fixed.*

The transition from two balls to three balls is essential because the first system is a Sinai (dispersing) billiard while the second is a semi-dispersing one (see e.g. [S9]). The simplest non-trivial example of the semi-dispersing billiard where one encounters some of the main difficulties was studied in

[KSS1]. Recently the proof of ergodicity was obtained for the gas of four balls but in dimension $d \geq 3$ only (see [KSS4]).

Nevertheless, the ergodicity problem of the gas of an arbitrary number of hard balls is still open. More generally the following general problem was upto recently unsolved. Does there exist a Hamiltonian system of an arbitrary finite number of interacting particles that is ergodic? This problem was solved in [BLPS]. The corresponding ergodic system is the gas of N hard balls in a vessel which contains as well some immovable scatterers (infintely heavy particles). This model is in some sense an intermediate one between the gas of hard balls and the Lorentz gas.

The situation with ergodicity of billiards with focusing components of the boundary is more complicated even in the two-dimensional case. The point is that the mechanism of defocusing is much more delicate than the mechanism of dispersing. That can be illustrated for instance by the examples of integrable billiards in circles and ellipses. Therefore the following natural problem was formulated in [B9]:

Given a focusing component Γ^f, does there exist a domain Q which generates the ergodic billiard and the boundary $\partial Q \supset \Gamma^f$?

In [B9] the property of strong focusing was considered which was called later (see [B11]) the property of absolutely focusing.

Definition A focusing component Γ^f is called alsolutely focusing if any infinitesimal beam of parallel trajectories (rays) which fall upon Γ^f after a series of consecutive reflections from Γ^f will leave Γ^f as a convergent (focusing) beam.

Another property of defocusing was considered in [D3], [B10] (condition (G4)) which looks on the first sight much more restrictive than absolute focusing but it is in fact equivalent (see [B11]). The results of [B13] (see also [B12]) give the strong evidence for the conjecture that in two-dimensional ergodic billiards all focusing components of the boundary have to be absolutely focusing.

Following the approach of Sinai and Chernov (see [SC2]) the general variant of the fundamental theorem was proven in [B10] which can be applied to billiards with focusing components in domains Q of arbitrary dimension d. Upto now it was applied yet to the case $d = 2$ only [B10], [B13].

The general theory of hyperbolic systems (see [An], [AS], [S2], [KS], [S9]) gives the possibility to derive immediately from ergodicity much more stronger statistical properties like mixing, K-property etc. In particular all examples of ergodic billiards considered above are K-systems. Moreover two-dimensional Sinai billiards are metrically isomorphic to Bernoulli shifts (see [GO]).

6 Motion of a Particle in a Compound Central Field and Ergodic Geodesic Flows on S^2 and Tor^2

After the appearance of Sinai's fundamental paper [S4] there were started different attempts to perturb Sinai billiards. There are two natural possibilities to consider such perturbations. One is to perturb the boundary ∂Q which was discussed above. The other one is to consider some potential field that differs from zero only in some convex bounded nonintersecting domains. In other words, to change the hard wall by some soft potential. This approach was developed first by Kubo who proved ergodicity and B-property for some class of repelling potentials (see [Ku], [KM], [T]). Then the case of Coulomb and some other attracting potentials was treated in [Kn1].

The most general results were obtained by Donnay and Liverani [DL]. They proved that for any $\alpha \in (0,2)$ there exist attracting potentials with singularity of order $-r^{-\alpha}$ for which the system is completely hyperbolic and ergodic. (In case of singularity of the type $-r^{-\alpha}, \alpha \geq 2$, a positive measure set of trajectories will be pulled into it.)

Again the same mechanisms of chaos are acting in these models. It is the mechanism of dispersing in case of attracting potentials and the mechanism of defocusing in case of repelling potentials. That is the reason in particular that potentials U considered in [Ku], [DL] are continous but not C^1. The discontinuities of U' occured at the boundaries of potential regions. In [DL] there were also constructed some potentials without singularities for which the motion of a particle is completely hyperbolic and ergodic as well. These are the first examples of a smooth Hamiltonian of the form H = Kinetic Energy + Potential Energy that generates an ergodic flow.

The natural problem arised after the appearance of examples of ergodic billiards in domains with focusing components (see [B1], [B2]) and especially in convex domains (like the stadium [B3]). To our knowledge it was V. M. Alexeev who formulated at first in 1973 in the Seminar on Dynamical Systems in Moscow University the problem to construct ergodic flows on S^2 and Tor^2. The first example however was introduced much later by Osserman [O] (see its exposition in [D1]). According to the Gauss-Bonnet theorem the average curvature over the two-dimensional surface of genus g equals $2\pi(2-2g)$. Therefore only for $g \geq 2$ the curvature can be negative everywhere in a surface (with respect to the natural metrics). Osserman has attached to S^2 in some circles (equators) three caps in the form of semispheres. The geodesic flow outside the caps is the same as in the case when we removed the caps and identified the opposite points on the boundary circle of a half-sphere. It is known [BBE] that the resulting geodesic flow is ergodic. This C^1-example was smoothed by using other caps up to C^∞ by Donnay (see [D1], [D2] which contain as well the analogous example on Tor^2) and then by Burns and Gerber [BG] who using some perturbation technique have constructed a real analytic Bernoulli geodesic flow on S^2. It is important to mention that in all these examples the mechanism of defocusing was explored: the geodesics entering a cap pass through conjugate points and leave the cap being defocused.

7 Systems of Falling Balls

A new interesting system of interacting particles was considered recently that demonstrate hyperbolic behavior. First Lehtihet and Miller [LM] studied numerically the motion of a billiard ball in a wedge with gravity. Then Wojtkowski in [W3] showed using the approach of [CFS] that this system is equivalent to the motion of two balls in a vertical line subjected to a constant vertical acceleration (balls fall down). Besides he considered a system of an arbitrary number of particles. The particles collide elastically between each other and the lowest one collides elastically with the floor as well. For equal masses this Hamiltonian system is completely integrable. However it was shown in [W3] that it has at least one non-zero Lyapunov exponent almost everywhere if the mass is a nondecreasing function of the number (starting from the floor) and not all masses are equal. Hence this system is partially hyperbolic. In the next paper [W4] it was shown that this system is completely hyperbolic if the potential $V(q)$ satisfies the conditions $V'(q) > 0$ and $V''(q) < 0$ where q is the vertical coordinate. For instance, one can take as $V(q)$ the standard potential $V(q) = -1/q$ of the gravitational field.

The condition on the masses is essential. If the masses above are bigger than the masses below the system has a linearly stable periodic orbit (see [CW]). Chernov [C2] proved recently that the system of two falling balls is actually ergodic on the constant energy surface under the same conditions as in [W3] and additionally a) $0 < const \leq V'(q) \leq const < \infty$, b) $| V''(q) | \leq const < \infty$. The natural conjecture is that it has to be the case for an arbitrary number of particles.

Most of the results of this and of the previous section were obtained using the so called cones techniques. It is the method to prove hyperbolicity of the system by constructing an invariant family of cones in the tangent space. This method was used for a long time (see [A], [S5]) to show the positivity of Lyapunov exponents in various dynamical systems. It became popular recently after the publishing of the paper by Wojtkowski [W1], where was stressed its efficiency for the proof of positivity of the largest Lyapunov exponent. He also introduced a general approach to the proof that in any $d > 2$ all Lyapunov exponents differ from zero (see e.g. [W5]). A similar technique was developed by Lewowicz (see [Le1] - [Le3]). Some further developments of this method were proposed in [BG] and the most general one in [Ka].

8 Markov Partitions and Strong Chaotic Properties of Two-Dimensional Hyperbolic Billiards

To study more subtle statistical properties of dynamical systems like the central limit theorem and the rate of decay of time-correlations one has to use some symbolic representation of them. In a sense the best such representations are given by the so called Markov partitions of the phase space (see [Bo], [BR], [R1], [R2], [S3], [S6]). Here and further we shall deal with the induced map T. Take a point $x \in \partial M^+, LUM\gamma^{(u)}(x), LSM\gamma^{(s)}(x)$ and choose subsets $\sqcap^{(u)} \subset \gamma^{(u)}(x), \sqcap^{(s)} \subset \gamma^{(s)}(x)$.

Definition A subset $\sqcap \subset \partial M^+$ is called a parallelogram with basic sets $\sqcap^{(u)}, \sqcap^{(s)}$ if for any $y_1 \in \sqcap^{(u)}, y_2 \in \sqcap^{(s)}$ the set $\gamma^{(s)}(y_1) \cap \gamma^{(u)}(y_2)$ consists of one point also belonging to $\sqcap$.

If $\sqcap$ is a parallelogram then for any $x' \in \sqcap$ the sets $\gamma^{(u)}(x') \cap \sqcap, \gamma^{(s)}(x') \cap \sqcap$ can be taken as basic sets. Each parallelogram has two natural partitions $\xi^{(u)}(\sqcap)$ and $\xi^{(s)}(\sqcap)$ with the elements of the form $C_{\xi^{(u)}(\sqcap)}(y) = \sqcap \cap \gamma^{(u)}(y), C_{\xi^{(s)}(\sqcap)}(y) = \sqcap \cap \gamma^{(s)}(y)$ respectively. If a partition η of the phase space is given whose elements are parallelograms then its partitions $\xi^{(u)}$ and $\xi^{(s)}$ are naturally defined.

Definition (see [Bo],[S3],[S6]) The partition η is called Markov partition if

(i) the elements of η are parallelograms $\sqcap_1, \sqcap_2, ...$
(ii) for almost each x

$$T^{-1}C_{\xi^{(u)}}(x) \subseteq C_{\xi^{(u)}}(T^{-1}x), TC_{\xi^{(s)}}(x) \subseteq C_{\xi^{(s)}}(x).$$

In case of smooth hyperbolic systems Markov partitions are finite. For billiards and other hyperbolic systems with singularities such partitions are by necessity countable. If $\mu(\sqcap) > 0$ then the basic sets $\sqcap^{(u)}, \sqcap^{(s)}$ are usually Cantor-like subsets of $\gamma^{(u)}(x), \gamma^{(s)}(x)$ of positive measure. The reason is again that in systems with singularities LUM and LSM can be arbitrarily small.

Having an arbitrary partition η of the phase space into measurable subsets we can construct its symbolic representation in the set of double-infinite sequences by the rule $x \to i = \{i_n\}$, where $x \in \partial M^+, T^n x \in \sqcap_{i_n}, \sqcap_{i_n}$ is an element of η. If a sequence i corresponds to x then $T\sqcap_{i_{n-1}} \cap \sqcap_{i_n} \neq \emptyset$. For Markov partitions the inverse is also true: if i is any sequence for which $\mu(T\sqcap_{i_{n-1}} \cap \sqcap_{i_n}) > 0$ for all $n, -\infty < n < \infty$, then there exists a point x of the phase space for which i is the symbolic representation.

First countable Markov partitions for two-dimensional Sinai billiards were constructed in [BS2]. In [BSC1] using the same ideas a much more simple construction was proposed that works as well for hyperbolic billiards with focusing components studied in [B1] - [B4]. Recently Krüger and Troubetzkoy (see [KT]) succeeded to construct countable Markov partitions for a large class of (not necessary two-dimensional) hyperbolic systems with singularities. The idea of their construction is the same as in [Bo]. It is based on the shadowing lemma for such systems proven in [KT].

The central limit theorem and the rate of decay of time correlations in two-dimensional Sinai billiards were obtained in [BS3] using the Markov partitions constructed in [BS2]. Besides in [BS3] was derived the diffusion equation for the two-dimensional periodic Lorentz gas with bounded free path.

The central limit theorem and the estimation of the rate of decay of time correlations were proved recently for a wider class of two-dimensional billiards (see [BSC2]). It is important to mention that following Chernov's idea we used in [BSC2] not the Markov partition but some its approximation which was called there Markov sieve.

9 Coupled Map Lattices and Space-Time Chaos

In this section we consider not Hamiltonian but some dissipative systems. The new class of dynamical systems with infinitely many degrees of freedom was introduced in the physical literature a few years ago. These new extended dynamical systems were called coupled map lattices (CML). The main reason to introduce CML was an attempt to understand some features of the fully developed turbulence, that is a basically space-time chaotic type of motion in extended dynamical systems.

Consider a smooth compact d-dimensional Riemannian manifold M. We assume that some transformation $f : M \to M$ is given or there exists an open subset $V \subset M$ such that $f : V \to M$ has an attractor Λ. The latter means that there exists such a neighbourhood $U \subset V$ of a set Λ that $\overline{f(U)} \subset U, \Lambda = \cap_{n \geq 0} f^n(U)$ and Λ is invariant with respect to f. One always assumes that f belongs to some space of transformations where some topology ρ is defined. Hence we can consider distances between such transformations.

Take a countable number of copies of the sets M, V, U, Λ and f and let

$$M_i \equiv M, V_i \equiv V, U_i \equiv U, f_i \equiv f, \Lambda_i \equiv \Lambda, i \in \mathbb{Z}^k, k \in \mathbb{Z}_+ ,$$

$$M = \otimes_{i \in \mathbb{Z}^k} M_i, V = \otimes_{i \in \mathbb{Z}^k} V_i, U = \otimes_{i \in \mathbb{Z}^k} U_i, F = \otimes_{i \in \mathbb{Z}^k} f_i, \Lambda_F = \otimes_{i \in \mathbb{Z}^k} \Lambda_i$$

Then F forms naturally a lattice of noninteracting (or uncoupled) maps. For sake of simplicity of the notations we shall consider the case $k = 1$.

A transformation $G : V \to U$ is called an interaction if it can be written as

$$G(x) = G(..., x_{-k}, ..., x_{-1}, x_0, x_1, ..., x_n, ...)$$
$$= (..., G_{-k}(x), ..., G_{-1}(x), G_0(x), G_1(x), ..., G_n(x), ...).$$

Definition A Coupled Map Lattice is a dynamical system generated on U by the tranformation $\Phi = G \circ F$.

Consider in U the shift S (to the left) that is defined by the relation $(Sx)_i = x_{i+1}, x = (x_i) \in U, i \in \mathbb{Z}$. The map S generates the group $\{S^k\}, k \in \mathbb{Z}$, of space translations in the phase space $\mathcal{M}$. It is clear that Φ and S commute.

Denote by $\{R^Z\}$ a group of space-time shifts that act in a phase space $\mathcal{M}$ of CML, where $z = (n, m), R^Z = \Phi^n \circ S^m, z \in \mathbb{Z}_+ \times \mathbb{Z}$. Let μ be a probability measure which is invariant with respect to $\{R^Z\}$.

Definition A dynamical system $(\mathcal{M}, \Phi, \mu)$ has the property of space-time chaos if for any two bounded measurable functions h_1, h_2

$$\lim_{|Z| \to \infty} \int h_1(R^Z u) h_2(u) d\mu = \int h_1(u) d\mu(u) \int h_2(u) d\mu(u) \quad .$$

The first example of extended dynamical systems that have the property of space-time chaos was considered in [BS4]. The corresponding class of CML is generated by expanding one-dimensional (local) maps with the short-range space interactions. It was shown there that in case of weak (short-range) space interaction in $\mathcal{M}$ there exists the unique Gibbs measure μ that has the property of space-time chaos. Moreover, under this condition correlations in time as well as in space decay exponentially fast. In [V] was constructed some class of nonequilibrium distributions on $\mathcal{M}$ which converge under dynamics to μ.

Recently the measures on Λ_Φ were considered for lattices of smooth hyperbolic transformations with short-range interactions (see [PS]). The main tool to prove the existence of space-time chaos in [BS4] and [PS] (see also [GR], [BS5]) is the construction of the thermodynamic formalism for the corresponding CML. Making use of this one can show the existence of Sinai-Bowen-Ruelle (SBR-) measures (see e.g. [S9]). These are the limits of iterations under the dynamics of absolutely continuous measures. The crucial point in this approach is the construction of Markov partitions that contain a continuum number of elements because CML are infinite dimensional dynamical systems.

10 Concluding Remarks

All dynamical systems discussed in this review upto Sect.9 belong to the class of hyperbolic systems with singularities. It is a customary situation that it is easier to study ergodic properties of such a system than of (in some sense) close to it smooth ones. (The typical example forms by the tent map and the logistic map.) That is also the reason why we still do not have structurally stable examples with coexistence of integrable and chaotic behavior in Hamiltonian systems. For instance the example of a smooth ergodic potential treated in [DL] is not structurally stable as well. (One has to mention the essential difference existing between geodesic flows and billiards. It allows Donnay to smooth the example by Osserman but the same approach can not be applied for the smoothing of stadium.) The remarkable results obtained recently by Benedics, Carleson and Young (see [BC], [BY]) for the Henon map give a hope for the advance in that problems.

The last remark refers to CML. The symbolic dynamics constructed in [BS4] and [PS] represents a d-dimensional CML ($d \geq 1$) as a space of configurations of a $(d+1)$-dimensional lattice spin system. Thus the minimal dimension of the spin systems that correspond to CML is equal to 2 and in these systems there can be phase transitions. Then it may be that at least in some cases by such phase transitions there can be generated coherent structures in the corresponding extended dynamical systems (CML). There are some computer simulations that confirm this idea, but basically this problem seems to be completely open.

I would like to thank C. Liverani and S. Troubetzkoy who read the text for useful remarks.

References

[A] Alekseev, V. M.: Quasirandom dynamical systems. Math.Sbornik **7**, 1-43 (1969)

[An] Anosov, D. V.: Geodesic flows on closed Riemannian manifolds of negative curvature. Proc.Steklov Institute **90**, 1-235 (1967)

[AS] Anosov, D. V., Sinai, Ya. G.: Some smooth dynamical systems. Russian Math.Surveys **22**, 107-152 (1967)

[B1] Bunimovich, L. A.: Candidate Dissertation. Moscow University, 123p. (1972) (in Russian)

[B2] Bunimovich, L. A.: On billiards close to dispersing. Math.Sbornik **94**, 49-73 (1974)

[B3] Bunimovich, L. A.: On ergodic properties of some billiards. Funct.Anal. Appl. **8**, 254-255 (1974)

[B4] Bunimovich, L. A.: On the ergodic properties of nowhere dispersing billiards. Comm.Math.Phys. **65**, 295-312 (1979)

[B5] Bunimovich, L. A.: Decay of correlations in dynamical systems with chaotic behavior. J.Exp.Theor.Phys. **89**, 1452-1471 (1985)

[B6] Bunimovich, L. A.: On stochastic dynamics of rays in resonators. Radiofizika **28**, 1601-1602 (1985)

[B7] Bunimovich, L. A.: Dynamical systems with chaotic behavior. Doctor of Sci. Dissertation. Moscow-Kiev. 336p. (1985) (in Russian)

[B8] Bunimovich, L. A.: Two mechanism of dynamical chaos: permanent stochasticity and intermittency. In: Nonlinear and Turbulent Processes in Physics (V. G. Bar'yakhtar et al. eds) Naukova Dumka, Kiev 190-193 (1988)

[B9] Bunimovich, L. A.: Many-dimensional nowhere dispersing billiards with chaotic behavior. Physica **D33**, 58-63 (1988)

[B10] Bunimovich, L. A.: A theorem on ergodicity of two-dimensional hyperbolic billiards. Comm.Math.Phys. **130**, 599-621 (1990)

[B11] Bunimovich, L. A.: On absolutely focusing mirrors. In: "Ergodic theory and related topics" (ed. by U. Krengel et al.), Springer (to be published)

[B12] Bunimovich, L. A.: Conditions of stochasticity of 2-dim billiards. Chaos **1**, (1991)

[B13] Bunimovich, L. A.: On criterium of mixing for 2-dim billiards (in preparation)

[BBE] Ballman, W., Brin, M., Eberlein, P.: Structure of manifolds of nonpositive curvature I. Ann.Math. **12**, 171-203 (1985)

[BC] Benedicks, M., Carleson, L.: The dynamics of the Henon map. Ann.Math. **133**, 73-169 (1991)

[BG1] Burns, K., Gerber, M.: Continuous invariant cone families and ergodicity of flows in dimension three. Ergod.Theory.Dyn.Syst. **9**, 19-25 (1989)

[BG2] Burns, K., Gerber, M.: Real analytic Bernoulli geodesic flows on S^2. Ergod.Theory.Dyn.Syst. **9**, 27-45 (1989)

[BLPS] Bunimovich, L. A., Liverani, C., Pellegrinotti, A., Sukhov, Yu. M.: Ergodic systems of N balls in a billiard table. Preprint 1991

[Bo] Bowen, R.: Equilibrium states and ergodic theory of Anosov diffeomorphisms. (Lecture Notes in Math., vol. 470) Springer, Berlin-Heidelberg-New York 1975

[BR] Bowen, R., Ruelle, D.: The ergodic theory of Axiom A flows. Invent.Math. **29**, 181-202 (1975)

[BS1] Bunimovich, L. A., Sinai, Ya. G.: On a fundamental theorem in the theory of dispersing billiards. Math.Sbornik **19**, 407-423 (1973)

[BS2] Bunimovich, L. A., Sinai, Ya. G.: Construction of Markov partitions for dispersing billiards. Comm.Math.Phys. **73**, 247-280 (1980)

[BS3] Bunimovich, L. A., Sinai, Ya. G.: Statistical properties of Lorentz gas with periodic configuration of scatterers. Comm.Math.Phys. **78**, 479-497 (1981)

[BS4] Bunimovich, L. A., Sinai, Ya. G.: Space-time chaos in coupled map lattices. Nonlinearity **1**, 491-516 (1981)

[BS5] Bunimovich, L. A., Sinai, Ya. G.: Statistical mechanics of coupled map lattices. In: Coupled map lattices (K. Kaneko, ed.) World Scientific, Singapore (in press)

[BSC1] Bunimovich, L. A., Sinai, Ya. G., Chernov, N. I.: Markov partitions for two-dimensional hyperbolic billiards. Russian Math.Surveys **45**, 105-152 (1990)

[BSC2] Bunimovich. L. A., Sinai, Ya. G., Chernov, N. I.: Decay of correlations and central limit theorem for two-dimensional hyperbolic billiards, Russian Math.Surveys **46** (1991)

[BY] Benedicks, M., Young, L.-S.: SBR-measures for certain Henon maps (in preparation)

[C1] Chernov, N. I.: Construction of transversal fibers for multi-dimensional semidispersing billiards. Funct.Anal.Appl. **16**, 35-46 (1982)

[C2] Chernov N. I.: The ergodicity of a Hamiltonian system of two particles in an external field. Preprint 1991

[CFS] Cornfeld, I. P., Fomin, S. V., Sinai, Ya. G.: Ergodic theory. Springer, Berlin-Heidelberg-New York, 1981

[CW] Cheng, I., Wojtkowski, M. P.: Linear stability of a periodic orbit in the system of falling balls. In: Proceedings of the Workshop on the Geometry of Hamiltonian Systems, MSRI, June 1989 (to appear)

[D1] Donnay, V.: Geodesic flow on the two-sphere. I: Positive measure entropy. Ergod. Theory Dyn.Syst. **8**, 531-553 (1988)

[D2] Donnay, V.: Geodesic flow on the two-sphere. II: Ergodicity. In: Springer Lecture Notes in Math. **1342** 112-153 (1988)

[D3] Donnay, V.: Using integrability to produce chaos: Billiards with positive entropy. Comm.Math.Phys. (to be published)

[DL] Donnay, V., Liverani, C.: Potentials on the two-torus for which the Hamiltonian flow is ergodic. Comm.Math.Phys. **135**, 267-302 (1990)

[Do] Douady, R.: Applications du théoréme des tores invariants. Thesis, University of Paris VII (1982)

[G] Gallavotti, G.: Lectures on the billiards. (Lecture Notes in Physics, vol. 38. Moser, J. ed.) Springer, Berlin-Heidelberg-New York, pp. 236-295 (1975)

[GO] Gallavotti, G., Ornstein, D.: Billiards and Bernoulli schemes. Comm. Math.Phys. **38**, 83-101 (1974)

[GR] Gundlach, V. M., Rand, D. A.: Spatio-temporal chaos. 1. Hyperbolicity, structural stability, shadowing and symbolic dynamics. Preprint 1991

[H1] Hopf, E.: Statistik der geodätischen Linien in Mannigfaltigkeiten negativer Krümmung. Ber.Verhandl.Sächs.Akad.Wiss. **91**, 261-304 (1939)

[H2] Hopf, E.: Statistik der Lösungen geodätischer Probleme vom unstabilen Typus. II. Math.Ann. **117**, 590-608 (1940)

[Ha] Hadamard, I.: Les surfaces á courbures opposées et leur lignes geodesiques. J.Math.Pures Appl. **4**, 27-73 (1989)

[He] Hedlund, G. A.: The dynamics of geodesic flows. Bull.Amer.Math.Soc. **45**, 241-260 (1939)

[K] Krylov, N. S.: Works on the foundations of statistical physics. Princeton University Press. Princeton 1979

[Ka] Katok, A. B.: Infinitesimal Lyapunov functions, invariant cone families and stochastic properties of smooth dynamical systems. Preprint 1991

[KM] Kubo, I., Murata, H.: Perturbed billiards systems. II. Nagoya Math.J. **81**, 1-25 (1981)

[Kn1] Knauf, A.: Ergodic and topological properties of coulombic periodic potentials. Comm.Math.Phys. **110**, 89-112 (1987)

[Kn2] Knauf, A.: On soft billiard systems. Physica **D36**, 259-262 (1989)

[KS] Katok, A. B., Strelcyn, I.-M., with collaboration of F. Ledrappier and F. Przytycki: Invariant manifolds, entropy and billiards, smooth maps with singularities (Lecture Notes in Mathematics, vol. 1222) Springer, Berlin-Heidelberg-New York, 1986

[Le1] Lewowicz, I.: Lyapunov function and topological stability. J.Diff. Equations **38**, 192-209 (1980)

[Le2] Lewowicz, I.: Persistence in expansive systems. Ergodic Theory Dyn.Syst. **3**, 567-578 (1983)

[Le3] Lewowicz, I.: Lyapunov functions and stability of geodesic flows. In: Springer Lecture Notes in Math. **1007** (1981)

[LM] Lehtichet, H. E., Miller, B. N.: Numerical study of a billiard in a gravitational field. Physica **D21**, 94-104 (1986)

[KSS1] Kramli, A., Simany, N., Szász, D.: Ergodic properties of semidispersing billiards. Two cylindric scatterers in the 3-D torus. Nonlinearity **2**, 311-326 (1989)

[KSS2] Kramli, A., Simany, N., Szász, D.: A "transversal" fundamental theorem for semi-dispersing billiards. Comm.Math.Phys. **129**, 535-560 (1990)

[KSS3] Kramli, A., Simany, N., Szász, D.: Three billiard balls on the ν-dimensional torus is a K-flow. Ann.Math. **133**, 37-72 (1991)

[KSS4] Kramli, A., Simany, N., Szász, D.: The K-property of four billiard balls. Preprint 1991

[KT] Krüger, T., Troubetzkoy, S.: Markov partitions and shadowing for non-uniformly hyperbolic systems with singularities. Preprint BiBoS 1991

[Ku] Kubo, I.: Perturbed billiard systems. I. Nagoya Math.J. **61**, 1-57 (1976)

[L] Lazutkin, V. F.: The existence of caustics for the billiard problem in a convex domain. Math.USSR Izvestija **37**, 188-223 (1973)

[M1] Markarian, R.: Billiards with Pesin region of measure one. Comm.Math. Phys. **118**, 87-97 (1988)

[M2] Markarian, R.: Non-uniform hyperbolicity, quadratic forms and billiards. Preprint. 1990

[O] Oseledec, V. I.: A multiplicative ergodic theorem: characteristic Lyapunov exponents of dynamical systems. Trans.Moscow Math.Soc. **19**, 197-231 (1968)

[Os] Osserman, R.: Lecture at geodesic flow workshop. Cal Tech, January 1985 (unpublished, see the description in [D1])

[P] Pesin, Ya. B.: Lyapunov characteristic exponents and smooth ergodic theory. Russian Math. Surveys **32**, 55-112 (1977)

[PS] Pesin, Ya. B., Sinai, Ya. G.: Space-time chaos in chains of weakly-coupled hyperbolic maps (in press)

[R1] Ruelle, D.: A measure associated with Axiom *A* attractors. Amer.J.Math. **98**, 619-654 (1976)

[R2] Ruelle, D.: Thermodynamic formalism. (Encyclopedia of Math. and its Appl.5) Addison-Wesley, 1978

[S1] Sinai, Ya. G.: A "physical" system with positive entropy. Vestnik MGU **5**, 6-12 (1963)

[S2] Sinai, Ya. G.: Classical dynamical systems with countable multiplicity Lebesgue spectrum. II. Math.USSR Izvestija **30**, 15-68 (1966)

[S3] Sinai, Ya. G.:Markov partitions and U-diffeomorphisms. Funct.Anal. Appl. **2**, 61-82 (1968)

[S4] Sinai, Ya. G.: Dynamical systems with elastic collisions. Russian Math. Surveys **25**, 141-192 (1970)

[S5] Sinai, Ya. G.: Appendix to the first edition of G. M. Zaslavsky's book "Statistical irreversibility in non-linear systems" ("Nauka" Press, Moscow) 114-139 (1970)

[S6] Sinai, Ya. G.: Gibbs measures in ergodic theory. Russian Math.Surveys **27**, 21-64 (1972)

[S7] Sinai, Ya. G.: Development of Krylov ideas. An addendum to the book [K]

[S8] Sinai, Ya, G.: Ergodic properties of Lorentz gas. Funct.Anal.Appl. **13**, 192-202 (1980)

[S9] Sinai, Ya. G. (ed.): Dynamical Systems II, Encyclopaedia of Math. Sci. vol **2**, Springer, Berlin-Heidelberg-New York (1989)

[S10] Sinai, Ya. G.: Hyperbolic billiards. In: Proceedings of ICM-90, Kyoto, 1990

[SC1] Sinai, Ya. G., Chernov, N. I.: Entropy of hard spheres gas with respect to the group of space-time translations. Trudy Semin. im. I. G. Petrovskogo **8**, 218-238 (1982)

[SC2] Sinai, Ya. G., Chernov, N. I.: Ergodic properties of some systems of 2-D discs and 3-D spheres. Russian Math.Surveys **42**, 153-174 (1987)

[Sm] Smale, S.: Differentiable dynamical systems. Bull.Amer.Math.Soc. **73**, 747-817 (1967)

[Sz] Szász, D.: On the K-property of Wojtkowski's planar billiards. Preprint, 1991

[T] Troubetzkoy, S. E.: A comparison of elastic and inelastic billiards. Nonlinearity **3**, 947-960 (1990)

[V] Volevich, D. L.: Kinetics of coupled map lattices. Nonlinearity **4**, 37-45 (1991)

[W1] Wojtkowski, M. P.: Invariant families of cones and Lyapunov exponents. Erg.Th.Dyn.Syst. **5**, 145-161 (1985)

[W2] Wojtkowski, M. P.: Principles for the design of billiards with nonvanishing Lyapunov exponents. Comm.Math.Phys. **105**, 391-414 (1986)

[W3] Wojtkowski, M. P.: A system of one dimensional balls with gravity. Comm.Math.Phys. **126**, 505-534 (1990)

[W4] Wojtkowski, M. P.: The system of one dimensional balls in an external field. II. Comm.Math.Phys. **127**, 425-432 (1990)

[W5] Wojtkowski, M. P.: Systems of classical interacting particles with nonvanishing Lyapunov exponents. Preprint 1990

Statistical Physics and Spectral Theory of Disordered Systems: Some Recent Developments

L.A.Pastur

Institute for Low Temperature,
Ukrainian Academy of Sciences, Kharkov

I Introduction

Disorder is one of the paradigms of modern physics. It suffices to look through issues of leading physical journals and conference proceedings to see how numerous the papers on the disordered systems are, treating an extremely wide variety of problems, from highly conceptual questions of quantum physics and neurobiology to crack propagation in concrete and spray painting. This certainly reflects the richness and freshness of the field and the continuous "production" of new problems and results.

Of course one should not expect that mathematical physicists, who, sharing the playing ground with theoretical physicists, use more restrictive playing rules, will produce the same amount and variety of papers and results per unit time as theoretical physicists, whose imagination is supported by modern computers. Nevertheless, the number of recent rigorous results on disordered systems is fairly large. Therefore, to give "... a generally accessible review of the field..." as was suggested by the Organizing Committee, I have chosen naturally the branches that are most familiar to me. These are phase transitions in some models of disordered statistical mechanics and spectral properties of random operators and matrices.

II Some General Facts in the Theory of Disordered Systems

A Models. A substantial part of the spectral problems that I am going to discuss emerged in the condensed matter theory, where the widely accepted one-body approximation allows us to express important thermodynamic and kinetic quantities through eigenvalues and eigenfunctions of the Schrödinger operator with a random potential. Thus, the spectral theory of disordered systems is in many aspects a part of statistical physics. Therefore, I shall discuss some general and, in fact, introductory facts on disordered systems in the context of statistical physics.

Disorder is a synonym of randomness. The randomness which is used in the theory of disordered systems satisfies some general physical requirements. Namely, it should be translationally invariant in the mean (homogeneous) and the statistical correlations between regions of a disordered system, which are far enough from one another, should vanish. Typical examples are a independent identically distributed (i.i.d.) for all $x \in \mathbb{Z}^d$ random potential $V(x)$ in the discrete analogue of the Schrödinger operator

$$-\Delta + V(x) \tag{1}$$

or a i.i.d. for all $(x, y) \in \mathbb{Z}^d \times \mathbb{Z}^d$ interaction J_{xy} and a i.i.d. for all $x \in \mathbb{Z}^d$ external field h_x in the classical Heisenberg Hamiltonian

$$-\frac{1}{2} \sum_{x,y \in \Lambda} J_{xy} s_x s_y - \sum_{x \in \Lambda} h_x s_x, \tag{2}$$

where Λ is a finite set in $\mathbb{Z}^d$ and s_x are unit n-dimensional vectors.

Besides, the randomness can be either annealed (in the thermodynamic equilibrium with the environment) or quenched (frozen-in). In other words annealed randomness is just an additional amount of dynamical degrees of freedom, perhaps with a somewhat unusual Hamiltonian. Annealed models are mathematically more tractable, while quenched ones are more realistic and much more popular in disordered systems. However, under the influence of the quantum field theory (large$-n$ expansions, two-dimensional gravity, low dimensional strings) a variety of annealed models have been introduced and solved in the last years [24].

Returning to the formulated general properties of disordered systems (translation invariance in the mean and decay of correlations) and bearing in mind the Schrödinger operator (1) as an archetype object of the random spectral theory and the Hamiltonian (2) as an archetype object of the disordered statistical physics, an educated reader can easily identify the first of these properties as a homogeneity of random fields $V(x)$ in (1) and h_x in (2) and the second property as a good mixing (or at least ergodicity) of these fields. Thus, if (Ω, F, P) is the corresponding probability space and T_x, $x \in \mathbb{Z}^d$ (or $\mathbb{R}^d$), is the corresponding ergodic group then for some measurable functions $\Phi_1(\omega)$ and $\Phi_2(\omega)$,

$$V(x) = \Phi_1(T_x\omega), \qquad h_x = \Phi_2(T_x\omega). \tag{3}$$

As for the interaction J_{xy} in (2), it can be represented as

$$J_{xy} = \Phi_3(x - y, T_y\omega), \tag{4}$$

where $\Phi_3(x, \omega) = \Phi_3(-x, \omega)$. The simplest cases of (4) are

$$J_{xy} = J_{x-y} c_x c_y, \tag{5}$$

where $J_x = J_{-x}$ is a rapidly decreasing function and c_x are i.i.d. random variables assuming values 0 and 1 with probabilities p and $1-p$ respectively (site disorder) and $J'_{xy}s$ which are i.i.d. for all nearest neighbours (bond disorder).

B Selfaveraging Property. Consider the free energy

$$F_\Lambda = -\ln Z_\Lambda \tag{6}$$

of the Hamiltonian (2), where Z_Λ is its partition function. If a large macroscopic volume Λ is split into a large number of congruent subvolumes Λ_i (macroscopic but small compared to Λ), then

$$F_\Lambda = \sum_i F_{\Lambda_i} + C_\Lambda,$$

where the correction term C_Λ is due to the interaction of subvolumes across their boundaries. For short-range interactions, i.e. for instance, if

$$\sum_{x \in \mathbb{Z}^d} | \Phi_3(x,\omega) | < \infty \tag{7}$$

with probability 1, then $C_\Lambda = o(| \Lambda |)$. Therefore, in the macroscopic limit $\Lambda \to \mathbb{Z}^d$ (we shall write $\Lambda \to \infty$) the specific free energy $f_\Lambda = N^{-1} F_\Lambda$, $N = | \Lambda |$, is the limit of the arithmetic mean of $| \Lambda_i |^{-1} F_{\Lambda_i}$, and, owing to the ergodic structure embedded, as it was explained above, into any disordered system, $\lim_{\Lambda \to \infty} f_\Lambda$ exists and is nonrandom. We say that the free energy of a disordered system is a selfaveraging quantity. This property of physical quantities (not necessarily the free energy) guarantees reproducibility of their measurements performed on a large number of macroscopic samples and justifies the quenched approach in the theory of disordered systems (see e.g. [34]).

Let us formulate now three natural forms of the selfaveraging property, that appeared in the theory of disordered systems. Consider some quantity a_Λ, defined for a disordered system, whose coefficients satisfy (3) and (4), and denote by $\mathbf{E}\{\cdot\}$ the respective expectation. We say that a_Λ has:

(i) the weak selfaveraging property, if

$$V\{a_\Lambda\} \equiv \mathbf{E}\{(a_\Lambda - \mathbf{E}\{a_\Lambda\})^2\} \to 0, \text{ as } \Lambda \to \infty; \tag{8}$$

(ii) the selfaveraging property, if

$$a_\Lambda \to a, \text{ as } \Lambda \to \infty, \tag{9}$$

where a is a nonrandom quantity and convergence is to be understood in some conventional probabilistic sense, i.e. in probability, in variance, with probability 1, etc;

(iii) the strong selfaveraging property, if

$$V\{a_\Lambda\} = o(|\Lambda|^{-1}). \tag{10}$$

Property (i) just tells us that fluctuations of a_Λ vanish in the macroscopic limit. Property (ii) is the most often used one and differs from (i) in that it includes the existence of $\lim_{\Lambda\to\infty} E\{a_\Lambda\} = a$ which for some mean field models is not too simple to prove. Property (iii) is one of the simplest that guarantees coincidence of the annealed and the quenched free energies. It is valid e.g. for the high temperature phase in some mean field models [5],[16], for the integrated density of states of some matrix ensembles [7],[27] and for the transparency of some toy models of wave propagations in random media [45].

The notion of selfaveraging is rather old (see [34] and references therein). I discuss it here for the following reasons.

First, weak selfaveraging (8) of the free energy of the important mean field model of spin glasses has been proved recently [46].

Second, it has been found [4] that the study of the fluctuations of some selfaveraging quantities allows us to rule out some phase transitions in model (2).

Third, it has been proved recently that some important quantities (the transparency of a random barrier [36], the order parameter in spin glass models [46]) are not selfaveraging in the macroscopic limit and therefore to describe respective models completely enough one should study their fluctuations, that do not vanish in the macroscopic limit.

Fourth, in the recent years an interesting concept has been clearly emerged in theoretical and experimental studies of quantum transport. It is the concept of non-selfaveraging of the conductance g_Λ of samples in the mesoscopic size range. These are sizes large on the atomic scale but small enough for the electron wave function to have well defined phase over the entire sample. Due to this quantum coherence, statistical fluctuations become much more long correlated and do not average to zero on the mesoscopic (submicron at millikelvin temperatures) scale. It turns out that these fluctuations show a remarkable universality: the variance of g_Λ is independent of the sample size and the amount of disorder. This universality was discovered by perturbative calculations [3] and supported by non-perturbative arguments, based on the nontrivial hypotheses on the properties of random matrices [22]. Mathematical physicists may find there a lot of new beautiful problems on limit theorems and intermediate asymptotes for random matrices, multichannel scattering and various expansion techniques.

III Phase Transitions in Disordered Systems

A Disordered Ferromagnets. Let us start with a very simple example. Assume that the interaction in (2) is translationally invariant and ferromagnetic and the external field is random but very slowly varying. This means, that $h_x = h + b(x/R)$, where h is a constant, $b(t), t \in \mathbb{R}^d$, is an ergodic field and the limit $R \to \infty$ should be taken after the macroscopic limit $\Lambda \to \infty$. Then, by using the above picture with a cube Λ split into subcubes Λ_i, whose size is much smaller than R, we can easily find the following expression for the free energy of our model

$$f(h) = \int f_{\mathrm{or}}(h + b)P(db), \tag{11}$$

where $f_{\mathrm{or}}(h)$ is the free energy of the translationally invariant (ordered) system with the same interaction and uniform field h, and $P(db)=$ $Prob\{b(t) \in db\}$. Since $f_{\mathrm{or}}(h)$ is an even function, then, if $b(x)$ is symmetrically distributed, the spontaneous magnetization $m = \mp\frac{df}{dh}\mid_{h=\pm 0}$ is zero, even if $m_{\mathrm{or}} \neq 0$.

Thus, the ferromagnetic ordering is unstable under perturbation by a slowly varying external field of an arbitrarily small amplitude. This is a very simple illustration of the general tendency of disorder: it smoothes over singularities of the free energy and increases the order of (or destroys completely) the phase transitions in ordered systems. Thus we come to the fairly natural problem of disordered statistical mechanics: study the stability of phase transitions in ordered systems under random perturbations, or, using the terminology of the theory of the critical phenomena, find out which random perturbations are relevant.

A fairly instructive example of such a study and much less trivial than the above example (11) of this tendency is the translationally invariant ferromagnet with a short correlated external field, i.e. with h_x in (2) independent, identically and symmetrically distributed. Then, according to Imry and Ma [23], ferromagnetic ordering is unstable under such a perturbation of an arbitrarily small amplitude in the Ising model ($s_x = \pm 1$) if $d \leq d_*^I = 2$ and in the classical Heisenberg model (s_x is the n-dimensional unit vector for $n \geq 2$) if $d \leq d_*^H = 4$. These arguments were made rigorous in the framework of the so-called disordered spherical model [44]. At that time, however, this model looked somewhat artificial. In the translationally invariant case the spherical model is the $n \to \infty$ limit of the classical Heisenberg (n-vector model). Recently it has been proved [25] that the model considered in [44] is also the $n \to \infty$ limit of the model defined by (2), in which $h_x^\alpha, \alpha = 1, \ldots, n, x \in \mathbb{Z}^d$ are i.i.d.

The controversial case of the Ising model was settled by J.Bricmont and A.Kupiainen [9], who established the ferromagnetic order for $d = 3$ if the random field is small enough, i.e. $d_*^I \leq 2$. The authors developed a powerful technique combining the renormalization group ideology with the contour technique, which for many years was a fairly important ingredient of all Ising model studies. These arguments however have no sufficiently good analogue in the case of continuous symmetry, $n \geq 2$. Recently M.Aizenman and J.Wehr [4] proposed a beautiful proof of the inequalities $d_*^I \leq 2, d_*^H \leq 4$, that is based on analysis of the size dependence of some free energy differences and rules out the discontinuities in the thermal averages of quantities conjugated to the random parameters. These arguments are of a very general nature and can be without any doubt applied to many interesting and important problems.

These arguments however give almost no information on the nature of the spin ordering for $d \leq d_*$. For the Ising model ($n = 1$), by using the correlation inequalities, one can show that two-dimensional ferromagnetism is completely destroyed and the Gibbs state is unique with probability 1 [4]. The same is valid for the disordered spherical model, i.e. for the $n \to \infty$ limit [25],[44]. What happens for finite $n \geq 2$? Further, what is the picture of the phase transitions for $d > d_*$? It is natural to expect, and the mean field calculations support this expectations, that we shall have a ferromagnet for small fields, however for large fields spins should follow the direction of the field. Thus, there should be a phase transition which can be reached by varying the magnitude of the random field.

The situation is similar in other cases of random perturbations. Consider, for example, the site disorder ferromagnet (5). Here, by decreasing the concentration p of ferromagnetic atoms, we can destroy ferromagnetism and produce the so-called concentration phase transition with its own criticality, which is closely related to the percolation transition. Physicists know a lot of things about this and similar transitions (see e.g. [11]) but mathematical physicists not so much (see however [18]).

Thus, even the simplest model of spin orderings, i.e. the ferromagnetic one, has many new very interesting features in the disordered case. I have not mentioned here such important topics as the Griffith singularities (C^∞-smoothness of thermodynamic functions in some temperature interval), stability of the ferromagnet under perturbation $D\sum_x(s_x n_x)^2$, where n_x are independent uniformly distributed unit vectors, and many others, whose rigorous study is in the initial stage, if any. Perhaps, as an illustration, showing that we have a lot of work to do, let me mention the absence of the proof of the existence of a ferromagnet for the Hamiltonian (2) with non-negative J_{xy}. The proof of this fact in the translationally invariant case uses the reflection positivity and infrared bounds, relies heavily on this invariance and, as is implied by example of the concentration phase transition, should be supplemented by new ideas.

A wide variety of stability results for quantum spin models has been obtained recently by A.Klein and coworkers. These results concern both, the stability of translation invariant spin orderings under small random perturbations and the absence of such orderings for large disorder, where these authors prove exponential decay with probability 1 of the respective correlation functions. These results are discussed in the lecture by A.Klein at this Congress. Here I only note, that their technique is based on the so-called multiscale analysis, which is, roughly speaking, a highly nontrivial version of the perturbation theory for operators with a dense point spectrum. This technique takes its origin in another branch of disordered mathematical physics, the spectral theory of random operators, and was introduced by J.Fröhlich and T.Spencer [19] to prove the Anderson localization for the Schrödinger operator with a large random potential. Recently, substantial improvement of this technique has been achieved by T.Spencer [49], H.von Dreifus and A.Klein [12],[13]. Its proper modification turns out to be the efficient method of analysis of disordered quantum spin models.

New interesting stability results have been obtained also by C.Albanese [1].

B Spin Glasses. The stability results which we have mentioned, being very important and quite difficult to prove in many cases (see e.g. [18]), are based in one way or another on the terminology and notions created in the translation invariant statistical physics. However, in the disordered case there exists a spin ordering that is hard to imagine in the translationally invariant setting. Let me begin with an extremely oversimplified definition of this ordering. Assume that the disordered Hamiltonian (2) has a unique Gibbs state and denote by $< \, . \, >$ the expectation in this state. Consider the so-called local magnetization $m_x \equiv < s_x >$. This quantity is independent of x or periodic in the translationally invariant case but it is a random field in the disordered case. Consider the two simplest characteristics of this random field: its mean value $m = \mathbf{E}\{m_x\}$ and variance $q = \mathbf{E}\{(m_x - \mathbf{E}\{m_x\})^2\}$. Then, except for the trivial case $m_x \equiv 0$ (paramagnet), we should expect that in the disordered case, $q > 0$. This leaves an alternative: $m \neq 0$ or $m = 0$. The former one corresponds to the disordered ferromagnet or rather the ferrimagnet, while the latter corresponds to a nontrivial magnetic structure of frozen but completely chaotically oriented spins with zero magnetization per site, because, referring again to the ergodic structure, we should expect that

$$m = \lim_{\Lambda \to \infty} | \Lambda |^{-1} \sum_{x \in \Lambda} m_x. \tag{12}$$

We shall call this frozen spin ordering a spin glass. Of course, the physical content of spin glass is much richer and includes notions of frustrations, broken ergodicity, multivalley energy landscape, macroscopic multiplicity of almost energetically equivalent states, remanence and dynamical effects and many others. However, even in the framework of the above, rather simplified definition, there are many open questions. Do sufficiently realistic Hamiltonians, say those with nearest neighbour symmetrically distributed (competing) interactions admit spin glass ordering for sufficiently low temperatures? Does this ordering result from some phase transition or does it develop gradually, like ordinary glass? If there is a phase transition, then what is the respective criticality, i.e. the number and properties of phases, critical indices, their dependence on the dimensionality of the lattice and the spin space, etc.?

However, in spite of many insights, suggestions and calculations, that can be found in the abundant physical literature on spin glasses, these and others important questions are still heavily debated. In this situation it is natural to consider the mean-field models that are fairly useful in the study of various qualitative and even quantitative aspects of the phase transitions. Since the simplest Curie-Weiss model in the translationally invariant case is generated by the infinite-range interaction $J_{xy} = J/N, N = \mid \Lambda \mid$, let us consider as its disordered counterpart the separable interaction

$$J_{xy} = J/N\, \xi_x \xi_y, \tag{13}$$

where ξ_x is an ergodic sequence and $\mathbf{E}\{\xi_x^2\} = 1$ (normalization condition). Then standard calculations give the free energy in the Ising case:

$$f = \max_t \left[\frac{Jt^2}{2} - \frac{1}{\beta} \mathbf{E}\{\ln 2 \operatorname{ch} \beta(Jt\xi_x + h_x)\} \right]. \tag{14}$$

Besides,

$$m_x = \operatorname{th} \beta(Jt_0\xi_x + h_x), \tag{15}$$

where t_0 is a maximum in (14), that is

$$t_0 = \mathbf{E}\{\xi_x \operatorname{th}\beta(Jt_0\xi_x + h_x)\}, \tag{16}$$

and

$$m = \mathbf{E}\{\operatorname{th}\beta(Jt_0\xi_x + h_x)\}. \tag{17}$$

For $h_x \equiv 0$ and $T < T_f = J$, (16) admits a nonzero solution, that plays the role of the order parameter here and we have phase transition at $T = T_f$ with criticality usual for the Curie-Weiss model. However, a new feature is that if ξ_x is symmetrically distributed, then according to (15) and (17), $m_x \neq 0$, but $m = 0$ even for $T < T_f$, and the magnetic susceptibility has a cusp at $T = T_f$, as is expected for the spin glass transition.

One can also consider a more general version of (13) and (14) [16],[35] containing any $N-$independent number p of modes:

$$J_{xy} = \frac{1}{N} \sum_{\mu=1}^{p} J_\mu \xi_x^\mu \xi_y^\mu + J_{xy}^{\text{s.r.}}, \tag{18}$$

where $\xi_x^\mu, \mu = 1,..p$, are ergodic sequences and $J_{xy}^{\text{s.r.}}$ is the short-range part of the interaction, satisfying (4). Then [16],

$$f = \operatorname*{ext}_{(t_\mu)} \left[\frac{1}{2} \sum_{1}^{p} J_\mu t_\mu^2 + \mathbf{E}\{f_{\text{s.r.}}(\sum_{1}^{p} t_\mu \xi_x^\mu + h_x)\} \right], \tag{19}$$

where ext denotes max for t_μ's with $J_\mu > 0$ and min for t_μ's with $J_\mu < 0$ and $f_{\text{s.r.}}(h_x)$ is the free energy, corresponding to $J_{xy}^{\text{s.r.}}$ and an external field h_x.

This model demonstrates a variety of phase transitions and other properties [2],[6],[16]. Besides, as the translation invariant Curie-Weiss model, the model (18) is an infinite range limit of the site-disordered-like interaction (cf. (5))

$$J_{xy}^{(R)} = \sum_{\mu=1}^{p} J_\mu \xi_x^\mu \xi_y^\mu R^{-d} \varphi(\frac{x-y}{R}) + J_{xy}^{\text{s.r.}} \tag{20}$$

(here $\varphi(x)$ is a nonnegative rapidly decreasing smooth function on $\mathbb{R}^d$ and $\int \varphi dx = 1$) in the sense that if f_R is the free energy of (20) in the macroscopic limit, then

$$f = \lim_{R\to\infty} f_R, \tag{21}$$

where f is given by (18).

However, according to the widely accepted opinion [8],[38], the model (18) with a finite p does not reflect many important features of the spin glass physics, in particular, because its dynamical properties are rather different from expected or experimentally observed. This model however arose considerable interest as a model of associative memory (see e.g. [6],[38]). In particular, it was found that the retrieval ability of learned memory patterns is good enough as long as $p \leq \alpha_c(T)N$, where $\alpha_c(0) = 0.14$, and beyond this limit the network suffers from "blackout" or, in terms of statistical mechanics, the system admits a spin glass phase. These and many other very interesting results were obtained by using the so-called replica trick, which, being a most popular and pragmatically efficient technical mean of theoretical physics of disordered systems (see e.g. [6],[38]), is very poorly understood rigorously.

Nevertheless, one of the numerous suggestions that are implied by these studies is that spin glass behaviour is to be expected when the number p of modes in (18) is macroscopically large: $p = \alpha N$. Let us therefore consider the extreme case of a very large number of modes, i.e. $p \gg N (\alpha \to \infty)$. Then, invoking the central limit theorem, we can expect that after replacing in (18) the normalization factor N^{-1} by $(pN)^{-1/2}$ this term will take the form

$$N^{-1/2} J_{xy}, \tag{22}$$

where $J_{xy}, x, y = 1, ...N$, are independent (except the symmetry condition $J_{xy} = J_{yx}$) Gaussian variables with

$$\mathbf{E}\{J_{xy}\} = 0, \quad \mathbf{E}\{J_{xy}^2\} = J^2. \tag{23}$$

This interaction defines the Sherrington-Kirkpatrick (SK) model that has been most popular in spin-glass studies of the last decades. It was proposed [47] as an infinite range version of the mean field theory of the spin glass, developed in the seminal paper by S.Edwards and P.Anderson [15].

According to [47], the free energy of the SK model obtained by the simplest version of the replica trick (the so-called symmetric replicas) is (for the Ising spins):

$$f_{SK} = -\frac{K^2}{4}(1-q)^2 - \frac{1}{\sqrt{2\pi}}\int_{-\infty}^{+\infty} \ln 2\,\mathrm{ch}\,(Kq^{1/2}z)e^{-z^2/2}dz, \tag{24}$$

where $K = \beta J$, q is a solution of the equation

$$q = \frac{1}{\sqrt{2\pi}}\int_{-\infty}^{+\infty} \mathrm{th}(Kq^{1/2}z)e^{-z^2/2}dz, \tag{25}$$

and plays the role of the order parameter. These expressions provide a quite reasonable description of the critical behaviour in the vicinity of the freezing temperature $T_f = J$, but give wrong results for low temperature, in particular, the entropy corresponding to (24), (25) is negative for $T < T_1 < T_f$.

The efforts to overcome these difficulties resulted in the solution, proposed by G.Parisi [40] and based on a version of the replica trick which is highly nontrivial and hard to follow using the known mathematical techniques. The corresponding free energy is given by an extremum of a rather complicated functional, depending on a nondecreasing function $0 \le q(x) \le 1$, $0 \le x \le 1$ (and coinciding for $q(x) \equiv$const with (24)). Thus, the order parameter in this solution is a function, not a number as in (19) and (24). This has no analogues in the vast family of the translationally invariant mean field models but is not too unexpected in view of the facts explained above, according to which (22) is the "limiting" form of (18) for $p \to \infty$ and the model defined by (18) has a p-dimensional order parameter.

One of the serious consequences of this important property of the Parisi solution is that the seemingly natural (in view of (24), (25)) candidate

$$\hat{q}_N = \frac{1}{N}\sum_{x=1}^{N} m_x^2 \tag{26}$$

for the order parameter (cf. (12)) is not a selfaveraging quantity. On the other hand, for the model (18) with a finite p, the p-dimensional order parameter $t_\mu, \mu = 1, ..., p$, is with probability 1 $\lim_{N\to\infty} N^{-1}\sum_1^N \xi_x^\mu m_x$, i.e. it is selfaveraging.

The Parisi solution received a beautiful physical interpretation, has numerous applications [8],[38] and is widely believed to be exact. This is, however, an open problem. Thus, let me list rigorous results on the SK model. The high temperature behaviour of this model was thoroughly studied in [5],[20] using versions of the cluster expansion method. In [20] it was proved that for high temperature the infinite volume Gibbs state of the SK model is unique and is the $R \to \infty$ limit of the Gibbs states corresponding to the interaction

$$J_{xy} R^{-d/2} \varphi(\frac{x-y}{R}), \tag{27}$$

where J_{xy} are the same as in (22) and (23) and $\varphi(x)$ is the same as as in (20). In [5] it was found that for $T > J$ the free energy of the SK model coincides with (24) for $q = 0$ (which is the annealed free energy), that the remainder (fluctuations) has the form φ_N/N (not $O(N^{-1/2})$!), where φ_N converges in distribution to a shifted Gaussian random variable and some bounds were obtained implying nontrivial behaviour for low temperatures.

In the recent paper [46] we proved first that the free energy f_N of the SK model admits the weak selfaveraging property (8). We used a version of the martingale ideology, since the above outlined scheme of the proof of selfaveraging does not work here. We also proved by using a technique similar to that developed in the random matrix studies [41] that if the variance of (26) vanishes as $N \to \infty$, then $\lim_{N\to\infty} \mathbf{E}\{f_N\} = f_{SK}$ where f_{SK} is given by (24), (25). Since this form of the free energy can not be true for low temperatures, we concluded that unlike the free energy the quantity (26) cannot be selfaveraging even in the weakest sense (8). This result is a first rigorous support of the Parisi solution.

Mean field models can be viewed as the infinite interaction radius ($R \to \infty$) limits of more realistic disordered spin models, as (5), (20) and (27). This limit and the largely similar limit of infinite dimensionality ($d \to \infty$) are the oldest and simplest "non-perturbative limits" in studies of strongly interacting systems of statistical mechanics and quantum field theory. Another rather popular non-perturbative limit is the limit of the infinite number of components ($n \to \infty$). It was introduced in statistical mechanics by M.Kac and H.Stanley and, unlike the $R \to \infty$ or $d \to \infty$ limits, it "feels" the dimensionality of space. In the translationally invariant case it is rather simple and reduces to the well-known spherical model. In the disordered case, the respective limiting model is much more complicated and includes the following problem [30]. Define the random operator A in $l^2(\mathbb{Z}^d)$ by the matrix $A_{xy} = J_{xy} - \delta_{xy} z_x(T)$, where $z_x, x \in \mathbb{Z}^d$, are to be found from the infinite system of equations for $G_{xy} = (A^{-1})_{xy}$

$$-TG_{xx} + (\sum_y G_{xy} h_y)^2 = 1, \qquad x \in \mathbb{Z}^d.$$

For large T or $d = 1, 2$, A is invertible and the thermodynamics is analytic in the temperature and the field. By lowering T, we should reach a point, where A ceases to be invertible, and this is a phase transition point.

Thus, to construct the thermodynamics in the $n \to \infty$ limit, we should solve a rather complicated inverse problem for the random operator A. This problem is solved for the translationally invariant interaction with a periodic or random field, random anisotropy and mean field interactions and gives quite reasonable results, that are discussed in detail in our review [26]. However, the interesting and important problems of the existence and nature of phase transitions for the random short-range ferromagnet ($J_{xy} \geq 0$) or the spin glass (J_{xy} is symmetrically distributed) are open.

The physical theory of spin glasses rises many rather fundamental questions. A number of its concepts and results are of fairly general character and have been successfully applied to a wide range of complex systems in computer science, neurobiology, biochemistry and economics. Therefore rigorous studies of this field would be very interesting and important.

IV Spectral Properties of Random Operators

A Localization and Related Problems. Let us consider the Schrödinger operator H in $L^2(\mathbb{R}^d)$ or its discrete analogue in $l^2(\mathbb{Z}^d)$ whose potential is an ergodic field. Using this property it is not hard to prove that the spectrum σ and all its components (point component σ_p , absolutely continuous component σ_{ac}, etc.) are nonrandom closed sets and if H_Λ is the restriction of H of (1) to a finite region Λ with some boundary conditions on $\partial\Lambda$ and $N_\Lambda(\lambda) = \mid \Lambda \mid^{-1} \#\{\lambda_i(\Lambda) \leq \lambda\}$ is the normalized counting function of eigenvalues $\lambda_i(\Lambda)$ of H_Λ, then with probability 1 [42]

$$\lim_{|\Lambda|\to\infty} N_\Lambda(\lambda) = N(\lambda), \tag{28}$$

where $N(\lambda)$ is known as the integrated density of states (IDS). Thus, the IDS is a spectral characteristic of H and selfaveraging in the sense of (9). It was studied by many authors (see e.g. reviews [28],[42]). To go further in the spectral analysis of H, assume that the potential is random, for instance, in the discrete case, $V(x)$ in (1) are i.i.d. P.Anderson argued in 1958 that the spectrum of this operator in the neighbourhoods of its edges is pure point, the respective eigenfunctions are localized at large fluctuations (wells) of the potential, decaying exponentially away from these wells and that for a large random potential all the spectrum is of this type. Since then this phenomenon, known as the Anderson localization, has become one of the central ones in condensed matter physics, having numerous applications.

The development of the rigorous spectral theory of the random Schrödinger operator began 10 years later, with the study of the IDS (28) (see e.g.[42]). As the next step, it was proved in [21] that the whole spectrum of the one-dimensional Schrödinger operator with smoothly distributed exponentially mixing Markov potential is pure point. In the middle eighties S.Kotani developed a fairly complete spectral theory of the one-dimensional case [31]. Recently, in the framework of this theory he has proved that the absolutely continuous spectrum is empty ($\sigma_{ac} = \emptyset$) if an ergodic potential assumes only finitely many values [32]. This criterion proved to be fairly useful in the recent studies of the one-dimensional Schrödinger equation with an almost periodic potential that have been actively developed during the last decade [17]. The nature of this criterion is rather different from the well known one [17], guaranteeing the absence of σ_{ac} if the Lyapunov exponent of the corresponding equation is positive in a given energy range. One more criterion has been given by B.Simon and T.Spencer [48]. It guarantees the absence of σ_{ac} if $\limsup_{x\to\pm\infty} | V(x) | = 0$ (discrete case). The proof is based on the representation of the operator as an orthogonal sum of finite blocks that are restrictions of the operator to the intervals between "peaks" of the potential and the remainder that proved to belong to the trace class. Improving this idea of splitting the operator into blocks in the spirit of the cluster expansions and adding new arguments induced by the random operator theory, Kirsch et al. [28] found recently new sufficient conditions on the unbounded (not necessarily random) potential guaranteeing pure point spectrum of the Schrödinger operator on the semiaxis for almost all boundary conditions. This is an example of fruitful influence of the random operator spectral studies on the conventional spectral analysis.

The mathematical history of multidimensional localization began with the paper by J.Fröhlich and T.Spencer [19], who proved exponential decay of the Green function of the Schrödinger operator in arbitrary dimension for a large smoothly distributed potential or for energies close enough to the spectrum edges. The subsequent progress in this direction was described in the review [37]. Recently this approach has been substantially improved and simplified, especially in its probabilistic part. This allows the consideration of new classes of random potentials, in particular strongly correlated ones (e.g. the Gaussian potential whose correlation function has power law decay [14]) and the applications of this approach to various fields (see, e.g., the lecture by A.Klein at this Congress). Thus, the case of strong disorder is rather well understood rigorously.

It is then natural to inquire what will happen in the case of small disorder. It is believed that in this case the part of the spectrum in the middle of the unperturbed band $| \lambda | \leq 2d$ belongs to σ_{ac} if $d \geq 3$. During the last decade physicists have developed efficient perturbation techniques, known as the weak localization theory (see e.g. the review [33]) which allows the calculation of the quantum correction of important physical quantities and, being supplemented by some renormalization group arguments, predicts the complete localization for $d = 1, 2$ and the mixed spectrum for $d \geq 3$.

B Limits of Infinite R and n and the Theory of Random Matrices. The reader familiar with statistical mechanics will recognize an analogy: the pure point spectrum (single phase) in low dimensions and the mixed one (several phases) in high dimensions. In the latter case we may expect, that moving from the edges of the spectrum, we should reach a nonrandom point that separates the pure point and the absolutely continuous components (it is believed, that these components do not overlap). This is an analogue of the phase transition known as the metal-insulator or the Anderson transition and this point is called the mobility edge, because when the Fermi energy of electrons passes through this point, the static conductivity at zero temperature becomes nonzero. Similar phenomena are expected in the case, where the energy is fixed, but the amplitude of the random potential or some other measure of the disorder varies.

However, since the problem of rigorous understanding of the weak disorder regime and the Anderson transition is still open, it is natural to try, just as for difficult phase transition problems, to construct analogues of the mean field model and the spherical model in the context of the spectral theory, i.e. analogues of the infinite interaction radius and the infinite number of components limits.

To this end, let us consider the random operator $H^{(R)}$, defined in $l^2(\mathbb{Z}^d)$ by the matrix (cf. (27))

$$H^{(R)} = H^{(0)}(x-y) + R^{-d/2} J_{xy}\phi(\frac{x-y}{R}), \tag{29}$$

where $H(x) = H^*(-x), \mid H(x) \mid \leq C\exp(-\rho x), C, \rho > 0, J_{xy} = J_{yx}$ are i.i.d.,

$$\mathbf{E}\{J_{xy}\} = 0, \quad \mathbf{E}\{J_{xy}^2\} = J^2, \quad \mathbf{E}\{J_{xy}^4\} < \infty \tag{30}$$

and $\phi(t), t \in \mathbb{R}^d$, is a rapidly decreasing function. Denote by $N_R(\lambda)$ the IDS of $H^{(R)}$. Then [41] $\lim_{n\to\infty} N_R(\lambda) = N(\lambda)$ in probability (cf. (21)), and the Stieltjes transform

$$r(z) = \int (\lambda - z)^{-1} dN(\lambda),\ Im\, z \neq 0 \tag{31}$$

of the limiting IDS $N(\lambda)$ is a unique solution of the functional equation

$$r(z) = r_0(z + J^2 r(z)), \tag{32}$$

in which $r_0(z)$ is the Stieltjes transform of IDS $N_0(\lambda)$ of the "unperturbed" operator $H^{(0)}$ in (29) that is equal to $\mathrm{mes}\{p \in \mathbf{T}^d : \sum_x H^{(0)}(x)\, e^{2\pi i p x} \leq \lambda\}$.

To explain this result, consider the finite $n \times n$ random matrix of the form

$$h_{ik}^{(n)} = h_{ik}^{(0)} + n^{-1/2} J_{ik}, \qquad i,k = 1,\ldots,n \tag{33}$$

where $h_{ik}^{(0)}$ is a matrix having the IDS $N_0(\lambda)$ and J_{ik} are the same as J_{xy} in (22) and (27). Then $h^{(n)}$ has also the nonrandom IDS $N(\lambda)$ that is related to $N_0(\lambda)$ by the same equation (32). In particular, if $h^{(0)} = 0$, then $r_0(z) = -z^{-1}$ and

$$N'(\lambda) = (2\pi J^2)^{-1}(4J^2 - \lambda^2)^{1/2}, \quad |\lambda| \leq 2J. \tag{34}$$

This distribution is known as the semicircle or the Wigner law. It was found by E.Wigner in the fifties and has been extensively studied (see the reviews [10],[42]). Formula (32) was obtained in my paper [41] and is true without the third condition (30) and even for non-identically distributed J_{ik}'s, satisfying the Lindeberg condition, well known in the probability theory. The IDS corresponding to (32) is known as the deformed semicircle law [10]. We see, that this eigenvalue distribution is as general and universal as the central limit theorem.

Let us outline the proof of (32). By using the resolvent identity we derive the infinite system of equations for the moments $\mathbf{E}\{\prod_l^1 G_R(x_i, y_i)\}$, $l = 1, 2, \ldots$, where $G_R(x,y) = (H^{(R)} - z)^{-1}(x,y)$, which is similar to the Kirkwood-Salzburg system in statistical mechanics. Some terms of this system contain the small parameter R^{-1} in front of them. Neglecting these terms, we can solve the truncated system by the ansatz $\prod_l^1 g(x_i - y_i)$, where $g(x-y)$ is the Green function of the nonrandom operator $H^{(0)}(x-y) - J^2\delta(x-y)r(z)$ differing from (29) in that the random part is replaced by the uniform "effective" potential $J^2 r$. The latter is to be found from the additional relation $r(z) = g(0)$, which is equivalent to (32). After that it can be shown that the remainder containing R^{-1} is indeed small if $\mathrm{Im} z$ is large enough. But, according to (31), the IDS is determined uniquely by $r(z)$ known outside of an arbitrarily wide strip.

Let us consider now the analogue of the $n \to \infty$ limit. Take the n-component discrete Schrödinger operator acting in $l^2(Z^d) \otimes \mathbf{C}_n$ defined by the matrix

$$H_{ik}^{(n)}(x,y) = H^{(0)}(x-y)\delta_{ik} + n^{-1/2} J_{ik}(x)\delta(x-y),$$

where $J_{ik}(x)$, $i, k = 1, 2, \ldots, n$, $x \in \mathbb{Z}^d$, are independent for $i \leq k$, identically distributed and satisfy (30) uniformly in x. This operator was introduced by F.Wegner [50] who proved on physical level of rigor that in the limit $n \to \infty$ the IDS of $H^{(n)}$ satisfies (32). A rigorous proof given recently [27] uses again the asymptotic analysis for $n \to \infty$ of the system of equations for some infinite family of the moments of the Green function of $H^{(n)}$.

Thus, at the level of the IDS, the $R \to \infty$ and $n \to \infty$ limits in the spectral theory coincide. This is rather different from statistical mechanics, where these limits yield the mean field model and the spherical model, respectively. However, it is believed that the IDS, unlike the free energy, contains little information on the Anderson transition. But according to Wegner's calculations [50] (which can be justified in the framework of our method), in the limit $n \to \infty$ the static conductivity is strictly positive in the whole IDS support. Since nonzero conductivity is possible only if the states of the respective operator are delocalized (supposedly belong to the σ_{ac}) it is tempting to expect that both limits ($R \to \infty$ and $n \to \infty$) remove completely the point spectrum conserving the absolutely continuous one.

A difficulty in the rigorous study of this conjecture is that in these limits there are no well-defined limiting operators (29) and (35). Thus, it seems that the proper form of the conjecture is that for large enough but finite R or n, the operators (29) and (35) have a sufficiently massive absolutely continuous component. We hope that using our technique, which is similar to the cluster expansion technique in statistical mechanics, it will be possible to proceed in justifying the above conjecture. By the way, the same conjecture can be formulated for large dimensionality d, because it can be shown that some natural analogue of (29) and (35) has in the $d \to \infty$ limit the IDS determined by the same equation (32).

Let me mention in conclusion that some support of this conjecture can also be found in the study of level repulsion, which seems strongly correlated with the existence of delocalized states.

References

1. Albanese, C.: Comm.Math.Phys. **134**, 1, 237 (1990)
2. Albeverio, S., Tirozzi, B., Zegarlinski, B.: Preprint, Bochum, 1990
3. Altshuler, B.: JETP Lett., **41**, 648 (1985); Lee, P.A., Stone, A.D.: Phys.Rev. Lett. **55**, 1622 (1985)
4. Aizenman, M., Wehr, J.: Comm.Math.Phys. **150**, 489 (1990); J.Statist.Phys. **60**, 287 (1990)
5. Aizenman, M., Lebowitz, J.L., Ruelle, D.: Comm.Math.Phys. **113**, 3 (1987)
6. Amit, D.J.: *Modelling Brain Functions.* Cambridge University Press, London, 1989
7. Berezin, F.A.: Teor.Mat.Fiz. **17**, 305 (1973)
8. Binder, K., Young, A.P.: Rev.Modern Phys. **58**, 801 (1986)
9. Bricmont, J., Kupiainen, A.: Comm.Math.Phys. **116**, 539 (1988)
10. Brody, T.A., Flores, J., French, J.B., Mello, P.A., Pandey, A., Wong, S.: Rev.Modern Phys. **53**, 385 (1980); Wigner, E.: SIAM Reviews J. **9**, 1 (1967); Mehta, M.L.: *Random Matrices.* Academic Press, N.Y., 1967
11. Dotzenko, Vik.S., Dotzenko, Vl.S.: Adv.in Phys. **32**, 129 (1983); Korenblit, I., Shender, E.: Soviet Phys.Uspekhi **126**, 233 (1978)
12. Dreifus, H.: Ph.D. thesis. N.Y.University, 1987
13. Dreifus, H., Klein, A.: Comm.Math.Phys. **124**, 285 (1989)
14. Dreifus, H., Klein, A.: Preprint, Irvine, 1990
15. Edwards, S., Anderson, P.W.: J.Phys. **F6**, 1927(1976)
16. Figotin, A., Pastur, L.: Teoret Mat.Fiz. **35**, 193 (1978)
17. Figotin, A.L., Pastur, L.A.: *Spectra of Random and almost Periodic Operators.* Springer, Heidelberg, 1991
18. Fröhlich, J.: In: Phenomen Critiques, North Holland, Amsterdam, 1986

19. Fröhlich, J., Spencer, T.: Comm.Math.Phys. **88**, 151 (1983)
20. Fröhlich, J.,Zegarlinski, B.: Comm.Math.Phys. **112**, 553 (1987)
21. Goldsheid, I.Ya., Molchanov, S.A., Pastur, L.A.: Functional Anal.Appl. **11**, 1 (1977); Molchanov, S.A.: Math. USSR, Izvestija **12**, 69 (1978)
22. Imry, Y.: Europhys.Lett. **1**, 249 (1986)
23. Imry, Y., Ma, S.: Phys.Rev.Lett. **35**, 1399 (1975)
24. Kazakov, V.: In: Fields, Strings. North Holland, Amsterdam, 1990
25. Khoruzhenko, B.: J.Statist.Phys. **62**, 21 (1991)
26. Khorunzhy, A., Khoruzhenko, B., Pastur, L., Shcherbina, M.: In: Phase Transit., Crit.Phen.(in press)
27. Khorunzhy, A., Pastur, L.: Preprint, Kharkov, 1989
28. Kirsch, W., Molchanov, S.A., Pastur, L.A.: Functional Anal.Appl. **24**, 14 (1990)
29. Klein, A.: Proc.Sympos. Pure Math. **51**, 315 (1990)
30. Knops, H.: J.Math.Phys. **14**, 1918 (1973); Shcherbina, M.: Teoret.Mat.Fiz. **77**, 460 (1988)
31. Kotani, S.: In: Proc.of ICM-90, Tokyo (in press)
32. Kotani, S.: Rev. in Math.Phys. **1**, 129 (1989)
33. Lee, P.A., Ramakrishnan, T.: Rev.Modern Phys. **57**, 287 (1985)
34. Lifshitz, I.M., Gredescul, S.A., Pastur, L.A.: Introduction to the Theory of Disordered System. Wiley and Sons, N.Y., 1988
35. Luttinger, J.M.: Phys.Rev.Lett **37**, 778 (1976); van Hemmen: J.Phys.Rev. Lett. **49**, 409 (1982)
36. Marchenko, A.V., Molchanov, S.A., Pastur, L.A.: Teoret.Mat.Fiz. **81**, 120 (1989)
37. Martinelli, F., Scoppola, E.: Riv.Nuovo Cimento, **10**, 2 (1987)
38. Mezard, M., Parisi, G., Virasoro, M.: Spin Glasses and Beyond. World Science, Singapore, 1987
39. Nattermann, T., Rujan, P.: Int.J.Mod.Phys. **B3**, 1597 (1989)
40. Parisi, G.: J.Phys. A **13**, 1101, 1887, L115 (1980)
41. Pastur, L.A:. Teoret.Mat.Fiz. **10**, 102 (1972)
42. Pastur, L.A.: Soviet Math.Usp. **28**, 1 (1973)
43. Pastur, L.A.: Comm.Math,Phys. **75**, 179 (1980)
44. Pastur, L.A.: J.Statist.Phys. **27**, 119 (1982)
45. Pastur, L.A.: Mat.Zametki, **46**, 50 (1989)
46. Pastur, L.A., Shcherbina, M.V.: J.Statist.Phys. **62**, 1 (1991); Shcherbina, M.V.: Preprint, Rome, 1991
47. Sherrington, D., Kirkpatrick, S.: Phys.Rev.Lett. **35**, 1792 (1975)
48. Simon, B, Spencer, T.: Comm.Math.Phys. **125**, 113 (1989)
49. Spencer, T.: J.Statist.Phys. **51**, 1009 (1988)
50. Wegner.,F.: Phys.Rev. B **19**, 783 (1979)

Low Temperature Stochastic Spin Dynamics: Metastability, Convergence to Equilibrium and Phase Segregation

Fabio Martinelli

Dipartimento di Matematica, Universita' "La Sapienza", Pz.A.Moro 2,Roma, Italy

Abstract: I will report on recent progresses made in the rigorous study of the long time behaviour of random dynamics of physical and mathematical interest for ferromagnetic spin systems in the low temperature regime. In particular I will discuss metastability and its connection with the problem of the convergence to equilibrium for a standard Glauber dynamics for the Ising model, dynamical phase transition, dynamics of the interface and phase segregation for conservative spin flips dynamics.

1 Metastability for the Stochastic Ising Model

Let us consider the d-dimensional nearest neighbor Ising system in a cubic box

$$\Lambda = \{x \in \mathbb{Z}^d,\ x = (x_1, \dots, x_d) : |x_i| \leq L, \quad i = 1, ..., d\}$$

of side $2L+1$ whose Hamiltonian is written as:

$$H_\Lambda^b(\sigma) = -\frac{1}{2} \sum_{\substack{x,y \in \Lambda \\ |x-y|=1}} \sigma(x)\sigma(y) - \frac{h}{2} \sum_x \sigma(x) - \frac{1}{2} \sum_{\substack{x \in \Lambda, z \notin \Lambda \\ |x-z|=1}} \sigma(x) b(z) \quad , \qquad (1)$$

where $\sigma \in \{-1,1\}^\Lambda$, h$\geq$ 0 is the external magnetic field and b is a fixed boundary configuration. The Gibbs measure associated to H at inverse temperature β is given by :

$$\mu(\sigma) \;=\; \frac{1}{Z(\beta,h)} \exp(-\beta H(\sigma)) \quad ,$$

where $Z(\beta, h)$ is the partition function.

Given the above Ising model one constructs an associated Markovian stochastic dynamics on the space of configurations $\{-1,+1\}^{\Lambda}$ by letting the spins change their value in a random way with the spin at time t at the site x flipping at a rate c(x,σ_t), where σ_t denotes the configuration at time t. The rates c(x,σ) should be such that the unique invariant measure of the corresponding Markov process coincides with the Gibbs state of the above Ising model. There are infinitely many choices of the rates satisfying the above requirements, but in this exposition for simplicity we will discuss only a particular one known as the Heat Bath (HB) dynamics. The rates c(x,σ) take in this case the value:

$$c(x,\sigma) = (1+\exp\{\beta \Delta_x H(\sigma)\})^{-1} \quad , \tag{2}$$

where $H(\sigma)$ is the Ising Hamiltonian and

$$\Delta_x H(\sigma) = H(\sigma^x) - H(\sigma) \tag{3}$$

with $\sigma^x(y) = \sigma(y)$ if $x \neq y$, $\sigma^x(x) = -\sigma(x)$. One can easily verify that the above defined rates satisfy the so-called "detailed balance " or reversibility condition with respect to the Gibbs measure $\mu(\sigma)$:

$$\mu(\sigma)c(x,\sigma) \;=\; \mu(\sigma^x)c(x,\sigma^x) \quad . \tag{4}$$

This assures that μ is the only invariant measure for the process and that starting from this measure one obtains a time-reversible process.

An explicit construction of the process goes as follows:

a) With rate $|\Lambda|$ one chooses a random site $x \in \Lambda$. We will then say that "x is visited".
b) Given x one extracts a number $\xi_x \in (0,1)$ with uniform distribution.
c) Given a configuration σ one sets:

$$\begin{aligned}
&\sigma_\tau(x) = +1 \quad \text{if} \quad \xi_x < \left[1+\exp(-\beta\left(h + \sum_{|y-x|=1} \sigma(y)\right))\right]^{-1} \\
&\sigma_\tau(x) = -1 \qquad \text{otherwise} \\
&\sigma_\tau(y) = \sigma(y) \quad \forall y \neq x \; .
\end{aligned}$$

It is clear that the above explicit construction of the process based on a), b) and c) above provides a random flow ϕ_t on $\{-1,1\}^{\Lambda}$ and by explicit computation one checks that:

$$\begin{aligned}
&\text{if } \sigma \le \eta \text{ i.e. } \sigma(x) \le \eta(x) \;\forall\, x \in \Lambda \\
&\qquad \text{then } \phi_t(\sigma) \le \phi_t(\eta) \quad .
\end{aligned} \tag{5}$$

Thus in this case if $\phi_t(\pm 1)$ denote the evoluted at time t of the configurations identically equal to ± 1, we get

$$\phi_t(-1) \leq \phi_t(\sigma) \leq \phi_t(+1) \qquad \forall\, \sigma \in \{-1,1\}^{\Lambda} \quad . \tag{6}$$

The analysis of the dynamical behaviour of the above Markov process, e.g. rate of its convergence to equilibrium, estimate of time correlations etc. have been mostly investigated in the small β regime where the spins behave very much as almost independent random variables (see e.g. the works by Holley [1],[2], Holley and Strook [3]). The case of large β and strong magnetic field h ($h > 2d$ in our case) also does not present any problem since the large majority of the spins will follow the magnetic field with high probability irrespectively of the sign of their nearest neighbors, with rare fluctuations against the field h. The case of large β and arbitrarily small h looks much more interesting from a physical point of view and it presents substantial new difficulties related to the phenomenon of metastability.

As it is well known the Ising model at low temperature and zero magnetic field exhibits a phenomenon of coexistence of two different phases with opposite non zero magnetization. The configurations of the system have a kind of "symmetric double well structure" even in the thermodynamic limit. The two wells are related to the typical configurations characterizing the two opposite phases. For small h this feature is in some sense preserved even though now the well corresponding to the phase parallel to h becomes deeper. The phase opposite to the field becomes metastable and does not correspond anymore to a true equilibrium state of the system on a macroscopic scale. The magnetic field h decides the phase no matter how small it is, but its effects become relevant only on a suitable scale diverging when $h \to 0$, in the sense that only on large scales the volume energy dominates the surface energy. It is clear that to describe the global approach to equilibrium, uniformly in the initial configuration and in the boundary condition, one needs to discuss the tunnelling from the metastable to the stable phase.

This "tunneling" is a local phenomenon consisting in the homogeneous formation, through the whole volume of many droplets (nuclei) of the new stable phase inside the metastable phase (homogeneous nucleation). The main physical feature of this transition is the existence of a critical value $l_c(h)$ for the size of the droplets. Droplets whose side is smaller than $l_c(h)$ have the tendence to shrink whereas the larger ones have the tendence to grow and there is an "activation energy" necessary to create them.

The existence of this "activation energy" can be deduced from simple heuristic arguments based on the dependence on L of the excess energy $E(L)$ associated to the formation of a cube of side L full of spins parallel to h in a sea of opposite spins:

$$E(L) \;=\; -|h|L^d \,+ 2dL^{d-1} \quad . \tag{7}$$

The critical size $l_c(h)$ is then taken as the value of L such that $E(L)$ attains its maximum i.e. $l_c(h) = \frac{2(d-1)}{|h|}$.

Since the formation of the critical droplet is a phenomenon of large fluctuation because it requires a local increases of the energy, on heuristic grounds one expects that its tipical time has to be of the order of $\exp(\beta E(l_c(h)) = \exp\frac{4\beta}{|h|}$. It is however clear that if we consider a very large volume then, due to stochastic fluctuations, critical droplets will be created at any time with a certain rate homogeneously in space and in some sense the system will evolve towards the final equilibrium in a smooth way. If instead the volume is kept fixed together with h with L $>> \frac{2(d-1)}{h}$ and β is subsequently taken large enough, then updatings of the spins that cause an increase of the energy will become very unlikely and they will play the role of a small random perturbation of a dynamics for which energy can either decrease or stay the same.

It is not surprising therefore that in this situation energy considerations are sufficient to determine the dynamics in a rather precise sense and in particular to give a rigorous dynamical characterization of the critical droplet and of the nucleation time (see below for a precise definition). The above setting, finite volume and large β depending on the volume, is not such a peculiar case as it may appear at first sight and in fact its analysis permits for example to obtain rigorous upper bounds on the time needed for the dynamics in the infinite volume limit starting from all minuses to reach locally (i.e. in region around the origin of size smaller than β) the true equilibrium which locally appears as a sea of pluses.

The above analysis was carried out for the first time by Jordao-Neves and R.Schonman for short range two dimensional Glauber dynamics as the one defined above (see [4],[5]) and, independently, by E.Olivieri, E.Scoppola and myself for the two dimensional Swendsen-Wang dynamics [6], a dynamical version of the Fortuin-Kastelein representation of the Ising model, whose invariant measure is again the Gibbs state of the Ising model. Such a dynamics has become very popular among people doing Monte Carlo simulations because of its very fast convergence towards equilibrium. We point out that for the Swendsen-Wang dynamics the monotonicity property (6) does not hold any more and that metastable behaviour appears only when boundary conditions are antiparallel to the external field. For the sake of exposition I will only describe now the results for the Heat Bath dynamics; the results for the Swendsen-Wang dynamics are very similar but their exposition is more involved. The main result is as follows.

If the external field h is small and positive the system, starting with all spins equal to -1 behaves as if it was in an equilibrium state for a very long time, with small and rare small nuclei of plus spins forming from time to time and rapidly disappearing. At a certain (random) time (nucleation time) a large fluctuation is able to create a single droplet of plus spins of a very precise shape and size. This droplet typically grows spontaneously under the dynamics and on a short time scale is able to invade the whole volume. Once the system has reached the configuration with all spins equal to +1 then it stays there for a time much longer then the nucleation time with small and rare islands of minus spins appearing and rapidly disappearing. The size and the shape of the critical droplet together with the order of magnitude of the nucleation time agree with the heuristic prediction: the critical droplet is a square of side $l_c = [\frac{2}{h}] + 1$ and the nucleation time τ, defined as the minum time needed to reach the configuration with all spins equal to +1 starting from the oppposite configuration, satisfies the following asymptotic result:

$$\lim_{\beta \to \infty} \frac{1}{\beta} log\, \tau \;=\; 4 l_c \;-\; (l_c^2 \;-\; l_c + 1) \text{ in probability.}$$

It is important to note that although the nucleation time is typically very large, of the order $\exp(\frac{4\beta}{h})$, its value for a single realization of the process is sharply defined. This is due to the fact that once the system decides to nucleate a critical droplet it does it in a rather abrupt way, on a time scale much smaller than the proper time scale of the nucleation time itself. On the other hand if one considers the nucleation time as a random variable then its fluctuations are "large" confirming the "unpredictability" of the nucleation phenomenon. More precisely if one rescales the nucleation time with its mean then, as $\beta \to \infty$, it converges to an exponential random variable of mean one.

A second very interesting series of results concerns the way the process nucleates the critical droplet. As a first step in this direction one observes that if the boundary conditions are periodic or -1 then configurations in which the plus spins form rectangles separated by more than one lattice spacing are local minima of the energy and therefore stable in the zero temperature limit. It is not so hard to show that the process starting from all spins equal to -1 will spend most of its time before reaching the equilibrium (all spins +1) or decaying back to all minuses in the subset of the above configurations for which the plus spins form a unique rectangle, and that with probability tending to one it will only make transitions between "nearest neighbors" rectangles, i.e. rectangles of the form $l_1 \times l_2$ and $(l_1 \pm 1) \times l_2$ or $l_1 \times (l_2 \pm 1)$. Then one proves that with probability tending to 1 as $\beta \to \infty$ the critical droplet, a square of side $l_c(h) = [\frac{2}{h}] + 1$, will grow through a sequence of squares $l \times l$, $(l+1) \times (l+1)$ and rectangles $l \times (l+1)$ or $(l+1) \times l$ of plus spins. Somehow this last fact is not so surprising since after all the square is the Wulff shape at zero temperature. One might therefore conjecture that for any Ising model (e.g. with anisotropic interaction or next nearest neighbors interaction) the critical droplet will be nucleated through a sequence of the corresponding Wulff shape of increasing sizes. Surprisingly enough this is not the case. R.Koteski and E.Olivieri [7] have shown for example that in the anisotropic case with different horizontal and vertical couplings, the growth of the critical droplet (in this case a rectangle with critical sides L_1^* and L_2^* determined by h and the strength of the couplings) occurs in a more complicated way: up to an intermediate critical length $L_o^* = min\{L_1^*, L_2^*\}$ the growth is along a sequence of squares like in the isotropic case; once the square of side equal to l_c has been formed then one side of the square stays fixed (up to small fluctuations) while the other one grows until it reaches the value $L^* = max\{L_1^*, L_2^*\}$. This fact must certainly be relevant for certain aspect of crystal growth.

So far the above results concern only the two dimensional case. In three dimensions the available results are much less precise than the ones exposed above. One can still prove that small enough droplets shrink under the dynamics with large probability, while sufficiently large ones grow, but the precise characterization of the critical droplet and of the way it is created are still missing a complete rigorous treatment. The main reason is the fact that in three dimensions the energy landscape is much more complicated than in the two dimensional case. There in fact the only local minima of the energy function are configurations in which the plus spins form rectangles separated by more than one lattice spacing, while here more complicated shapes appear. Therefore in order to reach the same level of detail reached in the two dimensional case it seems necessary to classify each local minimum of the energy according to its stability, that is according to the time scale needed for the process to leave it and jump to another minimum. Such a work has been initiated by Jordao-Neves [8] but there are no definite results in this direction.

Let us now examine some interesting consequences of the above results for the problem of the convergence of the dynamics in the thermodynamic limit (or arbitrarily large volume) to the equilibrium measure represented by the unique Gibbs state of the Ising model at inverse temperature β and external field h. It is a widespread folklore that, since the Gibbs state for the above choice of the parameters β and h is unique with very good mixing properties, then the dynamics should approach in a suitable sense exponentially fast its equilibrium measure. A proof of this fact has been given for the first time in [9] for the Heath bath case and in [10] for the Swendsen-Wang dynamics. We would like to emphasize here that the ideas and methods of [9] and [10] are genuinely dynamical in nature and that they could in principle applied also to other stochastic spin dynamics like stochastic cellular automata which exihibit the same type of metastable behaviour. Let us describe the results for the Heat Bath dynamics.

The theorem one proves (see [9]) is the following one:

Theorem 1 *Let d=2. Given $h > 0$ one can find $\beta_o = \beta_o(h)$ and $t_o(\beta, h)$ such that for any $\beta > \beta_o$ there exist $k = k(\beta) > 0$ with $k(\beta) \to \infty$ as $\beta \to \infty$, $m > 0$ independent of the volume Λ and of the b.c. b such that:*

a) $$\sup_{\sigma\in\{-1,1\}^{\Lambda}} P_\sigma(\sigma_t(x) = -sign\, h \quad \forall x \in A) \le e^{-k(\beta)|A|}$$

for all $A \subset \Lambda$ *with* $dist(A,\partial\Lambda)\, 1/|h|^3$, *if* $t > t_o(\beta,h)$

b) $$\sup_{\sigma\in\{-1,1\}^{\Lambda}} |\mu_\Lambda(f) - E_\sigma(f(\sigma_t))| \le C_f e^{-mt}$$

for arbitrary cylindrical function f, *where* C_f *is a positive constant depending on* f *and* $t > t_o(\beta,h)$.

Remark.

i) Clearly a) says that after time $t_o(\beta,h)$ we have reached the right phase parallel to the magnetic field h inside the bulk. We emphasize that this result holds in an arbitrarily large volume.
ii) It turns out that the critical time $t_o(\beta,h)$ after which exponential convergence takes place can be chosen of order

$$t_o(\beta,h) \approx \exp(\frac{4\beta}{h}) \quad .$$

This time coincides with the time necessary to nucleate a critical droplet of side $l_c(h) = 2/h$ of the stable phase (parallel to h) from the metastable phase antiparallel to h for the dynamics in a square L×L with e.g. L=$\frac{1}{h^3}$ with -1 boundary conditions.

Let us now explain the physical mechanism behind the proof of the theorem. Let us start with two opposite configurations, one with all spins minus and the other with all spins plus, and let us them evolve together under the same random flow ϕ_t (see (5)). Let us fix a square Λ_L around the origin of side L with e.g. L=$\frac{1}{h^3}$, and let us denote with $\tilde{\phi}_t(-1)$ the evoluted of the minus configuration with extra -1 boundary conditions at the boundary of Λ_L. Then using the monotonicity of the Heat Bath dynamics we get:

$$\tilde{\phi}_t(-1) \le \phi_t(-1) \le \phi_t(+1) \quad .$$

Due to the presence of the external field and using the results of [ ] , the process $\tilde{\phi}_t(-1)$ will become identically equal to +1 inside Λ_L in a time of the order of $\exp(\frac{4\beta}{h})$ (nucleation time) and a fortiori the same will happen to $\phi_t(-1)$. Obviously the origin here did not play any role and by the translation invariance of the dynamics the same will happen around most of the sites in the lattice. In conclusion at the nucleation time the differences between the two processes $\phi_t(-1)$ and $\phi_t(+1)$ will consist of small islands well isolated one from the other. These islands, under the action of the common sea of plus spins and of the field h, will disappear in a rather short time of the order log(L) where L is the size of the box in consideration thus producing a loss of memory of the initial condition. It is then easy to see that this produces in turn an exponentially fast convergence to the equilibrium.

Remark.

We emphasize that the above argument based on the introduction of extra negative boundary conditions together with the attractivity of the Heat Bath dynamics only gives an upper bound on the time necessary for the negatively magnetized configuration to reach the plus phase in the sense of part a) of Theorem 1. One might conceive that locally the spins reach the plus phase before they have been able to nucleate their own local critical droplet simply because they are hit by a growing supercritical droplet that has been created faraway in space by an anomalous large fluctuation in the past. There are actually some results that indicate that this is indeed possible and that the actual time for the nucleation of the right phase around the origin in the dynamics in the thermodynamic limit is much smaller than the nucleation time in a large but finite (w.r.t. β) volume ([11]).

2 Low Temperature Dynamics in the Phase Transition Region

A second very interesting but rather hard class of problems are those concerned with a spin dynamics (e.g. for the Ising model) in the phase transition region (zero external field and large enough β in the Ising model). Here one has to distinguish between dynamics that conserve the total magnetization, hereafter referred to as Kawasaki dynamics, and nonconservative ones like the Heat Bath or the Swendsen-Wang dynamics discussed above. The generator L_β of the Kawasaki evolution acts, on the cylindrical functions, as

$$L_\beta f(\sigma) = \frac{1}{2}\sum_{x,y} c_\beta(x,y;\sigma)\big[f(\sigma^{x,y}) - f(\sigma)\big] \tag{8}$$

$$c_\beta(x,y;\sigma) = (\exp\Big(-\frac{\beta}{2}\big[H(\sigma^{x,y}) - H(\sigma)\big]\Big) \quad , \tag{9}$$

where $\sigma^{x,y}$ is obtained from σ by interchanging the spins at x and y, and x and y in (8) are nearest neighbor sites and H is the Ising Hamiltonian (1) with h=0. Notice that the energy difference is always well defined also in the infinite volume. For the existence of the process we refer to Liggett, [12]. L_β is self-adjoint with respect to any Gibbs measure at temperature β i.e the detailed balance condition (4) is satisfied with σ^x substituted by $\sigma^{x,y}$.

Among the various questions that one may ask, the following ones appear to us of great relevance and only recently partial rigorous answers to some of them have started to appear:

i) Behaviour of the fluctuations inside a single phase, e.g. long time behaviour of the time autocorrelation function of the spin at the origin for $\beta > \beta_c$:

$$\int d\mu_\beta^+(\sigma)\sigma(0)\mathbf{E}(\sigma_t(0)) \;-\; m^2 \quad , \tag{10}$$

where β_c is the inverse critical temperature for the Ising model, $\mu_\beta^+(\sigma)$ denotes the plus phase and m is the spontaneous magnetization.

ii) In the nonconservative case study of the shape and of the dynamics of the interface between the plus and minus phases induced by fixed (e.g +) boundary condition along e.g. the coordinate axes in two dimensions.

iii) Dynamical proof of the existence of a phase transition without using the reversibility of the process with respect to the Gibbs state. This question is of great relevance if one hopes to extend techniques and results from random dynamics reversible with respect to Ising Gibbs measure to stochastic cellular automata whose equilibrium states are in general not known.

iv) Phase segregation and hydrodynamic limit when starting with a magnetization which does not correspond to any of the pure phases.

Let us analyse separately these various points:

i) We only discuss the nonconservative case represented by a Glauber dynamics e.g the Heat Bath dynamics. Few years ago an interesting paper by D.Fisher and D.Huse [13] appeared where the authors argued, among other things, that long lived rare droplet fluctuations of the opposite phase can dominate the long time behaviour of quantities like (10), yielding in the two dimensional case stretched exponential decay like $\exp(-k\sqrt{t})$. Such an anomalous behaviour attracted the attention of people doing numerical simulation. In a very interesting numerical work Ogielski [14] found that although the behaviour predicted in [13] may very well be the correct one, it is attained by the dynamics only after a time T_o so large that the autocorrelation function has already become negligeable small (something like 10^{-100}). Before T_o the behaviour is even slower and of the form $\exp(-kt^\alpha)$ with $\alpha < \frac{1}{2}$.

Let us shortly review the Fisher-Huse argument in order to make a contact with point ii) above.

First of all we rewrite (10) as :

$$\int d\mu_\beta^+(\sigma) \int d\mu_\beta^+(\eta)\sigma(0)[\mathbf{E}(\sigma_t(0)) \;-\; \mathbf{E}(\eta_t(0))] \quad , \tag{11}$$

where we have used the reversibility of the dynamics with respect to the Gibbs state. Let us now assume that σ is a typical configuration of the + phase, i.e. most spins around the origin are +1, while the configuration η is atypical in the sense that the origin belongs to a, roughly spherical, droplet of radius R of the opposite phase while outside the droplet one has again the plus phase.

Under the influence of the external sea of plus spins the boundary of the droplet is expected to shrink with a velocity proportional at a given site to the local curvature. Such a behaviour to my knowledge has not been proved rigorously for any of the dynamics introduced above at fixed low temperature, but it can be derived in a mean field approximation (Lebowitz-Penrose limit) at zero temperature. In this case the (deterministic) evolution of a magnetization profile m(r,t) can be described by a reaction diffusion equation of the form:

$$\frac{\partial m(r,t)}{\partial t} = \lambda^2 \Delta m(r,t) - [f'(m(r,t))] \quad , \tag{12}$$

where the reactive potential f is the local free energy which, at $\beta > \beta_c$, has two equal minima, $\pm m_\beta$. For physical reasons one is actually interested in the behavior of the equation when $\lambda \to 0$, which corresponds to take the range of the original Kac potential between the spins vanishingly small. Then in this case it is known that if the initial state $m(r,0)$ has value $-m_\beta$ inside a regular region Γ and m_β outside (some assumptions on the behavior at the boundaries are also needed) then the solution $m_\lambda(r, \lambda^{-2}t)$ of (12) converges when $\lambda \to 0$ to a function which has value $\pm m_\beta$ outside, respectively inside, the region Γ_t. The boundary of Γ_t evolves according to a motion by mean curvature: the velocity of any point of $\partial\Gamma_t$ equals a constant times the curvature of Γ_t at that point and it is directed toward the interior (see e.g. De Mottoni-Schatzmann [15]).

From the above discussion one concludes that the life time of the droplet before disappearing in the sea of pluses should be of the order of its area. Since in two dimensions the statistical weight given by the Gibbs state to configurations with the properties described above is $\approx \exp(-\beta R)$, if we take R = const. $\sqrt{t}$ we get that with probability of the order $\exp(-k\sqrt{t})$, up to time t, $[\mathbf{E}(\sigma_t(0)) - \mathbf{E}(\eta_t(0))] \approx 2$ in (11) and the Fisher-Huse result follows.

Of course the above argument is very rough since it does not include random fluctuations in the dynamics of the domain wall. These fluctuations have been simulated in [13] by adding a noise term with a suitable variance depending on the radius of curvature to the deterministic equation of motion of the domain wall, and it turned out that they do not modify the previous conclusion. It is clear that in order to implement rigorously the above ideas one needs to control the motion and the fluctuations of the interface between a droplet of the minus phase and the external plus phase at a fixed low temperature. To be more precise we first observe that (11) can be bounded by:

$$2\exp(-k(\beta)R) \quad + \quad 2P(\tilde{\phi}_t(+1)(0) \neq (\tilde{\phi}_t(-1)(0)) \quad , \tag{13}$$

where $(\tilde{\phi}_t(\pm 1)(0)$ is the evoluted with the H.B. dynamics at time t and at site x=0 of the configuration equal to ± 1, with + boundary conditions at the two lines x=-R and y=-R. In order to justify such a bound we argue as follows: if in (11) we add and substract the same expression but with extra plus boundary conditions at the above lines in the Gibbs states as well in the dynamics, then an easy computation using the Gibbs state shows that the difference between the two expressions, with and without extra boundary conditions, is estimated by the first term in (13). The remaining term is estimated using the monotonicity properties of the H.B. dynamics by the second term in (13). The value of R is totally arbitrary and it should be chosen in such a way that both terms of (12) becomes of the same order. Clearly according to Fisher and Huse such a choice should be: R $\approx \sqrt{t}$. In order to study the second term of (13) one is therefore led to the second question ii) posed above.

ii) At zero temperature it has been known since long time (see [12]) that the problem of the growth of the plus cluster induced by plus boundary conditions along the coordinate axes starting from an initial configuration of minus spins can be mapped exactly into a one dimensional stochastic particle system known as "the simple exclusion process". In such a process there is at most one particle per lattice site and particles jump to the right or to the left of their position with equal rate provided that the site they choose to jump to is empty. The analysis of the one dimensional process then proves that if one denotes by B_t the set of sites in the first quadrant of the lattice $\mathbb{Z}^2$ where $\phi_t(-1) = +1$ then $\frac{B_t}{\sqrt{t}}$ converges as $t \to \infty$ to a well precise region, symmetric around the y=x, enclosed by the coordinate axes and the graph of a certain positive function y=f(x) $x \geq 0$ with $\lim_{x\to\infty} f(x) = 0$. Thus the motion of the interface along the line y=x is proportional to $\sqrt{t}$, in agreement with the previous discussion. At positive temperature we can reformulate the problem as follows (see above): study the behaviour of F(x,t) = $P(\phi_t(+1)(x) \neq \phi_t(-1)(x))$ as a function of t and x. Although it is still a work in progress we have strong indications that it is possible to show by means of a suitable multiscale analysis in space-time, that the above probability is smaller than $\exp(t^\alpha)$ for a suitable $\alpha < \frac{1}{2}$ provided that the distance of x from the coordinate axes is smaller than t^γ with $\gamma < \frac{1}{2}$. Such a result, although quite interesting since it would provide a rigorous upper bound on time autocorrelation functions (see (13)), it is still far from being satisfactory since the exponents α and γ are not the right ones (see however the numerical work by Ogielski [14] above). Moreover a lower bound seems to us even harder to obtain.

iii) As it is well known Glauber dynamics, like the Heath Bath dynamics discussed in this report, are not ergodic in the infinite volume limit at low enough temperature in the absence of an external field. Absence of ergodicity is easily proved using the attractivity of the dynamics and its reversibility with respect to the Gibbs measure. Attractivity together with reversibility, is sufficient to prove for example that, starting from all pluses, the expected value of the spin at the origin at time t will always be greater or equal than the magnetization in the plus phase, while starting from all minuses the same average will always be smaller or equal than the magnetization in the minus phase. If the temperature is below the critical point there is spontaneous magnetization and therefore the system is no longer ergodic. It is however well known that it is very difficult to prove this result by purely dynamical methods, i.e. without using the attractivity and reversibility of the dynamics, and, to our knowledge, no rigorous results are available in this direction. A notable exception is represented by the beautiful work of Toom on stochastic cellular automata [16].

Although we have not made any progress in this direction for the short range case of the Heat Bath dynamics, we succeded in proving dynamically (i.e. without using any information about the Gibbs measure) the existence of a phase transition for the Swendsen-Wang algoritm [17]. The heart of the proof is a dynamical estimate of the probability of the presence at time of long Peierls contours uniformly in time(!). Such an estimate is non-trivial due to the presence of time correlations in a space-time picture of the dynamics (a long contour may be created at some time s and resists for a very long time with a probability independent of its length). Using the estimate we then show, among various things, that starting from a configuration identically equal to +1, the expected value of the spin at the origin stays positive uniformly in time, while starting from a configuration consisting of an infinite sequence of concentric rings with spins of alternating sign, the expected value of the spin at the origin is zero at any given t. Thus the breakdown of ergodicity.

iv) The situation that one faces when studying the kinetics of phase separation is well illustrated by the following example: one considers the Ising model at zero external field and inverse temperature β above the critical value β_c and one evolves in time with a reversible conservative (Kawasaki) dynamics an initial high temperature magnetization profile. This last condition can be implemented by taking for example the values of the spins at t=0 to i.i.d. Bernoulli random variables with mean m.

If the magnetization profile intersects the phase transition region $(-m_\beta, +m_\beta)$ (i.e. $-m_\beta < m < m_\beta$ in our example) then under the dynamics many interesting and complex phenomena take place, related to growth of the two phases. Since the initial magnetization density m does not correspond to any of the allowed values $\pm m_\beta$, in a short time the system changes structure by locally reaching values of the magnetization approximately equal to $\pm m_\beta$ while preserving the total magnetization density. This last constraint implies that the dynamics must be able to form clusters of both phases $(\pm)$. This is clearly observed in computer simulations and of great importance is the typical size L(t) of the clusters as a function of time t. In particular one may ask how much time is needed to separate phases on a macroscopic scale. In order to clarify this issue let us discuss the expected hydrodynamical behaviour of the system. Let ϵ be a small parameter eventually going to zero and let us consider the variable

$X_t^\epsilon(\phi) = \epsilon^d \sum_x \phi(\epsilon x)\sigma(x, \epsilon^{-2}t)$ where ϕ is a test function of compact support. Then in the limit $\epsilon \to 0$ (hydrodynamic limit) one expects that $X_t^\epsilon(\phi)$ converges to $\int drm(r,t)\phi(r)$, where m(r,t) is the solution of

$$\frac{\partial m}{\partial t} = \nabla(D(m)\nabla m) \quad . \tag{14}$$

It is also conjectured that the diffusion matrix D(m) is equal to the diffusion matrix D^{GK} given by the Green-Kubo theory. For such a quantity Spohn [18] has derived the following expression:

$$D_{l,l'}^{(GK)} = \frac{1}{2\chi(m)}\left(\frac{1}{2}\delta_{l,l'} \sum_{|e|=1} \mathbf{E}_{\mu_{\beta,m}}(c(0,e;\sigma)) - \right.$$

$$\left. -2\int_0^\infty ds \sum_x \mathbf{E}_{\mu_{\beta,m}}(j_\ell(x,s)j_{l'}(0,0))\right) \quad , \tag{15}$$

where

$$j_i(x,t) = [\sigma(x,t) - \sigma(x+e_i,t)]c(x,x+e_i;\sigma)$$

and $\chi(m)$ is the magnetic susceptibility at magnetization m.

Since the magnetic susceptibility χ diverges inside the phase transition region, this implies that if the initial value of the magnetization density is inside the phase transition interval, then (14) becomes degenerate and one does not observe any macroscopic change in the magnetization profile on a length scale $\approx \frac{1}{\epsilon}$ for time $\approx \frac{1}{\epsilon^2}$. In particular the size L(t) of the clusters of the growing phases should satisfy : L(t) $< t^\alpha$ with $\alpha < \frac{1}{2}$.

Unfortunately the above picture has not been proved so far for the short range ferromagnetic Ising model with a Kawasaki dynamics. There is an interesting work by Rezakhanlou [19] on the Kawasaki dynamics for a gradient system of the Ginzburg-Landau type, in which, using the techniques developed by Guo, Pappanicolau and Varadhan, the hydrodynamic limit in the sense explained above is proved together with the equality D(m)=$D^{GK}(m)$. The difficulties to extend such a result to the usual Ising model lies in the fact that, unlike the Ginzburg-Landau model, the nearest neighbors Ising model is not a gradient system.

Other available results concern the Ising model with a Kac potential in the so called Lebowitz-Penrose limit in one dimension. The Hamiltonian in this case is given by :

$$H(\sigma) = -\sum_{x,y} J_\gamma(x,y)\sigma_x\sigma_y \quad , \tag{16}$$

where $J_\gamma(x,y) = \gamma J(\gamma|x-y|)$ and J is a nonnegative smooth function of compact support and mass $\int dr J(r) = a$. For such an Hamiltonian it is possible to compute the free energy $F_o(\beta, m)$ in the so called Lebowitz-Penrose limit $\gamma \to 0$ and one obtains:

$$F_o(\beta, m) = CE\{-\frac{1}{2}am^2 - \beta^{-1}e(m)\} \quad , \tag{17}$$

where e(m) is the entropy of a Bernoulli process with mean m. The phase transition region is then given by the flat part of F_o, $(-m_\beta\,,\, m_\beta)$, where m_β is the positive solution of

$$2\beta a m_\beta = \log\,[\frac{1+m_\beta}{1-m_\beta}] \quad . \tag{18}$$

In order to study the hydrodynamic limit for the above model one has to choose the parameter ϵ^{-1} determining the spatial scale much larger than the typical scale of the interaction γ^{-1}. Of course when taking the hydrodynamic limit $\epsilon \to 0$ one is forced to take at the same time the local mean field limit $\gamma \to 0$ since otherwise one is back to the yet unsolved short range Ising model.

In [20] the following result has been proved:

Theorem *Let $\epsilon = \gamma^{1+\alpha}$, $\alpha > 0$, and let μ^ϵ be the product measure on $\{-1,1\}^{\Lambda_\epsilon}$, such that the average of the spin at x equals $\tilde{m}_0(\epsilon x)$: Λ_ϵ is the interval centered at 0 of length ϵ^{-2} with identification of the extremes; $\tilde{m}_0(r)$ is a C^∞ function on the torus of side ϵ^{-1} which coincides for $|r| \le \epsilon^{-1} - 1$ with $m_0(r)$, a fixed function on $\mathbb{R}$. It is assumed that each derivative of $\tilde{m}_0(r)$ is uniformly bounded. Assume finally that if $\beta > 1/a \equiv \beta_c$ the range of m_0 does not intersect the interval $[-m^\star_\beta, m^\star_\beta]$ where*

$$m^\star_\beta = \sqrt{1 - [\beta a]^{-1}} \quad . \tag{19}$$

Then, if α is small enough, for any $n \ge 1$ and $t > 0$

$$\lim_{\epsilon \to 0} \sup_x \left| bfE_{\mu^\epsilon}\Big(\prod_{i=1}^n \sigma(x_i, \epsilon^{-2}t)\Big) - \prod_{i=1}^n m(\epsilon x_i, t)\right| = 0 \quad ,$$

where $m(r,t)$ solves the equation

$$\frac{\partial m(r,t)}{\partial t} = \frac{1}{2}\frac{\partial}{\partial r}\Big(D(m(r,t))\frac{\partial m(r,t)}{\partial r}\Big)$$

$$D(m) = 1 - \beta a(1 - m^2)$$

and $m(r,0) = m_0(r)$.

It is important to note that the system behaves diffusively (i.e. $D(m) > 0$) outside of the region

$$\{(\beta, m) : \beta > 1/a,\; m \in [-m^\star_\beta, m^\star_\beta]\} \tag{20}$$

which is not the phase transition region; this last one is strictly larger than the interval in (20), for $\beta > 1/a$. The region in (20) is usually referred to as the spinodal region while its complement inside the phase transition region is the metastable region (see [21],[22]).

If in the above theorem one takes instead the hydrodynamic scale equal to the range of the Kac potential then one gets a different equation for the magnetization profile [20]:

$$\begin{aligned}\frac{\partial m(r,t)}{\partial t} &= \frac{1}{2}\frac{\partial}{\partial r}\left(\frac{\partial m(r,t)}{\partial r} + \beta\big(1-m(r,t)^2\big) \times \right.\\ &\left. \times \int_0^\infty dr' J'(r')[m(r+r',t) - m(r-r',t)]\right) \quad . \qquad (21)\end{aligned}$$

Here there is a very interesting result obtained by Dal Passo and De Mottoni [23] concerning the stationary solutions of (21). In [23] it is proved that there exists a stationary solution whose asymptotic values at $\pm\infty$ are just the $\pm m_\beta$ thus providing some evidence for phase separation in the model. If one assumes phase separation then a natural question arises on how do phases start to separate assuming that the initial magnetization profile is a small perturbation of e.g. m(r)$\equiv$ 0. If one linearizes (21) around the zero solution then one gets a set of eigenvectors, periodic in space, with positive eigenvalues;.therefore the system should leave $m_o(r)$ following the eigenvector with the largest eigenvalue. Computer simulations (see [24]) seem to confirm this picture fairly well.

Acknowledgments It is a pleasure to thank here the scientific committee for having invited me as a speaker in non-equilibrium statistical mechanics, a fascinating subject which I entered only recently. I am very much in debt with E.Olivieri and E.Scoppola for the long collaboration and for the many clarifying discussions on some of the subjects discussed in this note and also with R.Schonmann for communicating and explaining to me most of his results with Jordao-Neves before publication. Finally a special thank is due to E.Presutti who patiently explained to me the present state of affair concerning phase segregation in spin dynamics.

References

1. Holley, R.: Possible rates of convergence in finite range, attractive spin systems. Contemp.Math. **41**, (1985)
2. Holley, R.:, Rapid convergence to equilibrium in one dimensional stochastic Ising models. Ann.Probab. **13**, 72-89 (1985)
3. Holley, R., Strook, D.: L^2 theory of the stochastic Ising model. Z. Wahr. verw.Geb. **35** (1976)
4. Jordao Neves, E., Schonmann, R.H.: Critical Droplets and Metastability for a Glauber Dynamics at Very Low Temperatures. Preprint Univ. Sao Paulo

5. Jordao Neves, E., Schonmann, R.H.: Behaviour of droplets for a class of Glauber dynamics at very low temperatures. Comm.Math.Phys. **137**, 209 (1991)
6. Martinelli, F., Olivieri, E., Scoppola, E.: On the Swendsen and Wang dynamics II : Critical droplets and homogeneous nucleation at low temperature for the 2 dimensional Ising model. J.Statist.Phys. **62** N. 1/2, 135 (1991)
7. Koteski, R., Olivieri, E.: Private communication
8. Jordao Neves, E.: PhD thesis Istituto de Matematica e Estatistica da Universitade de Sao paulo, Brasil 1990
9. Martinelli, F., Olivieri, E., Scoppola, E.: Metastability and exponential approach to equilibrium for low temperature stochastic Ising models. J.Statist. Phys. **61** N.5/6, 1105 (1990)
10. Martinelli, F., Olivieri, E., Scoppola, E.: On the Swendsen and Wang dynamics I. Exponential convergence to equilibrium. J.Statist.Phys. **62** N. 1/2, 117 (1991)
11. Schonmann, R.: Private communication
12. Liggett, T.M.: Interacting Particle Systems. Grundlehren der mathematischen Wissenschaften, 276, (1985) Springer, Berlin-Heidelberg-New York
13. Huse, D.A., Fisher, D.S.: Phys.Rev. B **35**, 6841 (1987)
14. Ogielski: Dynamics of fluctuations in the order phase of kinetic Ising model. Phys.Rev. B **36** No. 13, 7315 (1987)
15. De Mottoni, Schatzmann: Geometrical evolutions of developed interfaces. Proc.Roy.Soc. Edinburgh, to appear
16. Toom, A.L.: Stable and attractive trajectories in multicomponents random fields. In: Multicomponenets Random Fields. R.L.Dobrushin and Ya.G.Sinai eds. New York, Marcel Dekker (1980)
17. Martinelli, F.: Low temperature behaviour of Swedsen-Wang Monte Carlo algorithm. Preprint Roma 1991
18. Spohn, H.: *The dynamics of systems with many particles.* Springer-Verlag New York (1991)
19. Rezakhanlou, F.: Hydrodynamic limit for a system with finite range interaction. Comm.Math.Phys. **137**, 209 (1991)
20. Lebowitz, J.L., Orlandi, E., Presutti, E.: A particle model for spinodal decomposition. J.Statist.Phys. **63**, 933-974 (1991)
21. Lebowitz, J.L., Penrose, O.: Rigorous treatment of the Van der Waals Maxwell theory of the liquid vapour transition. J.Math.Phys. **7**, 98 (1966)
22. Lebowitz, J.L., Penrose, O.: Rigorous treatment of metastable states in the Van der Waals Maxwell theory. J.Statist.Phys. **3**, 211 (1971)
23. Presutti, E.: Private communication
24. Giacomin: Van der Wals limit and phase separation in a particle model with Kawasaki dynamics. CARR report in math. phys. n.5/91 (1991)

Entropy Methods in Hydrodynamical Scaling

S. R. S. Varadhan

Courant Institute, NYU, New York 10012

We shall describe these methods by examining very closely an explicit model. Consider N lattice sites arranged periodically in one dimension with a lattice spacing of $1/N$. We have spin variables x_j attached to each site j/N, the sites being viewed as equally spaced points on the circle of unit circumference. The spins x_j vary in time in such a manner that they undergo a diffusion on $\mathbb{R}^N$ denoted by $\{x_1(t), \ldots, x_N(t)\}$. The diffusion process is described by

$$dx_i(t) = dz_{i-1,i}(t) - dz_{i,i+1}(t),$$

$$dz_{i,i+1}(t) = \frac{N^2}{2}\left[\phi'(x_i(t)) - \phi'(x_{i+1}(t))\right] dt + N\, d\beta_{i,i+1}(t).$$

Here $\phi'(x) = d\phi/dx$ for a suitable function $\phi(x)$. One can think of $\phi(x)$ as somewhat like $x^2/2$. If ϕ where $x^2/2$ we would have a linear model and our model should be thought of, as a nonlinear version. The random process $\beta_{i,i+1}(t)$ are N, independent Brownian motions. Equivalently one could write down the generator of the process

$$L_N = \frac{N^2}{2}\sum\left(\frac{\partial}{\partial x_i} - \frac{\partial}{\partial x_{i+1}}\right)^2 - \frac{N^2}{2}\sum\left[\phi'(x_i) - \phi'(x_{i+1})\right]\left[\frac{\partial}{\partial x_i} - \frac{\partial}{\partial x_{i+1}}\right].$$

One can also describe the process through its Dirichlet form

$$\frac{N^2}{2}\int\sum\left(\frac{\partial u}{\partial x_i} - \frac{\partial u}{\partial x_{i+1}}\right)^2 e^{-\Sigma\phi(x_i)}\prod dx_i\,.$$

We normalize ϕ so that $\int \exp[-\phi(x)]\,dx = 1$. We assume

$$M(\theta) = \int \exp\left[\theta x - \phi(x)\right] dx$$

is finite for all $\theta \in \mathbb{R}$. Then

$$h(x) = \sup_\theta[\theta x - \log M(\theta)]$$

is strictly convex and

$$\lambda = h'(x)$$

$$x = \frac{M'(\lambda)}{M(\lambda)} = [\log M(\lambda)]'$$

are inverse functions. The measure $\prod e^{-\phi(x_i)}\, dx_i$ is invariant for the diffusion. It is however not ergodic. $x_1 + \cdots + x_N$ is a conserved quantity. The ergodic pieces are restrictions of $\prod e^{-\phi(x_i)}\, dx_i$ to various hyperplanes $\frac{1}{N} \sum x_i = a$. The factor N^2 corresponds to speeding up of time. Since space has already been scaled by a factor of N, we are clearly anticipating a diffusive limit in our choice of parabolic scaling of space and time. We would like to think of

$$\nu_N(t) = \frac{1}{N} \sum x_j(t)\, \delta_{j/N}$$

as a randomly changing signed measure on the circle S. At time 0, we have a random initial configuration with density

$$f_N^0(x_1, \ldots, x_N) \exp\left[-\sum \phi(x_i)\right] \prod dx_i$$

on $\mathbb{R}^N$. The distribution of the configuration at time t will be given by a density

$$f_N^t(x_1, \ldots, x_N) \exp\left[-\sum \phi(x_i)\right] \prod dx_i\,,$$

where f_N^t satisfies the Kolmogorov equation

$$\frac{\partial f_N^t}{\partial t} = L_N f_N^t$$

$$f_N^t\big|_{t=0} = f_N^0\,.$$

Although in principle we should use L_N^*, because L_N is symmetric with respect to the weight $\exp[-\sum \phi(x_i)]$ we do not have to take the adjoint. Suppose at time 0 for some smooth deterministic profile $m_0(\theta)$

$$\nu_N(0) \sim m_0(\theta)\, d\theta\;,$$

or more precisely

$$\frac{1}{N} \sum J(\frac{i}{N}) x_i \to \int J(\theta)\, m_0(\theta)\, d\theta$$

in probability with respect to $f_N^0(x_1, \ldots, x_N) \exp[-\sum \phi(x_i)]\, \Pi\, dx_i$ for every nice test function $J(\theta)$ on S. Under suitable regularity conditions of a simple nature on f_N^0 we want to conclude that for some profile $m(\theta, t)$ we have

$$\frac{1}{N} \sum J(\frac{i}{N}) x_i \to \int J(\theta)\, m(\theta, t)\, d\theta$$

in probability with respect to $f_N^t(x_1,\ldots,x_N)\exp[-\sum\phi(x_i)]\prod dx_i$. In addition we should be able to describe how to compute $m(\theta,t)$ from the initial profile $m_0(\theta)$. Hopefully $m(\theta,t)$ does not depend on f_N^0 in any manner other than through the initial profile $m_0(\theta)$ determined by it. The main step is to compute

$$\begin{aligned} d\frac{1}{N}\sum J(\frac{i}{N})x_i(t) &= -\frac{N^2}{2}\cdot\frac{1}{N}\sum\left[J(\frac{i+1}{N})-J(\frac{i}{N})\right]\left[\phi'(x_{i+1}(t))\right. \\ &\quad \left.-\phi'(x_i(t))\right]dt \\ &\quad +N\cdot\frac{1}{N}\sum\left[J(\frac{i+1}{N})-J(\frac{i}{N})\right]d\beta_{i,i+1}(t) \\ &= \frac{1}{2N}\cdot N^2\sum\left[J(\frac{i+1}{N})-2J(\frac{i}{N})\right. \\ &\quad \left.+J(\frac{i-1}{N})\right]\phi'(x_i(t))+d(\text{“noise”}). \end{aligned}$$

Approximately “noise” $=\frac{1}{N}\sum J'(\frac{i}{N})\,d\beta_{i,i+1}(t)$ and goes to zero by the law of large numbers. We therefore have roughly

$$d(\frac{1}{N}\sum J(\frac{i}{N})\,x_i(t))\simeq\frac{1}{2N}\sum J''(\frac{i}{N})\,\phi'(x_i(t))\,dt\,.$$

The problem is that the measure

$$\frac{1}{2N}\sum\phi'(x_i)\delta_{i/N}$$

is not so easily expressed in terms of $\frac{1}{N}\sum x_i\delta_{i/N}$. We need to know how the spins arrange themselves so that a knowledge of

$$\frac{1}{N\epsilon}\sum_B x_j=m$$

determines

$$\frac{1}{N\epsilon}\sum_B\phi'(x_j)$$

over a block B of size $N\epsilon$ (ϵ is a small positive number).

On the infinite lattice the invariant measures for the diffusion are

$$\prod\frac{1}{M(\lambda)}e^{\lambda x_i-\phi(x_i)}\,dx_i$$

for various values of λ. These are the “Gibbs” measures in this context.

$$\begin{aligned} \text{The average spin } &= \int x\cdot\frac{1}{M(\lambda)}e^{\lambda x-\phi(x)}\,dx \\ &= \frac{M'(\lambda)}{M(\lambda)}=m\,. \end{aligned}$$

Therefore we know that λ has to be chosen so that

$$\lambda = h'(m) .$$

Moreover

$$\int \phi'(x) \frac{1}{M(\lambda)} e^{\lambda x - \phi(x)} \, dx = \lambda$$

so that, if the average spin is m and the spins are organized according to a Gibbs ensemble then the average of $\phi'(x)$ is $h'(m)$.

If we assume that locally the system is close to Gibbs ensembles then we get

$$\frac{\partial}{\partial t} \int J(\theta)\, m(\theta,t)\, d\theta = \frac{1}{2} \int J''(\theta)\, h'(m(\theta,t)\, d\theta ,$$

or equivalently the nonlinear heat equation

$$\frac{\partial m(\theta,t)}{\partial t} = \frac{1}{2} \left[h'(m(\theta,t)) \right]_{\theta\theta}$$

with initial condition

$$m(\theta,0) = m_0(\theta) .$$

This is the hydrodynamic equation. The problem is to prove that the density $f_N^t(x_1,\ldots,x_N) \exp\left[-\sum \phi(x_i) \right] \prod dx_i$ has the correct profile.

A version of the model on the infinite lattice was originally considered by J. Fritz [2]. The method we will outline now was developed by Guo-Papanicolaou-Varadhan in [3]. The basic step is to prove

$$\frac{1}{N} \sum_i \Big| \frac{1}{2N\epsilon} \sum_{|j-i|\le N\epsilon} \phi'(x_j) - h'\Big(\frac{1}{2N\epsilon} \sum_{|j-i|\le N\epsilon} x_j \Big) \Big| = \psi_{N,\epsilon}$$

is small most of the time. If we take a finite interval $0 \le t \le T$ for time and define

$$\bar{f}_N(x_1,\ldots,x_n) = \frac{1}{T} \int_0^T f_N^t(x_1,\ldots,x_N)\, dt$$

then we need

$$\lim_{\epsilon\to 0} \overline{\lim}_{N\to\infty} \int \psi_{N,\epsilon} \bar{f}_N \, e^{-\sum \phi(x_i)} \prod dx_i = 0 .$$

However, ϕ' is an unbounded function. If we assume

$$\int e^{\theta|\phi'(x)| - \phi(x)} \, dx < \infty$$

for all $\theta > 0$, then we can do effective truncation. For any bounded continuous function $g(x)$ on the line let us define

$$\hat{g}(m) = \frac{1}{M(\lambda)} \int e^{\lambda x - \phi(x)} \, dx$$

with $\lambda = h'(m)$. Then we need to prove that for

$$\tilde{\psi}_{N,\epsilon} = \frac{1}{N}\sum_i \Big|\frac{1}{2N\epsilon}\sum_{|j-i|\le N\epsilon} g(x_i) - \hat{g}\Big(\frac{1}{2N\epsilon}\sum_{|j-i|\le N\epsilon} x_j\Big)\Big|$$

$$\lim_{\epsilon\to 0}\limsup_{N\to\infty}\int \tilde{\psi}_{N,\epsilon}\bar{f}_N\, e^{-\sum\phi(x_i)}\prod dx_i = 0\,.$$

The key idea in [3] is the following. If we assume initially that

$$\int f_N^0 \log f_N^0\, e^{-\sum\phi(x_i)}\prod dx_i \le C\,N$$

for some constant C independent of N, then $H_N(t)$ defined by

$$H_N(t) = \int f_N^t \log f_N^t\, e^{-\sum\phi(x_i)}\prod dx_i$$

is decreasing in t and

$$\begin{aligned}\frac{dH_N(t)}{dt} &= -\frac{N^2}{2}\int\sum\left(\frac{\partial f_N^t}{\partial x_i} - \frac{\partial f_N^t}{\partial x_{i+1}}\right)^2 \frac{1}{f_N^t}\, e^{-\sum\phi(x_i)}\prod dx_i\\ &= -N^2\, I_N(t) \qquad \text{(by definition)}.\end{aligned}$$

Clearly

$$\int_0^T I_N(t)\,dt \le \frac{C}{N}\,.$$

If we define

$$I_N(f) = \frac{1}{2}\int\sum\left(\frac{\partial f}{\partial x_i} - \frac{\partial f}{\partial x_{i+1}}\right)^2\cdot\frac{1}{f}\, e^{-\sum\phi(x_i)}\prod x_i$$

then $I_N(t) = I_N(f_N^t)$. By convexity

$$\begin{aligned}I_N(\bar{f}_N) &\le \frac{1}{T}\int_0^T I_N(f_N^t)\,dt\\ &\le \frac{C}{T\,N}\,.\end{aligned}$$

Moreover, again by convexity and the monotonicity of $H_N(t)$,

$$\begin{aligned}H_N(\bar{f}_N) &= \int \bar{f}_N \log \bar{f}_N\, e^{-\sum\phi(x_i)}\prod dx_i\\ &\le \frac{1}{T}\int_0^T H_N(f_N^t)\,dt\\ &\le C\,N\,.\end{aligned}$$

The basic step in [3] is to show that for any densities f_N relative to $\exp[-\sum\phi(x_i)]\prod dx_i$ satisfying

$$H_N(f_N) \le C\,N$$

and

$$I_N(f_N) \le \frac{C}{N}$$

$$\lim_{\epsilon\to 0}\limsup_{N\to\infty}\int \tilde{\psi}_{N,\epsilon}\, f_N\, e^{-\sum\phi(x_i)}\prod dx_i = 0\,.$$

This is carried out in two steps. First we replace $N\epsilon$ by k and define

$$\tilde{\psi}_{N,k} = \frac{1}{N}\sum_i \Big|\frac{1}{2k+1}\sum_{|j-i|\le k} g(x_j) - \hat{g}\big(\frac{1}{2k+1}\sum_{|j-i|\le k} x_j\big)\Big|\,.$$

Then

$$\lim_{k\to\infty}\limsup_{N\to\infty}\int f_N\tilde{\psi}_{N,k}\, e^{-\sum\phi(x_i)}\, c\prod dx_i = 0\,. \qquad (1)$$

This is relatively easy and is proved in fair generality. If we define

$$f_k^N = \frac{1}{N}\sum_i (\text{marginal density of } x_{i-k},\ldots,x_{i+k})$$

then (1) is the same as

$$\lim_{k\to\infty}\limsup_{N\to\infty}\int \Big|\frac{1}{2k+1}\sum g(x_i) - \hat{g}\left(\frac{1}{2k+1}\sum x_i\right)\Big|\, f_N^k e^{-\sum\phi(x_i)}\prod dx_i$$
$$= 0\,.$$

We can let $N\to\infty$, keeping k fixed and see what the limit points of f_N^k are as $N\to\infty$. If we consider

$$I_k(f) = \frac{1}{2}\int\sum_{i=-k}^{k-1}\left(\frac{\partial f}{\partial x_i} - \frac{\partial f}{\partial x_{i+1}}\right)^2 \frac{1}{f}\, e^{-\sum\phi(x_i)}\prod dx_i$$

then by convexity

$$I_k(f_N^k) \le \frac{2k}{N^2}\cdot C\,.$$

By passing to the limit we conclude that for any limit point f^k, $I_k(f^k) = 0$. This forces f^k to be essentially a Gibbs measure and what we need is an easy consequence that is generally known as equivalence of ensembles.

To go from k large to $N\epsilon$ with ϵ small is harder. It can be done somewhat easily in our model, but can sometimes be very difficult. In the case of hyperbolic scaling one can see a priori that there have to be difficulties.

This method has worked successfully in various models, see for instance [3], [5], [7] and [8].

We will now discuss another method based again on entropy developed in [9] by H. T. Yau for this model. This method avoids the passage from k to $N\epsilon$ and therefore has wider applicability. See for instance [6], [10] where application is made to classical Hamiltonian systems with small noise.

Assume that $m(\theta, t)$ is given as a smooth solution of

$$\frac{\partial m}{\partial t} = \frac{1}{2}(h'(m(\theta,t)))_{\theta\theta}\,.$$

We go from $m(\theta, t)$ to $\lambda(\theta, t)$ where

$$\lambda(\theta,t) = h'(m(\theta,t))\,.$$

λ is also a smooth function. We expect f_N^t to be approximately of the form

$$g_N^t(x_1,\ldots,x_N) = \prod_{j=1}^{N} \frac{1}{M(\lambda(j/N,t))}\, e^{\lambda(j/N,t)x_j}\,.$$

Let us assume $f_N^0 = g_N^0$, a special locally Gibbsian ensemble. We want to see how close f_N^t is to g_N^t for $t > 0$. As a measure of closeness we take relative entropy

$$\overline{H}_N(t) = \int \log \frac{f_N^t}{g_N^t} \cdot f_N^t\, e^{-\sum \phi(x_i)} \prod dx_i\,,$$

$$\overline{H}_N(0) = 0\,.$$

One can show that

$$\lim_{N\to\infty} \frac{1}{N}\overline{H}_N(t) = 0 \quad \text{for} \quad t > 0\,.$$

This proves hydrodynamic scaling. The profile is correct for g_N^t with exponentially small error probabilities using the large deviation theory. The entropy inequality

$$Q(A) \le \frac{H(Q,P)+1}{\log 1/P(A)}$$

takes care of the rest. We need to compute $\frac{d\overline{H}_N(t)}{dt}$. Although it is somewhat of a mess

$$
\begin{aligned}
\frac{\partial}{\partial t}\overline{H}_N(t) &= -\frac{\partial}{\partial t}\int f_N^t \log\frac{g_N^t}{f_N^t}\, e^{-\sum \phi(x_i)} \prod dx_i \\
&= -\int \frac{\dot g_N^t}{g_N^t} f_N^t\, e^{-\sum \phi(x_i)} \prod dx_i \\
&\quad - \int \dot f_N^t \log\frac{g_N^t}{f_N^t}\, e^{-\sum \phi(x_i)} \prod dx_i \\
&= -\int \frac{\dot g_N^t}{g_N^t} f_N^t\, e^{-\sum \phi(x_i)} \prod dx_i \\
&\quad + \int f_N^t \bigl(L_N \log\frac{f_N^t}{g_N^t}\bigr)\, e^{-\sum \phi(x_i)} \prod dx_i \\
&\le \int \frac{(-\partial/\partial t + L_N^*) g_N^t}{g_N^t}\, f_N^t\, e^{-\sum \phi(x_i)} \prod dx_i\,.
\end{aligned}
$$

Based on the above general calculation in our special case, since we have

$$
g_N^t = \exp\bigl[-\sum \log M(\lambda(\tfrac{j}{N},t)) + \sum \lambda(\tfrac{j}{N},t)x_j\bigr]\,.
$$

We can compute

$$
\frac{1}{N}\frac{\partial}{\partial t}\overline{H}_N(t) = E^{f_N^t}\{\psi_N\}\,,
$$

where

$$
\begin{aligned}
\psi_N &= \frac{1}{N}\frac{(-(\partial/\partial t) + L_N^*) g_N^t}{g_N^t} \\
&= \frac{1}{N}\sum m(\tfrac{j}{N},t)\,\dot\lambda(\tfrac{j}{N},t) + \frac{1}{2N}\sum \lambda_{\theta\theta}(\tfrac{j}{N},t)\,\phi'(x_j) \\
&\quad - \frac{1}{N}\sum \dot\lambda(\tfrac{j}{N},t)x_j + \frac{1}{2N}\sum \bigl[\lambda_0(\tfrac{j}{N},t)\bigr]^2 \\
&\simeq \frac{1}{N}\sum m(\tfrac{j}{N},t)\,\dot\lambda(\tfrac{j}{N},t) + \frac{1}{2N}\sum \lambda_{\theta\theta}(\tfrac{j}{N},t)\,h'(\xi_{j,k}) \\
&\quad - \frac{1}{N}\sum \dot\lambda(\tfrac{j}{N},t)x_j + \frac{1}{2N}\sum \bigl[\lambda_0(\tfrac{j}{N},t)\bigr]^2 \\
&\simeq \frac{1}{N}\sum m(\tfrac{j}{N},t)\,\dot\lambda(\tfrac{j}{N},t) + \frac{1}{2N}\sum \lambda_{\theta\theta} h'(m(\tfrac{j}{N},t)) \\
&\quad + \frac{1}{2N}\sum \lambda_{\theta\theta}(\tfrac{j}{N},t)\,h''(m(\tfrac{j}{N},t))\,(\xi_{j,k} - m(\tfrac{j}{N},t)) \\
&\quad - \frac{1}{N}\sum \dot\lambda(\tfrac{j}{N},t)x_j + \frac{1}{2N}\sum \bigl[\lambda_\theta(\tfrac{j}{N},t)\bigr]^2 + \frac{1}{N}(\text{Error})\,.
\end{aligned}
$$

The equation $\partial m/\partial t = \frac{1}{2}(h'(m))_{\theta\theta}$ is equivalent to

$$
\frac{\partial\lambda}{\partial t} = h''(m)\frac{\partial m}{\partial t} = \frac{1}{2}h''(m)\,\lambda_{\theta\theta}\,.
$$

So only the error term survives.

$$\frac{d\frac{1}{N}\overline{H}_N(t)}{dt} \le E^{f_N^t}\{\frac{1}{N}(\text{error})\}$$
$$\le \frac{1}{\alpha N}\overline{H}_N(t) + \frac{1}{\alpha N}\log E^{g_N^t}\{e^{\alpha\{\text{error}\}}\}$$

One shows that

$$\frac{1}{N}\log E^{g_N^t}\{e^{\alpha(\text{error})}\} \to 0 \quad \text{as} \quad N \to \infty .$$

Then Gronwall's inequality will take care of the rest.

Large Deviation Theory

The method yields superexponential estimates in equilibrium for deviations from the Boltzmann-Gibbs principle and this allows one to calculate the large deviation rates for deviations from hydrodynamical limit rather explicitly because fluctuations can be averaged with respect to local Gibbs distributions without affecting the expnential rate for the probabilities of large deviation. See [1], [4] for examples.

Nongradient Systems

We can use ideas from large deviation theory to handle the following modification of the model that we just considered. We change the Dirichlet form to

$$\frac{N^2}{2}\int \sum a(x_i, x_{i+1})\left(\frac{\partial u}{\partial x_i} - \frac{\partial u}{\partial x_{i+1}}\right)^2 e^{-\sum \phi(x_i)} \prod dx_i .$$

The corresponding operator is

$$L_N = \frac{N^2}{2}\sum a(x_i, x_{i+1})\left(\frac{\partial}{\partial x_i} - \frac{\partial}{\partial x_{i+1}}\right)^2$$
$$- \frac{N^2}{2}\sum W(x_i, x_{i+1})\left(\frac{\partial}{\partial x_i} - \frac{\partial}{\partial x_{i+1}}\right),$$

where

$$W(x,y) = a(x,y)\left[\phi'(x) - \phi'(y)\right] - \left[a_x(x,y) - a_y(x,y)\right].$$

If we compute

$$d\left(\frac{1}{N}\sum J\left(\frac{i}{N}\right)x_i(t)\right)$$

and ignore the noise terms that are negligible, we now get

$$d\left(\frac{1}{N}\sum J\left(\frac{i}{N}\right)x_i(t)\right) \simeq \sum J'\left(\frac{i}{N}\right)W(x_i(t), x_{i+1}(t))\,dt .$$

There is a factor of $1/N$ missing because we could do summation by parts only once. On the other hand $E\{W(x_i, x_{i+1})\} = 0$ under every Gibbs measure so we need to do more analysis. By using ideas from large deviations one can establish hydrodynamical limits to an equation

$$\frac{\partial m}{\partial t} = \frac{1}{2}\frac{\partial}{\partial \theta}\left(\hat{a}(m(\theta,t))\right)\left(h'(m(\theta,t))_\theta\right),$$

where $\hat{a}(m)$ is the Green-Kubo diffusion coefficient. It turns out one can actually give a variational characterization of $\hat{a}(m)$.

References

1. Donsker, M.D., Varadhan, S.R.S.: Comm.Pure Appl.Math. **42**, 243 (1989)
2. Fritz, J.: IMA **9** Springer-Verlag, Berlin – Heidelberg – New York (1987)
3. Guo, M.Z., Papanicolaou, G.C., Varadhan, S.R.S.: Comm.Math.Phys. **118**, 31 (1988)
4. Kipnis, C., Olla, S., Varadhan, S.R.S.: Comm.Pure Appl.Math. **42**, 115 (1989)
5. Olla, S., Varadhan, S.R.S.: Comm.Math.Phys. **135**, 355 (1991)
6. Olla, S., Varadhan, S.R.S., Yau, H.T.: In preparation
7. Freydoun Rezhakhanlou: Comm.Math.Phys. **129**, 445 (1990)
8. Varadhan, S.R.S.: Comm.Math.Phys. **135**, 313 (1991)
9. Yau, H.T.: Lett.Math.Phys. **22**, 63 (1991)
10. Yau, H.T.: These proceedings

Quantum Groups and Non-commutative Differential Geometry

Yu. I. Manin

Steklov Math. Institute, 42 Vavilova, 117966, Moscow, USSR

0 Introduction

Recently it became common to identify the notion of a quantum group with that of a Hopf algebra. However, this does not quite agree with the experience gained in classical group theory. In fact, classically, Hopf algebras arise in the following framework. One starts with a category $\mathcal{S}$ of "spaces" (finite sets, schemes, differential manifolds, topological spaces ...) and a linear functor $\mathcal{F}$ on it, covariant or contravariant. The functor must satisfy certain formal properties, in particular, transform direct products into tensor products. The values of this functor upon group objects in $\mathcal{S}$ will then be Hopf algebras, commutative in the contravariant case, cocommutative in the covariant case.

There are four most important functors:

i) Functions,
ii) Distributions (variant: distributions supported at a marked point, e.g. identity),
iii) De Rham complexes,
iv) Cohomology.

In the same vein, when one drops the commutativity/cocommutativity assumption, one should keep imagining a linear functor applied to a quantum space.

This talk is devoted to a discussion of quantum de Rham complexes. Our viewpoint is close in spirit to that of Woronowicz [18], but the presentation is strongly motivated by the Wess-Zumino paper [17]. For more details and proofs, see [14].

1 Universal Coaction Principle

In [10]-[13], we suggested the following algebraic construction method of quantum automorphism groups of quantum spaces. Let (A_α, x_α) be a family of associative algebras over a field k, endowed with generators $x_\alpha = (x_\alpha^{(1)}, \dots, x_\alpha^{(n)})$. Consider a family of algebra maps $\delta_\alpha : A_\alpha \to E \otimes A_\alpha$ such that $\delta_\alpha(x_\alpha^{(i)}) = \sum_j z_j^i \otimes x_\alpha^{(j)}$. There is a "maximal" E with this property, satisfying an obvious universality condition. This E is endowed with a canonical structure of a bialgebra, such that δ_α make A_α comodules over E.

Intuitively, A_α are algebras of polynomial functions of quantum spaces, and E is a similar algebra on their common quantum endomorphism space. If we want to get an automorphism object, we must replace E by its Hopf envelope H, constructed in [11].

Recently I have extended this construction method replacing associative algebras by differential graded algebras ([14]). In this way, starting from de Rham complexes of quantum spaces, one arrives to the de Rham complexes of their endo- or automorphism spaces. Independently this generalization was suggested by G. Maltsiniotis ([8], [9]).

However, the universal coaction principle applied in the simplest situations often gives too large (exponential growth) algebras when we expect to obtain objects "of the same size" as the classical ones, for example, flat deformations of the latter. Therefore, it should be completed by additional algebraic requirements. In a down-to-earth language, we want to find some "missing relations" between the generators z_i^j.

2 Bounded Growth Principles

There are several more or less formalized prescriptions for pinpointing non-commutative rings of correct size.

In the Drinfeld's theory of deformations of universal enveloping algebras, one essentially imposes a flat deformation condition.

The notion of a *regular graded algebra* due to M. Artin and W. Schelter ([1]), furnishes a class of quantum spaces which may have interesting automorphism groups. This is certainly true for two-dimensional rings where this idea led to a discovery of a non-standard "Jordan" deformation of $GL(2)$. The next case, that of three-dimensional rings connected with elliptic curves is not yet understood.

In [17], a special role is played by a notion which we formalized in [14] and called a *skew product*. It probably was known to algebraists in different contexts, but its relation to quantum groups deserves a special study.

Let $A, B \subset C$ be two subalgebras of an associative k–algebra. We will say that they are *skew-commuting* if the multiplication map induces injections of linear spaces with the same image $m: A \otimes B \to C$, $m: B \otimes A \to C$. This image is a subalgebra $AB = BA$ of C. If it coincides with C, we say that C is a skew product of A and B.

The de Rham complexes, constructed in [17], [14], [8], [9] are all skew products of a subalgebra of functions and a subalgebra of forms with constant coefficients. This structure is crucial for two reasons. First, it furnishes a correct growth rate. Secondly, it allows us to define two important classes of operators on the function algebra: vector fields, and the Wess-Zumino operators O_i^j (in our notation, σ, τ) which govern the commutation laws between functions and differentials, and which are necessary for the closure of the differential operator rings with respect to comultiplication.

3 Structure of Skew Products

We shall state our structure theorem in the context of $\mathbb{Z}_2$–graded algebras. Let $\tilde{a}$ denote the parity of a. Assume that a skew product $C = AB = BA$ is given. Let B be also $\mathbb{Z}$–graded, and $AB^n = B^n A$ for all n. Let $r = r_C : B \otimes A \to A \otimes B$ be the isomorphism induced by multiplication. Consider for every $b \in B$ the operator $\sigma_b : A \to B \otimes A$ defined by

$$\sigma_b(a) = (-1)^{\tilde{a}\tilde{b}} r^{-1}(a \otimes b).$$

We may and will consider σ_b as elements of the k–algebra $B \otimes \mathbf{End}_{k-mod}(A)^{op}$.

One can prove that the map

$$\sigma: B \to B \otimes \mathbf{End}_{k-mod}(A)^{op}, \quad b \mapsto \sigma_b,$$

is a $\mathbb{Z}_2$–graded ring homomorphism.

Consider now the universal algebra morphism, which is automatically universal bialgebra coaction linear upon B^n :

$$\delta_r : B \to B \otimes E_r(B).$$

Using universality, one sees that instead of σ one can give an algebra homomorphism

$$\rho: \quad E_r(B)^{op} \to \mathbf{End}_{k-mod}(A).$$

This means that, starting with a skew product of A and B, we have constructed an action of $E_r(B)^{op}$ upon the left k–module A. We may call it the *structural action*.

The comultiplication map $\Delta_r : E_r(B) \to E_r(B) \otimes E_r(B)$ is also a comultiplication map for $E_r(B)^{op}$. Using it, we can define the action ϕ of $E_r(B)^{op}$ upon $A \otimes A$: for $e \in E_r(B)^{op}$, $c \in A \otimes A$, put

$$\phi(e)(c) = (\rho \otimes \rho)(\Delta(e))(c).$$

One can check that the multiplication morphism $m: A \otimes A \to A$ is a morphism of $E_r(B)^{op}$-modules.

3.1 Theorem: *This construction establishes a bijection between the following two sets:*

(SP) Skew products of A and B (compatible with the gradation of B).
(EA) Actions of $E_r(B)^{op}$ upon A, defined by an invertible operator-valued matrix such that $m: A \otimes A \to A$ is a morphism of $E_r(B)^{op}$-modules.

3.2 Skew Products and de Rham Complexes. In the examples below, we shall consider $\mathbb{Z}$-graded skew products $\Omega = AB = BA$, endowed with an odd differential d of degree 1 such that $A = \Omega^0$ (functions), $B \subset Ker(d)$ (forms with constant coefficients). Every cycle of the dual complex can be interpreted as a linear functional $\int : \Omega^n \to k$ such that $\int d\omega = 0$ for any $\omega \in \Omega^{n-1}$. However, the second property of Connes' cycles, $\int [\omega, \omega'] = 0$ generally does not hold. Looking at the simplest examples, one sees that the usual (super)commutator of forms must be replaced by a certain "quantized commutator". It has no universal definition. In our setting, quantized commutators are just corollaries of the cross-commutation rules hidden in the skew-product property $AB = BA$.

4 Examples

We start with (2,2)-quantum matrices. Choose two invertible even elements p, q in the ground field k. The ring of polynomial functions E on the noncommutative quantum space $M_{p,q}(2)$ is generated by even variables a, b, c, d with the following commutation relations:

$$ab = p^{-1}ba; \;\; ac = q^{-1}ca; \;\; ad = da + (q^{-1} - p)cb;$$

$$bc = pq^{-1}cb; \;\; bd = q^{-1}db; \;\; cd = p^{-1}dc. \tag{1}$$

This ring is a bialgebra with the standard comultiplication and counit. For $pq \neq -1$, it can be defined as universally coacting algebra on the function algebras of two quantum planes:

$$A_q: \;\; k\langle x, y\rangle/(xy - q^{-1}yx); \;\; x, y \text{ even}; \tag{2}$$

$$B_p: \;\; k\langle \xi, \eta\rangle/(\xi^2, \eta^2, \xi\eta + p\eta\xi); \;\; \xi, \eta \text{ odd}. \tag{3}$$

Adding the relation $ad - p^{-1}bc = 1$, we get the coordinate ring of $SL_{p,q}(2)$.

Order the generators by $d < b < c < a$. The Diamond Lemma is applicable, and shows that monomials $d^k c^l b^m a^n$ form a k-basis of E. (Of course, a similar statement is true and can be easily checked also for A_q, B_p.)

A slight generalization of the Wess-Zumino construction gives the following result.

4.1 Proposition: *There are precisely two skew products Ω of A_q and B_p, admitting a differential with $dx = \xi$, $dy = \eta$, such that all these structures are compatible with the coaction of E (with zero differential).*

Sketch of Proof. We actually prove slightly more: the skew product property should be postulated only on forms of degree two and three in (x, y, ξ, η); the full strength then follows from the Diamond Lemma.

Comparing dimensions, we see that we must find a system of four cross-commutation relations expressing $x\xi, x\eta, y\xi, y\eta$ linearly via $\xi x, \xi y, \eta x, \eta y$. (Actually, one relation is known: $d(xy - q^{-1}yx) = 0$). We have sixteen indeterminate coefficients which are successively constrained by the rest of conditions. It is convenient to use them in the following order.

i) Apply the differential to the cross-commutation relations. The result should be a consequence of (3). This eliminates about half of coefficients.

ii) Use compatibility with the coaction of E. This leaves one parameter free. Of course, this can also be explained conceptually, by analyzing the decomposition of the space

$$(kx \oplus ky) \otimes (k\xi \oplus k\eta) \oplus (k\xi \oplus k\eta) \otimes (kx \oplus ky)$$

with respect to the coaction of E. The result is:

$$x\xi = [(p + q^{-1})r - 1]\xi x; \;\; y\eta = [(p + q^{-1})r - 1]\eta y;$$
$$x\eta = (rq^{-1} - 1)\xi y + pq^{-1}r\eta x; \;\; y\xi = r\xi y + (pr - 1)\eta x \qquad (4)$$

where r is the last undetermined coefficient.

iii) Constraint r by checking linear relations between cubic monomials. In practice, it suffices to resolve the "overlap ambiguity" $x\eta\xi$, that is, to reduce this monomial first as $x(\eta\xi)$ and second as $(x\eta)\xi$. In both cases, the result is proportional to $\xi\eta x$, but the coefficients coincide iff $r^2 - (p^{-1} + q)r + qp^{-1} = 0$, that is, $r = p^{-1}$ or $r = q$. The net result is:

Variant 1: $\Omega_{p,q}^{(1)}$

$$x\xi = (pq)^{-1}\xi x; \;\; y\eta = \eta y;$$
$$x\eta = (p^{-1}q^{-1} - 1)\xi y + q^{-1}\eta x; y\xi = p^{-1}\xi y. \qquad (5)$$

Variant 2: $\Omega_{p,q}^{(2)}$

$$x\xi = pq\xi x; \;\; y\eta = pq\eta y;$$
$$x\eta = p\eta x; \;\; y\xi = q\xi y + (pq - 1)\eta x. \qquad (6)$$

Notice that (5) and (6) are connected by the isomorphism interchanging x and y, ξ and η, p and p^{-1}, which is compatible with the isomorphism interchanging a and d, b and c.

iv) Finally, one checks that the combined relations (2), (3) and either (5), or (6), satisfy all the conditions of the Diamond Lemma, that is, all ambiguities of cubic monomials are resolvable (with respect to any of the orders $\xi < \eta < x < y$ or $x < y < \xi < \eta$.

□

We now turn to the calculation of the de Rham complex of $M_{p,q}(2)$. We start with determining the universal differential quadratic algebra $W_{p,q}^{(i)}$ left coacting upon $\Omega_{p,q}^{(i)}$. Denote $\alpha, \beta, \gamma, \delta$ the differentials of a, b, c, d respectively. We assume $pq \neq -1$.

4.2 Proposition: *$W_{p,q}^{(1)}$ is defined by the relations (1) and the following ones:*

a) Relations between differentials of matrix coefficients:

$$\alpha^2 = \beta^2 = \gamma^2 = \delta^2 = 0;$$

$$\beta\alpha = -q^{-1}\alpha\beta; \; \gamma\alpha = -p^{-1}\alpha\gamma;$$

$$\delta\gamma = -q^{-1}\gamma\delta; \; \delta\beta = -p^{-1}\beta\delta; \tag{7}$$

$$pq\delta\alpha + p\gamma\beta = -\alpha\delta - q\beta\gamma. \tag{8}$$

b) Cross-commutation relations:

$$a\alpha = (pq)^{-1}\alpha a;$$

$$a\beta = (p^{-1}q^{-1} - 1)\alpha b + p^{-1}\beta a;$$

$$a\gamma = q^{-1}\gamma a + (p^{-1}q^{-1} - 1)\alpha c;$$

$$b\alpha = q^{-1}\alpha b;$$

$$b\beta = (pq)^{-1}\beta b;$$

$$b\delta = q^{-1}\delta b + (p^{-1}q{-}1 - 1)\beta d;$$

$$c\alpha = p^{-1}\alpha c;$$

$$c\gamma = (pq)^{-1}\gamma c;$$

$$c\delta = p^{-1}\delta c + (p^{-1}q^{-1} - 1)\gamma d;$$

$$d\beta = p^{-1}\beta d;$$

$$d\gamma = q^{-1}\gamma d;$$

$$d\delta = (pq)^{-1}\delta d; \tag{9}$$

$$a\delta + pc\beta = (p^{-1}q^{-1} - 1)\alpha d + (q^{-1} - p)\gamma b + p^{-1}\beta c + \delta a;$$
$$pqd\alpha + qb\gamma = \alpha d + p\gamma b;$$
$$pqd\alpha + pc\beta = \alpha d + q\beta c; \tag{10}$$

$W^{(2)}_{p,q}$ can be obtained from here by the involution described above.

The proof consists of direct calculations. The answer is not quite satisfactory because we want the de Rham complex of $M_{p,q}(2)$ to be of the same size as that of the commutative polynomial ring of four variables. For this to hold in degree two, we miss one quadratic relation between differentials and one cross-comutation relation. They are supplied by the following result, again stressing the role of skew products.

4.3 Theorem: *For $pq \neq 1$, $W^{(i)}_{p,q}$ has precisely three quadratic differential quotients for which quadratic and cubic components are freely spanned by the lexicographically ordered monomials in $\alpha < \beta < \gamma < \delta < d < c < b < a$. Every such quotient is a skew product of $k[a, b, c, d]$ and $k[\alpha, \beta, \gamma, \delta]$ (modulo relations), and their Hilbert series are the same as in the commutative case. For $pq = 1$, these quotients coincide.*

For $W^{(1)}_{p,q}$, the missing relations are as follows:
Variant 1:

$$d\alpha = \alpha d; \ \ \delta\alpha = -\alpha\delta. \tag{11}$$

Variant 2:

$$c\beta = p^2\beta c; \ \ \gamma\beta = -p^2\beta\gamma. \tag{12}$$

Variant 3:

$$b\gamma = q^2\gamma b; \ \ \beta\gamma = -q^2\gamma\beta. \tag{13}$$

The long calculations needed to check the Theorem 4.3 were made by D. Zhdanovich and the author. Proposition 4.2 was independently proved by G. Maltsiniotis [8], who has also discovered the missing relations (11), but not (12), (13). The reason was that Maltsiniotis was only looking for the DGA-quotients of $W^{(1)}_{p,q}$ compatible with the coalgebra structure, and thus leading to the de Rham complex of $GL_{p,q}(2)$. Only (11) satisfy this condition. (The existence of $W^{(2)}_{p,q}$ also is not mentioned by Maltsiniotis).

D. Zhdanovich remarked that (7), (8), and $(11)_{\alpha,\delta}$ determine the algebra of differential forms with constant coefficients which is naturally isomorphic to $M_{q,p}(2)^!$ (even quadratic dual). This does not hold for the other two quotients.

The following remark is also due to D. Zhdanovich. Relations (12) and (13) can be obtained by imitating the Wess-Zumino approach. More precisely, let us *start* with an algebra of functions determined by (1), and its differentials defined by (7), (8), and either $(12)_{\gamma,\beta}$ or $(13)_{\gamma,\beta}$. Let us then construct the universal bialgebra F coacting upon these algebras (by the same (4,4)-matrix on coordinates and their differentials). Then the cross-commutation relations (9), (10), and $(12)_{c,\beta}$ (resp. $(14)_{b,\gamma}$) are uniquely defined by the Wess-Zumino type conditions.

Question Is F a flat deformation of the polynomial algebra of four variables?

4.4 n-Dimensional Wess-Zumino de Rham Complex. The following generalization of the DGA (5) was considered by several people, among them G. Maltsiniotis, E. Demidov, E. Mukhin, after the discovery by M. Artin, J. Tate, A. Sudbery et. al. of the correct pair of $n-$dimensional quantum spaces leading to the construction of many-parametric quantum group $GL_{P,Q}(n)$ by the universal coaction method. This algebra is generated by $x_i,\ i=1,\ldots,n;\ \xi_i = dx_i$, subject to the following commutation relations:

$$x_i x_j = q_{ji}^{-1} x_j x_i, \qquad i<j; \tag{14}$$

$$\xi_i^2 = \xi_i\xi_j + \lambda_j q_{ji}^{-1}\xi_j\xi_i = 0, \qquad i<j; \tag{15}$$

$$\xi_i x_i = \lambda_i x_i \xi_i; \tag{16}$$

$$\xi_i x_j = \lambda_j q_{ji}^{-1} x_j \xi_i; \qquad i<j; \tag{17}$$

$$x_i \xi_j = q_{ji}^{-1}\xi_j x_i - (1-\lambda_j^{-1})\xi_i x_j; \qquad i<j. \tag{18}$$

Here $q_{ij}\,(i<j), \lambda_j$ are invertible quantization parameters. One can check without serious difficulties that this DGA $\Omega_{Q,\lambda}$ is well defined and is a flat deformation of the de Rham complex of the polynomial ring in n variables. In fact, the lexicographically ordered monomials in x, ξ form its free $k-$basis.

G.Maltsiniotis has applied the universal coaction construction in the DGA-category to $\Omega_{Q,\lambda}$ and established the following result ([9]):

4.5 Theorem: *The universal coacting DGA-algebra for $\Omega_{Q,\lambda}$ has a quotient which is a flat deformation of the standard de Rham complex of the polynomials in matrix entries iff $\lambda_j = \lambda$ does not depend on j. This quotient is then unique, and is a skew product of two subalgebras, generated by matrix entries, and their differentials, respectively.*

For a list of relations, see [9].

Question Do there exist other quotients, with the standard Hilbert function, generalizing (12) and (13) to general n? Up to now, problems of this type have been treated by a direct application of the Diamond Lemma. G. Maltsiniotis says it took him eighty pages of calculations to prove Theorem 4.5. About forty pages were spent to the discovery of (12) and (13). Clearly, a more intelligent approach is highly desirable.

These examples, together with other ones, like the de Rham complex of the Jordan plane and Jordan quantum $SL(2)$, should be a testing ground for developing a consistent framework of a new chapter of non-commutative differential geometry. Cf. [14] for further information.

5 Cohomology of Elementary Extensions

Let $U \subset V$ be an embedding of differential algebras, $t \in V$ an even element, dt its differential. We shall say that V is *an elementary extension of* U by (t, dt), if the following conditions are satisfied:

i) $\{t^i; dt.t^j\}$ for $i, j \geq 0$ form a free basis of V as a right U-module.
ii) $(dt)^2 = 0$; $tdt = v^2 dt.t$ for an invertible element v of a ground ring k.

Put

$$[i]_v = \frac{v^i - v^{-i}}{v - v^{-1}}.$$

The following result can be useful for an inductive calculation of the de Rham cohomology. It exhibits again the well known in the representation theory "roots of unity" effects.

5.1 Theorem: *Assume that all non-vanishing $[i]_v$ are invertible in k.*

a) If v is not a root of unity, then the embeding $U \rightarrow V$ defines the isomorphism of the cohomology spaces $H^*(U) \rightarrow H^*(V)$.
b) If v is a primitive root of unity of degree l, then $H^*(V)$ as a right $H^*(U)$-module is freely generated by the cohomology classes of cocycles t^{jl}, $j \geq 0$, and $dt.t^{jl-1}$, $j \geq 1$.

5.2 Application to the n-Dimensional Wess-Zumino-de Rham Complex. Let Ω_n be the DGA described in 4.4. One easily checks that the natural embedding $\Omega_{n-1} \subset \Omega_n$ (with compatible sets of quantization constants) makes Ω_n a left elementary extension of Ω_{n-1}, for which $v_n^2 = \lambda_n^{-1}$, which allows us to calculate $H^*(\Omega_n)$. From the Theorem 4.5 one infers that Ω_n is "quantum homogeneous" exactly when $\lambda_1 = \ldots = \lambda_n = \lambda$, so that the cohomology of Ω_n is then determined by this unique parameter.

Actually, Ω_n is a skew product of Ω_{n-1} and $k[x_n, dx_n]_{v_n}$.

5.3 Question: Can one apply a similar reasoning to the de Rham complexes of matrix bialgebras described in Section 3?

References

1. Artin, M., Schelter, W.F.: Graded algebras of global dimension 3. Adv. in Math. **66**, 171-216 (1987)
2. Artin, M., Schelter, W.F., Tate, J.: Quantum deformations of GL_n. Preprint MIT (1990)
3. Connes, A.: Non-commutative differential geometry. Publ.Math. de l'IHES **62**, 41-144 (1985)
4. Demidov, E.E., Manin, Yu.I., Mukhin, E., Zhdanovich, D.: Non-standard quantum deformations of $GL(n)$ and constant solutions of the Yang-Baxter equation. Prog.Theoret.Phys.Suppl. **102** (1990)
5. Drinfeld, V.G.: Quantum groups. Proc.Int.Math. Berkeley, 1986, Vol. 1, 798-820
6. Dubois-Violette, M., Kerner, R., Madore, J.: Non-commutative differential geometry of matrix algebras. J.Math.Phys. **31:2**, 316-322 (1990)
7. Lusztig, G.: Modular representations and quantum groups. In: Classical groups and related topics. Contemp.Math. **82**, 59-77 (1989)
8. Maltsiniotis, G.: Groupes quantiques et structures différentielles. C.R.Acad. Sci. **311**, Sér. I, 831-834 (1990)
9. Maltsiniotis, G.: Calcul différentiel sur le groupe linéaire quantique. Preprint 1990
10. Manin, Yu.I.: Some remarks on Koszul algebras and quantum groups. Ann.Inst.Fourier **37:4**, 191-205 (1987)
11. Manin, Yu.I.: Quantum groups and non-commutative geometry. CRM, Université de Montréal (1988)
12. Manin, Yu.I.: Multiparametric quantum deformation of the general linear supergroup. Comm.Math.Phys. **123**, 163-175 (1989)
13. Manin, Yu.I.: *Topics in non-commutative geometry.* Princeton University Press (1991)
14. Manin, Yu.I.: Notes on quantum groups and quantum de Rham complexes. Preprint Max-Planck-Institut, Bonn, MPI/91-60 (1991)
15. Reshetikhin, N.: Multiparametric quantum groups and twisted quasitriangular Hopf algebras. Lett.Math.Phys. **20**, 331-335 (1990)
16. A. Sudbery,A.: Consistent multiparameter quantization of $GL(n)$. J.Phys. A **23**, 697-704 (1990)
17. Wess, J., Zumino, B.: Covariant differential calculus on the quantum hyperplane. Preprint CERN-TH-5697/90
18. Woronowicz, S.: Differential calculus on compact matrix pseudogroups (quantum groups). Comm.Math.Phys. **122**, 125-170 (1989)

Hamiltonian Methods in Conformal Field Theory

L.D. Faddeev

Steklov Mathematical Institute, St.Petersburg and
Research Institute in Theoretical Physics, Helsinki University,
Helsinki

Conformal Field Theory (CFT) existed for a long time on the periphery of Quantum Field Theory (QFT). It got a very strong boost by the bootstrap programm of Belavin, Polyakov and Zamolodchikov [1]. Hundreds of papers were published since. CFT has conscripted new young and revolutionary devoters (call them the conformists to distinguish from die konformisten). New terminology and methodology, e.g. primary fields, vertex operators, operator expansion, mixing of states and operators is indispensable for the paper on CFT.

In this talk I would like to show, that the more traditional methods of QFT are quite suitable for CFT. In a sense, CFT provides a family of exactly soluble models of QFT in 1+1 dimensional space-time. I believe that the conservative and nonconformal approach will unable us to understand the reason for such a solvability. The pioneer paper in this direction was that of Alekseev and Shatashvili [2], who have shown that Polyakov's two-dimensional gravity has lagrangian formulation and that its phase space is a model space for Virasoro algebra.

I must admit, that the ideology of noncommutative geometry will manifest itself in my talk. Thus all the general invited lectures at this Congress are illustrating different aspects of this fascinating subject.

Following St.Petersburg tradition, coming back to Euler, I shall base my exposition on the concrete example — Wess-Zumino-Novikov-Witlen model (WZNW) for a compact group G. However most of the formulas will be quite general.

I shall use Minkowskian metric in space-time X, which is a cylinder

$$0 \leq x \leq 2\pi; \quad -\infty < t < \infty.$$

The field $g(x,t)$ has values in the group G, which will be examplified by $SU(2)$; naturally $g(x,t)$ is periodic in x

$$g(x+2\pi, t) = g(x,t)$$

The action is given by the expression, proposed by Witten [3]

$$A = \frac{1}{8\gamma} \int tr(\partial_\mu g g^{-1})^2 + \frac{1}{12\gamma} \int (d^{-1} tr(dgg^{-1})^3)_* .$$

The first term in the RHS is the usual chiral lagrangian; the second — so called Wess-Zumino term, introduced in [4] and interpreted by Novikov in the differentional-topological terms in [5], has the following meaning: the 3-form $\alpha_3 = tr(dgg^{-1})^3$ on G is closed but not exact. Thus it can be written as $\alpha_3 = d\alpha_2$ (or $\alpha_2 = d^{-1}\alpha_3$) with a singular 2-form α_2. The integrand in the second term is a pull-back of α_2 from G to X. The singularity of α_2 requires quantized values for the coupling constant γ

$$\gamma = \frac{\pi}{l}$$

where l is an integer.

The fixed time values of g and $J_0 = \partial_t g g^{-1}$ can be used as canonical (albeit nonlinear) variables. The canonical two-form (on the space of fields) looks as follows

$$\Omega = \frac{1}{4\gamma} \int tr(dJ_0 \wedge dgg^{-1} + J_1 dgg^{-1} \wedge dgg^{-1}) dx ,$$

where $J_1 = \partial_x g g^{-1}$. The second term in the RHS is generated by WZ-term. In derivation one can use the fact, that the variation of α_3 is exact.

The Poisson brackets are most transparent in terms of linear combinations of currents J_0 and J_1

$$L = \frac{1}{2}(J_0 + J_1);$$
$$R = \frac{1}{2} g^{-1}(J_0 - J_1) g .$$

Introducing components $L^a(x)$

$$L(x) = \sum_a L^a(x) \sigma^a ,$$

where σ^a, $a = 1, 2, 3$, are Pauli matrices, we have

$$\{L^a(x), L^b(y)\} = \gamma \left(\varepsilon^{abc} L^c(x) \delta(x-y) + \delta^{ab} \delta'(x-y) \right)$$

or in terms of matrices $L(x)$ themselves

$$\{L^1(x), L^2(y)\} = \gamma C \delta'(x-y) + \gamma \left[C, L^1(x) - L^2(y) \right] \delta(x-y).$$

Here we use suitable notations for matrices in $\mathbb{C}^2 \otimes \mathbb{C}^2$

$$L^1 = L \otimes I; \quad L^2 = I \otimes L$$

and

$$C = \sum_a \sigma^a \otimes \sigma^a .$$

The relations for R differ only in the sign in front of $\delta(x-y)$. Finally $L(x)$ and $R(x)$ commute,

$$\{L^1(x), R^2(y)\} = 0.$$

Thus we see, that the phase space of the WZNW model contains two exemplars of the Kac-Moody algebra; the coupling constant γ defines the corresponding central charge.

The dynamics is trivial in terms of L and R: the equations of motion are linear

$$\partial_t L - \partial_x L = 0; \quad \partial_t R + \partial_x R = 0;$$

so only the reformulation of dynamics in terms of original field $g(x,t)$ is needed. However the most interesting things happen in this passage. To get g we are to solve the differential equation

$$\partial_x g = Lg - gR .$$

The general solution is given by

$$g(x) = u_0(x)\, K\, v_0(x)$$

where

$$u_0(x) = \overleftarrow{\exp}\left\{\int_0^x L(y)dy\right\};$$

$$v_0(x) = \overrightarrow{\exp}\left\{-\int_0^x R(y)dy\right\}$$

are solutions of equations

$$\partial_x u_0 = L u_0; \quad \partial_x v_0 = -v_0 R$$

and

$$K = g(0)$$

plays the role of the integration constant. However it can not be arbitrary. Indeed, introducing the monodromies M_L and M_R for u_0 and v_0

$$u_0(x+2\pi) = u_0(x) M_L;$$

$$v_0(x+2\pi) = M_R v_0(x),$$

or

$$M_L = u_0(2\pi); \quad M_R = v_0(2\pi)$$

we have from the periodicity of $g(x)$

$$K = M_L\, K\, M_R$$

or

$$M_R^{-1} = K^{-1} M_L K.$$

Thus M_L and M_R could be diagonalized simultaneously

$$M_L = Z_1 \, D \, Z_L^{-1},$$
$$M_R = Z_R^{-1} D^{-1} Z_R,$$

where

$$D = \begin{pmatrix} e^{ip} & 0 \\ 0 & e^{-ip} \end{pmatrix}$$

and p must be confined to the half-period $0 \leq p \leq \pi$ to kill the Weyl invariance. Now we have the relation on K

$$[Z_L^{-1} K Z_R^{-1}, D] = 0,$$

so that

$$K = Z_L Q Z_R$$

and Q is diagonal

$$Q = \begin{pmatrix} e^{iq} & 0 \\ 0 & e^{-iq} \end{pmatrix}, \quad 0 \leq q \leq 2\pi.$$

The coordinate q is the only one which enters $g(x)$, but is absent in currents L and R. It is a cyclic coordinate. Its role is not evident untill both chiralities L and R, or truly canonical fields g and J_0 are considered.

This concludes the discussion of the maps

$$(g(x), J_0(x)) \leftrightarrow (L(x), R(x), q).$$

Alternatively we can use new chiral fields

$$u_F(x) = u_0(x) Z_L Q,$$
$$v_F(x) = Q Z_R v_0(x).$$

They have diagonal monodromy

$$u_F(x + 2\pi) = u_F(x) D,$$
$$v_F(x + 2\pi) = D^{-1} v_F(x)$$

and local field $g(x)$ is expressed through them as follows

$$g(x) = u_F(x) Q^{-1} v_F(x).$$

We have algebra of Poisson brackets for $L(x)$ and $R(x)$. To understand the underlining structure better one is to get the corresponding brackets for $u_F(x)$ and $v_F(x)$. To get an idea, how they could look, it is instructive to consider first the finite-dimensional analog of our model — namely classical isotropic top.

The phase space of the top is T^*G. The natural coordinates ω and a_L (or $a_R = \omega^{-1}a_L\omega$) have values in the group and Lie algebra, correspondingly. The lagrangian

$$l = tr(a_L\dot{\omega}\omega^{-1} - \frac{1}{2}\omega^2) = \\ = tr(a_R\dot{\omega}^{-1}\dot{\omega} - \frac{1}{2}\omega^2)$$

leads to the equations of motion

$$\dot{a}_L = 0; \quad \dot{\omega} = a_L\omega$$

and Poisson brackets

$$\{\omega^1, \omega^2\} = 0; \\ \{a_L^1, \omega^2\} = C\omega^2; \\ \{a_L^1, a_L^2\} = [C, a_L^1 - a_L^2].$$

Introducing

$$a_L = u_0 \begin{pmatrix} p & 0 \\ 0 & -p \end{pmatrix} u_0^{-1};$$

$$a_R = v_0^{-1} \begin{pmatrix} p & 0 \\ 0 & -p \end{pmatrix} v_0,$$

where $u_0, v_0 \in G/H$, H being the Cartan subgroup of diagonal matrices, we have

$$\omega = u_0 Q v_0,$$

where Q is diagonal

$$Q = \begin{pmatrix} e^{iq} & 0 \\ 0 & e^{-iq} \end{pmatrix}.$$

The analogy with the formulas above is striking. We are now for the Poisson brackets for u_0 and v_0. Let us parametrize them by means of Euler angles

$$u_0 = e^{i\alpha\sigma_3}e^{i\beta\sigma_2};$$

$$v_0 = e^{i\gamma\sigma_2}e^{i\delta\sigma_3}.$$

The absense of diagonal part on the right of u_0 and left of v_0 corresponds to the choice of the representatives in G/H. The canonical 1-form looks as follows

$$tr(a_L d\omega\omega^{-1}) = pdq + p\cos 2\beta d\alpha + p\cos 2\gamma d\delta,$$

so that we can calculate all the brackets we want. In particular, as was shown in [6], for the matrices

$$u = u_0 Q, \quad v = Q v_0$$

we get the quadratic algebra

$$\begin{aligned} \{u^1, u^2\} &= u^1 u^2 z_0(p); \\ \{v^1, v^2\} &= -z_0(p) v^1 v^2; \\ \{u_0^1, v_0^2\} &= 0; \\ \{u^1, D^2\} &= u^1 D^2 s; \\ \{v^1, D^2\} &= s D^2 v^1; \end{aligned}$$

where

$$r_0(p) = \frac{i}{p} \begin{pmatrix} 0 & 0 & 0 & 0 \\ 0 & 0 & 1 & 0 \\ 0 & -1 & 0 & 0 \\ 0 & 0 & 0 & 0 \end{pmatrix}$$

and

$$s = \operatorname{diag} \frac{1}{2}(1, -1, -1, 1).$$

This example shows, that one could expect the quadratic Poisson algebra for $u_F(x)$ and $v_F(x)$ with a suitable generalization of matrix $r_0(p)$. It is indeed the case and corresponding algebra was proposed in 1989 [7]–[9]. It was an exchange algebra with the classical R-matrix from the theory of quantum groups. The interrelation of CFT and quantum groups which was already seen previously in connection with the Knizhnik-Zamolodchikov equation [10]–[11], became more straightforward. However the derivations or reasonings in [7]–[9] were not completely satisfactory and looked rather misterious. Because of that Volkov and me in the beginning of 1990 made an effort to derive the exchange algebra directly from the lagrangian formulation. Evidently the same spirit prompted the recent publications [12],[13].

The important new idea appeared in the end of 1990, in the conversations between Alekseev, Volkov, Semenov-Tjan-Shansky and me. It led to the connection of the exchange algebra and R-matrix so strightforwardly that we called our publication [14] "The unraveling of the quantum group structure in WZNW model". Here follows the idea of our proposal.

If the coefficient $L(x)$ in the equation

$$u' = Lu$$

is gauge transformed

$$L(x) = T(x) s(x) T^{-1}(x) + \partial_x T(x) T^{-1}(x)$$

than the solution can be written as

$$u(x) = T(x) \overleftarrow{\exp} \left\{ \int_0^x s(y) dy \right\}.$$

The idea is to find the entries $T(x)$, $s(x)$ in such a way, that they have only ultralocal Poisson brackets (i.e. without $\delta'(x-y)$ terms) leading to the usual Kac-Moody brackets for $L(x)$. As soon as $s(x)$ is ultralocal, the Poisson brackets of transport matrix $\exp\left\{\int_0^x s(y)dy\right\}$ are easily calculatble, as is well known in the inverse scattering method [15].

The realization of this idea is quite simple and beautiful. Consider $T(x)$ — upper triagular matrix and $s(x)$ — lower triangular matrix with the Poisson brackets

$$\begin{aligned}\left\{T^1(x),T^2(y)\right\} &= 0;\\ \left\{T^1(x),s^2(y)\right\} &= \gamma T^1(x) r\delta(x-y);\\ \left\{s^1(x),s^2(y)\right\} &= \gamma\left[s^1(x)+s^2(y),r\right]\delta(x-y).\end{aligned}$$

Here r is a classical r-matrix

$$r=\frac{1}{2}\sigma^3\otimes\sigma^3+2\sigma^+\otimes\sigma^-,$$

or

$$r=\begin{pmatrix}1/2 & 0 & 0 & 0\\ 0 & -1/2 & 2 & 0\\ 0 & 0 & -1/2 & 0\\ 0 & 0 & 0 & 1/2\end{pmatrix}.$$

Then

$$L(x)=T(x)s(x)T^{-1}(x)+\partial T(x)T^{-1}(x)$$

has the Poisson brackets of the left current of WZNW model.

The relations for $T(x)$ and $s(x)$ have a nice geometrical meaning. As they are ultralocal we can drop the variable x, returning from loops to the finite dimensional objects. Then the relations are nothing but those on T^*B, where B is a Borel subgroup of G, namely the subgroup of upper triangular matrices. Thus $T(x)$ and $s(x)$ constitute the loops of T^*B.

Of course, T and s contain 4 variables whereas $L(x)$ has only 3. The resolution consists in the observation, that one can construct from the diagonal elements of T and s a current $l(x)$, such that

$$\begin{aligned}\{l(x),l(y)\} &= \gamma\delta'(x-y);\\ \{L(x),l(y)\} &= 0.\end{aligned}$$

In other words, we can write symbolically

$$KM=LT^*B/LH,$$

where LH is a loop of Cartan subgroup, generated by $l(x)$.

Note also the appearance of the r-matrix in an entirely classical setting — it is nothing but a structure constants matrix for the Borel subalgebra.

The ultralocal expression for $L(x)$ allows to calculate the Poisson brackets for $u(x)$. We shall present the corresponding formulas in already quantized form.

Quantization can be performed in the spirit of Quantum Inverse Scattering Method [16],[17]. The circle S^1, over which x runs is to be substituted by finite chain with running discrete variable n, $n = 1, \ldots, N$. The continuous limit $N \to \infty$, or $\Delta \to 0$, where

$$\Delta = 2\pi/N$$

is obtained in usual way

$$x = n\Delta.$$

The current $L(x)$ is substituted now by a set of matrices L_n; in continuous limit we have

$$L_n = I + \Delta L(x) + \ldots \quad .$$

The differential equation

$$u' = Lu$$

turns into

$$u_{n+1} = L_n u_n$$

with the solution

$$u_n = \overleftarrow{\prod_{k \leq n}} L_k .$$

The gauge transformation now looks as

$$L_n = T_{n+1} S_n T_n^{-1}$$

where in continuous limit

$$T_n \to T(x); \quad S_n \to I + \Delta s(x) + \ldots .$$

We introduce now the main commutation relations

$$\begin{aligned} R S_n^1 S_n^2 &= S_n^2 S_n^1 R \\ R T_n^1 T_n^2 &= T_n^2 T_1^n R \\ T_n^1 S_n^2 &= S_n^2 T_n^1 R \, . \end{aligned}$$

Here R is a quantum R-matrix

$$R = \begin{pmatrix} t^{1/2} & 0 & 0 & 0 \\ 0 & t^{-1/2} & t^{1/2} - t^{-3/2} & 0 \\ 0 & 0 & t^{1/2} & 0 \\ 0 & 0 & 0 & t^{1/2} \end{pmatrix}$$

where

$$t = e^{i\gamma\hbar}$$

and $\hbar$ is a Planck constant. In the classical limit $\hbar \to 0$

$$R = 1 + i\gamma\hbar r + \ldots ,$$

so that the quantum relations turn into the classical continuous ones in the limit $\Delta \to 0$, $\hbar \to 0$ if we identify

$$\frac{i}{\hbar}[\,,\,] = \{\,,\,\}.$$

Note that the group variables T_n are not commuting now — this is a first manifestation of the noncommutative geometry, more of which will be seen in what follows.

The commutation relations for T_n, S_n for a fixed n define the regular representation of "quantized" or rather derormed group B; In other words it is quantization of the deformed phase space $(T^*B)_\gamma$.

Now the problem of the calculation of the commutation relations for the chiral components u_n is reduced to pure algebra. We have

$$u_k = L_k L_{k-1} \dots L_1 = T_{k+1} S_k \dots S_1 T_1^{-1}.$$

In particular, for the monodromy M_L we have

$$M_L = T_1 \left(\overleftarrow{\prod_{n=1}^{N}} S_n \right) T_1^{-1}.$$

One can easily get the relations

$$\overset{1}{u}_i \overset{2}{u}_k = \overset{2}{u}_k \overset{1}{u}_i R^{\pm} \quad i \gtrless k$$

$$R^+ \overset{1}{u}_i \overset{2}{u}_i R^- = \overset{2}{u}_i \overset{1}{u}_i,$$

where

$$R^+ = R; \quad R^- = PR^{-1}P.$$

Note, that the fields with coinsided arguments are not commuting. This effect can be hardly seen in the continuous limit.

For monodromy we have

$$\overset{1}{M}_L (R^-)^{-1} \overset{2}{M}_1 R^- = (R^+)^{-1} \overset{2}{M}_L R^+ \overset{1}{M}_L$$

— this is nothing but the commutation relation for the quantized (deformed) Lie algebra in the form of Reshetikhin and Semenov-Tjan-Shansky [18].

Thus we see the proper place of the quantum group in CFT — it is a monodromy of chiral field with a trivial normalization at some point ($u_0(0) = I$). However it is not all. Let us turn our attention to the local field

$$g_i = u_i K v_i,$$

where $K = g_0 = g_N$.

The consistency of the periodicity and relations for monodromy leads to the relations

$$RK^1K^2 = K^2K^1R;$$
$$R^+K^1M^2 = M^2R^-K^1,$$

which together with relation for M constitute the set of relations for the quantized phase of the deformed top over G, or $(T^*G)_\gamma$ (see [9]).

Thus in CFT we have not only the quantum Lie algebra M, but rather a pair (K, M) — a local field at a fixed point and a monodromy of a chiral component around the circle from this point. This pair is nothing but a finite dimensional dynamical system, comprizing the zero models of the WZNW model. It essentially defines the full structure of the model.

Let us stress once more, that in the continuous limit this object is barely seen, because the notion of the value of a field $g(x)$ at one point loses sense.

The local field g_i for general i has relations

$$g_i^1 g_k^2 = g_k^2 g_i^1, \quad i \neq k,$$

and

$$Rg_i^1 g_i^2 = g_i^2 g_i^1 R.$$

Thus we see that the coordinates in the target space *do not* commute and natural geometry there is noncommutative. Not all implications of this observation are clearn yet.

Let us see now what one can learn from what was said about the quantum WZNW model.

First about the top. It is known that compact quantum Lie algebra $SU(2)_q$ for $q = e^{i\pi/l}$ has $l-1$ representations V_j, $j = 0, \frac{1}{2}, \ldots, \frac{l}{2} - 1$ of dimension $2j+1$ with highest weight and infinite series of cyclic representations of dimension l. One can forget about the latter in the quantization of a top, if one restricts values of p and q to the reals. Thus the Hilbert space of quantized (and deformed) top is

$$\mathcal{H}_{\text{top}} = \sum_j V_j \otimes V_j \, .$$

Now let us return to WZNW. The quantized $L(x)$ give a representation of Kac-Moody algebra of the level $k = l-2$ (the appearence of the famous shift is explained in terms nearest to this text in [19]). There are $k+1 = l-1$ such representation $\mathcal{H}_j$. In each $\mathcal{H}_j$ the monodromy M_L acts (with infinite multiplicity) so that (see [19])

$$\mathcal{H}_j = \mathcal{H}_0 \otimes V_j \, .$$

In other words the representation of quantum group V_j is kind of a label on the representation $\mathcal{H}_j$ of Kac-Moody algebra. In particular the fusion rules for $\mathcal{H}_j$ are generated by the tensor algebra of V_j. The Hilbert space of WZNW model has the form

$$\mathcal{H}_{\text{WZNW}} = \sum \mathcal{H}_j \otimes \mathcal{H}_j.$$

The local field at a fixed point can be used as an "intertwiner" between different $\mathcal{H}_j$ in the sense of the generalized approach to superselection sectors (see e.g. [20]).

The representation of the local field through the chiral components

$$g(x) = u_0(x) K v_0(x)$$

could prove to be very useful for the calculations of the correlation functions. Indeed, the reconstruction of the conformal block

$$\langle u_0^1(x_1 + t_1) \ldots u_0^n(x_n + t_n) \rangle$$

can be reduced to the Reimann-Hilbert problem similar (and simpler) to that considered by Smirnov [21], see also [22]. However this programm was not realized yet.

I turn now to the general conclusions.

1. Change of the nature of symmetry

The relations for L_n, which follow from those for S_n and T_n, are invariant to the external gauge transformation

$$L_n \to h_{n+1} L_n h_n^{-1}$$

only if h_n constitute a deformed loop group

$$h_n^1 h_m^2 = h_m^2 h_n^1, \quad n \neq m$$
$$R h_n^1 h_n^2 = h_n^2 h_n^1 R,$$

commuting with L_n

$$h_n^1 L_m^2 = L_m^2 h_n^1.$$

So the symmetry itself becomes deformed — it is a symmetry with respect to an action of the quantum loop group.

2. Noncommutative geometry

I have already made several comments to the effect, that the ideology of noncommutative geometry naturally enters my exposition. Let us reiterate that the natural object for the values of the local field is a deformed loop group in the sence just explained. Having in mind the connection of σ-model and string it is natural to speculate, that the target space for the string could be noncommutative.

3. Dimensional reduction

We have seen, that due to the monodromy map the full WZNW (1+1 dim) model was reduced to the finite-dimensional (1 dim) dynamical system on $(T^*G)_\gamma$, $\gamma = \frac{\pi}{k+2}$. In its turn WZNW is a reduction of a Chern-Simons model (1+2 dim) by transgression. Thus our example prompts the investigation of a more general situation, where such reductions are possible.

4. Equivalent deformations

The same thing can be put in a different fashion. Beginning with a finite dimensional Lie algebra G we can construct an affine algebra $(\hat{G})_k$ or quantized algebra G_q. The first is still Lie algebra, but infinite dimensional, the second is finite dimensional, but has only deformed Lie algebra structure. Each modification is parametrized by one parameter: k and q, correspondingly. We have seen, that $\hat{G}_k$ and G_q are essentially equivalent if $q = e^{i\pi/k+2}$ (for $SU(2)$).

One can draw a diagramm

$$\begin{array}{ccc} & G & \\ \swarrow & & \searrow \\ \hat{G} & \sim & G_q \end{array}$$

and ask natural question if this diagramm has a continuation (It was mentioned to me that I.Frenkel raised the same question). The continuation must be a two-parameter family, which may be a local Lie-algebra (in two dimensions), or deformed Lie algebra (in one dimension), or finite dimensional algebra with more complicated relations. There are candidates for all three — extended current algebra on a disk, deformed $K - M$ algebra and Sklyanin algebra. To establish the relation between their representations is an interesting unsolved problem.

References

1. Belavin, A., Polyakov, A., Zamolodchikov, A.: Nuclear Phys. B **241**, 333 (1984)
2. Alekseev, A., Statashvili, S.: Nuclear Phys. B **323**, 719 (1989)
3. Witten, E.: Comm.Math.Phys. **92**, 451 (1984)
4. Wess, J., Zumino, B.: Phys.Rev. B **37**, 95 (1971)
5. Novikov, S.: Uspekhi Mat.Nauk **37**, 3 (1982)
6. Alekseev, A., Faddeev, L.: MIT preprint CTP N1957 (1991) (to be published in Comm.Math.Phys.)
7. Blok, B.: Phys.Lett. B **233**, 359 (1989)
8. Faddeev, L.: Comm.Math.Phys. **132**, 131 (1990)

9. Alekseev, A., Statashvili, S.: Comm.Math.Phys. **133**, 353 (1990)
10. Tsuchiya, A., Kanie, Y.: Adv.Stud.Pure Math. **16**, 297 (1988)
11. Kohno, T.: Ann.Inst.Fourier (Grenoble) **37**, 4 (1987)
12. Gawedsky, K.: Preprint IHES (1990)
13. Chu, M., Goddard, P., Halliday, L., Olive, D., Schwimmer, A.: Phys.Lett. B **266**, 71 (1991)
14. Alekseev, A., Faddeev, L., Semenov-Tjan-Shansky, M., Volkov, A.: Preprint CERN-TH-5981/91 (1991)
15. Faddeev, L., Takhtajan, L.: *Hamiltonian Methods in the Theory of Solitons.* Springer (1987)
16. Sklyanin, E., Takhtajan, L., Faddeev, L.: Theoret.and Math.Phys. **40**, 688 (1980)
17. Faddeev, L.: Les Hauches lecture in: Recent Advances in Field Theory. R.Stora, J-B.Zuber (eds.), Amsterdam, North-Holland (1984)
18. Reshetikhin, N., Semenov-Tjan-Shansky, M.: Lett.Math.Phys. **19**, 133 (1990)
19. Alekseev, A., Faddeev, L., Semenov-Tjan-Shansky, M.: LOMI preprint E-5-91 (1991)
20. Mack, G., Schomerus, V.: Preprint DESY 91-037 (1991)
21. Smirnov, F.: *Formfactors in Completely Integrable Models of Quantum Field Theory.* World Scientific (to be published)
22. Smirnov, F.: Preprint RIMS-772 (1991)

On the General Theory of Quantized Fields

dedicated to the memory of Res Jost (†1990)

Klaus Fredenhagen

II. Institut für Theoretische Physik, Universität Hamburg

Content

1 The Aim of the General Theory of Quantized Fields

The term "General Theory of Quantized Fields", replacing the synonymous but somewhat misleading term "Axiomatic Field Theory", is to my knowledge due to Res Jost. He was one of the great pioneers in our field, and I dedicate this lecture to his memory. What is the aim of the general theory of quantized fields?

An answer may be found in Jost's famous book with the same title [1]. Jost formulates it as follows: "The general theory of quantized fields analyzes the notions which are at the basis of all previously analyzed specific models". So the general theory of quantized fields is concerned with model independent aspects of quantum field theory. Actually, the richness of the general theory of quantized fields is intimately connected with the notorious difficulties in constructing and evaluating examples of quantum field theories.

There has been a lot of progress in quantum field theory since the time the project of the general theory of quantized fields was started by the work of Wightman, Haag, Lehmann, Symanzik, Zimmermann and others. Non-abelian gauge theories became the most promising models for elementary particle physics, renormalization has much better been understood by now, interacting models in low dimensions were constructed , numerical approximation on lattices and semiclassical arguments supplement the perturbation theoretical approach, and experiments in elementary particle physics have supported much of the theoretical predictions. More recently, a rich class of interesting models was detected in 2 dimensional conformal field theory, integrable models with exact S-matrices were found, and string theory and topological field theory broadened the range of quantum field theory.

In spite of all these successes, quantum field theory remains a very difficult subject, and the general analysis of structural properties is often the most convenient way to derive predictions in combination with some input from the models or even directly from experiment.

In my lecture I want to describe the present stage of the general theory of quantized fields on the example of 5 subjects. They are ordered in the direction from large to small distances.

The first one is the by now classical problem of the structure of superselection sectors. It involves the behavior of the theory at spacelike infinity and is directly connected with particle statistics and internal symmetries. It has become popular in recent years by the discovery of a lot of nontrivial models in 2d conformal field theory, by connections to integrable models and critical behavior in statistical mechanics and by the relations to the Jones' theory of subfactors in von Neumann algebras and to the corresponding geometrical objects (braids, knots, 3d manifolds,...).

At large timelike distances the by far most important feature of quantum field theory is the particle structure. This will be the second subject of my lecture. It follows the technically most involved part which is concerned with the behavior at finite distances. Two aspects, nuclearity which emphasizes the finite density of states in phase space, and the modular structure which relies on the infinite number of degrees of freedom present even locally, and their mutual relations will be treated.

The next point, involving the structure at infinitesimal distances, is the connection between the Haag-Kastler framework of algebras of local and the framework of Wightman fields. Finally, problems in approaches to quantum gravity will be discussed, as far as they are accessible by the methods of the general theory of quantized fields.

2 Sectors, Statistics and Symmetry

The theory of superselection sectors has ever been a challenge since the observation of Wick, Wightman and Wigner [2] that the superposition principle in quantum physics does not hold unrestrictedly. It was formulated as a classification problem on the representations of the algebra of observables by Haag and Kastler [3], was initiated by Borchers [4] and reached a certain degree of completeness in the work of Doplicher, Haag and Roberts [5].

The basic structure is a net of von Neumann algebras

$$\mathcal{A} = (\mathcal{A}(\mathcal{O}))_{\mathcal{O}\in\mathcal{K}}, \qquad \mathcal{A}(\mathcal{O}) \subset \mathcal{B}(\mathcal{H}_0)\,, \tag{2.1}$$

where $\mathcal{K}$ is the set of open double cones ("diamonds") in Minkowski space. This net is assumed to satisfy the usual Haag-Kastler axioms,

$$(i) \quad \mathcal{A}(\mathcal{O}_1) \subset \mathcal{A}(\mathcal{O}_2) \quad \text{for} \quad \mathcal{O}_1 \subset \mathcal{O}_2 \quad \text{(isotony)} \tag{2.2}$$

$$(ii) \quad \mathcal{A}(\mathcal{O}_1) \subset \mathcal{A}(\mathcal{O}_2)' \quad \text{for} \quad \mathcal{O}_1 \subset \mathcal{O}_2' \quad \text{(locality)} \tag{2.3}$$

($\mathcal{A}(\mathcal{O})'$ is the commutant of $\mathcal{A}(\mathcal{O})$ and $\mathcal{O}'$ the spacelike complement of $\mathcal{O}$, $\mathcal{O} \in \mathcal{K}$)

$$(iii) \bigcap_{\mathcal{O}\in\mathcal{K}} \mathcal{A}(\mathcal{O})' = \mathbb{C}\mathbb{1} \quad \text{(irreducibility)}\,. \tag{2.4}$$

One then looks for representations π fulfilling the DHR selection criterion

$$\pi|_{\mathcal{A}(\mathcal{O}')} \simeq \pi_0|_{\mathcal{A}(\mathcal{O}')}, \quad \mathcal{O} \in \mathcal{K}\,, \tag{2.5}$$

where $\pi_0(A) = A$ is the defining representation of $\mathcal{A}$ ("vacuum representation") and $\mathcal{A}(\mathcal{O}')$ is the algebra generated by $\mathcal{A}(\mathcal{O}_1)$ with $\mathcal{O}_1 \subset \mathcal{O}', \mathcal{O}_1 \in \mathcal{K}$.

Crucial for the analysis is a strengthened version of locality which is supposed to hold in the vacuum representation

$$\mathcal{A}(\mathcal{O}) = \mathcal{A}(\mathcal{O}')' \quad \text{(Haag duality)}\,. \tag{2.6}$$

Haag duality is known to hold in many cases; there are, however, also situations where Haag duality fails, e.g. in the case of spontaneous breakdown of symmetry, a subject treated in the poster session by Roberts.

Provided Haag duality holds, there exists an endomorphism $\rho \in End(\mathcal{A})$ such that

$$(i) \qquad \pi \simeq \pi_0 \circ \rho \tag{2.7}$$

$$(ii) \quad \rho(A) = A \quad \text{for} \quad A \in \mathcal{A}(\mathcal{O}')\,. \tag{2.8}$$

Let $\Delta_\pi(\mathcal{O})$ be the set of all endomorphisms of $\mathcal{A}$ satisfying (i) and (ii), and let $\rho, \rho' \in \Delta_\pi(\mathcal{O})$. Then again by Haag duality there exists a unitary $U \in \mathcal{A}(\mathcal{O})$ intertwining ρ and ρ',

$$U\rho(A) = \rho'(A)U \quad , A \in \mathcal{A}\,. \tag{2.9}$$

(The set of all operators satisfying (2.9) will be denoted by (ρ, ρ').) Thus, composition of endomorphisms induces a composition law for unitary equivalence classes of representations satisfying (2.5),

$$[\pi_1] \times [\pi_2] := [\pi_0 \circ \rho_1\rho_2] \quad , \quad \rho_i \in \Delta_{\pi_i}(\mathcal{O})\,. \tag{2.10}$$

In cases where ρ is an automorphism, the irreducibility of π_0 implies the irreducibility of $\pi_o \circ \rho$; such sectors are called simple, and the product of simple sectors is again simple. If, however, the corresponding endomorphism is not surjective, the product of irreducible representations is, in general, reducible. This phenomenon is related to the violation of Haag duality in the pertaining representation. As a measure of the degree of violation one may use the Jones Index [6]

$$\mathrm{Ind}[\pi(\mathcal{A}(\mathcal{O}'))' : \pi(\mathcal{A}(\mathcal{O}))] \,, \tag{2.11}$$

which coincides with the square $d(\pi)^2$ of the statistical dimension $d(\pi)$ of π in the DHR theory [7].

The Jones Index may be illustrated most easily in the finite dimensional example of the inclusion

$$\begin{cases} M_n(\mathbb{C}) \to M_{nm}(\mathbb{C}), & n, m \in \mathbb{N}, \quad n > m \\ A \quad \mapsto A \otimes \mathbb{1}_m \end{cases} , \tag{2.12}$$

where $M_k(\mathbb{C})$ denotes the algebra of $(k \times k)$-matrices over $\mathbb{C}, k \in \mathbb{N}$. In this case one has

$$\mathrm{Ind}[M_{nm}(\mathbb{C}) : M_n(\mathbb{C})] = \frac{\dim(M_{nm}(\mathbb{C}))}{\dim(M_n(\mathbb{C}))} = m^2 \,. \tag{2.13}$$

In the infinite dimensional situation of quantum field theory an analytic characterization of the index due to Pimsner and Popa [8] is more useful. There is a conditional expectation $\mathcal{E}_\pi$ from $\pi(\mathcal{A}(\mathcal{O}'))'$ onto $\pi(\mathcal{A}(\mathcal{O}))$ (in the finite dimensional example above this is the partial trace) such that

$$\mathcal{E}_\pi(A^*A) \geq d(\pi)^{-2} A^*A \quad , \quad A \in \mathcal{A} \,, \tag{2.14}$$

and $d(\pi)$ is the smallest positive number for which the inequality is true. Inequality (2.14) already appears in [5]. Representations with $d(\pi) < \infty$ were said to have finite statistics.

The nice feature of such representations is that they are completely reducible. Moreover, one finds

$$(i) \quad d(\pi_1 \times \pi_2) = d(\pi_1) d(\pi_2) \tag{2.15}$$

$$(ii) \quad d(\oplus_i \pi_i) = \sum_i d(\pi_i) \tag{2.16}$$

hence the statistical dimensions behave like dimensions of vector spaces under tensor products and direct sums.

The main mean for the decomposition into irreducible components is the notion of statistics, very similar to the role permutation symmetry plays in the reduction of tensor products of group representations. Let (DHR) denote the class of all representations of $\mathcal{A}$ satisfying (2.5) and having finite statistics, and let

$$\Delta(\mathcal{O}) = \bigcup_{\pi \in (\mathrm{DHR})} \Delta_\pi(\mathcal{O}) \quad . \tag{2.17}$$

As a consequence of locality, endomorphisms $\rho_i \in \Delta(\mathcal{O}_i), i = 1,2$, with $\mathcal{O}_1 \subset \mathcal{O}_2'$ commute. Now let $\rho \in \Delta(\mathcal{O})$ and $\rho_1 \in \Delta_{\pi_0 \circ \rho}(\mathcal{O}_1)$ with $\mathcal{O}_1 \subset \mathcal{O}'$, and choose a unitary $U \in (\rho, \rho_1)$. Then one defines the "statistics operator"

$$\epsilon_\rho = \rho(U)^{-1}U \quad \in (\rho^2, \rho^2) = \rho^2(\mathcal{A})' \subset \mathcal{A}(\mathcal{O}) \,. \tag{2.18}$$

ϵ_ρ turns out to be independent of the choice of

$$\begin{aligned} &(i) \quad U \in (\rho, \rho_1) \\ &(ii) \quad \rho_1 \in \Delta_{\pi_0 \circ \rho}(\mathcal{O}_1) \\ &(iii) \quad \mathcal{O}_1 \text{ in a connected component of } \mathcal{O}' \,. \end{aligned} \tag{2.19}$$

Hence in $D \geq 3$ space time dimensions ϵ_ρ is uniquely determined, and in $D = 2$ dimensions there are at most two different operators ϵ_ρ dependent on whether $\mathcal{O}_1$ is in the right or left component of the spacelike complement of $\mathcal{O}$.

ϵ_ρ satisfies the braid (i.e. constant Yang Baxter) relation

$$\epsilon_\rho \rho(\epsilon_\rho) \epsilon_\rho = \rho(\epsilon_\rho) \epsilon_\rho \rho(\epsilon_\rho) \tag{2.20}$$

which implies that

$$\epsilon^{(\rho)} : \sigma_i \mapsto \rho^{i-1}(\epsilon_\rho) \tag{2.21}$$

induces a unitary representation of the braid group B_∞. (σ_i denotes the transposition of the i-th and (i+1)-st string.) In $D \geq 3$ dimensions one finds the additional relation

$$\epsilon_\rho^2 = \mathbb{1} \tag{2.22}$$

which says that $\epsilon^{(\rho)}$ factors through a representation of the symmetric group.

For an analysis of the occuring representations [5],[9]one uses a so called left inverse of ρ. Let $\rho \in \Delta(\mathcal{O})$ be irreducible. Then there exists a unique left inverse

$$\phi : \mathcal{A} \to \mathcal{A} \tag{2.23}$$

with $\phi \circ \rho = id$ and with $\rho \circ \phi$ being a conditional expectation from $\mathcal{A}$ onto $\rho(\mathcal{A})$ (which coincides essentially with the expectation $\mathcal{E}_\pi$ considered in the definition of the statistical dimension (2.14)).

Now for ρ irreducible, $\phi(\epsilon_\rho)$ is a multiple of the identity, and

$$\phi(\epsilon_\rho) = \frac{\kappa(\rho)}{d(\rho)} \mathbb{1}, \quad |\kappa(\rho)| = 1 \,, \tag{2.24}$$

where $\kappa(\rho)$ is called the statistical phase and $d(\rho) = d(\pi_0 \circ \rho)$ is the statistical dimension defined above. By iteration one obtains a function φ of positive type on the braid group B_∞,

$$\varphi(b)\mathbb{1} = \lim_{n \to \infty} \phi^n(\epsilon^{(\rho)}(b)) \,. \tag{2.25}$$

φ is a trace on B_∞,

$$\varphi(b_1 b_2) = \varphi(b_2 b_1) \quad , \quad b_1, b_2 \in B_\infty \, , \tag{2.26}$$

and fulfils

$$\varphi(b\sigma_n^{\pm 1}) = \frac{\kappa(\rho)^{\pm 1}}{d(\rho)} \varphi(b) \quad , \quad b \in B_n \, ; \tag{2.27}$$

such traces were called Markov traces by Jones [6], since due to Markovs Theorem [10] they induce ribbon invariants and, after renormalization, link invariants.

A partial classification can be performed according to the number of different eigenvalues of ϵ_ρ.

In the simplest case ϵ_ρ is a multiple of the identity. This happens if and only if ρ is an automorphism or, equivalently, if and only if ρ^2 is irreducible. In the next simple case ϵ_ρ has two different eigenvalues. This is always true when (2.22) holds [5] or when ρ^2 is a direct sum of two irreducible representations [9]. Then $\epsilon^{(\rho)}$ induces a representation of the Hecke algebra, hence is quasiequivalent to the braid group representations found by Jones, Ocneanu and Wenzl [7],[11],[12] tensored by a one dimensional representation, and φ leads to the Jones- and HOMFLY-invariants [7],[13].

In the presence of three different eigenvalues a general analysis was not yet possible. Only the case where ρ^2 is a direct sum of three irreducible representations one of which is a simple sector was completely analyzed [14],[15]. One obtains the Birman-Wenzl algebra [16], the Murakami-Wenzl representation of B_∞ [17],[18] and the Kauffman link invariant [19].

According to the general analysis each (DHR)-representation of a 2d quantum field theory satisfying Haag duality leads to a ribbon invariant. It is, however, not to be expected that every representation which describes particles is in the (DHR)-class. What can be proved is that (in $D \geq 3$ dimensions) every irreducible representation π with an isolated mass shell in the energy momentum spectrum satisfies the following weaker version of (DHR) [20],[21]:

(i) there exists a unique vacuum representation π_0 such that

$$\pi|_{\mathcal{A}(\mathcal{C}')} \simeq \pi_0|_{\mathcal{A}(\mathcal{C}')} \tag{2.28}$$

for all spacelike cones $\mathcal{C} = a + \bigcup_{\lambda>0} \lambda\mathcal{O} \quad , \mathcal{O} \in \mathcal{K}, \mathcal{O} \subset \{0\}'$.

(ii) provided π_0 satifies Haag duality, π has finite statistics.

One then can perform an analogous analysis of the set of sectors satisfying (2.28). The relation $\epsilon_\rho^2 = 1$ can be shown only in $D \geq 4$ dimensions, hence one may have braid group statistics in $D = 3$ dimensions [22],[23]. In $D = 2$ dimensions the mentioned result leads to the possible existence of two different vacua corresponding to the two different directions in which spacelike cones can point ("solitons"); composition of soliton sectors was defined in [24] but not yet analyzed in detail (see [25] for an earlier approach to this problem).

A finer classification of the set of sectors may be done in terms of a generalized internal symmetry. The most directly visible structure is the following category. Its objects are the endomorphisms $\rho \in \Delta(\mathcal{O})$, its arrows are the intertwiners $T \in (\rho, \rho')$, $\rho, \rho' \in \Delta(\mathcal{O})$, and the composition of arrows is the product of intertwiners. In addition this category has a tensor (monoidal) structure,

$$\begin{cases} \rho, \rho' \mapsto \rho\rho' \\ S, T \mapsto S \times T := S\rho(T) \end{cases} \tag{2.29}$$

$(S \in (\rho, \rho'), T \in (\rho_1, \rho_1'), S \times T \in (\rho\rho', \rho_1\rho_1'))$ with the statistics operators describing the permutation of factors in the tensor product. This category has all the properties postulated by Moore and Seiberg as "axioms for rational conformal field theory" [26], so this structure turns out to be a generic feature of quantum field theory and is by no means restricted to conformal invariant theories.

In that cases where permutation statistics holds, Doplicher and Roberts [27],[28] showed that this category is equivalent to the category of finite dimensional representations of a uniquely determined compact group G which may in turn be considered as the group of internal symmetries. Moreover, they construct an embedding of the observable algebra $\mathcal{A}$ into a so called field algebra $\mathcal{F} = (\mathcal{F}(\mathcal{O}))_{\mathcal{O} \in \mathcal{K}}$ which satisfies graded locality and on which G acts by automorphisms such that $\mathcal{A}(\mathcal{O})$ is the set of fixed points of $\mathcal{F}(\mathcal{O})$.

In the braid group case examples of these categories can be obtained from quantum groups at roots of unity. In the approach of Fröhlich and Kerler [29] one looks at the ring of representations and factors out the ideal of degenerate representations. In an alternative approach due to Todorov et al. [30] fields transforming covariantly under a quantum group act on an indefinite metric space from which the physical Hilbert space may be recovered. A recent construction of Mack and Schomerus [31] , explained in more detail in the poster session by Schomerus, avoids the use of indefinite metrics. Covariant fields were constructed which act on the physical Hilbert space, but the symmetry is not the quantum group itself but the quotient with respect to some ideal resulting in a modification of the Hopf algebra structure ("weak quasi quasitriangular Hopf algebra").

There is also a proposal for the general case with a finite number of sectors ("rational theories"). Rehren [32] constructs a field algebra similar to the field algebra of Doplicher and Roberts where now the role of the symmetry group is played by a bimodule over the hyperfinite von Neumann factor $\mathcal{R}$ of type II_1 which is the standard algebra in the Jones theory of subfactors. There are fields $\psi_i, i = 1,..n$, and a homomorphism

$$\rho : \begin{cases} \mathcal{R} \to M_n(\mathcal{R}) \\ r \mapsto (\rho_{ij}(r)) \end{cases} \tag{2.30}$$

into the algebra of $(n \times n)$-matrices with entries in $\mathcal{R}$ such that covariance takes the form

$$r\psi_i = \sum_j \psi_j \rho_{ji}(r) \quad , r \in \mathcal{R} . \tag{2.31}$$

For more details one may consult the contribution of Rehren to these proceedings.

A classification of symmetries may also be done directly in terms of the Jones theory of subfactors. So Longo analyzed the inclusion

$$\rho(\mathcal{A}(\mathcal{O})) \subset \mathcal{A}(\mathcal{O}) \quad , \rho \in \Delta(\mathcal{O}) , \tag{2.32}$$

and found additional restrictions on the statistical dimensions [14].

3 Particles and Infraparticles

At large timelike distances the dominant structure of quantum field theory is the occurence of particles. The relation between particles and fields is most elegantly expressed in terms of the LSZ reduction formulae; there is, however, in general no 1-1 correspondence between particles and fields.

According to Wigner particles are characterized as eigenstates of the mass operator. The Haag-Ruelle theory [33],[34] then succeeds in constructing the corresponding incoming and outgoing multiparticle scattering states provided the mass shell of the particle is isolated in the energy momentum spectrum and the particle is uncharged. This result has been generalized in several directions. Doplicher, Haag and Roberts performed the same construction for particles in (DHR)- representations [5], Buchholz proved (by different methods) corresponding results for massless particles [35], Buchholz and myself did the construction for charged particles in general massive 4d theories, Fröhlich, Gabbiani and Marchetti [36] and Rüger [37]generalized this construction to the case of braid group statistics in 3 dimensions ("anyons" (simple sectors), "plektons" (general case), resp.). The existence of scattering states could be proven even in euclidean lattice field theories [38] where the local structure of quantum field theory is strongly disturbed.

As a resumee the mathematically rather nontrivial fact of existence of scattering states is by now well understood and may be considered as the easy part of the problem. There are, however, really hard problems where progress has been very slow.

The first one to mention is the famous problem of asymptotic completeness. Even in quantum mechanics this problem is by no means simple, but has nevertheless been solved recently (see the contribution of Graf to these proceedings). In the case that all particles are massive and uncharged one may approach this question in terms of the analyticity properties of Wightman functions (see the contribution of Bros to these Proceedings).

A more general approach is due to Buchholz. It takes into account that Wigner's characterization of particles is too narrow since it excludes the so called infraparticles occuring in theories with massless particles which cannot be characterized by normalizable eigenvectors of the mass operator due to the presence of a cloud of massless particles. Also the usual description by improper momentum eigenstates meets problems since states with different momenta typically cannot be coherently superimposed. There is, however, a new proposal by Buchholz, Porrmann and Stein [39] (presented in the poster session by Buchholz) where improper momentum eigenstates are defined as linear mappings from a left ideal $\mathcal{L}$ of almost local vacuum annihilators into the physical Hilbert space $\mathcal{H}$,

$$|p\rangle : \begin{cases} \mathcal{L} \to \mathcal{H} \\ L \mapsto L|p\rangle \end{cases} \tag{3.1}$$

where $|p\rangle$ represents the bare particle ("particle weight") and $L|p\rangle$ is a normalizable state describing the particle together with some infrared cloud. In this framework, multiparticle scattering weights can be constructed as linear functionals on an algebra of particle counters, and cross sections can be directly defined without any infrared regularization.

Wigner's particle definition seems also to be too restrictive for situations at finite temperature. As first observed by Narnhofer, Requardt and Thirring [40], eigenstates of the mass operator at finite temperature describe noninteracting particles. In which sense the concept of interacting particles at finite temperature is meaningful is unclear at present; an interpretation in terms of resonances similar to the Stark effect in atomic physics was presented in the poster session by Landsman [41].

4 Nuclearity and Modular Structure

At finite distances, two competing principles are to be taken into account. The first one emphasizes that locally one has a situation similar to quantum mechanics with a finite number of degrees of freedom. Semiclassically, the density of states in phase space for n degrees of freedom is h^{-n}, h denoting Planck's constant. In order to incorporate this structure into quantum field theory, Haag and Swieca [42] introduced the compactness criterion, i.e.

$$\{P_E A\Omega, A \in \mathcal{A}(\mathcal{O}), ||A|| \leq 1\} , \tag{4.1}$$

where P_E denotes the spectral projection onto energies $\leq E$ and Ω is the vacuum supposed to be a relatively compact subset of Hilbert space (w.r.t. the strong topology).

The Haag-Swieca criterion was sharpened to nuclearity by Buchholz and Wichmann [43]. Namely, the set

$$\{e^{-\beta H} A\Omega, A \in \mathcal{A}(\mathcal{O}), ||A|| \leq 1\} \tag{4.2}$$

shall be nuclear (which roughly means that it is contained in the image of the unit ball under a trace class operator T) for $\beta > 0$, and the nuclear size $Z(\beta, \mathcal{O})$ of this set (roughly the trace norm of T) which plays a role similar to a partition function is supposed to satisfy the bound

$$Z(\beta, \mathcal{O}) \leq const \ \exp\{const \ \beta^{-n}\}, n > 0 \, . \tag{4.3}$$

Nuclearity implies the existence of product states, i.e. states having no correlations between observables in regions separated by a finite distance, equivalently, that inclusions $\mathcal{A}(\mathcal{O}) \subset \mathcal{A}(\mathcal{O}_1)$ are split if the closure of $\mathcal{O}$ is contained in $\mathcal{O}_1$, i.e. there is a von Neumann factor $\mathcal{N}$ of type I (i.e. isomorphic to $\mathcal{B}(\mathcal{H})$) such that $\mathcal{A}(\mathcal{O}) \subset \mathcal{N} \subset \mathcal{A}(\mathcal{O}_1)$ and, moreover, the existence of KMS-states for all $\beta > 0$ [44].

The contrary principle, that quantum field theory, strictly speaking, even locally has an infinite number of degrees of freedom relies on the Reeh-Schlieder theorem [45] which states that every vector which is in the image of $e^{-\beta H}$ for some $\beta > 0$ (in particular the vacuum Ω) is cyclic and separating for each algebra $\mathcal{A}(\mathcal{O})$, $\mathcal{O} \in \mathcal{K}$. Therefore the modular operator Δ and the modular conjugation J of the Tomita-Takesaki theory may be defined for the pair $(\mathcal{A}(\mathcal{O}), \Omega)$,

$$\begin{aligned} (\Omega, A\Delta B\Omega) &= (\Omega, BA\Omega) \\ J\Delta^{\frac{1}{2}} A\Omega &= A^*\Omega \quad , \quad A, B \in \mathcal{A}(\mathcal{O}) \, . \end{aligned} \tag{4.4}$$

A geometrical interpretation of the modular operators was found by Bisognano and Wichmann [46] for the case where $\mathcal{O}$ is replaced by the so called wedge $W = \{x \in \mathbb{R}^D, |x^o| < x^1\}$.Under the additional assumption that the local algebras are generated by Wightman fields they show that Δ^{it} is the Lorentz boost with velocity $\tanh 2\pi t$ in the 1-direction and J is the PCT operator with parity P replaced by the reflection on the plane $x^1 = 0$.

Recently Borchers proved a remarkable result [47](presented in the poster session) which shows that the structure discovered by Bisognano and Wichmann does in no means depend on special features of quantum field theory. Let $\mathcal{M}$ be a von Neumann algebra on a Hilbert space $\mathcal{H}$, let $\Omega \in \mathcal{H}$ be cyclic and separating for $\mathcal{M}$ and denote by Δ and J the operators of the Tomita-Takesaki theory associated to the pair $(\mathcal{M}, \Omega)$. Let in addition $U(x)$ be a 1-parameter group of unitaries acting on $\mathcal{H}$ with a positive generator such that Ω is invariant and $U(x)\mathcal{M}U(x)^{-1} \subset \mathcal{M}$ for $x \geq 0$. Then Borchers proves the commutation relation

$$\begin{aligned} \Delta^{it} U(x) \Delta^{-it} &= U(e^{-2\pi t} x) \\ JU(x)J &= U(-x) \end{aligned} \quad , \tag{4.5}$$

which are precisely the commutation relations of Lorentz boosts with light-like translations.

There are mutual relations between nuclearity and modularity [48],[49]. In particular, it is possible to replace the Hamiltonian H in the nuclearity condition by a function of the modular operator of a slightly larger region. This is based on the following inequality between the modular Hamiltonian $K = -\ln\Delta$ and H. As is easily seen the modular Hamiltonian increases with the algebra, hence $K \leq K_W$ for wedges W containing $\mathcal{O}$. Using the fact that the latter are boost generators Buchholz, D'Antoni and Longo find the relation [48]

$$K \leq \frac{\tan \pi\lambda}{\lambda} dH + \frac{\ln 2}{\lambda} \quad ,0 < \lambda < \frac{1}{4}, \tag{4.6}$$

where d is the timelike extension of $\mathcal{O}$.

5 Haag-Kastler Nets Versus Wightman Fields

There are two main frameworks for quantum field theory; Haag-Kastler nets and Wightman fields. The formulation in terms of Haag-Kastler nets is very well suited for the discussion of structural problems. It is, however, difficult to write down Haag-Kastler nets explicitely.

In this respect, Wightman fields are much better behaved, since they may be directly extracted from models, they have a simple connection to an euclidean formulation and they can be obtained in terms of formal perturbation series. Usually, one believes that both frameworks are physically equivalent. There is however no proof. Actually, there are some structural problems for Haag-Kastler nets for which the relations to Wightman fields are important. Such problems are

(i) the geometrical meaning of modular operators (see however the mentioned result of Borchers)
(ii) a general proof of Haag duality, a problem connected to (i)
(iii) the notion of a scaling limit.

The relations between both frameworks have been studied in several papers. For the construction of Haag-Kastler nets conditions were found by Borchers and Zimmermann [50](Ω analytic vector for smeared field), Driessler and Fröhlich [51](linear H bounds), Driessler, Summers and Wichmann [52], Buchholz [53] and Borchers and Yngvason [54](see below). The construction of Wightman fields associated to a given Haag-Kastler net was first investigated by Hertel and myself [55](fields with polynomial H bounds) and then generalized by Summers [56](generalized H bounds) and Rehberg and Wollenberg [57],[58] (pre-Wightman fields). Recently, Jörß[59] succeeded in deriving fields from a conformally covariant Haag-Kastler net.

Let me describe the result of Borchers and Yngvason in more detail. Let φ be a hermitean scalar Wightman field satisfying a generalized H bound

$$\|\varphi(f)e^{-H^\alpha}\| < \infty \quad \text{for some } \alpha, 0 < \alpha < 1. \tag{5.1}$$

Under this condition the weak commutants

$$\mathcal{A}_w(\mathcal{O}) \tag{5.2}$$
$$= \left\{ A \in \mathcal{B}(\mathcal{H}), (\Phi, A\varphi(f)\Psi) = (\varphi(f)^*\Phi, A\Psi)\,,\; f \in \mathcal{D}(\mathcal{O})\,,\; \Phi, \Psi \in \bigcup_{\alpha>0} e^{-H^\alpha}\mathcal{H} \right\}$$

are von Neumann algebras [52]. Now, following Powers [60], Borchers and Yngvason define a state ω on a *-algebra $\mathcal{A}$ to be centrally positive w.r.t. $A \in \mathcal{A}, A = A^*$ if

$$\omega(\sum_n B_n A^n) \geq 0 \tag{5.3}$$

whenever $\sum_n B_n \lambda^n \geq 0$ for all $\lambda \in \mathbb{R}$.

Theorem [54] *There exists a Haag Kastler net* $(\mathcal{A}(\mathcal{O}))_{\mathcal{O}\in\mathcal{K}}$ *such that for each real test function* $f \in \mathcal{D}(\mathcal{O})$ $\varphi(f)$ *has a selfadjoint extension affiliated to* $\mathcal{A}(\mathcal{O})$ *if and only if for each* f, ω_o *is centrally positive with respect to* $\varphi(f)$ *on the algebra generated by* $\varphi(f)$ *and* $\varphi(g)$, *supp* $g \subset \mathcal{O}'$.

A nice feature of this criterion is that it leads to testable predictions. Look for instance at the polynomial $p(x,y) = x^2y^2(x^2+y^2-1) + \frac{1}{27}$. p assumes positive values for real arguments, hence for each $x \in \mathbb{R}$ it is a sum of absolute squares of polynomials in y. Thus when φ is associated to a Haag Kastler net, ω_o must satisfy the inequality

$$\omega_0(\varphi(f)^2\varphi(g)^2) \leq \omega_0(\varphi(f)^4\varphi(g)^2) + \omega_0(\varphi(f)^2\varphi(g)^4) + \frac{1}{27}\,, \tag{5.4}$$

where f and g are real test functions, *supp* $f \subset \mathcal{O}$, *supp* $g \subset \mathcal{O}'$.

6 Conceptual Problems in Approaches to Quantum Gravity

The impact of quantum gravity on the framework of the general theory of quantized fields may be discussed on three levels. The first one is the semiclassical level where gravity is treated classically and quantum field theory lives on a curved space time. On a first sight this situation does not seem to lead to conceptual problems, since all local aspects of quantum field theory can be easily generalized. There is, however, one important global aspect of quantum field theory, namely the spectrum condition, and it is a major problem to find the appropriate generalization of this condition in a curved background. In general, there is no timelike Killing vector which could play the role of the generator of time translations; but even in stationary spacetimes where by definition a timelike Killing vector field exists the positivity of the corresponding generator is not always the good property required for a physical state. For instance, for a Schwarzschild black hole, it is generally believed that not the ground state with respect to the time translations of the Schwarzschild space time ("Boulware vacuum") is the right state but states which exhibit Hawking radiation. The problem was first observed by Fulling [61] and found a conceptually satisfactory formulation by Haag, Narnhofer and Stein [62] and, independently, by Kay [63]. According to these authors, in order to define the theory, one must in addition to the field equations and the commutation relations specify the set of admissible states such that the local algebras have a unique completion to von Neumann factors (principle of "local definiteness" [62]). It is, however, not known in general how the admissible states can be characterized. Physically most appealing is the idea of "local stability" [62],[64] which says that the Wightman functions should have a short distance scaling limit which then as a theory on Minkowski space should satisfy the spectrum condition. Local stability is a condition which is applicable also in interacting theories, unfortunately it does not fix the local algebras uniquely, in general. A related condition is the so called Hadamard condition [65] which can be formulated only for linear field theories; it states that the singularity of the 2-point function is the singularity of Hadamard's fundamental solution of the wave equation. The Hadamard condition implies local stability, and Kay and Wald [66] showed that for stationary states on the Kruskal space time it implies that the state is a KMS state with the Hawking temperature. It is to be expected that the Hadamard condition fixes the local von Neumann algebra but a proof is not yet available (see however [67]). A third approach to the determination of the local von Neumann algebras is the construction of "adiabatic vacuum states". This idea was carried out by Lüders and Roberts [68] for Robertson-Walker space times; it amounts to an approximation of a nonstationary space time by a stationary one with a nonsingular timelike Killing vector and to a construction of the admissible states in terms of corrections to the ground state of the stationary problem. Lüders and Roberts could show that this procedure fixes the local von Neumann algebras uniquely and that they are factors in accordance with the principle of local definiteness.

The second level in an approach to quantum gravity is the treatment of backreaction. As shown in [69], the scaling behavior of the state near the horizon of a spherically symmetric collapsing star is in 1-1 correspondence to the long time asymptotic radiation. One has tried to treat the change of the metric due the mass loss by radiation in terms of a semiclassical Einstein equation

$$G_{\mu\nu} = 8\pi\kappa\langle T_{\mu\nu,\mathrm{ren}}\rangle \,, \tag{6.1}$$

where $T_{\mu\nu,\mathrm{ren}}$ means the energy momentum tensor after subtraction of some diverging term which depends on the space time metric. The procedure looks somewhat arbitrary and actually the equation has no solution. Recently, a different approach related to ideas of Haag has been proposed by Salehi [70]. Salehi considers the singular part of $T_{\mu\nu}$ as that part which is responsible for the gravitational field and introduces a cutoff at length $\kappa^{\frac{1}{2}}$. His attempt results in an effective nonlinear equation for the matter field φ; whether his approach is free from inconsistencies is not known at the moment.

At the end of my lecture I want to add some remarks on full quantum gravity (level 3 of the above classification). There are several directions in which quantum gravity is approached at present; string theory, gravity in 2 and 3 dimensions (leading to topological field theories), noncommutative geometry and Ashtekar's Hamiltonian formulation of classical gravity. Unfortunately, general quantum field theory has not much to offer at the moment to these problems. The main obstacle in incorporating gravity into the framework of general quantum field theory is the fact that classical space time plays a distinguished role in quantum field theory. This becomes especially clear in the algebraic formulation in terms of Haag-Kastler nets, but is implicit in every formulation of field theory. The main role played by space time is in the interpretation of the theory. One possible route to quantum gravity is therefore the derivation of a space time structure from observables.An interesting idea how this could be done is due to Bannier [71]. Bannier starts from a family of von Neumann algebras $(\mathcal{A}_i)_{i\in I}$ representing the observable algebras of physical subsystems. The crucial ingredient is now the notion of a "good subsystem" $i < j$, $i,j \in I$, which is a partial ordering stronger than the set theoretical inclusion $\mathcal{A}_i \subset \mathcal{A}_j$. A subset $J \subset I$ is called overlapping if any finite subset $J_0 \subset J$ has a lower bound in I. Let M be the set of maximal overlapping subsets of I. Then M gets the structure of a topological Hausdorff space by defining as open sets in M those which contain with every $x \in M$ for some finite subset $J_0 \subset x$ all $y \in M$ with $J_0 \subset y$. If one applies this construction to the local net $(\mathcal{A}(\mathcal{O}))_{\mathcal{O}\in\mathcal{K}}$ and uses the partial ordering

$$\mathcal{O}_1 < \mathcal{O}_2 \iff \text{the inclusion } \mathcal{A}(\mathcal{O}_1) \subset \mathcal{A}(\mathcal{O}_2) \text{ is split} \tag{6.2}$$

one recovers Minkowski space [71]. In order to apply this idea in a more general situation one has to find an intrinsic characterization of the algebras $\mathcal{A}_i$ (candidates: hyperfinite factors of type III_1 [72]) and an ordering relation (candidate: split inclusion). The result will be a topological space which can be used as a device for an interpretation of the theory.

References

1. Jost, R.: The General Theory of Quantized Fields. AMS 1965, Providence, Rhode Island
2. Wick, G.C., Wightman, A.S., Wigner, E.P.: Phys.Rev. **88**, 101 (1952)
3. Haag, R., Kastler, D.: J.Math.Phys. **5**, 848(1964)
4. Borchers, H.-J.: Comm.Math.Phys. **1**, 281 (1965)
5. Doplicher, S., Haag, R., Roberts, J.E.: Comm.Math.Phys. **23**, 199 (1971); Doplicher, S., Haag, R., Roberts, J.E.: Comm.Math.Phys. **35**, 49 (1974)
6. Jones, V.F.: Invent.Math. **72**, 1 (1983)
7. Longo, R.: Comm.Math.Phys. **126**, 217(1989); Longo, R.: Comm.Math.Phys. **130**, 285 (1990)
8. Pimsner, M., Popa, S.: Ann. Sci. École Norm. Sup. **19**, 57 (1986)
9. Fredenhagen, K., Rehren, K.H., Schroer, B.: Comm. Math. Phys. **125**, 201 (1989)
10. Birman, J.: Braids, Links and Mapping Class Groups. Ann. Math.Studies **82**, Princeton University Press 1974
11. Ocneanu, A.: London Math.Soc., Lect. Notes Vol. **136**, Evans, Takesaki (eds.), 119-172 (1989)
12. Wenzl, H.: Invent. Math. **92**, 349 (1988)
13. Freyd, P., Yetter, D., Hoste, J., Lickorish, W., Millet,K., Ocneanu, A.: Bull.AMS **12**, 103 (1985)
14. Longo, R.: Minimal Index and Braided Subfactors (preprint)
15. Rehren, K.H.: Braid Group Statistics and their Superselection Rules. In: The Algebraic Theory of Superselection Sectors. Introduction and Recent Results. D. Kastler (ed.), World Scientific 1990
16. Birman, J., Wenzl, H.: Trans. AMS **313**, 249 (1989)
17. Murakami, J.: Osaka J.Math. **24**, 745 (1987)
18. Wenzl, H.: Comm. Math. Phys. **133**, 383 (1990)
19. Kauffman, L.H.: On Knots. Annals of Math. Study **115**, Princeton University Press (1987)
20. Buchholz, D., Fredenhagen, K.: Comm. Math. Phys. **84**, 1 (1982)
21. Fredenhagen, K., Comm. Math. Phys. **79**, 141 (1981)
22. Fröhlich, J., Gabbiani, F.: Rev.Math.Phys. **2**, 251 (1990)
23. Fredenhagen, K.: Structure of Superselection Sectors in Low Dimensional Quantum Field Theory. In: Differential Geometric Methods in Theoretical Physics. L.L. Chau, W. Nahm (eds.), Plenum Press 1990
24. Fredenhagen, K.: Generalizations of the Theory of Superselection Sectors. See ref.15
25. Fröhlich, J.: Comm. Math. Phys. **47**, 269 (1976)
26. Moore, G., Seiberg, N.: Phys.Lett. **212 B**, 451 (1988)
27. Doplicher, S., Roberts, J.E.: Ann.Math. **130**, 75 (1989)
28. Doplicher, S., Roberts, J.E.: Comm. Math. Phys. **131**, 51 (1990)
29. J.Fröhlich, Kerler, T. (to be published)
30. Hadjiivanov, L.K., Paunov, R.R., Todorov, I.T.: Quantum Group Extended Chiral p-Models. INRNE-TH-90-7 (preprint)
31. G. Mack, V. Schomerus: Quasi Quantum Group Symmetry and Local Braid Relations in the Conformal Ising Model. DESY 91-060 (preprint)

32. Rehren, K.H.: Field Operators for Anyons and Plektons. DESY 91-043
33. Haag, R.,: Phys.Rev. **112**, 669 (1958)
34. Ruelle, D.: Helv.Phys.Acta **35**, 147 (1962)
35. Buchholz, D.: Comm. Math. Phys. **52**, 147 (1977)
36. Fröhlich, J., Marchetti, P.: Nucl.Phys. B **356**, 533 (1991)
37. Rüger, S.: Streutheorie für Teilchen mit Zopfgruppenstatistik. Diplomarbeit FU Berlin 1990
38. Barata, J.C.A., Fredenhagen, K.: Comm. Math. Phys. **138**, 507 (1991)
39. Buchholz, D., Porrmann, M., Stein, U.: Dirac versus Wigner: Towards a Universal Particle Concept in Local Quantum Field Theory. DESY 91-062 (preprint) (to appear in Phys.Lett.B)
40. Narnhofer, H., Requardt, H., Thirring, W.: Comm. Math. Phys. **92**, 247 (1983)
41. Landsman, N.P.: Nucl.Phys. A **525**, 397 (1991)
42. Haag, R., Swieca, J.A.: Comm. Math. Phys. **1**, 308 (1965)
43. Buchholz, D., Wichmann, E.: Comm. Math. Phys. **106**, 106 (1986)
44. Buchholz, D., Junglas,P.: Lett. Math. Phys. **11**, 51 (1986)
45. Reeh, H., Schlieder, S.: Nuovo Cim. X **22**, 1051 (1961)
46. Bisognano, J.J., Wichmann, E.H.: J.Math.Phys **17**, 303 (1975)
47. Borchers, H.J.: The CPT Theorem in Two-Dimensional Theories of Local Observables (preprint 1991)
48. Buchholz, D., D'Antoni, C., Longo, R.: J.Funct.Anal. **88**, 233 (1990); Buchholz, D., D'Antoni, C., Longo, R.: Commun. Math. Phys. **129**, 115 (1990)
49. Buchholz, D., Yngvason, J.: Generalized Nuclearity Conditions and the Split Property in Quantum Field Theory (preprint)
50. Borchers, H.J., Zimmermann, W.: Nuovo Cim. **31**, 1047 (1964)
51. Driessler, W., Fröhlich, J.: Ann.Inst.H. Poincaré **27**, 221 (1977)
52. Driessler, W., Summers, S.J., Wichmann, E.H.: Comm.Math.Phys. **105**, 49 (1986)
53. Buchholz, D.: J.Math.Phys. **31**, 1839 (1990)
54. Borchers, H.J., Yngvason, J.: Comm. Math. Phys. **127**, 607 (1990)
55. Fredenhagen, K., Hertel, J.: Comm. Math. Phys. **80**, 555 (1981)
56. Summers, S.J.: Helv.Phys.Acta **60**, 1004 (1987)
57. Rehberg, H., Wollenberg, M.: Math. Nachr. **125**, 1 (1986)
58. Wollenberg, M.: Rep.Math.Phys. **22**, 409 (1985); Wollenberg, M.: Math.Nachr. **128**, 169 (1986)
59. Jörß, M.: Lokale Netze auf dem eindimensionalen Lichtkegel. Diplomarbeit FU Berlin 1991
60. Powers, R.T.: Comm.Math.Phys. **21**, 85 (1971); Powers, R.T.: Trans.AMS **187**, 261 (1974)
61. Fulling, S.A.: Phys.Rev. D **7**, 2850 (1973)
62. Haag, R., Narnhofer,H., Stein, U.: Comm. Math. Phys. **94**, 219 (1984)
63. Kay, B.S.: contribution to these proceedings and references therein
64. Fredenhagen, K., Haag, R.: Comm. Math. Phys. **108**, 91 (1987)
65. Adler, S., Liebermann, J., Ng, Y.J.: Ann.Phys. **106**, 279 (1978)

66. Kay, B.S., Wald, R.M.: Theorems on the Uniqueness and Thermal Properties of Stationary, Nonsingular, Quasi-Free States on Spacetime with a Bifurcate Killing Horizon (preprint)
67. Bernard, D.: Phys.Rev. D **33**, 3581 (1986)
68. Lüders, C., Roberts, J.E.: Comm. Math. Phys. **134**, 29 (1990)
69. Fredenhagen, K., Haag, R.: Comm. Math. Phys. **127**, 273 (1990)
70. Salehi, H.: Ph.D.Thesis Hamburg 1991
71. U. Bannier, U.: Ph.D.Thesis Hamburg 1987
72. Buchholz, D., D'Antoni, C., Fredenhagen, K.: Comm.Math.Phys. **111**, 123 (1987)

Atomic and Molecular Structure
– A Renormalized Picture –

Jan Philip Solovej

Department of Mathematics, Fine Hall, Princeton University,
Princeton, N.J. 08544–1000, USA

0 Introduction

It is my goal here to review some recent developments in the rigorous mathematical study of atomic and molecular structure.

We shall restrict ourselves entirely to the ground state structure. By structure we shall mean the following features of the ground state: *Total energy* (E), *Binding energy* (BE), *Ionization energy* (I), *Stability*, *Maximal positive and negative ionization charges* ($Q_\pm$), *Nuclear configuration (distance between R_n)*, *Electronic density distribution (radius R_e)*.

We shall first give the exact definitions of these quantities and concepts and then review the main known results about them. Most of these results are obtained by approximating the full quantum theory by simplified models, especially the famous Thomas–Fermi and Hartree–Fock models.

These models are certainly of independent interest and we shall, indeed, put our emphasis on a simplified Hartree–Fock model. In this model we can give a description of the ground state which is much more complete than what has so far been achieved for the full quantum theory [23]. It is this description that contains a renormalized picture of the atom.

Supported in parts by US-NSF grant DMS–9002416

1 Definitions

A molecule is described by giving the number of electrons N, the number of nuclei M and their charges $Z_1, \ldots, Z_M > 0$. The total nuclear charge is $Z = Z_1 + \ldots + Z_M$.

We shall here only consider the special Born–Oppenheimer approximation in which the nuclear coordinates $\mathcal{R}_1, \ldots, \mathcal{R}_M \in \mathbb{R}^3$ are parameters, i.e., we neglect the kinetic energy of the nuclei. Letting $x_1, \ldots, x_N \in \mathbb{R}^3$ denote the electronic coordinates the **Hamiltonian** is (we are using units with $\hbar = 2m_e = e = 1$)

$$H(N, \underline{Z}, \underline{\mathcal{R}}) = \sum_{i=1}^{N} [-\Delta_i - V(x_i)] + \sum_{i<j} |x_i - x_j|^{-1} + U(\underline{\mathcal{R}}),$$

where $\underline{\mathcal{R}} = (\mathcal{R}_1, \ldots, \mathcal{R}_M)$ and $\underline{Z} = (Z_1, \ldots, Z_M)$,

$$V(x) = \sum_{k=1}^{M} Z_k |x_i - \mathcal{R}_k|^{-1}$$

and

$$U(\underline{\mathcal{R}}) = \sum_{k<\ell} Z_k Z_\ell |\mathcal{R}_k - \mathcal{R}_\ell|^{-1}.$$

To break the translational invariance we assume $\mathcal{R}_1 + \ldots + \mathcal{R}_M = 0$. When $M = 1$ and then $U = 0$ we are in the case of an atom.

We shall consider the Hamiltonian $H(N, \underline{Z}, \underline{\mathcal{R}})$ as acting in one of the following two Hilbert spaces in which it is a self-adjoint operator bounded from below. The big space $\mathcal{H}_0 = L^2(\mathbb{R}^{3N})$ or the physical space of antisymmetric functions $\mathcal{H}_p = \bigwedge^N L^2(\mathbb{R}^3; \mathbb{C}^2)$ ($\mathbb{C}^2$ is the spin space, on which H acts trivially).

Since the ground state in $\mathcal{H}_0$ will be symmetric, this space corresponds to considering bosonic "electrons". The antisymmetric space $\mathcal{H}_p$ of course corresponds to fermionic electrons, i.e., the Pauli principle is satisfied. When we do not want to specify the space we simply write $\mathcal{H}$.

We define the **energy** as a function of $\underline{\mathcal{R}}$ to be

$$E(N, \underline{Z}, \underline{\mathcal{R}}) = \inf \operatorname{spec}_{\mathcal{H}} H(N, \underline{Z}, \underline{\mathcal{R}}),$$

and the **Born–Oppenheimer energy** of the molecule

$$E(N, \underline{Z}) = \inf_{\underline{\mathcal{R}}, \sum \mathcal{R}=0} E(N, \underline{Z}, \underline{\mathcal{R}}).$$

The lowest possible energy of a break up of the molecule into smaller independent molecules is the **threshold energy**

$$E_*(N,\underline{Z}) = \liminf_{\substack{\underline{\mathcal{R}}\to\infty \\ \sum \mathcal{R}=0}} E(N,\underline{Z},\underline{\mathcal{R}})\,.$$

The **binding energy** is then

$$BE(N,\underline{Z}) = E_*(N,\underline{Z}) - E(N,\underline{Z}) \geq 0\,,$$

and the **ionization energy** is

$$I(N,\underline{Z}) = E(N-1,\underline{Z}) - E(N,\underline{Z}) \geq 0\,.$$

The binding energy is the energy it costs to separate the molecule into smaller molecules, while the ionization energy is the cost of removing one electron.

We shall say that the molecule is **stable** if both the binding and ionization energies are strictly positive, i.e,

$$\text{(a)}\quad BE(N,\underline{Z}) > 0 \quad\text{and}\quad \text{(b)}\quad I(N,\underline{Z}) > 0\,.$$

The condition (a) implies that all nuclei are bounded and that there exists a finite **nuclear configuration** $\mathcal{R}_0$ such that

$$E(N,\underline{Z}) = E(N,\underline{Z},\underline{\mathcal{R}_0}).$$

For such a configuration we shall be particularly interested in the minimal distance $R_n = \min_{k\neq\ell}|\mathcal{R}_k - \mathcal{R}_\ell|$. The condition (a) is satisfied as long as N is not too small. We denote by $N_-(\underline{Z})$ the minimal electron number such that (a) is satisfied.

The condition (b), which is the only relevant condition in the case of an atom ($M = 1$), implies by the HVZ-Theorem (see [3]) that $E(N,\underline{Z})$ is a discrete eigenvalue for $H(N,\underline{Z},\underline{\mathcal{R}}_0)$. If we denote the corresponding normalized eigenfunction by $\psi \in \mathcal{H}$, we define the **electronic density** as

$$\rho_Q(x) = N\int \|\psi(x,x_2,\ldots,x_N)\|^2 dx_2\cdot\ldots\cdot dx_N\,.$$

Here $\|\cdot\|$ denotes the norm in $\mathbb{R}$ (for $\mathcal{H}_0$) or in $\bigotimes_1^N \mathbb{C}^2$ (for $\mathcal{H}_p$). The condition (b) is satisfied as long as N is not too big. We denote by $N_+(\underline{Z})$ the maximal electron number such that (b) is satisfied. The excess charges are defined as $Q_\pm = |Z - N_\pm|$.

For the electronic configuration there are two quantities of special interest. The **screened nuclear charges** at radius $R < R_n$

$$\nu_k(R) = Z_k - \int_{|x-\mathcal{R}_k|<R} \rho_Q(x)dx$$

and the **outer electronic radius** R_e defined by

$$1 = \int_{|x|<R_e} \rho_Q(x)dx\,.$$

2 Results

Obtaining good estimates for the quantities defined in the previous section for arbitrary values of N and $\underline{Z}$ is in general very difficult. However, in recent years there has been a great interest in studying the behavior as the total nuclear charge Z tends to infinity, or more precisely, as $\underline{Z}$ tends to infinity along a direction. The main reason for this is that this is essentially a semiclassical limit. It is then possible to formulate precise asymptotic statements that are tractable by a semiclassical analysis.

One can of course object that the limit $Z \to \infty$ is unphysical, since nuclei become unstable at about $Z \approx 100$, or even more important that there are relativistic corrections and instabilities for large Z values.

The hope is that the asymptotic analysis will give good estimates for Z not too large, say, Z of the order of 20.

Theorem 2.1 *(Energy, Fermions $\mathcal{H} = \mathcal{H}_p$). If $R_n \gg O(Z^{-2/3})$ then*

$$E(N = Z, \underline{Z}, \underline{R}) = -C_{TF}Z^{7/3} + \frac{1}{8}Z^2 - C_{DS}Z^{5/3} + o(Z^m).$$

In the atomic case ($M = 1$) $m = 5/3$, in the molecular case ($M > 1$), one can so far only prove $m = 2$.

Remark. The first term C_{TF} is the Thomas–Fermi term derived from Thomas–Fermi theory (see below). It was originally studied independently by Thomas [24] and Fermi [5] and was first rigorously proved by Lieb and Simon [12]. The second term $C_S = \frac{1}{8}$ was derived by Scott [17] and proved in [8], [18] in the atomic case and recently for $M > 1$ in the preprint [9]. The last term C_{DS} can really be considered as a sum of two conceptually different contributions which happen to be of the same order. One was studied by Dirac [4] the other by Schwinger [16]. The mathematical proof of this term has been announced in [7]. For $M > 1$ it has not yet been proved. The actual results proved are somewhat stronger than stated in the theorem.

In the case of bosons the knowledge is not as great. The following is essentially proved in [2].

Theorem 2.2 *(Energy, Bosons $\mathcal{H} = \mathcal{H}_0$). If $R_n \gg O(Z^{-1})$*

$$E(N = Z, \underline{Z}, \underline{R}) = -C_H Z^3 + o(Z^3).$$

Unfortunately, Theorem 2.1, although of great mathematical interest, does not directly give much information about the physically more interesting quantities BE and I. But for BE this is basically the only way to get an upper bound.

Theorem 2.3 *(BE, Fermions $\mathcal{H} = \mathcal{H}_p$)*

$$0 < BE(N = Z, \underline{Z}) < o(Z^2)\,.$$

Remark. The lower bound says that a neutral molecule will always be bounded if we neglect the kinetic energy of the nuclei.

This important effect which is due to the so-called van der Waals force was proved in [13]. The upper bound is a corollary of Theorem 2.1.

Theorem 2.4 *(I, Fermions $\mathcal{H} = \mathcal{H}_p$). For $N \geq Z$*

$$I(N, \underline{Z}) \leq O(Z^{20/21}) \ll O(Z^{4/3})\,.$$

Remark. For $M = 1$ this was proved in [19] and for $M > 1$ in [21] and [15]. We compare with $Z^{4/3}$ since this is the average energy per electron.

Theorem 2.5 *($N_\pm$, Fermions $\mathcal{H} = \mathcal{H}_p$, Non-asymptotic bounds).*

$$\begin{aligned} N_+ &\leq 2Z + M\,, \\ N_- &\geq \frac{2}{3}\frac{Z_1 Z_2}{Z_1 + Z_2} \quad \text{for} \quad M = 2\,. \end{aligned}$$

Remark. The upper bound is in [10], the lower bound in [21]. In [14] weaker bounds are given without use of the Born–Oppenheimer assumption.

Theorem 2.6 *($N_\pm$, Fermions $\mathcal{H} = \mathcal{H}_p$, Asymptotic neutrality).*

$$\lim_{Z\to\infty} \frac{N_+(\underline{Z})}{Z} = \lim_{Z\to\infty} \frac{N_-(\underline{Z})}{Z} = 1\,.$$

Remark. For $M = 1$ this is only a result about N_+ and has been proved in [12], [6] and [19]. For $M > 1$ where we have both N_- and N_+ this has been proved in [21] and [15].

Theorem 2.7 *($N_\pm$, Bosons $\mathcal{H} = \mathcal{H}_0$). For $M = 1$ we have*

$$\lim_{Z\to\infty} \frac{N_+(Z)}{Z} = 1 + \lambda,$$

with $\lambda \approx 0.21$.

Remark. An asymptotic lower bound was proved in [2] while a corresponding asymptotic upper bound is in [22]. The numerical value of λ was computed in [1] from the non-linear Hartree theory (see below).

Theorem 2.8 *(R_n, R_e, Fermions $\mathcal{H} = \mathcal{H}_p$).*

$$R_n,\, R_e \geq O(Z^{-2/7}) \gg O(Z^{-1/3}).$$

Remark. These estimates have been proved in [19], [21] and [15]. We compare with $Z^{-1/3}$, since this is the radius of the bulk of electrons.

For fermions I believe that the correct asymptotic is that all the quantities BE, I, $Q^{\pm}$, R_e, R_n are $O(1)$ as $Z \to \infty$. This belief relies on some experimental evidence and on the study of models that we shall now discuss.

We shall in the following restrict our attention to these models that are simpler than the full quantum theory. Within the scope of this review we can only sketch some of the ideas used in giving a very complete description of the ground state for our models. For simplicity and because we can prove stronger results in this case we shall mainly consider the atomic case $M = 1$. The results stated for the quantum theory above are all proved by approximating with these models and we shall indicate this along the way.

3 Hartree–Fock Models

In the standard Hartree–Fock model one neglects correlations between electrons and restrict to wave functions $\psi \in \mathcal{H}_p$ of the form

$$\psi = N^{-1/2}\varphi_1 \wedge \ldots \wedge \varphi_N$$

where φ_i, $i = 1, \ldots, N$ is an orthonormal family in $L^2(\mathbb{R}^3; \mathbb{C}^2)$. Let $\tilde{\gamma} = \sum_i \varphi_i \otimes \overline{\varphi}_i$ be the integral kernel for the projection in $L^2(\mathbb{R}^3; \mathbb{C}^2)$ onto the subspace spanned by this orthonormal set. If we contract this over $\mathbb{C}^2$, (i.e., take the trace over $\mathbb{C}^2$) we get an integral kernel for an operator in $L^2(\mathbb{R}^3)$ which we denote by γ.

$$\begin{aligned}\langle \psi \mid H(N, \underline{Z}, \underline{R}) \mid \psi \rangle =& \mathrm{Tr}[\gamma h] + \frac{1}{2}\int\int \rho_\gamma(x)|x-y|^{-1}\rho_\gamma(y)dxdy \\ &+ U(\underline{R}) - \frac{1}{2}\int\int \frac{|\gamma(x,y)|^2}{|x-y|}dxdy\,, \end{aligned} \tag{3.1}$$

where h is the one-particle operator

$$h = -\Delta - V(x)$$

and

$$\rho_\gamma(x) = \gamma(x,x) \equiv \sum_i \|\varphi_i(x)\|^2_{\mathbb{C}^2}$$

is the electronic density.

We can now generalize this model to consider (3.1) for all positive trace class operators $\gamma : L^2(\mathbb{R}^3) \to L^2(\mathbb{R}^3)$. Without any restrictions, except that the terms in (3.1) are finite, this will actually give a model for bosonic "electrons". We call these γ, *bosonic admissible.* To treat fermions in this framework we restrict to operators satisfying

$$0 \le \gamma \le 2\,\mathbf{Id}\,, \qquad (2 = \dim \mathbb{C}^2)\,. \tag{3.2}$$

We call operators that are trace class and satisfy (3.2) *fermionic admissible.* In [11] it was shown that minimizing (3.1) over all fermionic admissible operators rather than just projections does not lower the infimum.

Unfortunately (3.1) is too complicated for the detailed analysis we shall do of the atomic case $M = 1$. The problem is that the minimizing γ will not necessarily be spherically symmetric. One possibility which might, indeed, be preferable because it approximates the quantum energy to $o(Z^{5/3})$ (see [7]) is simply to restrict to spherically symmetric γ. Another, which we shall choose here simply because it was used in [23] is to remove the exchange term – the last term in (3.1). Then it is easy to prove that the minimizer is spherically symmetric, since the obtained functional is convex in γ.

Thus we define the reduced Hartree–Fock functional

$$\begin{aligned}\mathcal{E}^{RHF}(\gamma) = \mathrm{Tr}[h\gamma] + \frac{1}{2}\int\int \rho_\gamma(x)|x-y|^{-1}\rho_\gamma(y)dxdy\,, \\ + U(\underline{\mathcal{R}})\,,\end{aligned} \tag{3.3}$$

defined on either bosonic or fermionic admissible density matrices γ. The corresponding energy is

$$E^{RHF}(N, \underline{Z}, \underline{\mathcal{R}}) = \inf\{\mathcal{E}^{RHF}(\gamma) \mid \gamma \text{ admissible, } \mathrm{Tr}\gamma = N\}\,. \tag{3.4}$$

We can again as in Section 1, define BE^{RHF}, I^{RHF}, $N_\pm^{RHF}$, ν^{RHF}. We shall say that the molecule is stable if there exist minimizing configurations $\underline{\mathcal{R}}_0$ and γ. One can prove

Theorem 3.1 *For fermions*

$$|E(N, \underline{Z}, \underline{\mathcal{R}}) - E^{RHF}(N, \underline{Z}, \underline{\mathcal{R}})| = O(Z^{5/3})\,.$$

For bosons

$$|E(N, \underline{Z}, \underline{\mathcal{R}}) - E^{RHF}(N, \underline{Z}, \underline{\mathcal{R}})| = O(Z^{7/3})\,.$$

Main Theorem 3.2 [23]. *For neutral atoms $M = 1$, $N = Z$, and fermionic electrons we get for the screened nuclear charge*

$$\nu^{RHF}(R) = 324\pi^2 R^{-3} + o(R^{-3}) \text{ as } R \to 0 \tag{3.5}$$

uniformly in Z, for $O(Z^{-1/3}) \le R$.

The important fact here is that (3.5) holds out to length scales R independent of Z, i.e., $O(1)$. This result implies the $O(1)$ estimates conjectured at the end of the previous section, but only for the RHF model.

To explain the main theorem we must introduce Thomas–Fermi theory.

4 Thomas–Fermi Theory

The only term in $\mathcal{E}^{RHF}$ which does not depend only on ρ_γ is $\mathrm{Tr}[-\Delta\gamma]$. In Thomas–Fermi theory one replaces this term by $\frac{3}{5}(3\pi^2)^{2/3}\int \rho^{5/3}$. To understand that this is a semiclassical approximation recall that semiclassically if $W \geq 0$

$$\begin{aligned}&\inf\{\mathrm{Tr}[\gamma(-\Delta - W)] \mid \gamma \text{ fermionic admissible}\}\\ &= 2\cdot(2\pi)^{-3}\int_{p^2-W(x)<0}(p^2 - W(x))dpdx = -\frac{2}{15}\int W^{5/2}.\end{aligned}$$

However $\rho \mapsto \frac{3}{5}(3\pi^2)^{2/3}\rho^{5/3}$ is exactly the Legendre transform of $W \mapsto \frac{2}{15}W^{5/2}$. The Thomas–Fermi functional is thus

$$\begin{aligned}\mathcal{E}^{TF}(\rho) = \tfrac{3}{5}(3\pi^2)^{2/3}\int(\rho^{5/3} - \rho V) + \tfrac{1}{2}\int\int\frac{\rho(x)\rho(y)}{|x-y|}dxdy\\ + U(\underline{R}).\end{aligned} \tag{4.1}$$

In the bosonic case one gets the Hartree functional which is obtained by replacing $\mathrm{Tr}[-\Delta\gamma]$ by $\int(\nabla\sqrt{\rho})^2$.

We define as before

$$E^{TF}(N,\underline{Z},\underline{R}) = \inf\{\varepsilon^{TF}(\rho) \mid \int\rho = N\}, \tag{4.2}$$

and again, BE^{TF}, I^{TF}, ν^{TF},.... This has been studied in [12]. The TF theory has an exact scaling in Z

$$\text{Energy} \sim Z^{7/3}, \quad \text{Length} \sim Z^{-1/3}, \quad \text{Density} \sim Z^2.$$

The semiclassical analysis gives

Theorem 4.1 *If $R_n \geq O(Z^{-1/3})$ then*

$$|E^{RHF}(N,\underline{Z},\underline{R}) - E^{TF}(N,\underline{Z},\underline{R})| = O(Z^{7/3(1-\varepsilon)}),$$

where the optimal ε is $1/7$ (see Theorem 2.1).

$$\nu^{RHF}(R) = \nu^{TF}(R) + o(R^{-3}) \text{ for } O(Z^{-1/3}) \leq R \leq O(Z^{-1/3(1-\varepsilon)}).$$

Theorem 4.2 *There exist minimizers if and only if* $N \leq Z$, *i.e.,* $N_+^{TF} = Z$. *Furthermore* $BE^{TF} = 0$ *(Teller's No-binding Theorem). In fact (see* [21],[15]*), if* $R_n > O(Z^{-1/3})$ *then*

$$E^{TF}(N, \underline{Z}, \underline{R}) - E_*^{TF}(N, \underline{Z}) > CR_n^{-7}.$$

In case of the Hartree theory $N_+^H = (1+\lambda)Z$ *(see* [2]*).*

Most of the results in Section II are proved by approximating quantum theory by first RHF (Theorem 3.1) and then approximating RHF by TF (Theorem 4.1) and finally using Theorem 4.2. However for I, $N_\pm$ and ν this is not enough, since we need to control $\nu(R)$ for $R \gg O(Z^{-1/3(1-\varepsilon)})$. For ν^{TF} we have

Theorem 4.3 *(Sommerfeld formula). In the neutral atomic case* ($M = 1$) *we have as* $R \to \infty$

$$\nu^{TF}(R) = 324\pi^2 R^{-3} + o(R^{-3}).$$

Notice that in contrast to Theorem 3.2 the asymptotics here is at infinity.

Remark. The Sommerfeld formula [20] is proved by studying the asymptotic behavior of the solution to the non-linear Thomas–Fermi equation, i.e., the ordinary differential equation essentially obtained as the Euler–Lagrange equation for the minimization problem in (4.2).

4 Renormalized Picture

We now sketch the idea of the proof of the main theorem, in the neutral atomic case, and why it implies estimates on I and N_+. The idea is to bootstrap the Sommerfeld formula for $\nu^{RHF}(R)$ from $O(Z^{-1/3(1-\varepsilon)})$ to $O(1)$:

1) Assume $\nu^{RHF}(R) = 324\pi^2 R^{-3} + o(R^{-3})$ holds for $R < R_1$ for some length scale $R_1 < 1$.

2) Let γ be a minimizer and define γ_- to be the restriction of γ to a ball of radius R_1 (in both variables) and γ_+ to be the restriction outside (in both variables). Then it is not too difficult, using spherical symmetry, to show that

$$\mathcal{E}^{RHF}(\gamma) = \mathcal{E}_-^{RHF}(\gamma_-) + \mathcal{E}_+^{RHF}(\gamma_+) + O(R_1^{-5}),$$

where $\mathcal{E}_-^{RHF}$ denotes the RHF-functional on the ball of radius R_1 and $\mathcal{E}_+^{RHF}$ denotes the functional on the outside but with the nuclear charge Z renormalized to the screened charge $\nu(R_1)$.

We point out that in case of the full quantum theory this separation in inside and outside problems is more difficult, since one has to control correlations. It can be done (see [19]), but the error terms are too big for the refined analysis we are doing here.

3) We now only estimate the outside renormalized part of the energy by introducing an outside TF-theory. We control the outside energy which is of order R_1^{-7} to order $O(R_1^{-7(1-\varepsilon)})$.

4) Show that outside TF satisfies the original Sommerfeld formula and use this to show that

$$\nu^{RHF}(R_1) - \int_{R_1\leq|x|\leq R} \rho_\gamma(x)dx = 324\pi^2 R^{-3} + o(R^{-3})$$

out to length scale $O(R_1^{(1-\varepsilon)}) > O(R_1)$.

But for $R \geq R_1$ we have

$$\nu^{RHF}(R) = \nu^{RHF}(R_1) - \int_{R_1\leq|x|\leq R} \rho_\gamma(x)dx$$

so we have been able to bootstrap the estimate and prove the main theorem.

Having concluded that $\nu^{RHF}(R)$ is $O(1)$ as a function of Z when R is $O(1)$ we get estimates on N_+^{RHF} and I^{RHF} by controlling the final outside region $O(1) \leq R \leq \infty$. But everything is $O(1)$ here so it is not difficult to get $O(1)$ estimates.

As an example let us show that $Q_+ < O(1)$: If γ is the minimizer corresponding to the largest possible N then

$$0 > -Q_+ = Z_{tot} - \int \rho_\gamma = \nu(R) - \int_{|x|>R} \rho_\gamma,$$

where $R = O(1)$. But we know now that $\nu(R)$ is $O(1)$, i.e., bounded above and below independent of Z. We must bound the charge outside R. Using the methods leading to the non-asymptotic bounds on N_+ given in Theorem 2.5 but now on the outside problem, which has nuclear charge $\nu(R)$, we get

$$\int_{|x|>R} \rho_\gamma < C_1\nu(R) + C_2.$$

Which implies the bound on $Q_\pm$.

The main idea in the above proof is that one can localize and renormalize such that the energy coming from length scale R is of order R^{-7} and the charge there is of order R^{-3}, i.e., the density is of order R^{-6}.

A similar multiscale analysis, where the energy at different lengths scales is controlled to different orders depending on the scale is an essential ingredient in [9]. However, there one moves closer to the nucleus while we move away.

Acknowledgment I would like to thank Matematisk Institut, University of Copenhagen for their hospitality while this review was prepared.

References

1. Baumgartner, B.: On the Thomas–Fermi–von Weizsäcker and Hartree energies as functions of the degree of ionization. J.Phys. A **17**, 1592-1602 (1984)
2. Benguria, R., Lieb, E.H.: Proof of stability of highly negative ions in the absence of the Pauli principle. Phys.Rev.Lett. **50**, 1771-1774 (1983)
3. Cycon, H.L., Froese, R.G., Kirsch, W., Simon, B.: Schrödinger Operators. Springer-Verlag, Berlin, Heidelberg (1987)
4. Dirac, P.A.M.: Note on exchange phenomena in the Thomas–Fermi atom. Proc.Cambridge Phil.Soc. **26**, 376-385 (1930)
5. Fermi, E.: Un metodo statistico per la determinazione di alcune priorietá dell'atomo. Rend.Accad.Naz.Lincei. **6**, 602-607 (1927)
6. Fefferman, C.L., Seco, L.A.: Asymptotic neutrality of large ions. Comm. Math.Phys. **128**, 109-130 (1990)
7. Fefferman, C.L., Seco, L.A.: On the energy of a large atom. Bull.Amer.Math. Soc. **23**, 525-529 (1990)
8. Hughes, W.: An atomic energy lower bound that agrees with Scott's Correction. Adv.Math. **79**, 213-270 (1990)
9. Ivrii, V., Sigal, I.M.: Asymptotics of the ground state of large Coulomb systems. Preprint (1991)
10. Lieb, E.H.: Bound on the maximum negative ionization of atoms and molecules. Phys.Rev. A **29**, 3018-3028 (1984)
11. Lieb, E.H.: A variational principle for many-fermion systems. Phys.Rev.Lett. **46**, 457-459 (1981)
12. Lieb, E.H., Simon, B.: The Thomas-Fermi theory of atoms, molecules and solids. Adv.Math. **23**, 22-116 (1977)
13. Lieb, E.H., Thirring, W.E.: Universal nature of the van der Waals forces for Coulomb systems. Phys.Rev. A **34**, 40-46 (1986)
14. Ruskai, M.B.: Limits on stability of positive molecular ions. Lett.Math.Phys. **18**, 121-132 (1989);
    Ruskai, M.B.: Limit on the excess negative charge of a dynamic diatomic molecule. Ann.Inst.H.Poincaré A: Phys.Theor. **52**, 397-414 (1990)
15. Ruskai, M.B., Solovej, J.P.: Asymptotic neutrality of polyatomic molecules. In preparation.
16. Schwinger, J.: Thomas–Fermi model: The second correction. Phys.Rev. A **24**, 2353-2361 (1981)
17. Scott, J.M.C.: The binding energy of the Thomas–Fermi atom. Philos.Mag. **43**, 859-867 (1952)
18. Siedentop, H., Weikard, R.: On the leading energy correction for the statistical model of the atom: Interacting case. Commun.Math.Phys. **112**, 471-490 (1987);
    Siedentop, H., Weikard, R.: On the leading correction of the Thomas–Fermi model: Lower bound. Invent.Math. **97**, 159-193 (1989)

19. Seco, L.A., Sigal, I.M., Solovej, J.P.: Bound on the ionization energy of large atoms. Comm.Math.Phys.**131**, 307-315 (1990)
20. Sommerfeld, A.: Asymptotische Integration der Differentialgleichung des ThomasFermischen Atoms. Z.Phys.**78**, 283-308 (1932)
21. Solovej, J.P.: Asymptotic neutrality of diatomic molecules. Comm.Math. Phys.**130**, 185-204 (1990)
22. Solovej, J.P.: Asymptotics for bosonic atoms. Lett.Math.Phys. **20**(2), 165-172 (1990)
23. Solovej, J.P.: Proof of the ionization conjecture in a reduced Hartree–Fock model. Invent.Math. **104**, 291-311 (1991)
24. Thomas, L.H.: The calculation of atomic fields. Proc.Camb.Philos.Soc. **23**, 542-548 (1927)

Gauge Invariance in Non–Relativistic Many–Body Theory*

J. Fröhlich, U.M. Studer

Institute of Theoretical Physics
ETH–Hönggerberg
CH–8093 Zürich

Abstract: We review some recent results on the physics of two-dimensional, incompressible electron- and spin liquids. These results follow from Ward identities reflecting the $U(1)_{em} \times SU(2)_{spin}$–gauge invariance of non-relativistic quantum mechanics. They describe a variety of generalized quantized Hall effects.

Electromagnetic U(1)-gauge–invariance of quantum mechanics was discovered by Hermann Weyl, more than sixty years ago. It has played a basic role in understanding properties of quantum mechanical systems coupled to external electromagnetic fields and/or the quantized radiation field. A key effect reflecting the $U(1)_{em}$-gauge principle realized in quantum dynamics is the Aharonov-Bohm effect [1] which describes quantum interference phenomena caused by non-integrable $U(1)_{em}$-phases of wave functions in the presence of magnetic solenoids.

The non-relativistic quantum theory of particles with spin, but without an anomalous magnetic moment, provides an example of supersymmetric quantum mechanics, in the sense of Witten. This supersymmetry gives rise to degeneracies between eigenvalues of the Hamilton operator corresponding to states with distinct spin orientations. It was discovered, originally, and exploited by Novikov and others. However, in the presence of electromagnetic or tidal fields, it is broken if the particles have anomalous magnetic moments, or if spin-orbit couplings are included.

A deeper property of the non-relativistic quantum theory of spinning particles with arbitrary magnetic moments is its $SU(2)_{spin}$-gauge-invariance. In non-relativistic quantum mechanics, spin plays a role analogous to that of internal quantum numbers. Rotations in spin space can be interpreted as global internal symmetry operations. The corresponding conserved Noether current is the spin current. Just like one can gauge internal symmetries

---

* Notes to a lecture presented by J.F., IAMP congress, Leipzig 1991

one can gauge the $SU(2)_{spin}$-symmetry. What is meant by "spin up" or "spin down" then depends on the space-time point at which a spinning particle is located. Parallel displacement of quantum mechanical spins is described by an SU(2)-connection. It turns out that, in a fixed coordinate system and in an SU(2)-gauge in which spin parallelism is globally defined, the $SU(2)_{spin}$-connection can be expressed in terms of the external electric and magnetic fields acting on the system. A key effect reflecting the $SU(2)_{spin}$-gauge principle realized in non-relativistic quantum theory is the Aharonov-Casher effect [1] describing quantum interference phenomena in systems of neutral particles with non-vanishing magnetic moments in the presence of charged wires.

An effect – at least as spectacular as the Aharonov-Bohm- and Aharonov-Casher effect – which is intimately related to the gauge-invariance of non-relativistic quantum theory is the quantized Hall effect encountered in two-dimensional electronic systems, e.g. heterojunctures. The quantized Hall effect for the electric current is tied to the $U(1)_{em}$-gauge-invariance, the quantized Hall effect for the spin current is tied to the $SU(2)_{spin}$-gauge-invariance of the non-relativistic quantum mechanics of electrons. It is an experimental fact that when the Hall conductivities are on plateaux the electron fluid in a heterojuncture is dissipationless. The longitudinal resistance, R_L, vanishes. This means that there are no gapless, extended modes in such systems, and, as a consequence, connected Green functions of electric- and spin current densities at zero temperature have "good" cluster decomposition properties. These electron fluids are therefore "incompressible".

In these notes, we summarise some results of a recent analysis of two-dimensional, incompressible electron fluids and of the Hall effects encountered when such fluids are subject to external electromagnetic or tidal fields. Details of our analysis appear in [2].

A basic recent observation is that the large-scale, small-frequency physics of incompressible electron fluids exhibits universal features: One can identify interesting physical quantities, such as the Hall conductivity or the quantum numbers of excitations above the groundstate, which only depend on the large-scale, low-frequency properties of the system and which can therefore be predicted precisely without detailed knowledge of the microscopic dynamics. A related notion of universality is familiar from the theory of critical phenomena accompanying continuous phase transitions. The idea that incompressible quantum fluids exhibit universal large-scale, low-frequency behaviour plays an important role in the analysis of the quantum Hall effect reported in [3]; see also [2].

A heterojuncture is an essentially two-dimensional gas of electrons. If a strong, transverse magnetic field is turned on then, for sufficiently small electron density, the spins of the electrons in states of low energy are aligned in the direction of the magnetic field and can therefore be ignored. In this approximation, the electron can be described as a non-relativistic scalar fermion. Using the fact that, in 2+1 dimensions, the quantum mechanical electric current can be derived from a vector potential [4,3], one can show that the large-scale, low-frequency physics of such a system is described by an abelian Chern-Simons gauge theory, the gauge fields being the vector potentials of the electric current [3]. Insisting on the property that, among the physical excitations of the Chern-Simons theory, there be excitations with the quantum numbers of the electron or hole (charge $\mp e$, Fermi statistics) one finds that the set of possible values of σ_H is discrete, that the odd-denominator rule holds, and that the system exhibits, in general, fractionally charged excitations with fractional statistics.

It has become clear that, for certain values of σ_H, spin effects (possibly for $\sigma_H = \frac{e^2}{h}\frac{2}{2\ell+1}$, $\ell = 1, 2, \cdots$) or effects of approximate internal symmetries – "isospin effects" – (e.g., for $\sigma_H = \frac{e^2}{h}\frac{5}{2}$) may play an important role [5],[6],[7],[3],[8]. The analysis in [2] is motivated, in part, by a desire to understand the significance of spin- and internal degrees of freedom in the physics of 2d electron fluids. We determine the effective theory describing the large-scale, low-frequency properties of 2d electron fluids, taking into account spin- and internal degrees of freedom. From the effective theory we then derive the basic equations of linear response theory and current sum rules which describe a variety of well known and less well known effects, including generalized Hall effects. (Some of our current sum rules imply the Kubo formula for σ_H.)

Our derivation of the effective theory in the adiabatic and scaling limit relies on two basic observations:

(a) The general observation that systems of non-relativistic electrons have a local $U(1)_{em} \times SU(2)_{spin}$–symmetry, i.e. a $U(1) \times SU(2)$–gauge invariance, as described above.

(b) The observation that if a system of non-relativistic electrons is incompressible then its effective action, as a functional of external electromagnetic fields, is local, and its general form is computable in the adiabatic and scaling limit. This is a manifestation of universality.

In computing the general form of the effective action, $U(1) \times SU(2)$ Ward identities turn out to play a crucial role, (but the specific form of the microscopic dynamics is unimportant).

In the following, we intend to describe some physical consequences of observations **(a)** and **(b)**, in particular a Hall effect for spin currents and quantization of magnetic susceptibility. One main idea is to treat the dynamics of systems of non-relativistic matter in a geometrical way. The $U(1) \times SU(2)$-gauge invariance of such systems is related to a natural notion of parallel transport, or, equivalently, of covariant differentiation of non-relativistic spinors. Consider, for example, two-component Pauli spinors ψ. The covariant derivatives acting on Pauli spinors are given by

$$D_\mu = \partial_\mu + i a_\mu + i \sum_{A=1}^{3} w_{\mu A} \sigma_A , \tag{1}$$

for $\mu = 0$ (time), $1, 2, (3)$ (space), where $\vec{\sigma} = (\sigma_1, \sigma_2, \sigma_3)$ are the usual Pauli matrices. A geometrically natural form of an action functional for a system of non-relativistic electrons confined to a region Ω of two- (or three-) dimensional space, in the formalism of second quantization, is given by the following expression:

$$S_\Omega(\psi^*, \psi; a, w) = \int_{\mathbb{R}\times\Omega} dt\, d\underline{x} \left[i\hbar c\, \psi^* D_0 \psi + \frac{\hbar^2}{2m} (D_\ell \psi)^* (D^\ell \psi) \right] - \int_{\mathbb{R}} dt\, H_I(t) , \tag{2}$$

where ψ^* (the creation operator) is the adjoint of the Pauli spinor ψ (the annihilation operator), and the interaction term $H_I(t)$ is a $\mathrm{U}(1) \times \mathrm{SU}(2)$-gauge invariant functional of ψ^* and ψ. The $\mathrm{U}(1) \times \mathrm{SU}(2)$-gauge transformations of the gauge fields a and w and of the Pauli spinor ψ are as follows.

$$a_\mu \mapsto a_\mu + \partial_\mu \chi , \quad \psi \mapsto e^{-i\chi} \psi \quad (\mathrm{U}(1)_{\mathrm{em}}) , \tag{3}$$

where χ is an arbitrary, real-valued function on space-time $\mathbb{R} \times \Omega$, and

$$w_\mu \equiv i \sum_{A=1}^{3} w_{\mu A} \sigma_A \mapsto g w_\mu g^{-1} + g\, \partial_\mu g^{-1} ,$$

$$\psi \mapsto g \psi, \quad \left(\mathrm{SU}(2)_{\mathrm{spin}} \right) \tag{4}$$

where g is an arbitrary $\mathrm{SU}(2)$-valued function on $\mathbb{R} \times \Omega$. It follows from (1), (3) and (4) that the action S_Ω given in (2) is $\mathrm{U}(1) \times \mathrm{SU}(2)$-gauge invariant.

We must ask whether the equations of motion obtained by varying the action S_Ω with respect to the dynamical fields ψ and ψ^* are related to the Pauli equation for systems of interacting, non-relativistic electrons in an external electromagnetic field, [9],[10],[11]. It turns out that, for a certain natural choice of $\mathrm{U}(1) \times \mathrm{SU}(2)$-gauge, the components of the U(1)- and SU(2)-gauge potentials can be expressed in terms of the electromagnetic vector potential A_μ and the electric and magnetic fields $\vec{E}$ and $\vec{B}$ as follows:

$$a_\mu = \frac{e}{\hbar c} A_\mu , \tag{5}$$

$$w_{0A} = -\frac{\mu_e}{2c} B_A, \quad \text{and} \quad w_{\ell A} = -\frac{\mu_e}{4c} \varepsilon_{\ell AB} E_B , \tag{6}$$

$A, B = 1, 2, 3$, $\vec{B} = (B_1, B_2, B_3)$, $\vec{E} = (E_1, E_2, E_3)$, and $\mu_e \approx -\frac{e}{m_0 c}$ is the magnetic moment of the electron. With identifications (5) and (6), the equations of motion derived from the action S_Ω given in (2) reduce to the usual Pauli equations for ψ and its adjoint ψ^*, including the Zeeman term and conventional spin-orbit couplings.

This explains where observation **(a)** comes from.

To show that a given system is incompressible, e.g. in the sense that connected Green functions of quantum mechanical currents have "good" cluster properties – the form of the incompressibility assumption required in [2] – is a difficult analytical problem of many-body theory. We have not solved this problem, so far. Rather, we have elucidated the physical properties of incompressible systems, assuming incompressibility and using U(1) × SU(2)-gauge invariance. We have found that the behaviour of such systems on large distance scales and in the adiabatic limit, for short, in the scaling limit, is universal: Assuming that the system under the influence of a certain background magnetic field $\vec{B}_c = (0, 0, B_c)$ is incompressible, we have studied its response to small fluctuations in the external electromagnetic field $(\vec{\tilde{E}}, \vec{\tilde{B}})$. In order to find the basic equations of linear response theory, we determine the general form of the effective action $S^*_\Omega(a_c, w_c; \tilde{a}, \tilde{w})$ in the scaling limit. Here the potentials a_c, w_c describe the background field $\vec{B}_c$, while $\tilde{a}$ and $\tilde{w}$ describe the fluctuation field $(\vec{\tilde{E}}, \vec{\tilde{B}})$ and are calculated from formulas (5) and (6), (in a specific choice of gauge). The total gauge potentials are given by $a = a_c + \tilde{a}$, $w = w_c + \tilde{w}$. With the help of observation **(a)** (U(1) × SU(2)-gauge invariance) and assuming incompressibility, the most general form of $S^*_\Omega(a_c, w_c; \tilde{a}, \tilde{w})$ for a two-dimensional electron fluid confined to some domain Ω in the $x - y$ plane is found to be given by the following expression, independently of what the specific microscopic dynamics of the system is, (universality!):

$$\begin{aligned} -\frac{1}{\hbar} S^*_\Omega(a_c, w_c; \tilde{a}, \tilde{w}) &= \int_{M_3} (*j_c) \wedge \tilde{a} + \int_{M_3} (*m_3) \wedge \tilde{w}_3 \\ &+ \frac{\sigma}{4\pi} \int_{M_3} \tilde{a} \wedge d\tilde{a} + \frac{\chi}{2\pi} \int_{M_3} \tilde{a} \wedge d\tilde{w}_3 + \frac{\sigma_s}{4\pi} \int_{M_3} \tilde{w}_3 \wedge d\tilde{w}_3 \\ &+ \frac{k}{4\pi} \int_{M_3} tr\left(w \wedge dw + \frac{2}{3} w \wedge w \wedge w\right) \\ &+ \sum_{A=1}^{2} \int_{M_3} \tau_1^{\mu\nu} \tilde{w}_{\mu A} \tilde{w}_{\nu A} d^3\xi + \sum_{A,B=1}^{2} \int_{M_3} \tau_2^{\mu\nu} \varepsilon_{AB} \tilde{w}_{\mu A} \tilde{w}_{\nu B} d^3\xi \\ &+ \sum_{A,B,C=1}^{3} \int_{M_3} \eta^{\mu\nu\rho}_{ABC} \tilde{w}_{\mu A} \tilde{w}_{\nu B} \tilde{w}_{\rho C}\, d^3\xi + b.t.. \end{aligned} \tag{7}$$

This form holds in an SU(2)-gauge where $w_{c,\mu A} = -\delta_{\mu 0}\delta_{A3} \frac{\mu_e}{2c} B_{c,3}$. Once we have found S^*_Ω, it is a matter of functionally differentiating S^*_Ω with respect to the gauge fields $\tilde{a}$ and $\tilde{w}$, using that

$$\langle j^{\mu}(\xi)\rangle_{a,w} = \frac{\delta S^{*}_{\Omega}}{\delta \tilde{a}_{\mu}(\xi)}(a_c, w_c; \tilde{a}, \tilde{w}), \cdots, \tag{8}$$

where j^{μ} is the electric current operator, and ξ a rescaled space-time point, in order to find the linear response equations of an incompressible system in the scaling limit. Denoting the electric charge density in physical units by ρ and the electric current in physical units by $\mathcal{J}^{i}, i = 1,2$, we derive from (7) and (8)

$$\langle \rho(\xi)\rangle_{\vec{E},\vec{B}} = \rho_c(\xi) + \frac{\sigma_H}{c}\tilde{B}_3(\xi) - \ell_0 \underline{\nabla} \cdot \underline{\tilde{E}}(\xi) + \cdots, \tag{9}$$

and

$$\langle \mathcal{J}^{i}(\xi)\rangle_{\vec{E},\vec{B}} = \mathcal{J}^{i}_c(\xi) + \sigma_H \varepsilon^{ij}\tilde{E}_j(\xi) + \ell_0 \delta^{ij}\partial_{\tau}\tilde{E}_j(\xi) + c\ell_1 \varepsilon^{ij}\partial_j \tilde{B}_3(\xi) + \cdots, \quad i = 1,2. \tag{10}$$

Here $\rho_c = \frac{e}{c} j^0_c$ describes the background charge density of the system and $\mathcal{J}^i_c = e\, j^i_c$ a possible persistent current circulating in the system. Moreover, $\vec{\tilde{E}} = \vec{E}$ is the electric field (we have set $\vec{E}_c = 0$) and $\vec{B} = \vec{B}_c + \vec{\tilde{B}}$ the magnetic field. The second terms on the r.h.s. of (9) and (10) describe the Hall effect for the electric charge density and current. One finds that

$$\sigma_H = \sigma \frac{e^2}{h}, \tag{11}$$

where σ is the coefficient of the third term in S^{*}_{Ω}. The remaining terms on the r.h.s. describe effects of the spin degrees of freedom of electrons. The dots stand for "irrelevant" terms.

Besides the electric current density $\mathcal{J}^{\mu}$, one can define spin current densities $\mathcal{S}^{\mu}_{A}$, $A = 1,2,3$, in a natural way. By differentiating S^{*}_{Ω} with respect to the components $w_{\mu A}$ of the SU(2)-gauge potential, using an analogue of (8), we discover a Hall effect for the spin current.

For example, for the expectation value of the 3-component $\mathcal{S}^{\mu}_{3}$ of the spin current density we find the equations

$$\langle \mathcal{S}^{0}_{3}(\xi)\rangle_{\vec{E},\vec{B}} = M^{0}(\xi) + \left(\sigma^{\text{spin}}_{H1} - \sigma^{\text{spin}}_{H2}\right)\frac{1}{2c}\underline{\nabla} \cdot \underline{\tilde{E}}(\xi) + \chi_{\perp}\frac{1}{\mu_e}\tilde{B}_3(\xi) + \ldots, \tag{12}$$

and, for $i = 1,2$,

$$
\begin{aligned}
\langle S_3^i(\xi)\rangle_{\vec{\tilde{E}},\vec{\tilde{B}}} = {} & M^i(\xi) + \left(\sigma_{H1}^{\text{spin}} - \sigma_{H2}^{\text{spin}}\right) \varepsilon^{ij}\partial_j \tilde{B}_3(\xi) + \sigma_{H1}^{\text{spin}}\, \varepsilon^{ij}\partial_j B_{c,3}(\xi) \\
& - \left(\sigma_{H1}^{\text{spin}} - \sigma_{H2}^{\text{spin}}\right) \delta^{ij}\, \frac{1}{2c}\, \partial_\tau \tilde{E}_j(\xi) \\
& + \chi_\perp \frac{c}{\mu_e} \varepsilon^{ij} \tilde{E}_j(\xi) + \cdots . \qquad (13)
\end{aligned}
$$

Here M^0 is proportional to the magnetization of the system when $\vec{\tilde{E}} = \vec{\tilde{B}} = 0$, and M^i is a possible persistent spin current. Furthermore, in terms of the coefficients of the effective action S_Ω^*, one finds that

$$
\sigma_{H1}^{\text{spin}} = \frac{k}{2\pi}\frac{\hbar}{2}\mu_e\,, \quad \sigma_{H2}^{\text{spin}} = \frac{\sigma_s}{4\pi}\frac{\hbar}{2}\mu_e\,, \qquad (14)
$$

and

$$
\chi_\perp = \frac{\chi}{4\pi}\frac{\mu_e e}{c}. \qquad (15)
$$

We note that $\mathcal{J}^\mu$ and $\mathcal{S}_3^\mu$ are conserved currents, while the other components $\mathcal{S}_A^\mu$, $A = 1,2$, are not conserved, for our choice of the background gauge potential $\vec{w}_0 = -\frac{\mu_e}{2c}\vec{B}_c$, $\vec{w}_\ell = 0$, $\vec{B}_c = (0,0,B_c)$. These components are, however, covariantly conserved. This implies relations between various distinct coefficients in the effective action S_Ω^*; see [2]. Their expectation values obey equations analogous to (12) and (13) describing a Hall effect for spin currents.

So far, we have not said anything about the possible values of the constants σ_H, $\sigma_{H1}^{\text{spin}}$, $\sigma_{H2}^{\text{spin}}$ and $\chi_\perp$ appearing in the equations above. One can show that they belong to certain discrete sets; for example k is an integer. This "quantization" is a consequence of the following observation: If we consider the transformation properties of S_Ω^* under U(1) × SU(2)-gauge transformations not vanishing on the boundary of Ω we realize that S_Ω^* is gauge-invariant only up to boundary terms! Since non-relativistic quantum mechanics is fully U(1) × SU(2)-gauge invariant, violations of U(1) × SU(2)-gauge invariance due to bulk terms of S_Ω^* must be compensated by corresponding violations of gauge invariance due to anomalous boundary terms in S_Ω^*. These anomalous boundary terms are uniquely determined by the Chern-Simons bulk terms in S_Ω^*. They correspond to chiral electric and spin currents, coupled to the gauge potentials $\tilde{a}$ and $\tilde{w}$, respectively, which circulate around the system and are localized near its boundary. Chiral electric currents circulating around the boundary of quantum Hall systems were originally found by Halperin [12] in a simple quantum mechanical analysis. For a general analysis see [13],[3],[14]. The study of the algebras of chiral boundary currents and of their representation theory (along with some input from the physics of quantum Hall fluids) provides us with fairly precise information on the possible, discrete values of the basic constants σ_H, $\sigma_{H1}^{\text{spin}}$, etc.; see also [13,3].

While the classical Hall effect for the electric current is standard knowledge, generalized classical Hall effects, e.g. (9), or the Hall effects for spin currents, do not appear to have been discussed in the literature; but see [15]. Surprizingly, one could have predicted them starting from classical physics. This is shown in [2].

Thus, besides the standard Hall effect for the electric current, as described in (10), quantum mechanics and classical physics predict several generalizations of the Hall effect, see (9),(12), (13). Of course, only quantum mechanics enables us to understand the quantization of the coefficients σ, χ, σ_s and k.

A corollary of our derivation of the effective action S^*_Ω in the scaling limit, see (7), using gauge invariance and assuming incompressibility, is the Goldstone theorem: If the magnetization M^0 (see (12)) does not vanish, as $\vec{B}_c \to 0$, then the system cannot be incompressible at $\vec{B}_c = 0$, i.e., there are gapless extended modes, the Goldstone bosons, coupled to the groundstate by the spin current; see [2].

The methods and results sketched in these notes (see [2],[3] for more details) can be extended to describe the effect of tidal (centrifugal and Coriolis) forces, rather than of electrostatic and Lorentz forces, on incompressible quantum fluids, e.g. on superfluid films or neutron stars. As a corollary one finds the usual quantization of superfluid vorticity (in units of $\frac{h}{m}$), tidal Hall effects and an analogue of the Aharonov-Bohm-Casher effects in superfluids. This is worked out in [16].

Acknowledgements

We thank Y. Avron and A. Wipf for very helpful discussions.

References

1. Aharonov, Y., Bohm, D.: Phys.Rev. **115**, 485 (1959)
   Aharonov, Y., Casher, A.: Phys.Rev.Lett. **53**, 319 (1984)
2. Fröhlich, J., Studer, U.M.:U(1)×SU(2)–Gauge Invariance of Non-Relativistic Quantum Mechanics and Generalized Hall Effects. ETH–TH/91-13, submitted to Phys.Rev. B
3. Fröhlich, J., Kerler, T.: Nuclear Phys. B **354**, 369 (1991)
   Fröhlich, J., Zee, A.: Nuclear Phys. B (1991), in press
4. Wilczek, F., Zee, A.: Phys.Rev. Lett. **51**, 2250 (1983)
5. Halperin, B.I.: Helv.Phys.Acta **56**, 75 (1983)
6. Haldane, F.D.M., Rezayi, E.H.: Phys.Rev.Lett. **60**, 956 (1988)
7. Eisenstein,J.P. et al.: Phys.Rev.Lett. **62**, 1540 (1989)
8. Greiter, M., Wen, X.G., Wilczek, F.: Phys.Rev.Lett. **66**, 3205 (1991)
9. Baym, G.: *Lectures on Quantum Mechanics.* Benjamin/Commings Publishing Company, Reading, Massachusetts, 1969

10. Messiah, A.: *Quantum Mechanics.* North-Holland Publishing Company, Amsterdam, 1961
11. Piron, C.: Mécanique Quantique (Base et Applications) (Presses Polytechniques et Universitaires Romandes, Lausanne, 1990)
12. Halperin, B.I.: Phys.Rev. B **25**, 2185 (1982)
13. Wen, X.G.: Phys.Rev. B **43**, 11025 (1991)
14. MacDonald, A.H.: Phys.Rev.Lett. **64**, 220 (1990)
    Blok, B., Wen, X.G.: Phys.Rev. B **42**, 8133 (1990)
    Stone, M.: Internat.J.Modern Phys. B **5**, 509 (1991)
15. Volovik, G.E. Yakovenko, V.M.: J.Phys.: Condens.Matter **1**, 5263 (1989)
16. Fröhlich, J. Studer, U.M.: preprint, to appear

Asymptotic Completeness for N-Body Quantum Systems

Gian Michele Graf

Division of Physics, Mathematics and Astronomy
California Institute of Technology
Pasadena, CA91125

Abstract: We give a sketch of a geometrical proof of asymptotic completeness for an arbitrary number of quantum particles interacting through short-range pair potentials.

1 Introduction

The first task of quantum scattering theory is to give a classification of the possible large time behaviours of Schrödinger orbits $e^{-itH}\psi$. In this contribution we study this problem for an arbitrary number N of particles interacting through pair potentials of short range, i.e. decaying at infinity faster than the Coulomb potential. In the intuitive picture of the scattering process, this system is well described at large times by a number of independently moving, bound clusters, i.e. by free composite particles. This statement is called asymptotic completeness. The subject has spurred an enormous amount of literature (see [1],[20],[21] for reviews) dealing with a variety of techniques. Here we mention the works of Povzner [19] and of Faddeev [9], where asymptotic completeness was first proved for $N = 2$, respectively for $N = 3$. Enss [5]-[7] and Mourre [15]-[17] introduced new geometric methods which strongly influenced the more recent developments. For arbitrary N the proof was first given by Sigal and Soffer [22].

The physical space is $\mathbb{R}^\nu$, $\nu \geq 1$. The configuration space of N mass points $m_i > 0$ in the center of mass frame is the real vector space

$$X := \left\{ x = (x^1, \ldots, x^N) \mid x^i \in \mathbb{R}^\nu,\ \sum_{i=1}^N m_i x^i = 0 \right\}$$

equipped with the metric

$$x \cdot y := \sum_{i=1}^{N} m_i x^i \cdot y^i \,, \tag{1}$$

where $x^i \cdot y^i$ is the scalar product on $\mathbb{R}^\nu$. We will also use the notation $x^2 = x \cdot x$. The Hilbert space of the quantum mechanical N-body system is $L^2(X)$, where the volume element of X is defined by the metric.

The Hamiltonian of the N-particle system is

$$H = -\frac{1}{2}\Delta + V := -\frac{1}{2}\Delta + \sum_{(ij)} V_{ij}(x^i - x^j) \tag{2}$$

on $L^2(X)$, where Δ is the Laplacian on X. We will come back to this and to other statements concerning the Hamiltonians at the end of this section. A cluster decomposition is a partition of $\{1, \ldots, N\}$ into disjoint, nonempty subsets, called clusters. The intercluster interactions for a cluster decomposition a is defined as the sum of all potentials linking different clusters in a, i.e. as

$$I_a := \sum_{(ij) \not\subset a} V_{ij}(x^i - x^j) \,,$$

where $(ij) \not\subset a$ means that i, j belong to different cluster in a. Noninteracting clusters of a are the described by the Hamiltonian $H_a := H - I_a$. It decomposes naturally into a part describing the center of mass motion of the clusters in a, and an internal part H^a, whose bound state projection is denoted P^a. States in the range of P^a represent, physically speaking, a set of composite particles, each of them consisting of the 'elementary' particles of some cluster of a.

We say that the N-body short-range system above is *asymptotically complete*, if for any state $\psi \in L^2(X)$ and any $\varepsilon > 0$ there are states $\psi^a = P^a \psi^a \in L^2(X)$ such that

$$\left\| e^{-itH}\psi - \sum_a e^{-itH_a}\psi^a \right\| \leq \varepsilon \tag{3}$$

for $t > 0$ large enough, where the sum ranges over all cluster decompositions. Given the existence of the wave operators $\Omega_a := \mathrm{s} - \lim_{t\to\infty} e^{itH} e^{-itH_a} P^a$, this definition is equivalent to the usual one:

$$\bigoplus_a \operatorname{Ran} \Omega_a = L^2(X) \,.$$

We assume that the pair potentials are real and of short-range in the sense that

$$(1 + |y|)^\mu V_{ij}(y) \,, \qquad (1 + |y|)^\mu |\nabla V_{ij}(y)| \tag{4}$$

are Δ_y-compact (on $L^2(\mathbb{R}^\nu)$) for some $\mu > 1$.

Theorem *The quantum N-body system* (2) *satisfying* (4) *is asymptotically complete.*

Let us now introduce some further notion concerning N-body systems. For any cluster decomposition a we introduce the external configuration space

$$X_a := \left\{ x \in X \mid x^i = x^j \text{ if } i, j \in C \text{ for some } C \in a \right\},$$

and its orthogonal complement, the internal configuration space

$$X^a := \left\{ x \in X \mid \sum_{i \in C} m_i x^i = 0 \text{ for } C \in a \right\}.$$

The direct sum $X = X_a \oplus X^a$ induces an orthogonal decomposition $x =: x_a + x^a$. Note that $\mathbb{1}_a : x \mapsto x_a$ replaces the position of each particle by the one of the center of mass of the cluster of a to which it belongs, while $\mathbb{1}^a : x \mapsto x^a$ replaces it with the position of the particle relative to that center of mass.

The particle momenta are the covariant components of the vector operator p acting as $p\psi = (1/\mathrm{i})d\psi$, where d is the derivative. The velocities $p_i/m_i = p^i$ are its contravariant components. The kinetic energy then is $\sum_i p_i^2/2m_i = \sum_i p_i p^i/2 = p^2/2 = -\Delta/2$.

The sum $X = X_a \oplus X^a$ also induces the factorization $L^2(X) = L^2(X_a) \otimes L^2(X^a)$ with respect to which we have

$$H_a = \frac{1}{2}(p_a)^2 \otimes \mathbb{1} + \mathbb{1} \otimes H^a$$

with

$$H^a = \frac{1}{2}(p^a)^2 + \sum_{(ij) \subset a} V_{ij}(x^i - x^j).$$

2 Propagation Estimates

To each configuration of the particles one can associate a cluster decomposition in such a way that particles in the same cluster are close together, and that clusters are very distant from each other. If at some large time t one groups the particles as explained, then the motion of the centers of mass of these clusters should be almost free, i.e. $p_a \approx x_a/t$. In other words, the quantum dynamics is concentrating on states where these two quantities coincide. This is the main intermediate result of the proof. It will be cast into precise mathematical terms by means of a *propagation estimate*. Several sorts of propagation estimates have been used in scattering theory. The type we are going to describe goes back to [12]-[14], except for the use of time-dependent observables, which is due to [23]. There, as well as in [3],[4],[8],[11],[22],[24], examples of propagation estimates of this kind can be found.

Let H be a self-adjoint operator on a Hilbert space $\mathcal{H}$. Let $P(t)$ be a bounded, time-dependent positive operator, i.e.

$$\begin{gathered} P: [1,\infty) \to \mathcal{L}(\mathcal{H}),\ t \to P(t)\,, \\ P(t) \geq 0\,. \end{gathered}$$

A propagation estimate is an estimate of the form

$$\int_1^\infty dt\,(\psi_t, P(t)\psi_t) \leq \text{const}\,\|\psi\|^2\,,$$

where $\psi_t := \mathrm{e}^{-\mathrm{i}tH}\psi$. It tells us that the trajectory $\mathrm{e}^{-\mathrm{i}tH}\psi$ concentrates where $P(t)$ is small.

Propagation estimates can be generated by a *propagation observable*

$$\Phi: [1,\infty) \to \mathcal{L}(\mathcal{H}),\ t \to \Phi(t),$$

where $\Phi(t)$ is uniformly bounded in t. We assume the existence (as a bounded operator) of the Heisenberg derivative $D\Phi(t)$, i.e.

$$\frac{d}{dt}(\psi_t, \Phi(t)\psi_t) = (\psi_t, D\Phi(t)\psi_t)\,,$$

which is formally given by $D\Phi(t) = \mathrm{i}[H, \Phi(t)] + \partial\Phi/\partial t$. Assume it is positive in the sense that

$$D\Phi(t) \geq P(t) + R(t)$$

in $\mathcal{L}(\mathcal{H})$, where

$$P(t) \geq 0$$

and $R(t)$ satisfies the *remainder estimate*

$$\int_1^\infty dt\,|(\psi_t, R(t)\psi_t)| \leq \text{const}\,\|\psi\|^2\,.$$

Then

$$(\psi_t, \Phi(t)\psi_t)\big|_1^T \geq \int_1^T dt\,(\psi_t, P(t)\psi_t) + \int_1^T dt\,(\psi_t, R(t)\psi_t)\,,$$

where both the l.h.s. and the last term on the r.h.s. are bounded by const $\|\psi\|^2$. Since the first integrand is nonnegative, we conclude that $P(t)$ satisfies a propagation estimate.

We now mention two ways to prove a remainder estimate. We assume one of the following

1) $$\int_1^\infty dt\, \|R(t)\| < \infty\,.$$

2) $$R(t) = R_1(t)^* B(t) R_2(t) + R_2(t)^* B(t)^* R_1(t)\,,$$

where $R_i(t)^* R_i(t)$, $i = 1, 2$, satisfy a propagation estimate and $B(t)$ is uniformly bounded. In fact

$$|(\psi_t, R(t)\psi_t)| \le \text{const}\, (\psi_t, R_1(t)^* R_1(t)\psi_t)^{1/2} (\psi_t, R_2(t)^* R_2(t)\psi_t)^{1/2}$$

from which the remainder estimate follows by the Cauchy inequality.

Coming back to (2), we will give a number of examples of propagation estimates, propagation observables, and remainder estimates. Actually they constitute the core of the proof.

Example 1 [23] Let $\Omega \subset \mathbb{R}$ be a bounded interval, and $\lambda > 0$ large enough. Then

$$\int_1^\infty \frac{dt}{t} (\psi_t, E_\Omega(H) F(\lambda \le |x/t| \le 2\lambda) E_\Omega(H)\psi_t) \le \text{const}\, \|\psi\|^2$$

for all $\psi \in L^2(X)$, where $E_\Omega(H)$ is the spectral projection for H associated with Ω and the constant depends on Ω, λ.

The idea is that a particle with finite energy and hence speed cannot propagate into regions moving away still faster. This propagation estimate is generated by the propagation observable

$$\Phi(t) = -E_\Omega(H) h(|x/\lambda t|) E_\Omega(H),$$

where $h' \in C_0^\infty(0, \infty)$, $h' \ge 0$ and $h'(y) \ge 1$ for $1 \le y \le 2$. One checks in fact that

$$D\Phi \ge \frac{\text{const}}{t} E_\Omega(H) F(\lambda \le |x/t| \le 2\lambda) E_\Omega(H) + O(t^{-2})\,.$$

Example 2 This example deals with a free particle, $H = p^2/2$, and is therefore not directly applicable to N-body systems. The basic observation is that a free particle has the classical trajectory $p = p_0$, $x = p_0 t + x_0$, where p_0, x_0 are the initial data. It follows that $(p - x/t)^2 = x_0^2/t^2$ is decreasing. This suggests that one can get a positive Heisenberg derivative by considering the propagation observable

$$-K_0(t) := -\frac{1}{2}\left(p - \frac{x}{t}\right)^2$$

(we ignore that $K_0(t)$ is not uniformly bounded for a while). In fact

$$-DK_0(t) = \frac{1}{t}\left(p - \frac{x}{t}\right)^2 \geq 0\,, \tag{5}$$

which 'proves'

$$\int_1^\infty \frac{dt}{t}\,(\psi_t, \left(p - \frac{x}{t}\right)^2 \psi_t) \leq \text{const}\,\|\psi\|^2\,. \qquad (wrong)$$

This means that the time evolved quantum state is concentrating on the region in phase-space where $(p - x/t)^2$ is small, i.e. around the classical trajectories. But the equation above is wrong as it stands, since $K_0(t)$ is not even a bounded operator. The next example will deal with this problem.

Example 2 (revisited) We replace $K_0(t)$ by a bounded operator by introducing some cutoffs, namely an energy cutoff $E_\Omega(H)$, where $\Omega \subset \mathbb{R}$ is some bounded interval, and a volume cutoff $f(x/\lambda t)$, where $f \in C_0^\infty(X)$, $f(x) = 1$ for $|x| \leq 1$, $f(x) = 0$ for $|x| \geq 2$, and $\lambda > 0$ is large. $-E_\Omega f K_0 f E_\Omega$ is now a suitable propagation observable, since it is uniformly bounded in t. Then

$$D(E_\Omega f K_0 f E_\Omega) = E_\Omega \left(f(DK_0)f + fK_0(Df) + (Df)K_0 f\right) E_\Omega\,.$$

Since Df is supported in $\lambda \leq |x/t| \leq 2\lambda$, it is easily checked that the terms above arising from Df satisfy a remainder estimate by Example 1. Hence

$$\int_1^\infty \frac{dt}{t}\,(\psi_t, E_\Omega f \left(p - \frac{x}{t}\right)^2 f E_\Omega \psi_t) \leq \text{const}\,\|\psi\|^2\,.$$

Example 3 We extend the previous result to the two-body case. Clearly $K_0(t)$ as defined above is not going to be non-increasing if the system is in a bound state or close to the scatterer. On the other hand the Hamiltonian H, being conserved, is trivially non-increasing (but, as a propagation observable, it does not provide any dynamical information). Let $v(x,t)$ be a vector field with $v = x/t$ for $|x| \gtrsim R$, resp. $v = 0$ for $|x| \lesssim R$ and with a smooth transition in between. Then

$$K(t) := \frac{1}{2}\left(p - v(x,t)\right)^2 + V(x)$$

reduces to $K_0(t)$, resp. H in those two regions and should thus be non-increasing. One computes

$$-DK(t) =$$

$$(p-v)\frac{v_* + v_*^\dagger}{2}(p-v) + \frac{1}{2}[(p-v)\cdot\mathcal{D}_v v + \mathcal{D}_v v\cdot(p-v)] - \frac{1}{4}\Delta(\nabla\cdot v) - v\cdot\nabla V\,, \tag{6}$$

where $v_*(x,t) : X \to X$ is the x-derivative of v, and $v_*^\dagger$ its adjoint with respect to the scalar product (1), and $\mathcal{D}_v = v\cdot\nabla + \partial/\partial t$ is the material derivative for the vector field v. If v is chosen appropriately, the first term on the r.h.s. of (6) is nonnegative and reduces to (5) far from the the origin. For $R := t^{1-\delta}$, $\delta > 0$, one gets $\mathcal{D}_v v = O(t^{-(1+\delta)})$, $\Delta(\nabla \cdot v) = O(t^{-(3-2\delta)})$ and $v \cdot \nabla V = O(t^{-(1-\delta)\mu})$. Thus all terms on the r.h.s. of (6), except for the first one, satisfy a remainder estimate if δ is taken small enough. After adding the same cutoffs to $K(t)$ as in the free case, its Heisenberg derivative then satisfies

$$-D(E_\Omega f K f E_\Omega) \geq \frac{1}{t} E_\Omega f j \left(p - \frac{x}{t}\right)^2 j f E_\Omega + R(t)$$

for $j = j(x/t)$ bounded and supported away from the origin, with $R(t)$ satisfying a remainder estimate. Hence

$$\int_1^\infty \frac{dt}{t} (\psi_t, E_\Omega f j \left(p - \frac{x}{t}\right)^2 j f E_\Omega \psi_t) \leq \text{const} \|\psi\|^2 , \tag{7}$$

i.e. the conclusion of the previous example still holds except for a region close to the origin.

Example 4 Next we extend the previous example to the N-body case. Consider a configuration x. Group the particles into clusters in such a way that particles in the same cluster are close together, and that clusters are far from each other (the scale being $R = t^{1-\delta}$ as before). Call a the resulting cluster decomposition. We express this by saying that the particles are a-clustered at x. Let $v = x_a/t$ in this case. Then $v_*(x)$ reduces to $\mathbb{1}_a$ and the first term on the r.h.s. of (6) to

$$(p - v)\frac{v_* + v_*^\dagger}{2}(p - v) = \frac{1}{t}(p - v)_a^2 = \frac{1}{t}\left(p - \frac{x}{t}\right)_a^2 .$$

Thus (7) goes over to

$$\int_1^\infty \frac{dt}{t} (\psi_t, E_\Omega f j_a \left(p - \frac{x}{t}\right)_a^2 j_a f E_\Omega \psi_t) \leq \text{const} \|\psi\|^2 \tag{8}$$

for $j_a = j_a(x/t)$ bounded and supported on a-clustered configurations. For large times, this estimate asserts that whenever the particles are a-clustered, the velocities of the centers of mass of the clusters of a are approximately given by their positions, divided by t. This is the result mentioned at the beginning of this section.

3 Asymptotic Completeness

The propagation estimates of the preceding section can be used to prove asymptotic completeness. Let us first investigate its consequences heuristically. Suppose that at some large time t the particles are a-clustered. By (8) the velocities of the clusters of a are $p_a \approx x_a/t$. Hence the motion still to come will separate them even more from each other, making the intercluster interactions I_a negligible. In other words, H_a will account for this motion almost as well as H does. Mathematically, if j_a is supported on a-clustered configurations, given ψ there should be a state ψ^a such that

$$\lim_{t\to\infty} \|j_a(x/t)\mathrm{e}^{-\mathrm{i}tH}\psi - \mathrm{e}^{-\mathrm{i}tH_a}\psi^a\| = 0 \,. \tag{9}$$

In particular, let the $\{j_a\}_a$ form a partition of unity on X indexed by all cluster decompostions, i.e. $\sum_a j_a = 1$. Then (9) implies

Lemma 1 *The Deift-Simon [2] wave operators*

$$W_a(v) := \mathrm{s} - \lim_{t\to\infty} \mathrm{e}^{\mathrm{i}tH_a} j_a(x/vt)\mathrm{e}^{-\mathrm{i}tH}$$

exist for any $v > 0$.

Thus

$$\mathrm{e}^{-\mathrm{i}tH}\psi = \sum_a j_a \mathrm{e}^{-\mathrm{i}tH}\psi = \sum_a \mathrm{e}^{-\mathrm{i}tH_a} W_a(v)\psi + \mathrm{o}(1) \tag{10}$$

as $t \to \infty$, showing that the system is well described by independently moving clusters. Yet for asymptotic completeness each cluster should also be bound, i.e. $\psi^a = P^a\psi^a$ in (3). Consider for simplicity the trivial cluster decomposition $a = (1 \ldots N)$. Then $(\psi, W_a(v)\psi)$ is the asymptotic probability for a cluster where all particles are close together on the scale vt. The probability for being in a bound state is included in $(\psi, W_a(v)\psi)$ but does not exhaust it: If the energy of ψ is sufficiently close to a threshold, then the particles can break apart (thus not forming a bound state) with speed less than v. However no other possibilities for $W_a(v)\psi$ exist:

Lemma 2 *Let $\Omega \subset \mathbb{R}$ be a compact set, which does not contain any eigenvalues or thresholds. Then*

$$W_{(1\ldots N)}(v)E_\Omega(H) = 0 \,, \tag{11}$$

provided $v > 0$ is small enough.

This result is based on the Mourre estimate [10],[16],[18]. These lemmas easily imply asymptotic completeness: Since (3) is trivial if ψ is a bound state, we may assume $\psi \perp \operatorname{Ran} P^{(1\ldots N)}$. Given $\varepsilon > 0$, we can find a compact set $\Omega \subset \mathbb{R}$ which does not contain any eigenvalues or thresholds, such that

$$\|(1 - E_\Omega(H))\psi\| \leq \varepsilon$$

and then $v > 0$ such that (11) holds. Thus, by (10)

$$e^{-itH} E_\Omega(H)\psi = \sum_{\#(a)\geq 2} e^{-itH_a} W_a(v) E_\Omega(H)\psi + o(1) ,$$

where the sum now ranges only over decompositions with at least two clusters. Since each of these clusters consists of less than N particles, by induction we can apply (3) to the terms in the last sum above. We end up with asymptotic completeness.

Acknowledgement: This work was partially supported by NSF grant DMS-9101715.

References

1. Cycon, H.L., Froese, R.G., Kirsch, W., Simon, B.: *Schrödinger Operators*, Springer (1987)
2. Deift, P., Simon, B.: A time-dependent approach to the completeness of multiparticle quantum systems. Comm.Pure Appl.Math. **30**, 573-583 (1977)
3. Dereziński, J.: A new proof of the propagation theorem for N-body quantum systems. Comm.Math.Phys. **122**, 203-231 (1989)
4. Dereziński, J.: Algebraic approach to the N-body quantum long range scattering, Warsaw University preprint (1990)
5. Enns, V.: Asymptotic completeness for quantum-mechanical potential scattering, I. Short-range potentials. Comm.Math.Phys. **61**, 285-291 (1978)
6. Enns, V.: Completeness of Three-Body Quantum Scattering. In: *Dynamics and Processes*. Ed. by P. Blanchard, L. Streit, Lecture Notes in Mathematics, Vol. 1031, pp. 62-88, Springer (1983)
7. Enns, V.: Introduction to asymptotic observables for multi-particle quantum scattering. In: *Schrödinger Operators. Aarhus 1985*. Ed. by E. Balslev, Lecture Notes in Mathematics, Vol. 1218, pp. 61-92, Springer (1986)
8. Enns, V.: Two- and three-body quantum scattering: Completeness revisited. In: *Symposium "Partial Differential Equations" Holzhau 1988*. Ed. by B.-W. Schulze and H. Triebel, Texte zur Mathematik, Vol. 112, pp. 108-120, Teubner (1989)
9. Faddeev, L.: *Mathematical Aspects of the Three Body Problem in Quantum Scattering Theory* (Steklov Institute 1963), Israel Program for Scientific Translation (1965)
10. Froese, R., Herbst, I.: A new proof of the Mourre estimate. Duke Math.J. **49**, 1075-1085 (1982)
11. Graf, G.M.: Asymptotic completeness for N-body short-range quantum systems: A new proof. Comm.Math.Phys. **132**, 73-101 (1990)

12. Kato, T.: Wave operators and similarity for some non-self-adjoint operators. Math.Ann. **162**, 258-279 (1966)
13. Kato, T.: Smooth operators and commutators. Stud.Math. **31**, 535-546 (1968)
14. Lavine, R.: Absolute continuity of Hamiltonian operators with repulsive potentials. Proc.Amer.Math.Soc. **22**, 55-60 (1969)
15. Mourre, E.: Link between the geometrical and the spectral transformation approaches in scattering theory. Comm.Math.Phys. **68**, 91-94 (1979)
16. Mourre, E.: Absence of singular continuous spectrum for certain selfadjoint operators. Comm.Math.Phys. **78**, 391-408 (1981)
17. Mourre, E.: Operateurs conjugés et propriété de propagation. Comm.Math. Phys. **91**, 279-300 (1983)
18. Perry, P., Sigal, I.M., Simon, B.: Spectral analysis of N-body Schrödinger operators. Ann.Math. **114**, 519-567 (1981)
19. Povzner, A.J.: On eigenfunction expansions in terms of scattering solutions. Dokl.Akad.Nauk SSSR **104**, (1955)
20. Reed, M., Simon, B.: *Methods of Modern Mathematical Physics, III. Scattering Theory*. Academic Press (1979)
21. Sigal, I.M.: *Scattering Theory for Many-Body Quantum Mechanical Systems*. Lecture Notes in Mathematics, Vol. 1011, Springer Verlag (1983)
22. Sigal, I.M., Soffer, A.: The N-particle scattering problem: asymptotic completeness for short-range systems. Ann.Math. **126**, 35-108 (1987)
23. Sigal, I.M., Soffer, A.: Local decay and propagation estimates for time-dependent and time-independent Hamiltonians. Princeton University preprint (1988)
24. Sigal, I.M., Soffer, A.: Long-range many-body scattering. Asymptotic clustering for Coulomb-type potentials. Invent.Math. **99**, 115-143 (1990)

Self-Dual Chern–Simons Solitons

R. Jackiw [1]

Center for Theoretical Physics, Laboratory for Nuclear Science and Department of Physics, Massachusetts Institute of Technology, Cambridge, Massachusetts 02139 U.S.A.

1 Introduction

These lectures are devoted to the two-year old topic of Chern–Simons solitons. Solitons have interested particle theorists since the mid-1970's[1]; Chern–Simons dynamics, since the early 1980's [2]. Now in the decade of the 1990's, we are combining the two subjects.

The Chern–Simons theories that we consider exist in three-dimensional space-time. They possess pedagogical interest — teaching us [we hope] useful lessons that may be relevant to the physical four-dimensional world. Also they are mathematically fascinating, giving rise to effects and structures that invite further study. Finally, they can have a physical role in descriptions of actual, laboratory processes that are confined to a spatial plane. In a very precise sense, Chern–Simons models are field theoretical relatives of Landau's work on charged particles in an external magnetic field and of the Aharonov–Bohm [Ehrenberg–Siday] effect. These days, the quantum Hall effect and high-T_c superconductivity — two planar phenomena — are analyzed in terms of Chern–Simons dynamics.

The concepts surrounding the Chern–Simons term thus lie at the intersection of physics and mathematics. They provide a topic that is most appropriate for discussing at this conferencee whose logo is $M \cap \Phi$. Indeed Chern–Simons theory has already been described at a previous meeting of this Congress — there the focus was mathematical. In contrast, my emphasis is on the physics, and in my lecture I shall speak about the structures that are supported by Chern–Simons interactions, specifically the soliton solutions to the appropriate equations of motion.

[1] This work is supported in part by funds provided by the U. S. Department of Energy (D.O.E.) under contract #DE-AC02-76ER03069.

Although we shall work with Chern–Simons theories in three-dimensional space-time, to set the stage for our calculations we start from a different place in mathematical physics — the (1 + 1)-dimensional non-linear Schrödinger equation, given in (1.1).

$$i\frac{\partial}{\partial t}\psi(t,x) = -\frac{1}{2}\,\frac{\partial^2}{\partial x^2}\psi(t,x) - g\psi^*(t,x)\psi(t,x)\psi(t,x) \tag{1.1}$$

This equation enjoys two honorable and important roles.

First, we can take ψ to be a classical, c-number, complex function of its arguments, and (1.1) is a non-linear partial differential equation that determines ψ. In this context (1.1) is one of those famous non-linear partial differential equations that is integrable and possesses a complete set of soliton solutions. Understanding the complete integrability and soliton structure of the non-linear Schrödinger equation was an important milestone along the road that now has led to the elaborate theory of completely integrable systems.

In the second, alternative view, ψ is a quantum field operator, satisfying an equal-time commutation relation with its Hermitian conjugate,

$$\left[\psi(t,x),\psi^\dagger(t,y)\right] = \delta(x-y) \tag{1.2}$$

and (1.1) is the Heisenberg equation of motion for the operator [$\psi^\dagger$ replaces ψ^* and $\hbar$ is scaled to unity]. This then is a non-relativistic quantum field theory and because the number operator N is conserved,

$$N = \int dx\,\psi^\dagger(t,x)\psi(t,x) \tag{1.3}$$

$$\frac{d}{dt}N = 0\ , \tag{1.4}$$

the quantum Hilbert space may be spanned by energy and number eigenstates $|E,N\rangle$. In a well-known way, it follows that in the N-sector, one is dealing with a N-body quantum-mechanical [linear] Schrödinger problem where the potential [coded in the non-linearity of (1.1)] is the pairwise δ-function interaction. Specifically, by defining

$$e^{-iEt}u_E(x_1,\ldots,x_n) = \langle 0|\psi(t,x_1)\ldots\psi(t,x_n)|E,N\rangle\ , \tag{1.5}$$

where $|0\rangle$ is the no-particle, zero-energy state, one finds that the wave function u_E satisfies

$$\left(-\frac{1}{2}\sum_i\frac{\partial^2}{\partial x_i^2} - g\sum_{ij}\delta(x_i-x_j)\right)u_E(x_1,\ldots,x_N) = E\,u_E(x_1,\ldots,x_N)\,. \tag{1.6}$$

This many-body problem is also completely solvable and the energy spectrum and S-matrix are known.

A final remarkable fact about the model is that the same quantum mechanical spectrum is obtained by quantizing the classical soliton solutions of the first approach, using the methods of soliton quantization. Thus the above two aspects of the non-linear Schrödinger equation are two sides of the same coin.

In view of the remarkable properties exhibited by the non-linear Schrödinger equation in (1 + 1)-dimensional space-time, it is natural to try extending the model to higher dimensionality, for example to (2 + 1)-dimensions, by positing, instead of (1.1)

$$i\partial_t\psi(t,\mathbf{r}) = -\frac{1}{2}\nabla^2\psi(t,\mathbf{r}) - g\psi^*(t,\mathbf{r})\psi(t,\mathbf{r})\psi(t,\mathbf{r})\,. \tag{1.7}$$

The single coordinate x has been replaced by the two-vector $\mathbf{r}$ and the second derivative with respect to x becomes the two-dimensional Laplacian. Unfortunately, nothing particularly interesting has been found for this planar generalization of (1.1).

The new suggestion that we make, and which does produce an interesting dynamical model, consists of introducing a gauge potential $A^\mu(t,\mathbf{r}) = (A^0(t,\mathbf{r}), \mathbf{A}(t,\mathbf{r}))$ by promoting all derivatives in (1.7) to gauge covariant derivatives and thereby allowing ψ to interact with A^μ in a gauge-invariant fashion. [When writing expressions involving gauge fields we shall frequently use Lorentz covariant notation, even though matter dynamics is non-relativistic. We use the Minkowski metric, with signature $(1,-1,-1)$ and set c to unity.]

2 Non-Relativistic Abelian Model

The equation that we propose to study is

$$iD_t\psi = -\frac{1}{2}\mathbf{D}^2\psi - g\psi^*\psi\psi \tag{2.1}$$

[3], where the ordinary derivatives of (1.7) are replaced by gauge covariant derivatives.

$$\begin{aligned} D_\mu &\equiv \partial_\mu + iA_\mu \\ D_t &= \partial_t + iA^0 \\ \mathbf{D} &= \boldsymbol{\nabla} - i\mathbf{A}\,. \end{aligned} \tag{2.2}$$

Equation (2.1) is also an evolution equation

$$\begin{aligned} &i\partial_t\psi(t,\mathbf{r}) \\ &= -\frac{1}{2}\left(\boldsymbol{\nabla} - i\mathbf{A}(t,\mathbf{r})\right)^2\psi(t,\mathbf{r}) + A^0(t,\mathbf{r})\psi(t,\mathbf{r}) - g\psi^*(t,\mathbf{r})\psi(t,\mathbf{r})\psi(t,\mathbf{r}) \end{aligned} \tag{2.3}$$

but to close the system we still need an equation specifying the time-evolution of A^μ. Owing to the gauge invariant coupling of A^μ to ψ, we expect that the equation determining A^μ should also be gauge-invariant, *i.e.* involve only the field strength $F_{\mu\nu}$.

$$F_{\mu\nu} = \partial_\mu A_\nu - \partial_\nu A_\mu \tag{2.4}$$

The field components, in non-relativistic notation are,

$$B = \nabla \times \mathbf{A} \tag{2.5a}$$

$$\mathbf{E} = -\nabla A^0 - \partial_t \mathbf{A} \,. \tag{2.5b}$$

The natural and obvious equation for gauge field dynamics is the Maxwell equation, relating a derivative of $F^{\mu\nu}$ to the conserved matter current j^μ.

$$j^\mu = \begin{cases} j^0 = \rho = \psi^* \psi \\ \mathbf{j} = \Im \psi^* \mathbf{D} \psi \end{cases} \tag{2.6}$$

$$\partial_\mu F^{\mu\nu} = j^\nu \tag{2.7a}$$

However, as is by now well-known, in three-dimensional space-time there exists another term that may be added to (2.7a): the modified gauge field equation of topologically massive electrodynamics could be adopted [2].

$$\partial_\mu F^{\mu\nu} + \frac{1}{2}\kappa \epsilon^{\nu\alpha\beta} F_{\alpha\beta} = j^\nu \tag{2.7b}$$

Here κ has dimensions of mass and measures the strength of the modification. It is therefore also true that at low energies [large distances] the modification dominates the Maxwell term, so if one is considering a low energy phenomenological model, it makes sense to drop the Maxwell term altogether, giving rise to a field-current identity [4], which we in fact adopt as the gauge field equation, complementing the gauged, non-linear Schrödinger equation.

$$\frac{1}{2}\epsilon^{\mu\alpha\beta} F_{\alpha\beta} = \frac{1}{\kappa} j^\mu \tag{2.7c}$$

This is of course the equation of the Abelian Chern–Simons gauge theory; in components it reads

$$B = -\frac{1}{\kappa}\rho \tag{2.8}$$

$$E^i = \frac{1}{\kappa}\epsilon^{ij} j^j \,. \tag{2.9}$$

Equation (2.8) — the time-component of (2.7c) — is the Gauss law of Chern–Simons dynamics. It shows that $\mathbf{A}$ is determined by the matter density, and upon integrating (2.8) over all space one arrives at the important consequence that

$$\Phi = \int d^2\mathbf{r}\, B = -\frac{1}{\kappa}\int d^2\mathbf{r}\,\rho \equiv -\frac{1}{\kappa}N\,. \tag{2.10}$$

Therefore, in this theory excitations carrying charge $Q = N$ also possess magnetic flux Φ. Indeed all that is interesting about the present model is encapsulated in (2.10), which in fact also holds in the topologically massive theory (2.7b): the space integral of the time component in that equation still gives (2.10). Thus, the adoption of (2.7c), (2.8) and (2.9) may be viewed as the promotion of the spatially global formula (2.10) to the spatially local relation (2.8).

The equations (2.1), (2.3), (2.7c), (2.8) and (2.9) are the Euler–Lagrange equations of the Lagrange density

$$\mathcal{L} = \frac{\kappa}{4}\epsilon^{\alpha\beta\gamma}A_\alpha F_{\beta\gamma} + i\psi^* D_t\psi - \frac{1}{2}|\mathbf{D}\psi|^2 + \frac{g}{2}(\psi^*\psi)^2\,. \tag{2.11}$$

The first term is the Chern–Simons density. It is not gauge invariant, even though the equations that follow from varying it are. [In the quantized theory, which relies on the action not merely on the equations of motion, the issue of gauge invariance has to be re-examined [2]. One finds that in the Abelian theory there are no quantum-mechanical constraints on κ, regardless of base-space topology — contrary statements appear occasionally, but they are wrong.]

The dynamical equations may also be given a Hamiltonian formulation, involving solely the matter variables. The Hamiltonian H is

$$H = \frac{1}{2}\int d^2\mathbf{r}\left(|\mathbf{D}\psi|^2 - g(\psi^*\psi)^2\right)\,, \tag{2.12}$$

where the vector potential $\mathbf{A}$ occurring in $\mathbf{D}\psi$ is not an independent variable, but is expressed in terms of $\rho = \psi^*\psi$ by

$$\mathbf{A}(t,\mathbf{r}) = \frac{1}{\kappa}\int d^2\mathbf{r}'\,\mathbf{G}(\mathbf{r}-\mathbf{r}')\rho(t,\mathbf{r}')\,, \tag{2.13}$$

where $\mathbf{G}$ is the Greens function

$$G^i(\mathbf{r}) = \frac{1}{2\pi}\epsilon^{ij}\partial_j \ln r = \frac{\epsilon^{ij}\hat{r}^j}{2\pi r} \tag{2.14a}$$

satisfying

$$\nabla\times\mathbf{G}(\mathbf{r}) = -\delta^2(\mathbf{r}) \tag{2.14b}$$

so that $\mathbf{A}$ solves (2.5a), (2.8). Equation (2.3) then is given as

$$i\partial_t\psi(t,\mathbf{r}) = \frac{\delta H}{\delta\psi^*(t,\mathbf{r})}\,. \tag{2.15}$$

Note that A^0 does not appear in H; A^0 arises in (2.15) when varying with respect to ψ^* the vector potential $\mathbf{A}$, which is present in the spatial covariant derivative occurring in H and depends on ψ^* through (2.13). A^0 enters (2.15) as

$$A^0(t,\mathbf{r}) = \frac{1}{\kappa}\int d^2\mathbf{r}'\, \mathbf{G}(\mathbf{r}-\mathbf{r}')\cdot \mathbf{j}(t,\mathbf{r}) \tag{2.16}$$

which is easily shown to satisfy (2.5b), (2.9).

Note that the Hamiltonian formulation, with gauge fields expressed in terms of matter variables, achieves a fixing of the gauge freedom

$$A_\mu \to A_\mu + \partial_\mu \omega \tag{2.17a}$$

$$\psi \to e^{-i\omega}\psi \tag{2.17b}$$

present in the formulation based on the Lagrange density (2.11). This is because the potentials are uniquely prescribed by the integral representations (2.13) and (2.16). Alternatively one may state the gauge choice as the Coulomb condition, $\nabla \cdot \mathbf{A} = 0$, supplemented by large distance fall-off requirements on the differential equations satisfied by A^μ; the integral representations imply

$$\lim_{r\to\infty} rA^i(t,\mathbf{r}) = \frac{1}{2\pi\kappa}\epsilon^{ij}\hat{r}^j N \tag{2.18a}$$

$$\lim_{r\to\infty} A^0(t,\mathbf{r}) = 0\,. \tag{2.18b}$$

The gauge fixed Lagrangian becomes

$$L = \int d^2\mathbf{r}\, i\psi^*\partial_t\psi - H \tag{2.19a}$$

and this defines the gauge-fixed action I.

$$I = \int dt\, d^2\mathbf{r}\, i\psi^*\partial_t\psi - \int dt\, H = \int dt\, L \tag{2.19b}$$

While issues of gauge invariance, the form of the action, *etc.* are largely irrelevant in the classical theory, which is fully determined by the equations of motion, these matters play a significant role in the quantum theory, and will concern us below.

Just as the non-linear Schrödinger equation on a line, so also our gauged, planar non-linear Schrödinger equation may be considered to be a classical field theory, or alternatively a non-relativistic quantum field theory. In the former context, we shall explore soliton solutions. In the latter, we again observe number conservation

$$\frac{d}{dt}N = 0 \tag{2.20}$$

and recognize that we are dealing with the second-quantized N-body anyon problem, where the anyons possess, in addition to their statistical [Chern–Simons] interaction, a further two-body δ-function attraction of strength g, coming from the $(\psi^*\psi)^2$ non-linearity.

In what follows we are principally concerned with the classical theory. We shall choose a particular value of g, which allows further explicit analysis.

3 Transformations on the Model

Before presenting classical solutions to our dynamical equations, we record various useful transformations that can be performed. Some of these are symmetries of the equations; others map our system into other interesting dynamical models. Familiarity with these transformations assists in understanding the properties of the solutions.

The Chern–Simons contribution to the action $\int d^3x\,\mathcal{L}$ of (2.11) is a topological invariant, hence invariant against all coordinate transformations, which we present in space-time notation.

$$x^\mu \to X^\mu \tag{3.1}$$

$$A_\mu \to \tilde{A}_\mu \tag{3.2a}$$

$$\tilde{A}_\mu(x) = A_\nu(X)\frac{\partial X^\nu}{\partial x^\mu} \tag{3.2b}$$

Owing to the field-current identity (2.7c), the matter current must then transform as a contravariant density,

$$\tilde{j}^\mu(x) = \frac{1}{\Delta} j^\nu(X)\frac{\partial x^\mu}{\partial X^\nu} \tag{3.3}$$

with $\Delta \equiv \det\left(\partial x^\mu / \partial X^\nu\right)$.

But of course, the action for the non-relativistic matter dynamics is not invariant under this general coordinate transformation. Nevertheless, it is interesting to consider a subclass of transformations for which the time coordinate is transformed into a function only of time and the spatial coordinate is rescaled by a time-dependent factor.

$$x^\mu = \begin{pmatrix} t \\ \mathbf{r} \end{pmatrix} \longrightarrow \begin{pmatrix} T \\ \mathbf{R} \end{pmatrix} = \begin{pmatrix} T(t) \\ \sqrt{\dot{T}(t)}\,\mathbf{r} \end{pmatrix} \tag{3.4}$$

$$\psi \to \tilde{\psi} \tag{3.5a}$$

$$\tilde{\psi}(t,\mathbf{r}) = \sqrt{\dot{T}}\, e^{-ir^2\ddot{T}/4\dot{T}}\,\psi(T,\mathbf{R}) \tag{3.5b}$$

[The vector potential continues to transform as in (3.2).]

For arbitrary $T(t)$, (3.4) and (3.5) is not a symmetry transformation; rather, our dynamical system is mapped into one with an external harmonic force, with time-dependent frequency given by [5]

$$\omega^2(t) = \frac{\dddot{T}}{2\dot{T}} - \frac{3}{4}\left(\frac{\ddot{T}}{\dot{T}}\right)^2 = -\sqrt{\dot{T}}\frac{d^2}{dt^2}\frac{1}{\sqrt{\dot{T}}} . \tag{3.6}$$

This feature will be exploited below. However, for three particular forms of $T(t)$, the matter action remains invariant, and the transformations form the $SO(2,1)$ group.

3.1 Symmetry Transformations

The three invariances comprise

$$T(t) = t + a \qquad \text{time translation} \tag{3.7}$$

$$T(t) = at \qquad \text{dilation} \tag{3.8}$$

$$T(t) = \frac{t}{1-at} \qquad \text{conformal transformation.} \tag{3.9}$$

With these, $\tilde{\psi}$ of (3.5b) is a solution to our equations provided ψ solves them. The time translations are of course generated by the Hamiltonian (2.12). The generator of dilations D is

$$D = tH - \frac{1}{2}\int d^2\mathbf{r}\,\mathbf{r}\cdot\mathbf{j}(t,\mathbf{r}) \tag{3.10}$$

while the conformal transformation is generated by K.

$$K = -t^2 H + 2tD + \frac{1}{2}\int d^2\mathbf{r}\,r^2\rho(t,\mathbf{r}) \tag{3.11}$$

All three generators are constants of motion [6]: $\frac{d}{dt}H = \frac{d}{dt}D = \frac{d}{dt}K = 0$.

Additionally there are the conventional, expected symmetries: spatial variables may be translated, rotated or boosted: $t \to t$, $\mathbf{r} \to \mathbf{R}$.

$$\mathbf{R} = \mathbf{r} + \mathbf{a} \qquad \text{space translation} \tag{3.12}$$

$$R^i = \mathcal{R}^{ij}(\varphi)R^j \qquad \text{rotation} \tag{3.13}$$

$$\mathbf{R} = \mathbf{r} - \mathbf{v}t \qquad \text{boost} \tag{3.14}$$

Here $\mathcal{R}^{ij}(\varphi)$ is the two-dimensional matrix implementing a rotation by angle φ: $\mathcal{R}^{ij}(\varphi) = \delta^{ij}\cos\varphi - \epsilon^{ij}\sin\varphi$, while $\mathbf{v}$ is the velocity of the boost. The transformed field takes the obvious form for translations and rotations,

$$\tilde{\psi}(t,\mathbf{r}) = \psi(t,\mathbf{r}+\mathbf{a}) \tag{3.15}$$

$$\tilde{\psi}(t,\mathbf{r}) = \psi(t,\mathcal{R}\mathbf{r}) \tag{3.16}$$

while boosts require the familiar 1-cocycle.

$$\tilde{\psi}(t,\mathbf{r}) = e^{i\left(\mathbf{r}\cdot\mathbf{v} - \frac{v^2}{2}t\right)}\psi(t,\mathbf{r}-\mathbf{v}t) \tag{3.17}$$

The time-independent generators of these symmetry transformations are, respectively, the momentum $\mathbf{P}$,

$$\mathbf{P} = \int d^2\mathbf{r}\, \mathbf{j}(t,\mathbf{r}) \tag{3.18}$$

the angular momentum J,

$$J = \int d^2\mathbf{r}\, \mathbf{r} \times \mathbf{j}(t,\mathbf{r}) \tag{3.19}$$

and the Galilean boost $\mathbf{B}$,

$$\mathbf{B} = t\mathbf{P} - \int d^2\mathbf{r}\, \mathbf{r}\rho(t,\mathbf{r}) \tag{3.20}$$

with $\frac{d\mathbf{P}}{dt} = \frac{dJ}{dt} = \frac{dB}{dt} = 0$. Finally phase invariance insures number conservation (2.20).

The constants of motion/symmetry generators can be given in terms of a non-relativistic energy-momentum tensor. We define the energy density $\mathcal{H}$

$$\mathcal{H} = T^{00} = \frac{1}{2}\left(|\mathbf{D}\psi|^2 - g\rho^2\right) \tag{3.21}$$

which together with the energy flux T^{i0}

$$T^{i0} = -\frac{1}{2}\left((D_t\psi)^* D_i\psi + (D_i\psi)^*(D_t\psi)\right) \tag{3.22}$$

satisfies the continuity equation,

$$\partial_t T^{00} + \partial_i T^{i0} = 0 \tag{3.23}$$

thus ensuring that energy

$$E = H = \int d^2\mathbf{r}\, T^{00} \tag{3.24}$$

is time-independent. The momentum density $\mathcal{P}$

$$\mathcal{P}^i = T^{0i} = j^i \tag{3.25}$$

satisfies a continuity equation with the momentum flux — the stress tensor T^{ij} —

$$\begin{aligned} T^{ij} &= \frac{1}{2}\left((D_i\psi)^*(D_j\psi) + (D_j\psi)^*(D_i\psi) - \delta^{ij}(D_k\psi)^*(D_k\psi)\right) \\ &+ \frac{1}{4}\left(\delta^{ij}\nabla^2 - 2\partial_i\partial_j\right)\rho + \delta^{ij}\mathcal{H} \end{aligned} \tag{3.26}$$

$$\partial_t T^{0i} + \partial_j T^{ji} = 0 \tag{3.27}$$

which renders the momentum

$$P^i = \int d^2\mathbf{r}\, T^{0i} \tag{3.28}$$

conserved. Also of course there is the current continuity equation

$$\partial_t \rho + \nabla \cdot \mathbf{j} = 0 \tag{3.29}$$

by virtue of which N in (2.20) is time-independent.

Note that the energy-momentum tensor is not symmetric: energy flux does not equal momentum density $T^{i0} \neq T^{0i}$; the theory is not Lorentz invariant. However, the stress-tensor is symmetric $T^{ij} = T^{ji}$; this is a consequence of rotational invariance, and the conservation of angular momentum

$$J = \int d^2\mathbf{r}\, \epsilon^{ij} r^i T^{0j} \tag{3.30}$$

follows from the continuity equation (3.27) and the symmetry of T^{ij}.

The current continuity equation (3.29), together with (3.25) and (3.28) insure the conservation of the boost generator (3.20).

Finally, the dilation and conformal generators may be presented as

$$D = \int d^2\mathbf{r} \left[tT^{00} - \frac{1}{2} r^i T^{0i} \right] \tag{3.31}$$

$$K = \int d^2\mathbf{r} \left[t^2 T^{00} - t r^i T^{0i} + \frac{1}{2} r^2 \rho \right] . \tag{3.32}$$

Their time-independence follows from the previously stated continuity equations and also from the fact that the trace of the stress tensor is twice the energy density.

$$\sum_{i=1}^{2} T^{ii} = 2T^{00} \tag{3.33}$$

[Recall that in a relativistic theory the trace relation for $T^{\mu\nu}$ that encodes conformal invariance is without the factor 2.]

That the above conformal invariance is respected in classical field theory by the $(\psi^*\psi)^2$ interaction is easily understood from the quantum viewpoint. As mentioned already, this field theoretic contact term is the second-quantized form of a $\delta^2(\mathbf{r})$ interaction. But on the plane, $\delta^2(\mathbf{r})$ scales as $1/r^2$, *i.e.* in the same way as the kinetic Laplacian operator. Hence, the interaction is scale-free. However, this consideration also puts into evidence a quantum mechanical complication. The $\delta^2(\mathbf{r})$ potential in quantum mechanics requires renormalization or self-adjoint extension, which in fact spoils the scale invariance — there occurs anomalous scale symmetry breaking [7]. Correspondingly in the quantum field theoretic formulation, the coupling strength of the $(\psi^*\psi)^2$ term needs renormalization; the scale symmetry and the trace relation (3.33) acquire anomalies and must be analyzed by renormalization group-type arguments. This is especially striking when it is recalled that the *relativistic* $(\psi^*\psi)^2$ interaction is super-renormalizable in three-dimensional space-time; it requires no coupling constant renormalization and has been "constructed" by axiomatists. Consequently, the pathologies that beset the non-relativistic theory, both in the field theoretic and the quantum mechanical formulations, must arise as the non-relativistic limit is taken [8].

Since here we are concerned with the classical theory, these problems do not arise, but it is important to keep them in mind for an eventual quantum mechanical application.

3.2 Other Transformations

We have remarked that for general $T(t)$ in (3.4) and (3.5) our problem is mapped onto one with an external harmonic force, where the frequency is the time-dependent quantity (3.6). This observation allows constructing solutions $\tilde{\psi}$ to the external field problem from solutions ψ of the original problem [5].

Within this approach, it is useful to view the dynamics as involving solely matter variables, with the potentials given by (2.13) and (2.16). It then follows that the transformation rules for A_μ (3.2b) and j^μ (3.3) are consequences of the transformation law (3.5) for ψ and the formulas (2.6), (2.13) and (2.16) for these quantities in terms of the fields.

If we take $T(t) = \frac{1}{\omega}\tan\omega t$, the harmonic frequency (3.6) is constant: $\omega(t) = \omega$. Therefore from (3.5) we know that if $\psi(t, \mathbf{r})$ solves the equations in the absence of external forces, then

$$\begin{gathered} \psi_\omega(t,\mathbf{r}) = \frac{1}{\cos\omega t} e^{-i\frac{1}{2}\omega r^2 \tan\omega t} \psi(T, \mathbf{R}) \\ T = \frac{1}{\omega}\tan\omega t \quad , \qquad \mathbf{R} = \frac{\mathbf{r}}{\cos\omega t} \end{gathered} \tag{3.34}$$

solves the problem in the presence of an additional harmonic force with frequency ω. [The subscript ω indicates quantities in the external harmonic field.]

$$i\partial_t\psi_\omega = -\frac{1}{2}\mathbf{D}^2\psi_\omega + \left(A^0_\omega - g\rho_\omega + \frac{1}{2}\omega^2 r^2\right)\psi_\omega \tag{3.35}$$

where $\mathbf{A}_\omega$, from which $\mathbf{D}$ is constructed, and A^0_ω are given by (2.13) and (2.16), with density of charge ρ_ω and current $\mathbf{j}_\omega$ constructed from ψ_ω by (2.6). The conserved energy and angular momentum now read

$$H_\omega = \frac{1}{2}\int d^2\mathbf{r}\left(|\mathbf{D}\psi_\omega|^2 - g\rho^2_\omega + \omega^2 r^2 \rho_\omega\right) \tag{3.36}$$

$$J_\omega = \int d^2\mathbf{r}\,\mathbf{r}\times\mathbf{j}_\omega \quad , \tag{3.37}$$

where ρ_ω and $\mathbf{j}_\omega$ are constructed from ψ_ω according to (2.6), or equivalently are related to ρ and $\mathbf{j}$ by (3.3). Substituting (3.34) for ψ_ω gives

$$E_\omega = E + \omega^2 K \tag{3.38}$$

$$J_\omega = J\,. \tag{3.39}$$

The angular momentum is unchanged, while the energy acquires a contribution proportional to the value in the force-free problem of the conformal generator K, (3.11) [9].

The above ideas and methods can be extended to the construction of a solution in the presence of an external, constant, magnetic field from a solution without the magnetic field [5]. The equation of motion in the presence of the constant magnetic field $\mathcal{B}$, described in the radial gauge by

$$\mathcal{A}^i = -\frac{1}{2}\epsilon^{ij} r^j \mathcal{B} \tag{3.40}$$

is

$$\begin{aligned} i\partial_t \psi_{\mathcal{B}} &= -\frac{1}{2}\mathbf{D}_{\mathcal{B}}^2 \psi_{\mathcal{B}} + \left(A_{\mathcal{B}}^0 - g\rho_{\mathcal{B}}\right)\psi_{\mathcal{B}} \\ \mathbf{D}_{\mathcal{B}} &= \boldsymbol{\nabla} - i\mathbf{A}_{\mathcal{B}} - i\mathcal{A}\,, \end{aligned} \tag{3.41}$$

where $A_{\mathcal{B}}^0$ and $\mathbf{A}_{\mathcal{B}}$ are still given by (2.13) and (2.16), with density of charge $\rho_{\mathcal{B}}$ and current $\mathbf{j}_{\mathcal{B}}$, which are constructed from $\psi_{\mathcal{B}}$ by formulas like (2.6), except that the covariant derivative in the definition of $\mathbf{j}_{\mathcal{B}}$ involves the external magnetic field. [The subscript $\mathcal{B}$ indicates quantities in the presence of the external magnetic field.] One may verify that the solution to (3.41) is constructed from the solution $\psi(t, \mathbf{r})$ in the absence of $\mathcal{B}$ by

$$\begin{aligned} \psi_{\mathcal{B}}(t,\mathbf{r}) &= \frac{1}{\cos\frac{\mathcal{B}}{2}t} e^{-i\frac{\mathcal{B}}{4}r^2 \tan\frac{\mathcal{B}}{2}t}\, e^{i(N/4\pi\kappa)\mathcal{B}t}\psi(T,\mathbf{R}) \\ T &= \frac{2}{\mathcal{B}}\tan\frac{\mathcal{B}}{2}t\,, \qquad R^i = \frac{1}{\cos\frac{\mathcal{B}}{2}t}\mathcal{R}^{ij}\left(\frac{\mathcal{B}}{2}t\right) r^j\,. \end{aligned} \tag{3.42}$$

The transformation exhibited in (3.42) is a time-dependent dilation [like in (3.34)] supplemented by a time-dependent rotation by angle $\frac{\mathcal{B}}{2}t$ [a generalization of (3.13) and (3.16)] and finally a gauge transformation with the gauge function $-\frac{\mathcal{B}}{4\pi\kappa}Nt$ [like in (2.17)].

The gauge transformation is needed to assure that the gauge potential $A_{\mathcal{B}}^0$ satisfy (2.18), or what is equivalent, that the equation of motion (3.41) follow from the matter Hamiltonian $H_{\mathcal{B}}$.

$$H_{\mathcal{B}} = \frac{1}{2}\int d^2\mathbf{r}\left(|\mathbf{D}_{\mathcal{B}}\psi_{\mathcal{B}}|^2 - g\rho_{\mathcal{B}}^2\right) \tag{3.43}$$

As a consequence of the gauge transformation, the vector potential transformation formula (3.2) is modified,

$$A_{\mu\mathcal{B}}(x) = A_\nu(X)\frac{\partial X^\nu}{\partial x^\mu} - \partial_\mu\left(\frac{\mathcal{B}}{4\pi\kappa}Nt\right)\,, \tag{3.44}$$

but the transformation law (3.3) for the gauge invariant current remains unchanged.

In addition to the energy, the momentum $\mathbf{P}_{\mathcal{B}}$ and angular momentum $J_{\mathcal{B}}$ are conserved. Their expressions, in the presence of the external magnetic field, are

$$P^i_{\mathcal{B}} = \int d^2\mathbf{r}\, j^i_{\mathcal{B}} - \mathcal{B} \int d^2\mathbf{r}\, \epsilon^{ij} r^j \rho_{\mathcal{B}} \tag{3.45}$$

$$J_{\mathcal{B}} = \int d^2 r\, \mathbf{r} \times \mathbf{j}_{\mathcal{B}} + \frac{1}{2}\mathcal{B} \int d^2\mathbf{r}\, r^2 \rho_{\mathcal{B}} \,. \tag{3.46}$$

On (3.42) these constants of motion are expressed in terms of the $\mathcal{B} = 0$ constants, and one finds

$$E_{\mathcal{B}} = E + \frac{\mathcal{B}^2}{4} K - \frac{\mathcal{B}}{2} J \tag{3.47}$$

$$P^i_{\mathcal{B}} = P^i + \frac{\mathcal{B}}{2} \epsilon^{ij} B^j \tag{3.48}$$

$$J_{\mathcal{B}} = J \,. \tag{3.49}$$

The energy acquires contributions from the conformal (K) and angular momentum (J) generators; the momentum is supplemented by a term involving the boost generator ($\mathbf{B}$); while the angular momentum remains unchanged.

The above transformations can be used to generate soliton solutions in the presence of external fields from those in their absence [10].

We conclude this discussion by remarking that the symmetry transformations of the force-free problem, which lead to eight conserved generators (H, D, K. $\mathbf{P}$, J, $\mathbf{B}$) and are discussed in Subsection 3.1 above, can be "imported" to the external harmonic and magnetic field problems by the transformations (3.34) and (3.42). In this way, one finds conserved generators/constants of motion that are additional to those in (3.36), (3.37) and (3.43), (3.45), (3.46) [11].

4 Static Solutions

4.1 General Properties

We consider static solutions to the model described in Section 2; *viz.* we seek solutions to

$$-\frac{1}{2}(\boldsymbol{\nabla} - i\mathbf{A})^2 \psi + A^0\psi - g\rho\psi = 0 \tag{4.1}$$

$$\boldsymbol{\nabla} \times \mathbf{A} = -\frac{1}{\kappa}\rho \tag{4.2}$$

$$\boldsymbol{\nabla} \times A^0 = \frac{1}{\kappa}\mathbf{j} \,. \tag{4.3}$$

These are the time-independent version of (2.1), (2.3), (2.5), (2.8) and (2.9).

Before proceeding, let us observe the following important consequences of the symmetries in our problem. Galileo invariance insures that the boost generator $\mathbf{B}$ in (3.20) is time-independent. But for static solutions $\int d^2\mathbf{r}\,\mathbf{r}\rho$ is also time-independent. It therefore follows that the momentum $\mathbf{P} = \int d^2\mathbf{r}\,\mathbf{j}$ vanishes on static solutions — a natural and not unexpected fact. An additional, remarkable vanishing theorem follows from conformal invariance, which requires that K in (3.11) be time-independent. Since static solutions lead to time-independent $\int d^2\mathbf{r}\,r^2\rho$, it must be that H and D also vanish, which from (3.10) further implies that $\int d^2\mathbf{r}\,\mathbf{r}\cdot\mathbf{j} = 0$. Thus static solutions carry zero energy and momentum [12]. [It should be stated that implicitly we are making various regularity assumptions, some of which will be made explicitly later.]

In order to explore the implication of vanishing energy [for static solutions] we make use of the following identity

$$|\mathbf{D}\psi|^2 = |(D_x - i\epsilon(\kappa)D_y)\,\psi|^2 - \epsilon(\kappa)\,(B\rho + \boldsymbol{\nabla}\times\mathbf{j}) \quad , \quad \epsilon(\kappa) = \operatorname{sign}\kappa \tag{4.4}$$

to present the Hamiltonian (2.12) as

$$H = \frac{1}{2}\int d^2\mathbf{r}\,\left\{|(D_x - i\epsilon(\kappa)D_y)\,\psi|^2 - \left(g - \frac{1}{|\kappa|}\right)\rho^2\right\} \; , \tag{4.5}$$

where $\int d^2\mathbf{r}\,\boldsymbol{\nabla}\times\mathbf{j}$ has been dropped with the hypothesis that $\mathbf{j}$ is sufficiently well-behaved. Since the first term in the integrand of (4.5) is non-negative and so is ρ^2, a non-trivial ($\psi \neq 0$) zero-energy configuration is possible only for $g \geq \frac{1}{|\kappa|}$, since otherwise the energy is positive for non-zero ψ.

Henceforth we take $g = \frac{1}{|\kappa|}$ so that the field non-linearity [δ-function interaction] is attractive and with strength correlated to the Chern–Simons coupling, κ. Also, without loss of generality κ may be taken positive, so the Hamiltonian becomes

$$H = \frac{1}{2}\int d^2\mathbf{r}\,|(D_x - iD_y)\,\psi|^2 \; . \tag{4.6}$$

Since time-independent solutions stationarize H and moreover H must also then vanish, we see that static solutions require a self-duality condition,

$$(D_x - iD_y)\,\psi = 0 \tag{4.7a}$$

$$\mathbf{D}\psi = i\mathbf{D}\times\psi \tag{4.7b}$$

and of course the vector potential $\mathbf{A}$ in the above covariant derivatives also satisfies

$$\boldsymbol{\nabla}\times\mathbf{A} = -\frac{1}{\kappa}(\psi^*\psi) \tag{4.8}$$

Thus the system of first-order equations (4.7), (4.8) is completely equivalent [with regularity assumptions] to the second-order (4.1), (4.2), (4.3) at $g = \frac{1}{\kappa}$.

That the first-order equations imply the second order ones is easily verified; that the implication also goes in the other direction is a consequence of conformal symmetry.

Note that self-duality (4.7) has the consequence that

$$\mathbf{j} = \frac{1}{2}\boldsymbol{\nabla} \times \rho\,. \tag{4.9}$$

Hence according to (4.3)

$$A^0 = \frac{1}{2\kappa}\rho \tag{4.10}$$

and also it is obvious from (4.9) that $\int d^2\mathbf{r}\,\mathbf{j}$ and $\int d^2\mathbf{r}\,\mathbf{r}\cdot\mathbf{j}$ do both indeed vanish, as was earlier established by a general argument. The angular momentum (3.19) is expressed in terms of N, which need not be an integer.

$$J = \int d^2\mathbf{r}\,\mathbf{r} \times \left(\frac{1}{2}\nabla \times \rho\right) = \int d^2\mathbf{r}\,\rho = N \tag{4.11}$$

This is related to flux Φ by the Chern–Simons constraint.

$$\Phi = \int d^2\mathbf{r}\,B = -\frac{1}{\kappa}\int d^2\mathbf{r}\,\rho = -\frac{1}{\kappa}N \tag{4.12}$$

It will be seen that Φ is quantized. The remaining, non-vanishing symmetry generators/constants of motion are

$$\mathbf{B} = -\int d^2\mathbf{r}\,\mathbf{r}\rho \equiv -N\langle\mathbf{r}\rangle \tag{4.13}$$

and

$$K = \frac{1}{2}\int d^2\mathbf{r}\,r^2\rho \equiv \frac{1}{2}N\left\langle r^2\right\rangle\,. \tag{4.14}$$

The values of $\langle\mathbf{r}\rangle$ and $\left\langle r^2\right\rangle$ depend on the form of the solution.

Finally we discuss the particular choice $g = 1/\kappa$ for the non-linearity. To be sure, this choice is made to facilitate the subsequent self-dual formulation, which leads to explicit solution, see below. Nevertheless, the choice has several "natural" features that are worth noting.

Observe that the Pauli equation

$$\boldsymbol{\sigma}\cdot\left(\frac{1}{i}\boldsymbol{\nabla} - \mathbf{A}\right)\chi = \epsilon\chi \tag{4.15}$$

for a two-spinor χ, and in two dimensions involves two Pauli matrices. Upon iteration one gets

$$\left\{\sigma\cdot\left(\frac{1}{i}\boldsymbol{\nabla} - \mathbf{A}\right)\right\}^2\chi = \left\{-(\nabla - iA)^2 - \sigma^3 B\right\}\chi = \epsilon^2\chi\,. \tag{4.16a}$$

One may choose the χ spinor to be an eigenstate of σ^3, $\sigma^3\chi_\pm = \pm\chi_\pm$, and also owing to the Chern–Simons constraint, $B = -\frac{1}{\kappa}\chi^*_\pm\chi_\pm$. Hence (4.16a) becomes

$$\left\{-\frac{1}{2}(\nabla - i\mathbf{A})^2 \pm \frac{1}{2\kappa}\chi^*_\pm\chi_\pm\right\}\chi_\pm = \frac{\epsilon^2}{2}\chi_\pm\,. \tag{4.16b}$$

With the lower sign, the left-hand side coincides with (2.3) and (4.1) at $g = 1/\kappa$ and A^0 evaluated from (4.10). Thus we see that the non-linearity may be alternatively viewed as a magnetic moment interaction with a magnetic field that is given self-consistently by the Chern–Simons equation. Moreover, when $g = 1/\kappa$ the magnetic moment is the "minimal" one in a spinorial formulation, and therefore there is a "hidden" supersymmetry [3].

When we later discuss a relativistic soliton model, we shall demonstrate that $g = 1/\kappa$ corresponds to the non-relativistic limit of that theory, which also possesses supersymmetric features.

4.2 Self-Dual Solutions

In fact the system (4.7) and (4.8) may be solved completely and explicitly. Equation (4.7) implies that

$$A^x - iA^y = i(\partial_x - i\partial_y)\ln\psi\,. \tag{4.17}$$

When this form for $\mathbf{A}$ is inserted in (4.8) one finds that $\rho = \psi^*\psi$ satisfies the Liouville equation.

$$\nabla^2 \ln\rho = -\frac{2}{\kappa}\rho \tag{4.18}$$

Since

$$\psi = \rho^{1/2}\, e^{i\omega} \tag{4.19}$$

we see that there is no equation determining the phase ω; rather it is fixed by regularity requirements: At the zeroes of ψ, equivalently at the zeroes of ρ, $\ln\psi$ and $\ln\rho$ may be singular. Demanding that $\mathbf{A}$, determined in (4.17) by the holomorphic derivative of $\ln\psi$, be non-singular fixes the phase of ψ, which is then further constrained by the requirement that ψ be single-valued near its zeroes. This is illustrated below in detail. It is also understood that (4.18) holds away from the zeroes of ρ, and the non-singular ρ is determined near its zeroes by continuity.

The general solution to (4.18) is known. Let us consider first in (r,θ) variables the radially symmetric, θ-independent solutions, which can be constructed by quadrature from the second-order r-differential equation that is implied by (4.18). One finds the most general regular, positive and θ-independent ρ to be [3]

$$\rho(\mathbf{r}) = \frac{4\kappa\mathcal{N}^2}{r^2}\left(\left(\frac{r_0}{r}\right)^{\mathcal{N}} + \left(\frac{r}{r_0}\right)^{\mathcal{N}}\right)^{-2}, \tag{4.20}$$

where r_0 and $\mathcal{N}$ are two positive constants of integration. Clearly r_0 is a scale, reflecting the scale invariance of the dynamics. To fix $\mathcal{N}$, let us observe that regularity at the origin, where $\rho \underset{r\to 0}{\sim} r^{2\mathcal{N}-2}$, and at infinity, where $\rho \underset{r\to\infty}{\sim} r^{-2\mathcal{N}-2}$, require $\mathcal{N} \geq 1$. Moreover, since $\rho^{1/2}$ has a zero at the origin of order $\mathcal{N}-1$, singularities in $\mathbf{A}$ are avoided if the phase of ψ is chosen to be $\omega = (1-\mathcal{N})\theta$. Finally for single-valued ψ, $\mathcal{N}$ must be an integer. In this way we conclude that [3]

$$\psi(\mathbf{r}) = \frac{2\sqrt{\kappa}\mathcal{N}}{r}\left(\left(\frac{r_0}{r}\right)^{\mathcal{N}} + \left(\frac{r}{r_0}\right)^{\mathcal{N}}\right)^{-1} e^{i(1-\mathcal{N})\theta} \quad , \quad \mathcal{N} = 1, 2, \ldots \tag{4.21}$$

This is interpreted as an $\mathcal{N}$-soliton/vortex solution, where the solitons are superimposed at the origin, and each has the same scale r_0. Note that with this solution

$$\Phi = -\frac{1}{\kappa}\int d^2\mathbf{r}\,\rho(\mathbf{r}) = -2\pi(2\mathcal{N})\,. \tag{4.22}$$

In units of 2π, which is the natural unit of flux with our conventions, our solution carries an even number, $2\mathcal{N}$, flux units. Also we get

$$J = N = -\kappa\Phi = 2\pi\kappa(2\mathcal{N}) \tag{4.23}$$

which is not an integer for arbitrary κ. Owing to the rotational symmetry, the Galileo boost generator vanishes.

$$\mathbf{B} = -N\langle\mathbf{r}\rangle = 0 \tag{4.24}$$

The conformal charge is

$$K = \frac{1}{2}N\langle r^2\rangle = \frac{1}{2}Nr_0^2\left(\frac{\pi/\mathcal{N}}{\sin\pi/\mathcal{N}}\right)\,. \tag{4.25}$$

This diverges for $\mathcal{N} = 1$, and tends to $\frac{1}{2}Nr_0^2$ for $\mathcal{N}\to\infty$.

The general solution to the Liouville equation, producing a physically acceptable ρ, involves a holomorphic function $f(z)$, $z = r\,e^{i\theta}$, and its complex conjugate

$$\rho(\mathbf{r}) = \frac{4\kappa\,|f'(z)|^2}{\left(1+|f(z)|^2\right)^2}\,. \tag{4.26}$$

The previous solution, with $\mathcal{N}$ solitons superimposed at the origin, corresponds to $f(z) = c_0\,z^{-\mathcal{N}}$. Separated solitons are obtained by taking [3]

$$f(z) = \sum_{n=1}^{\mathcal{N}}\frac{c_n}{z-z_n}\,; \qquad z_n = x_n + iy_n\,. \tag{4.27}$$

This describes $\mathcal{N}$ solitons at $2\mathcal{N}$ real positions z_n and $2\mathcal{N}$ real scales and phases coded in the complex numbers c_n. Thus the solution depends on $4\mathcal{N}$ real parameters. That this is the most general solution is confirmed by index theory [13].

It is easy to evaluate the flux carried by the general solution (4.26), (4.27); it is given by asymptotic data and coincides with (4.22).

$$\Phi = -2\pi(2\mathcal{N}) \tag{4.28}$$

Angular momentum as well as number follow as before.

$$J = N = 2\pi\kappa(2\mathcal{N}) \tag{4.29}$$

The boost generator is given by the center-of-mass, which is now non-vanishing.

$$\mathbf{B} = -4\pi\kappa\sum_{n=1}^{\mathcal{N}}\mathbf{r}_n \tag{4.30}$$

The conformal generator cannot be in general evaluated explicitly. But it should be remarked that since $f(z) \underset{z\to\infty}{\longrightarrow} \sum_{\kappa=1}^{\mathcal{N}} c_n/z$, the large distance behavior of the generic $\mathcal{N}$-soliton solution is the same as that of a single soliton, unless $\sum_{\kappa=1}^{\mathcal{N}} c_n = 0$. Consequently, in the generic case K diverges.

The flux in our static solution is quantized by even integers ($2\mathcal{N}$), as in (4.22) and (4.28) A solution involving an arbitrary integer is available, but it describes a soliton/vortex condensate. To achieve this one takes $f(z)$ in (4.26) to be doubly periodic, producing a periodic charge density. Flux must now be computed over a unit cell in space. With a particular choice for f, ρ possesses a single zero in the unit cell, leading to a single unit of flux [14].

Finally let us remark that it has been occasionally stated that the Chern–Simons interaction has no dynamical effect save modifying spin and statistics in the quantum theory. This assertion appears to be based on the observation that the Green's function $\mathbf{G}$ occurring in (2.13) and given by (2.14) can be presented as a gradient of the azimuthal angle $\theta(\mathbf{r}) \equiv \tan^{-1} y/x$,

$$\mathbf{G}(\mathbf{r}) = -\frac{1}{2\pi}\boldsymbol{\nabla}\theta \tag{4.31}$$

so that

$$\mathbf{A}(t,\mathbf{r}) = -\frac{1}{2\pi\kappa}\int d^2\mathbf{r}'\boldsymbol{\nabla}\theta\,(\mathbf{r}-\mathbf{r}')\,\rho(t,\mathbf{r}')\,. \tag{4.32}$$

If one could interchange the $\mathbf{r}'$-integral with the $\mathbf{r}$-gradient, one would conclude that $\mathbf{A}$ is a [multi-valued] pure gauge, removable by a singular gauge transformation, which only affects phases of quantum mechanical wave functions, and has no other dynamical role. However, the integral *cannot* be exchanged with the derivative, because the integrand is multi-valued; $\mathbf{A}$ is *not* a pure gauge [3]. That this is indeed true is put into evidence by the existence of non-trivial soliton solutions: if the gauge field interaction is removable, the only dynamics would be provided by the scalar field self-interaction, but it is known that this cannot bind static solitons on the plane. [Consistency between (2.14b) and (4.31) is achieved by virtue of the unexpected formula $\nabla \cdot \nabla \times \theta = 2\pi\delta^2(\mathbf{r})$.]

5 Time-Dependent Solutions

A complete analysis of the time-dependent, gauged non-linear Schrödinger equation on the plane has not been achieved thus far, and it is not known whether the time-dependent system is as completely integrable in $(2+1)$-dimensional space-time, as is the static system on the two-dimensional plane. Our model is tantalizingly similar to the Davey–Stewartson equation, which is known to be integrable in $(2+1)$ dimensions, for a range of its parameters [15]. But the similarity has not been elevated to an identity.

Some time-dependent solutions have been found. All of them are related to our previously described static solutions, and they arise both in the absence and in the presence of external [harmonic, magnetic] interactions. We describe these in turn.

5.1 Solutions in the Absence of External Fields

In the absence of external fields, the time-dependent solutions that have been constructed thus far are symmetry transforms of the static solutions. Both a Galilean boost [with velocity $\mathbf{v}$] and a conformal boost [with parameter a] will introduce time-dependence in the static soliton/vortex profile, while preserving the validity of the equation of motion. The transformation can also be a combination of the two boosts [K and $\mathbf{B}$ commute in the Lie algebra]. Therefore the boosted function [described by subscript b] is [16]

$$\psi_b(t,\mathbf{r}) = \frac{1}{1-at} e^{iv^2/2a}\, e^{-i(a/2(1-at))(\mathbf{r}-\mathbf{v}/a)^2}\, \psi(\mathbf{R}) \tag{5.1}$$

$$\mathbf{R} = \frac{\mathbf{r}-\mathbf{v}t}{1-at}\,, \tag{5.2}$$

where $\psi(\mathbf{r})$ is one of our static solutions. The transformation laws for the potentials and current are as in (3.2) and (3.3), with $X^\mu = \left(\frac{t}{1-at}, \mathbf{R}\right)$.

The constants of motion on these boosted generators differ from their static forms [save for K and $\mathbf{B}$, which remain unchanged since they commute]. The momentum and energy no longer vanish; rather one finds boosted expressions.

$$\mathbf{P}_b = N\mathbf{v}_a \tag{5.3}$$

$$E_b = \frac{1}{2}Nv_a^2 + \frac{1}{2}Na^2 \langle \Delta\mathbf{r}\rangle^2 \tag{5.4}$$

The angular momentum is also boosted.

$$J_b = \langle \mathbf{r}\rangle \times N\mathbf{v}_a + N \tag{5.5}$$

Here $\mathbf{v}_a \equiv \mathbf{v} - a\langle \mathbf{r}\rangle$, $(\Delta\mathbf{r})^2 \equiv \langle r^2\rangle - \langle \mathbf{r}\rangle^2$. With a pure Galileo boost ($a = 0$), we see that the soliton moves as a particle with mass $M = N$. [Note that the mass m in the Schrödinger equation has been scaled to unity.] A conformal boost ($a \neq 0$) shifts the velocity $\mathbf{v} \to \mathbf{v}_a$ and shows that there is internal structure, since $(\Delta\mathbf{r})^2 \neq 0$, Finally, the dilation generator reads

$$D_b = -\frac{1}{2}\langle \mathbf{r}\rangle \cdot \mathbf{P}_b + \frac{1}{2}Na(\Delta\mathbf{r})^2 \ . \tag{5.6}$$

Specific forms for the static solution are needed to evaluate $\langle \mathbf{r}\rangle$ and $\langle r^2\rangle$, defined by (4.13) and (4.14). For the rotationally symmetric profile, $\langle \mathbf{r}\rangle$ vanishes while $\langle r^2\rangle$ converges only when $\mathcal{N} > 1$; see (4.25). For the generic $\mathcal{N}$-vortex solution $\langle r^2\rangle$ diverges, unless the previously mentioned sum rule holds. There is no physical answer to the question why infinite energy is required to boost conformally the generic vortex.

5.2 Solutions in the Presence of External Fields

Transforming according to (3.5) the solutions in the absence of external fields, both the static solutions and the above-described, time-dependent boosted ones, produces time-dependent solutions [10] in the presence of (i) a harmonic force [17] with frequency ω and/or (ii) a constant magnetic field [10] $\mathcal{B}$. We describe these new solutions, but restrict the discussion to transforms of static solutions; transforms of boosted solutions are discussed in the literature, [10],[17], but do not appear to provide new information.

In the presence of an external harmonic force with frequency ω,

$$\psi_\omega(t, \mathbf{r}) = \frac{1}{\cos\omega t} e^{-i\frac{1}{2}\omega r^2 \tan\omega t}\psi\left(\frac{\mathbf{r}}{\cos\omega t}\right) \tag{5.7}$$

is a periodic solution to (3.35), with period $2\pi/\omega$, where $\psi(\mathbf{r})$ is a static, zero-energy solution at $\omega = 0$, see (3.34). The energy of this solution, from (3.38), is

$$E_\omega = \frac{1}{2}N\omega^2 \langle r^2\rangle \tag{5.8}$$

and may diverge, as explained before. According to (3.39) the angular momentum does not see the harmonic force.

$$J_\omega = N \tag{5.9}$$

The momentum, which is *not* a constant of motion, is found to be

$$P_\omega(t) = \int d^2\mathbf{r}\, \mathbf{j}_\omega(t,\mathbf{r}) = -N\langle \mathbf{r}\rangle\, \omega \sin\omega t\ . \tag{5.10}$$

Because (5.7) is periodic in time it may be quantized by the semiclassical [Bohr–Sommerfeld] method. To this end we integrate the canonical one-form $\int d^2\mathbf{r}\, i\psi_\omega^* \partial_t \psi_\omega$ over the periods $2\pi/\omega$, and equate this to $2\pi n$, where n is a "principal quantum number." Since $\int d^2\mathbf{r}\, i\psi_\omega^*\partial_t\psi = L_\omega + H_\omega$ where L_ω and H_ω are the Lagrangian and Hamiltonian [see (2.19) and (3.36)], an energy quantization condition follows.

$$E_\omega = \omega\left(n - \frac{1}{2\pi}\int_0^{2\pi/\omega} dt\, L_\omega\right) \tag{5.11}$$

When the Lagrangian is evaluated on a solution to (3.35) only the non-quadratic terms survive.

$$L_\omega = \int d^2\mathbf{r}\left(A_\omega^0 - \frac{1}{2\kappa}\rho_\omega\right)\rho_\omega \tag{5.12}$$

Substituting the solution (5.7) into (5.12) shows that L_ω vanishes on the solution. Thus the semi-classical quantization rule (5.11) becomes [18]

$$E_\omega = \omega n \tag{5.13}$$

which of course coincides with that for a point-particle in a harmonic potential on a plane.

A similar analysis can be performed for the problem with external magnetic field $\mathcal{B}$ [10],[18]. The solution to (3.41), which is transformed from the static $\mathcal{B}=0$ solution, is given by (3.42) as

$$\begin{aligned} \psi_{\mathcal{B}}(t,\mathbf{r}) &= \frac{1}{\cos\frac{\mathcal{B}}{2}t} e^{-i\frac{\mathcal{B}}{4}r^2 \tan\frac{\mathcal{B}}{2}t}\, e^{i(N/4\pi\kappa)\mathcal{B}t}\psi(\mathbf{R}) \\ R^i &= r^i - \tan\left(\frac{\mathcal{B}}{2}t\right)\epsilon^{ij}r^j\ . \end{aligned} \tag{5.14}$$

The kinematical characteristics of this solution are coded in the constants of motion. The energy, from (3.47), is

$$E_{\mathcal{B}} = \frac{1}{8}N\mathcal{B}^2\langle r^2\rangle - \frac{\mathcal{B}}{2}N \tag{5.15}$$

and this may diverge. The angular momentum of (3.49) again does not see the external field.

$$J_{\mathcal{B}} = N \tag{5.16}$$

For the momentum one gets from (3.48)

$$P_{\mathcal{B}}^{i} = -\frac{1}{2} N \mathcal{B} \epsilon^{ij} \left\langle r^{j} \right\rangle \tag{5.17}$$

Note that

$$E_{\mathcal{B}} + \mathcal{B} J_{\mathcal{B}} = \frac{P_{\mathcal{B}}^{2}}{2N} + \frac{\mathcal{B}}{2} N + \frac{N}{8} \mathcal{B}^{2} (\Delta \mathbf{r})^{2} . \tag{5.18}$$

This shows that apart from the internal energy, proportional to $(\Delta\mathbf{r})^2$, the kinematics of our vortex-soliton is that of a point particle, with mass $M = N$, spin N and unit g-factor, moving in the external magnetic field $\mathcal{B}$.

The semiclassical quantization condition may be imposed on $\psi_{\mathcal{B}}$ [18],[19]. Note that $\tan \frac{\mathcal{B}}{2} t$ has period $2\pi/\mathcal{B}$. The Jacobian factor in (5.14), *viz.* $1/\cos \frac{\mathcal{B}}{2} t$, really enters with absolute value, hence the period of $\psi_{\mathcal{B}}(t, \mathbf{r})$ is $2\pi/\mathcal{B}$, and the quantization condition, analogous to (5.11) is

$$E_{\mathcal{B}} = \mathcal{B} \left(n - \frac{1}{2\pi} \int_{0}^{2\pi/\mathcal{B}} dt\, L_{\mathcal{B}} \right) , \tag{5.19}$$

where the Lagrangian, evaluated on the solution reads

$$L_{\mathcal{B}} = \int d^{2}\mathbf{r} \left(A_{\mathcal{B}}^{0} - \frac{1}{2\kappa} \rho_{\mathcal{B}} \right) \rho_{\mathcal{B}} . \tag{5.20a}$$

Evaluating $L_{\mathcal{B}}$ with the explicit solution (5.4) leaves,

$$L_{\mathcal{B}} = -\frac{\mathcal{B} N^{2}}{8\pi\kappa} . \tag{5.20b}$$

Therefore the quantization condition becomes

$$E_{\mathcal{B}} = \mathcal{B} \left(n + \frac{N^{2}}{8\pi\kappa} \right) . \tag{5.21}$$

Note that the divergence in the classical harmonic oscillator and magnetic energy, evaluated on solutions with slow large-distance fall off, for which have no physical interpretation, does not appear to affect the semiclassical quantization results (5.13) and (5.21).

Finally let us remark that static solutions in the presence of a magnetic field have also been considered [10] — only numerical results are available.

6 Relativistic Abelian Model

Another Abelian Chern–Simons theory, closer to the interests of particle physicists, makes use of relativistic dynamics for the matter degrees of freedom and supports self-dual solitons if the field potential is taken in a particular form. We consider the Lagrange density

$$\mathcal{L} = \frac{\kappa}{4}\epsilon^{\alpha\beta\gamma}A_\alpha F_{\beta\gamma} + (D_\mu\varphi)^* (D_\mu\varphi) - V(|\varphi|^2) \tag{6.1}$$

with a potential function [24],[25]

$$V(|\varphi|^2) = \frac{1}{\kappa^2}|\varphi|^2 \left(|\varphi|^2 - v^2\right)^2 \tag{6.2}$$

that exhibits two, zero-energy and degenerate minima: one at

$$|\varphi| = v \tag{6.3}$$

which breaks the $U(1)$ symmetry, and another, symmetry preserving one, at

$$\varphi = 0\ . \tag{6.4}$$

The form of V is dictated by our desire to arrive at a static, self-dual system. However, there is also a supersymmetry connection: the Lagrangian (6.1), (6.2) is the bosonic partner in an $N = 2$ supersymmetry of a fermionic Lagrangian [26].

$$\mathcal{L}_F = i\bar{\psi}\gamma^\mu D_\mu\psi + \frac{1}{\kappa}\left(3|\varphi|^2 - v^2\right)\bar{\psi}\psi \tag{6.5}$$

The supersymmetry is preserved at both minima (6.3) and (6.4).

At the symmetric minimum, there are two bosonic excitations of mass

$$m = \frac{v^2}{\kappa} \tag{6.6}$$

associated with the charged scalar field; the fermions carry the same mass as is seen from (6.5), while the gauge field does not support any excitations. At the symmetry breaking minimum, the Higgs phenomenon sets in [in spite of the purely Chern–Simons kinetic term], the Higgs particle carries mass

$$m_H = 2\frac{v^2}{\kappa} \tag{6.7}$$

which is also the mass of the single degree of freedom now associated with the gauge field — for a total of two bosonic excitations as before; the fermion mass in the symmetry breaking case also is (6.7). The mass degeneracy in the two cases is of course a consequence of the supersymmetry, which remains unbroken.

Before proceeding with the analysis of the relativistic model, it is interesting to discuss its non-relativistic limit, which we consider in the symmetric sector. Upon writing φ as

$$\varphi = \frac{1}{\sqrt{2m}}e^{-imt}\psi + \frac{1}{\sqrt{2m}}e^{imt}\tilde{\psi}^* \tag{6.8}$$

ψ becomes in the non-relativistic ($m \to \infty$) limit the annihilation operator for particles and $\tilde{\psi}^*$ the creation operator for anti-particles. Upon substituting (6.8) in (6.2), and dropping all oscillatory and sub-dominant terms at large m, we arrive at a model in which particles and anti-particles are separately conserved, and the particle, zero anti-particle sector coincides with our non-relativistic model [in which m is set to unity], where the non-linearity is precisely $1/\kappa$. Thus the previously considered self-dual $U(1)$ theory of (4.6) is the non-relativistic limit of the relativistic self-dual $U(1)$ theory (6.1), with a symmetric realization of the $U(1)$ gauge symmetry.

Static equations stationarize the static energy functional, which is the negative of the Lagrangian with all time-derivatives suppressed. Moreover, with the help of the Gauss law constraint

$$B = \nabla \times \mathbf{A} = -\frac{1}{\kappa}\rho\,, \tag{6.9}$$

where the relativistic charge density for static fields reads,

$$\rho = i\left(\varphi^* D_0 \varphi - \varphi D_0 \varphi^*\right) = -2A^0|\varphi|^2 \tag{6.10}$$

A^0 may be eliminated.

$$A^0 = \frac{B}{2\kappa|\varphi|^2} \tag{6.11}$$

Note that in contrast to Nielsen–Olesen vortices, here the magnetic field must vanish whenever φ does. With (6.11) the static energy functional is manifestly positive and becomes

$$E = \int d^2\mathbf{r}\left(\frac{\kappa^2}{4}\frac{B^2}{|\varphi|^2} + |\mathbf{D}\varphi|^2 + V\left(|\varphi|^2\right)\right) \tag{6.12}$$

which is further rewritten by using the identity

$$|\mathbf{D}\varphi|^2 = |(D_x \pm iD_y)\varphi|^2 \pm B|\varphi|^2 \pm \nabla \times \mathbf{j} \tag{6.13}$$

and dropping the last, total derivative when integrating (6.12) over all space.

$$E = \int d^2\mathbf{r}\left\{|(D_x \pm iD_y)\varphi|^2 + \left|\frac{\kappa}{|\varphi|}B \pm \frac{1}{\kappa}|\varphi|\left(|\varphi|^2 - v^2\right)\right|^2\right\} \pm v^2\Phi\,. \tag{6.14}$$

Here Φ is the total flux, which as always equals $-\frac{1}{\kappa}N$, by virtue of the Chern–Simons condition (6.9). Thus we conclude that for a fixed value of flux there is a lower bound on energy

$$E \geq v^2|\Phi|\,. \tag{6.15}$$

Since static configurations that are stationary points of the energy are also stationary points of the action, the Euler-Lagrange equations of the theory will be satisfied by static configurations obeying the self-duality equation,

$$(D_x \pm iD_y)\varphi = 0 \tag{6.16a}$$

$$\mathbf{D}\varphi = \mp i\mathbf{D} \times \varphi \tag{6.16b}$$

and the Chern–Simons constraint,

$$eB = \pm\frac{m_H^2}{2}\frac{|\varphi|^2}{v^2}\left(1 - \frac{|\varphi^2|}{v^2}\right), \tag{6.17}$$

where the upper [lower] sign corresponds to a positive [negative] value of Φ; these solutions achieve the lower bound in Eq. (6.15).

Equations (6.16) and (6.17) possess topologically stable vortex solutions for which $|\varphi| \to v$ at large distances and Φ is quantized [24],[25]. But there also exist non-topological soliton solutions for which $\varphi \to 0$ asymptotically [25],[27]. For these the flux is not quantized but rather is an arbitrary parameter describing the solution. This parameter can be continuously varied; therefore it is evident, since the energy $E = v^2|\Phi|$, that these solutions are not stationary points of the energy. This is understood by recalling that the energy is stationary provided that the field variations vanish faster than $1/r$. Such variations are sufficient to establish the Euler–Lagrange equations, so the solutions of the self-duality equations are indeed solutions of the full field equations. The above-mentioned non-stationary variations have $\delta\mathbf{A} \sim 1/r$.

Since the relativistic theory is not conformally invariant no argument has been constructed which would imply that [regular] static solutions are always self-dual, as in the non-relativistic case.

The energy of a non-topological soliton with a given charge $Q = N$ is

$$E = v^2|\Phi| = \frac{v^2}{\kappa}|Q| = mQ\,, \tag{6.18}$$

where m is the scalar mass (6.6) in the symmetric vacuum. Thus the energy per unit charge is identical to that of the elementary excitations in the symmetric phase. This indicates that the collective, non-topological excitations are just at the threshold of stability against emission of elementary particles. Consequently, stability does not impose an upper bound on the non-topological soliton charges. The topological solitons are of course stable for topological reasons; their flux is quantized.

Equation (6.16) implies for $\varphi = |\varphi|e^{i\omega}$ that

$$A^i = \partial_i\omega \mp \epsilon^{ij}\partial_j \ln|\varphi|\,. \tag{6.19}$$

When substituted into (6.17) this gives, away from zeroes of $|\varphi|$

$$\nabla^2 \ln|\varphi|^2 = -\frac{4v^2}{\kappa^2}|\varphi|^2\left(1 - \frac{|\varphi|^2}{v^2}\right). \tag{6.20}$$

Both topological $\left(|\varphi|^2 \underset{r\to\infty}{\longrightarrow} v^2\right)$ and non-topological $\left(|\varphi|^2 \underset{r\to\infty}{\longrightarrow} 0\right)$ solitons satisfy the same equation, which is the generalization to the relativistic context of the Liouville equation, the latter being regained in the non-relativistic limit, where $|\varphi|^2/v^2$ is dropped relative to 1.

Unfortunately (6.20) does not possess closed-form solutions. Only numerical integration has been performed [25],[28] and a mathematical existence proof has been constructed [29].

The nature of the solutions is as follows. [We discuss the rotationally symmetric case.]

For topological solitons/vortices in the broken symmetry phases, the constant large distance asymptote is approached exponentially and at the origin there is a zero, which can be multiple.

$$|\varphi|^2 \underset{r \to 0}{\longrightarrow} r^{2\mathcal{N}-2} \quad , \qquad \mathcal{N} = 2, 3, \ldots \qquad \text{(topological)} \tag{6.21}$$

The integrality restriction on $\mathcal{N}$ comes from familiar arguments on the absence of singularities in the vector potential, together with a single valuedness requirement on φ. The flux is quantized

$$\Phi = \int d^2\mathbf{r}\, B = 2\pi(\mathcal{N} - 1)\,. \tag{6.22}$$

The value of the integral is determined by the strength of the zero at the origin.

The non-topological solitons can possess the same zeroes as in (6.21) at the origin, but now there is also a solution which does not vanish at the origin, *i.e.*

$$|\varphi|^2 \underset{r \to 0}{\longrightarrow} r^{2\mathcal{N}-2} \quad , \qquad \mathcal{N} = 1, 2, \ldots \qquad \text{(non-topological)}\,. \tag{6.23}$$

[The $\mathcal{N} = 1$ solution is not available in the topological case.] The approach to the vanishing large-distance asymptote is power-law,

$$|\varphi|^2 \underset{r \to \infty}{\longrightarrow} r^{-2(\alpha+1)} \tag{6.24}$$

and one can prove that

$$\alpha \geq \mathcal{N}\,. \tag{6.25}$$

The flux is no longer quantized,

$$\Phi = \int d^2\mathbf{r}\, B = 2\pi(\mathcal{N} + \alpha) \tag{6.26}$$

with [quantized] contributions from the origin combining with [unquantized] contributions from infinity. In the non-relativistic limit, α approaches $\mathcal{N}$ and (6.26) coincides with the non-relativistic result. A useful point of view about the arbitrary-$\mathcal{N}$ solution is that the true non-topological soliton corresponds to $\mathcal{N} = 1$, while higher-$\mathcal{N}$ solutions describe $\mathcal{N} - 1$ vortices threading through the non-topological soliton.

Index theory may be used to count parameters and associate them with physical characteristics of the solitons. A picture consistent with the above description emerges [25]. In particular the $\mathcal{N}$-vortex/topological soliton solution depends on $2\mathcal{N}$ parameters, identified as the locations of the vortices on the plane.

We record formulas for the angular momentum. From the energy momentum tensor

$$T_{\mu\nu} = (D_\mu\varphi)^*(D_\nu\varphi) + (D_\nu\varphi)(D_\mu\varphi)^* - g_{\mu\nu}\left[|D_\lambda\varphi|^2 - V(|\varphi|^2)\right] \quad (6.27)$$

we infer that the angular momentum, which is constructed from the momentum density $\mathcal{P}^i = T^{0i}$ as,

$$J = \int d^2\mathbf{r}\,\mathbf{r} \times \boldsymbol{\mathcal{P}} \quad (6.28)$$

becomes after some manipulation

$$J = -\frac{\kappa}{4\pi}\Phi^2 + \kappa\alpha\Phi\,. \quad (6.29)$$

This formula holds for both the non-topological and topological solitons $[\alpha = -1]$.

One may consider the slow motion of the topological solitons/vortices and derive an effective Lagrangian from the underlying field theory. One finds a statistical interaction term with the value of the statistical factor in agreement with the generalized spin-statistic relation [30].

7 Conclusions

Our understanding of the gauged, planar, non-linear Schrödinger equation is not as thorough as of the $(1+1)$-dimensional model: the only classical solutions that are completely cataloged [in the Abelian theory] are the static ones; but the complete integrability of the time-dependent equations remains an open question; in the quantum theory only partial results on the N-anyon problem are available — we have no Hans Bethe guessing the N-body solution as he did for the $(1+1)$-dimensional model.

The quantum meaning of the classical solution is of course well-known for the relativistic theory: upon quantization, the classical solutions give rise to quantum particle states [1]. In the non-relativistic model quantization yields a spectrum of soliton states, which may be expressed in terms of coherent states in the usual multi-particle Hilbert space [31].

Quantum and thermal fluctuations have been computed in the relativistic theory, with the fermions included so that supersymmetry holds. The pleasing result is that renormalization does not change the shape of the classical potential; only v^2 in (6.2) acquires an uncalculable correction. The remaining corrections are calculable and small [in the perturbative regime]. Moreover, they do not remove the supersymmetry and both the $U(1)$ symmetric and asymmetric vacua persist. At high temperature only the $U(1)$ symmetry is realized; symmetry breaking configurations are not stable [32].

Various further generalizations, which however retain self-duality, have been examined: adding the Maxwell term to both the non-relativistic and relativistic models so that the topologically massive equation(2.76) holds [33], adding non-Abelian group structure to the relativistic theory [34], using unconventional matter dynamics [35]. Such modifications complicate the dynamics significantly, explicit solution is never possible, but qualitative behavior is similar to what happens in the models that we have described.

Although the static, non-relativistic Abelian equations are completely integrated, a question remains. We have seen that this theory is the non-relativistic limit of the relativistic model in its $U(1)$ symmetric realization. Correspondingly, the Schrödinger theory solitons/vortices are the non-relativistic limits of the non-topological, relativistic solitons — this is also confirmed by the fact that the non-relativistic charge density vanishes at large distance, as it does for relativistic, non-topological solutions. Nevertheless, the non-relativistic solitons/vortices appear to possess topological characteristics, *e.g.* the flux is quantized. So one is left wondering where is the topology in the gauged, planar non-linear Schrödinger equation, and how does it arise as a non-relativistic limit is performed on non-topological excitations.

References

1. For reviews see Jackiw, R.: Rev.Modern Phys. **49**, 681 (1977); Rajaraman, R.: *Solitons and Instantons.* North Holland, Amsterdam, 1982
2. Jackiw, R., Templeton, S.: Phys.Rev. D **23**, 2291 (1981); Schonfeld, J.: Nuclear Phys. B **185**, 157 (1981); Deser, S., Jackiw, R., Templeton, S.: Phys.Rev.Lett. **48**, 975 (1982); Deser, S., Jackiw, R., Templeton, S.: Ann. Physics (NY) **140**, 372 (1982); (E) **185**, 406 (1988)
3. Jackiw, R., Pi, S.-Y.: Phys.Rev.Lett. **64**, 2969 (1990); (C) **66**, 2682 (1991); Phys.Rev. D **42**, 3500 (1990)
4. Hagen, C.: Ann.Physics (NY) **157**, 342 (1984); Hagen, C.: Phys.Rev. D **31**, 848, 2135 (1985)
5. Takagi, S.: Progr.Theoret.Phys. **84**, 1019 (1990); **85**, 463, 723 (1991); Tohoku University preprint (June 1991)
6. This conformal symmetry of non-relativistic dynamics was discussed in the early seventies, when conformal symmetry of relativistic theories was explored in connection with deep inelastic scattering. Jackiw, R.: Physics Today **25**, Nov. 1, 23 (1972); Hagen, C.: Phys.Rev. D **5**, 377 (1972); Niederer, U.: Helv.Phys.Acta **45**, 802 (1972); **46**, 191 (1073); **47**, 119, 167 (1974); **51**, 220 (1978). In the context of Chern–Simopns dynamics, the symmetry was identified in Jackiw, R.: Ann.Physics (NY) **201**, 83 (1990).

7. Jackiw, R.: *Bég Memorial Volume* (World Scientific, Singapore). In press, MIT preprint CTP#1937 (January, 1991)
8. These issues are under further investigation by O. Bergman (in preparation)
9. de Alfaro,V., Fubini, S., Furlan, G.: Nuovo Cimento A **A34**, 569 (1976); Jackiw, R.: Ann.Physics (NY) **129**, 183 (1980) and last entry Ref. [6]; Takagi, Ref. [5]
10. The idea of generating time-dependent solitons from static solitons by these transformations is due to Ezawa, Z., Hotta, M., Iwazaki, A.: Phys.Rev.D **44**, 452 (1991)
11. Hotta, M.: Tohoku University preprint, TU-372 (July 1991)
12. Freedman, D., Newell, A.: (unpublished)
13. Kim, S., Soh, K., Yee, J.: Phys.Rev. D **42**, 4139 (1990)
14. Olesen, P.: Phys.Lett. B **265**, 361 (1991)
15. Davey, A., Stewartson, K.: Proc.Roy.Soc. London Ser.A **338**, 101 (1974); Anker, D., Freeman, N.: Proc.Roy.Soc. London Ser.A **360**, 529 (1978)
16. Jackiw, R., Pi, S.-Y.: Phys.Rev.D **44**, xxx (1991) and Ref. [3]
17. Jackiw, R., Pi, S.-Y.: Phys.Rev.D **44**, xxx (1991)
18. Jackiw, R., Pi, S.-Y.: Phys. Rev. Lett. **67**, 415 (1991) and Ref. [16]
19. Ezawa, Z., Hotta, M., Iwazaki, A.: Phys.Rev.Lett. **67**, 411 (1991) and Ref. [10]
20. Grossman, B.: Phys.Rev.Lett. **65**, 3230 (1990)
21. Dunne, G., Jackiw, R., Pi, S.-Y., Trugenberger, C.: Phys.Rev.D **43**, 1332 (1991)
22. Ward, R.: Philos.Trans.Roy.Soc. London Ser.A **A315**, 451 (1985); Hitchin, N.: Proc.Math.Soc. London **55**, 59 (1987); Donaldson, S.: Proc.Math.Soc. London **55**, 127 (1987)
23. Yoon, Y.: Ann.Physics (NY) **211**, 316 (1991)
24. Hong, J., Kim, Y., Pac, P.Y.: Phys.Rev.Lett. **64**, 2230 (1990)
25. Jackiw, R., Weinberg, E.: Phys.Rev.Lett. **64**, 2234 (1990); Jackiw, R., Lee, K., Weinberg, E.: Phys.Rev. D **42**, 3488 (1990)
26. Lee, C., Lee, K., Weinberg, E.: Phys.Lett. B **243**, 105 (1990)
27. Khare, A.: Phys.Lett. B **225**, 393 (1991)
28. Spruck, J. Yang, Y.: (preprint 1991); Ren, H., Weinberg, E.: (in preparation)
29. Wang, R.: Comm.Math.Phys. **137**, 587 (1991) Integrability of this equation and its generalizations when the spatial geometry is not flat have been examined by Schiff, J.: J.Math.Phys. **32**, 753 (1991)
30. Kim, S., Min, H.: Seoul National University preprint SNU TP 91-41 (1991)
31. Kabat, D.: MIT preprint CTP#2016 (September 1991)
32. İpekoğlu, Y., Leblanc, M., Thomaz, M.T.: Ann.Physics (NY) (in press)
33. a. Non-relativistic matter dynamics: Dunne, G., Trugenberger, C.: Phys. Rev. D **43**, 1323 (1991)
    b. Relativistic matter dynamics: Lee, C., Lee, K., Min, H.: Phys.Lett. B **252**, 79 (1990);
    Lee, C., Min, H., Rim, C.: Seoul National University preprint SNU TP 90-39 (1990); Lee,J., Nam, S.: Phys.Lett. B **261**, 437 (1991)
34. Lee, K.: Phys.Lett. B **255**, 381 (1991); Lee, K.: Phys.Rev.Lett. **66**, 553 (1991)
36. Lee,J., Nam, S.: Ref. [33]; Yang, Y.: Lett.Math.Phys. (in press)

Mathematical Theory of Classical Fields and General Relativity

Sergiu Klainerman

Princeton University, Princeton, NJ 08544

The subject of Classical Field Theory has received relatively little attention from both mathematicians and theoretical physicists. The former look at it primarily from the perspective of Quantum Mechanics, namely, as subject of the elusive task of quantization. Mathematicians, on the other hand, never quite took the underlying geometric structure of Classical Field Theory, the space-time, seriously. The geometers, though greatly emboldened by the success of Riemann's visionary ideas in the formulation of General Relativity, have stayed away, with few notable exceptions, from the fundamental new twist given to them by Einstein who replaced the positive definite metric of Riemannian Geometry by a Lorentzian, or more appropriate, Einsteinian metric.

The other group of mathematicians who might have naturally been drawn to the mathematical theory of classical fields, with its rich examples of interesting nonlinear partial differential equations of hyperbolic type, are the analysts. Their attention nevertheless, and I mean here those analysts whose interest lies in understanding the mathematical structure of the nonlinear partial differential equations of Mathematical Physics, has been, in the past, almost entirely focused on the mathematics of Continuum Mechanics[1].

The aim of my lecture is to draw the attention of the community of mathematical physicists to the beautiful mathematics of Classical Field Theory, to some of its wealth of, largely unexplored, fundamental questions as well as to some of the progress made in the last few years.

As I have mentioned above, the underlying geometry of Classical Field Theory, which itself becomes a dynamical variable in General Relativity, is that given by a space-time. This consists of a pair $(\mathcal{M}, \mathbf{g})$ with $\mathcal{M}$ a smooth $(n+1)$-dimensional manifold and $\mathbf{g}$ an Einsteinian metric i.e. a nondegenerate quadratic form of signature $(-1, 1, \ldots, 1)$, defined on the tangent space $T_p(\mathcal{M})$ and varying smoothly with $p \in \mathcal{M}$. Thus, locally, a

[1] As an exception to this one ought to mention the work of K. Jörgens [19], I. Segal [31], W. Strauss [35], C. Morawetz-W. Strauss [26] on nonlinear scalar wave equations and C. Bruhat [5] on General Relativity.

space-time looks Minkowskean. The Minkowski space-time itself, which will be denoted by $\mathbf{M}^{n+1}$, consists of a copy of $\mathbb{R}^{n+1}$ together with a metric $\mathbf{m}$ and a distinguished class of global coordinates x^α, $\alpha = 0, 1, \dots, n$, relative to which the line element of the metric takes the form $-(dx^0)^2 + (dx^1)^2 + \dots + (dx^n)^2$.

A classical field theory on $\mathcal{M}$ is specified by a Lagrangean L which depends on the metric $\mathbf{g}$ and a collection of fields $\psi = (\psi_1 \dots, \psi_m)$ corresponding to some matter-fields acting in our space-time. One defines the action integral $\mathcal{S}[\psi, \mathbf{g}, \mathcal{U}] = \int_{\mathcal{U}} L d\vartheta_{\mathbf{g}}$ where $\mathcal{U}$ is a given relatively compact domain in $\mathcal{M}$ and $d\vartheta_{\mathbf{g}}$ is the volume element of $\mathbf{g}$. A compact variation of the fields ψ and metric $\mathbf{g}$ are defined to be smooth 1-parameter families $(\psi(s), \mathbf{g}(s))$ with $s \in (-\epsilon, \epsilon)$ verifying the properties $\psi(0) = \psi, \quad \mathbf{g}(0) = \mathbf{g}$ everywhere in $\mathcal{M}$ and $\psi(s) = \psi, \quad \mathbf{g}(s) = \mathbf{g}$ at all points p in $\mathcal{M} \setminus \mathcal{U}$. The field equations of a particular theory are obtained by requiring that the acceptable physical solutions of the theory are those stationary relative to arbitrary compact variations of the corresponding action integral. Here are some simple examples:

Example 1. Scalar Field Equations

These are obtained by considering a real scalar ϕ defined on $(\mathcal{M}, \mathbf{g})$ with Lagrangean density $L[\phi] = -\frac{1}{2}\, \mathbf{g}^{\mu\nu}\partial_\mu\partial_\nu\phi + V(\phi)$ with V a given real function of ϕ. The corresponding Euler-Lagrange equations, obtained by varying ϕ alone[2], are given by

$$\Box_g\phi + V'(\phi) = 0\,, \qquad (S.W.)$$

where $\Box_g$ is the D' Alembertian operator, $\Box_g = \frac{1}{\sqrt{-g}}\, \partial_\mu(\mathbf{g}^{\mu\nu}\sqrt{-g}\; \partial_\nu\phi)$. The equation (S.W.) is called the scalar wave equation with potential $V(\phi)$. It has an energy-momentum tensor given by $\mathbf{T}_{\alpha\beta} = \frac{1}{2}\Big(\phi_{,\alpha}\phi_{,\beta} - \frac{1}{2}\mathbf{g}_{\alpha\beta}(\mathbf{g}^{\mu\nu}\phi_{,\mu}\phi_{,\nu} - 2V(\phi))\Big)$. It verifies the conservation law, $\mathbf{D}^\beta\mathbf{T}_{\alpha\beta} = 0$.

Example 2. Wave Maps

Consider a space-time $(\mathcal{M}, \mathbf{g})$ and a Riemannian manifold $\mathcal{N}$ with metric h. Let ϕ be mappings from $\mathcal{M}$ to $\mathcal{N}$. We define the Lagrangean density for wave maps to be $L = -\frac{1}{2}\; Tr_{\mathbf{g}}(\phi^* h)$ where $Tr_{\mathbf{g}}(\phi^* h)$ denotes the trace relative to $\mathbf{g}$ of the pull-back $\phi^* h$ of the metric h. In local coordinates x^α, $\alpha = 0, 1, \dots, n$, on $\mathcal{M}$ and y^a, $a = 1, \dots, m$ the Lagrangean density takes the form $L = -\frac{1}{2}\mathbf{g}^{\mu\nu} h_{ab}(\phi)\frac{\partial\phi^a}{\partial x^\mu}\frac{\partial\phi^b}{\partial x^\nu} = -\frac{1}{2}\mathbf{g}^{\mu\nu}\langle\phi_\mu, \phi_\nu\rangle_h$ with $\langle\ ,\ \rangle_h$ the Riemannian metric on $\mathcal{N}$. The corresponding Euler-Lagrange equations, obtained by variations of $\mathcal{S}$ with respect to ϕ alone, are given by

---

[2] while the metric $\mathbf{g}$ is kept fixed

$$\Box_{\mathbf{g}}\phi^a + \Gamma^a_{bc}(\phi)\mathbf{g}^{\mu\nu}\partial_\mu\phi^b\partial_\nu\phi^c = 0\,, \qquad (W.M.)$$

where Γ^a_{bc} are the Christoffel symbols of $(\mathcal{N}, h)$.

The e-m tensor of (W.M.) is given by $\mathbf{T}_{\alpha\beta} = \frac{1}{2}\Big(\langle\phi_{,\alpha},\phi_{,\beta}\rangle - \frac{1}{2}\mathbf{g}_{\alpha\beta}(\mathbf{g}^{\mu\nu}\langle\phi_{,\mu}\phi_{,\nu}\rangle)\Big)$. It verifies the conservation law, $\mathbf{D}^\beta\mathbf{T}_{\alpha\beta} = 0$.

Example 3. Yang-Mills Equations

We first recall the Maxwell equations. Given a $3+1$ dimensional space-time $(\mathcal{M},\mathbf{g})$ an electromagnetic field F is a 2-form on $\mathcal{M}$ which is exact, i.e. $F = dA$ with A a 1-form on $\mathcal{M}$. The vector potential A is unique up to a gauge transformation $A \mapsto A + d\chi$ with χ an arbitrary scalar function. The Lagrangean density of F is defined by, $L[F] = -\frac{1}{2}F_{\mu\nu}F^{\mu\nu}$ and the Euler-Lagrange equations are, $\mathbf{D}^\nu F_{\mu\nu} = 0$ with $\mathbf{D}$ denoting the covariant differentiation on $(\mathcal{M},\mathbf{g})$. Together with the equations $dF = 0$, which are a consequence of $F = dA$, they form the Maxwell equations. We can write the Maxwell equations in a more symmetric form by setting ${}^*F_{\mu\nu} = \frac{1}{2}\epsilon_{\mu\nu\alpha\beta}F^{\alpha\beta}$ the Hodge dual of F with $\epsilon_{\alpha\beta\gamma\delta}$ the components of the volume form on $(\mathcal{M},g)$. With this notation the Maxwell equations take the form

$$dF = 0, \quad d^*F = 0\,. \qquad (M)$$

The energy-momentum tensor of (M) is given by

$$\mathbf{T}_{\alpha\beta} = F^\mu_\alpha F_{\beta\mu} - \tfrac{1}{4}g_{\alpha\beta}\,F_{\mu\nu}F^{\mu\nu}.$$

The Yang-Mills equations give a simple generalization of the Maxwell equations where the vector potential $A = A_\mu dx^\mu$ becomes a $\mathcal{G}$-valued 1-form with $\mathcal{G}$ the Lie algebra of a Lie group G. Let $[\cdot,\cdot]$ the Lie bracket on $\mathcal{G}$ and $\langle\cdot,\cdot\rangle$ its Killing scalar product which one assumes positive definite. For any $\mathcal{G}$-valued tensor ψ one defines the gauge covariant derivative $D_\mu\psi = \mathbf{D}_\mu\psi + [A_\mu,\psi]$ where $\mathbf{D}$ is the covariant derivative on $\mathcal{M}$. The definition is invariant with respect to "gauge transformations". These are transformations of the type

$$\psi(p) \mapsto O^{-1}(p)\psi(p)O(p)$$
$$A_\alpha(p) \mapsto O^{-1}(p)A_\alpha(p)O(p) + O^{-1}(p)O_\alpha(p)\,.$$

One can easily check that for any $\mathcal{G}$-valued scalar ψ, $D_\mu D_\nu\psi - D_\nu D_\mu\psi = [F_{\mu\nu},\psi]$ where the 2-form $F_{\mu\nu}dx^\mu dx^\nu$, given by the formula $F_{\mu\nu} = \partial_\mu A_\nu - \partial_\nu A_\mu + [A_\mu, A_\nu]$, is called the curvature of the connection. It is manifestly invariant under a gauge transformation $F \to O^{-1}FO$ and satisfies the Bianchi equation

$$D_\alpha F_{\beta\gamma} + D_\beta F_{\gamma\alpha} + D_\gamma F_{\alpha\beta} = 0\,. \qquad (Y-M')$$

The Lagrangean of the Yang-Mills equations is defined by $L[A] = -\frac{1}{4}\langle F_{\mu\nu}, F^{\mu\nu}\rangle$. The corresponding Euler-Lagrange equations are

$$D^{\nu}F_{\mu\nu} = 0 \qquad (Y - M"),$$

which together with (Y-M') form the Yang-Mills equations. The energy momentum for (Y-M) is, $\mathbf{T}_{\alpha\beta} = \langle F^{\mu}_{\alpha}, F_{\beta\mu}\rangle - \frac{1}{4}g_{\alpha\beta}\langle F_{\mu\nu}, F^{\mu\nu}\rangle$.

Example 4. The Einstein Field Equations

According to the general relativistic variational principle the space-time metric $\mathbf{g}$ itself is stationary with respect to an action $\mathcal{S} = \int_{\mathcal{U}} L dv_g$ whose Lagrangean $L = L_G + L_M$ consists of a purely gravitational part $L_G = \mathbf{R}$, given by the scalar curvature $\mathbf{R}$ of the metric $\mathbf{g}$, and the matter Lagrangean L_M which depends on the matter fields ψ present in the space-time. The corresponding Euler-Lagrange equations are the so called Einstein field equations,

$$\mathbf{R}_{\mu\nu} - \frac{1}{2}\mathbf{g}_{\mu\nu}\mathbf{R} = \mathbf{T}_{\mu\nu}, \qquad (E)$$

where $\mathbf{R}_{\mu\nu}$ is the Ricci curvature of $\mathbf{g}$ and $\mathbf{T}_{\mu\nu}$ the energy momentum tensor of matter, $\mathbf{T}_{\mu\nu} = \left(\frac{\partial L_M}{\partial \mathbf{g}^{\mu\nu}} - \frac{1}{2}\mathbf{g}_{\mu\nu}L_M\right)$ obtained from the variation of $S_M = \int L_M dv_{\mathbf{g}}$ with respect to the metric. Thus the energy momentum tensors of the matter fields discussed before arise from the variations of their action integrals with respect to the background metric. From this perspective we remark that the purely gravitational part of the Einstein field equations has no energy-momentum tensor, as defined for the other field theories, since the left hand side of the equations (E) have been derived precisely by varying the action $\mathcal{S}_G$ relative to $\mathbf{g}$. This formal difficulty is connected to the well known difficulty of principle of defining local concepts of energy, linear and angular momenta for General Relativity as we have them for the other field theories.

As is well known Einstein proposed General Relativity as a unified theory of space, time and gravity. In fact, according to it, gravity is a manifestation of the space-time. An important consequence of this is that the Einstein field equations (E) admit nontrivial solutions even in the absence of external matter fields. This amounts to setting $\mathbf{T} \equiv 0$ on the right-hand side of (E). The resulting equations, called the Einstein-Vacuum equations, take the form,

$$R_{\alpha\beta} = 0 . \qquad (E - V)$$

Most of the fundamental difficulties of the general Einstein field equations are present in the form (E-V). From a physical point of view the equation (E-V) are perfectly adequate to describe the propagation of gravitational waves at large distances from a source.

Going back to the general case (E) we recall that the Riemann curvature tensor satisfies the Bianchi identities,

$$\mathbf{D}_{\epsilon}\mathbf{R}_{\alpha\beta\gamma\delta} + \mathbf{D}_{\alpha}\mathbf{R}_{\beta\epsilon\gamma\delta} + \mathbf{D}_{\beta}\mathbf{R}_{\epsilon\alpha\gamma\delta} = 0 . \qquad (B)$$

Contracting (B) twice we derive, $\mathbf{D}^\nu(\mathbf{R}_{\mu\nu} - \frac{1}{2}\mathbf{g}_{\mu\nu}\mathbf{R}) = 0$. This implies that the right-hand side of (E) also satisfies

$$\mathbf{D}^\nu \mathbf{T}_{\mu\nu} = 0\,,$$

which is the law of conservation for the energy-momentum tensor of the matter-field.

In view of our definition, the energy-momentum tensor is automatically symmetric. The energy-momentum tensor of the scalar wave equation satisfies the positive energy condition[3] if V is nonpositive. The energy momentum tensors for the wave maps, Maxwell equations and Yang-Mills also satisfy the positive energy condition.

The energy-momentum tensor of a field theory is intimately connected with the Noether's Principle which asserts the following:

Noether's Principle. To any one parameter group of transformation which preserves the action there corresponds a conservation law.

We illustrate this fundamental principle as follows. Let $\mathcal{S} = \mathcal{S}[\psi, \mathbf{g}]$ be the action integral of the fields ψ. Consider χ_t to be a 1-parameter group of isometries of $\mathcal{M}$, i.e. $(\chi_t)_*\mathbf{g} = \mathbf{g}$.

Then $\mathcal{S}[(\chi_t)_*\psi, \mathbf{g}] = \mathcal{S}[(\chi_t)_*\psi, (\chi_t)_*\mathbf{g}] = \mathcal{S}[\psi, \mathbf{g}]$. In view of Noether's Principle we ought to find a conservation law of the corresponding Euler-Lagrange equations.

Let X be an arbitrary vectorfield and P be the 1-form obtained by contracting the energy-momentum tensor of ψ with X i.e. $P_\alpha = \mathbf{T}_{\alpha\beta}X^\beta$. Then, since $\mathbf{T}$ is symmetric and divergence-less,

$$D^\alpha P_\alpha = \frac{1}{2}\mathbf{T}^{\alpha\beta}\,\pi_{\alpha\beta}\,, \tag{1}$$

where

$$\pi_{\alpha\beta} = D_\beta X_\alpha + D_\alpha X_\beta \tag{2}$$

is the deformation tensor of X. We now integrate (1) on a lens shaped domain $\mathcal{D}$ be bounded by two space-like hypersurfaces $\mathcal{H}_0$, $\mathcal{H}_1$. In view of the divergence theorem we derive the integral identity

$$\int_{\mathcal{H}_0} \mathbf{T}(X,T)da_g - \int_{\mathcal{H}_1} \mathbf{T}(X,T)da_g = -\int_{\mathcal{D}} \mathbf{T}^{\alpha\beta}\,\pi_{\alpha\beta}\,, \tag{3}$$

where T is the future oriented unit normal, g the induced metric and da_g the area element of $\partial\mathcal{D}$.

In the particular case when X is Killing[4] its deformation tensor π vanishes identically and we derive the conservation law

[3] $\mathbf{T}$ is said to satisfy the positive energy condition if, for any causal future directed vectors X, Y, we have $\mathbf{T}(X,Y) \geq 0$.

[4] i.e. it generates a one parameter group of isometries

$$\int_{\mathcal{H}_0} \mathbf{T}(X,T)da_g = \int_{\mathcal{H}_1} \mathbf{T}(X,T)da_g \ . \tag{4}$$

Remark that the result (4) remains true if X is a conformal Killing vector field, i.e. it generates conformal isometries, and $\mathbf{T}$ is traceless. Indeed, if X is conformal Killing, $\pi_{\alpha\beta} = \Lambda \mathbf{g}_{\alpha\beta}$ and hence $\pi_{\alpha\beta}\mathbf{T}^{\alpha\beta} = \Lambda \mathrm{tr}(\mathbf{T}) = 0$. One can easily show that if the action integral $\mathcal{S}[\psi, \mathbf{g}]$ is invariant under conformal rescalings of the metric, i.e. $\tilde{\mathbf{g}} = \Omega^2 \mathbf{g}$, then the corresponding energy-momentum tensor is traceless. This is the case of the Yang-Mills Theory in $3+1$ dimensions.

The identities (3) and (4) are usually applied to time-like future oriented vectorfields X, in which case, in view of the positive energy condition, the integrand $\mathbf{T}(X,T)$ is positive. This is the essence of what we call the energy method. The method allows us to obtain apriori bounds, typically L^2 on space-like hypersurfaces, of the field under consideration. This provides us with a very powerful analytic tool. The fundamental mathematical difficulty in the study of interesting nonlinear field equations is due to the fact that these L^2 bounds, given to us for free by the theory, do not suffice to understand the equations. We thus have to scramble to derive additional information every way we can.

One interesting extension of the energy method has become an important tool in the study of linear field equations in Minkowski space-time as well as in nonlinear perturbation theory ([21],[22],[12],[13]). The extension is based on the invariance properties of the linear theory relative to the groups of motion of $\mathbf{M}^{n+1}$. More precisely, one uses the Killing and conformal Killing vector fields generated by the conformal group of the Minkowski space-time to define a global energy norm which is invariant relative to the linear evolution. Precise asymptotic information, including uniform bounds on different components of the field, are then an immediate consequence of a global version of the Sobolev inequalities ([18],[21],[23]).

To illustrate the method consider the Maxwell equations in $\mathbf{M}^{3+1}$. As mentioned above the Maxwell theory is conformal invariant. This means that whenever F is a solution to (M) and X is an arbitrary conformal Killing vectorfield the Lie derivative of F with respect to X is also a solution to (M). Combining this fact with the standard energy method describe above we can now derive L^2 estimates for various vectorfields applied to F. For example, consider the following quantities,

$$\mathbf{Q}_0(t) = \int_{\mathcal{H}_t} \mathbf{T}(F)(K_0, T_0)dx$$

$$\mathbf{Q}_1(t) = \sum_{a=1}^{3} \int_{\mathcal{H}_t} \mathbf{T}(\mathcal{L}_{\Omega(a)}F)(\overline{K_0}, T_0)dx + \int_{\mathcal{H}_t} \mathbf{T}(\mathcal{L}_S F)(\overline{K_0}, T_0)dx$$
$$+ \int_{\mathcal{H}_t} \mathbf{T}(\mathcal{L}_{T_0}F)(\overline{K_0}, T_0)dx$$

$$\mathbf{Q}_2(t) = \sum_{a,b=1}^{3} \int_{\mathcal{H}_t} \mathbf{T}(\mathcal{L}_{\Omega(a)}\mathcal{L}_{\Omega(b)}F)(\overline{K_0}, T_0)dx$$
$$+ \sum_{a=1}^{3} \int_{\mathcal{H}_t} \mathbf{T}(\mathcal{L}_S\mathcal{L}_{\Omega(a)}F)(\overline{K_0}, T_0)dx$$
$$+ \int_{\mathcal{H}_t} \mathbf{T}(\mathcal{L}_S\mathcal{L}_S F)(\overline{K_0}, T_0)dx + \int_{\mathcal{H}_t} \mathbf{T}(\mathcal{L}_{T_0}\mathcal{L}_{T_0}F)(\overline{K_0}, T_0)dx \; . (5)$$

Here $\mathcal{H}_t$ are the level hypersurfaces of the time function $t = x^0$ with future directed unit normal $T_0 = \partial_0$. T_0 is also the generator of time translations. The vectorfields $\Omega(a) = \epsilon_{abc}x_b\partial_c$, $a = 1,2,3$ are the usual angular momentum operators. $S = x^\alpha\partial_\alpha$ is the generator of dilations while $\overline{K}_0 = K_0 + T_0$ with $K_0 = (t^2 + r^2)\partial_0 + 2tx^i\partial_i$ the generator of inverted time translations.

In view of the above discussion the quantities $\mathbf{Q}_0(t), \mathbf{Q}_1(t), \mathbf{Q}_2(t)$ are all time independent and therefore bounded for all $t \in \mathbb{R}$ if $\mathbf{Q}_i(0)$, $i = 0,1,2$ are all bounded. These L^2 bounds can be combined with some global version of the classical Sobolev inequalities to derive very sharp uniform estimates for F.

To do this we decompose F relative to the pair of conjugate null vectors $e_+ = \partial_t + \frac{x^i}{r}\partial_i$, $e_- = \partial_t - \frac{x^i}{r}\partial_i$ and an arbitrary orthonormal frame $(e_A)_{A=1,2}$ on the 2-spheres $S_{u,v}$ obtain by the intersection of the level hypersurfaces of $u = t - r$ with those of $u = t + r$, into the vectors $\alpha_A = F(e_+, e_A)$, $\underline{\alpha}_A = F(e_-, e_A)$ and scalars $\rho = \frac{1}{2}F(e_+, e_-)$, $\sigma = \frac{1}{2}{}^*F(e_+, e_-)$. This is called the null decomposition of F. We can now state the following,

Proposition 1 ([11]) *Let F be a solution of (M) and assume that the initial conditions on the hyperplane $t = 0$ are such that the quantities $\mathbf{Q}_i(0)$, $i = 0,1,2$ are all bounded. Consider the asymptotic behaviour of F for $t > 0$. In the space-time region $r \le 1 + \frac{t}{2}$ all components of F behave in the same way, $|F(t,x)| \le c(1+t)^{-\frac{5}{2}}$. Outside that region[5] we have, $|\alpha(t,x)| \le (1+t+r)^{-\frac{5}{2}}$, $|\rho(t,x), \sigma(t,x)| \le (1+t+r)^{-2}(1+|t-r|)^{-\frac{1}{2}}$, $|\underline{\alpha}(t,x)| \le (1+t+r)^{-1}(1+|t-r|)^{-\frac{3}{2}}$.*

We have thus been able to derive the asymptotic properties of (M)-equations relying only on their geometric properties. This feature is crucial in applications to nonlinear problems. A variant of this method has been used recently to prove the nonlinear stability of the Minkowski space-time (see our discussion below). At the heart of the proof lies the derivation of L^2 estimates for the modified Lie derivatives of the curvature tensor $\mathbf{R}$ of a

[5] The crucial region, where different null components of F decay in different ways, is the wave zone region where $u = t - r$ is bounded.

space-time $\mathcal{M}, \mathbf{g}$ verifying the Einstein-Vacuum equations (E-V) relative to some modification of the vectorfields $T_0, \Omega, S, \overline{K_0}$, as above. This is based on the crucial observation that the Bianchi identities (B), if interpreted as equations satisfied by $\mathbf{R}$ relative to a fixed background metric $\mathbf{g}$, are conformally covariant. Moreover, despite the fact that the equations are not derived from a Lagrangean, one can define a four-covariant tensor, called the Bell-Robinson tensor, $\mathbf{T}_{\alpha\beta\gamma\delta} = \frac{1}{2}(\mathbf{R}_{\alpha\mu\beta\nu}\mathbf{R}_{\gamma}{}^{\mu}{}_{\delta}{}^{\nu} + {}^{\star}\mathbf{R}_{\alpha\mu\beta\nu}{}^{\star}\mathbf{R}_{\gamma}{}^{\mu}{}_{\delta}{}^{\nu})$ where, ${}^{\star}\mathbf{R}_{\alpha\beta\gamma\delta} = \frac{1}{2}\epsilon_{\alpha\beta\mu\nu}\mathbf{R}^{\mu\nu}{}_{\gamma\delta}$, which exhibits all the properties of an energy-momentum tensor[6]. Thus $\mathbf{T}$ is symmetric relative to all pairs of indices and satisfies the conservation law, $\mathbf{D}^{\delta}\mathbf{T}_{\alpha\beta\gamma\delta} = 0$. Moreover it verifies the positive energy condition, i.e. $\mathbf{T}(X, Y, X, Y)$ is positive whenever X, Y are future directed time-like vectors. The conformal covariance of (B) translates into that $\mathbf{T}$ is traceless. Also, if we define the modified Lie derivative of $\mathbf{R}$ relative to a vectorfield X by the formula, $\hat{\mathbf{L}}_X\mathbf{R} = \mathbf{L}_X\mathbf{R}_{\alpha\beta\gamma\delta} - \frac{1}{2}(\pi^{\mu}_{\alpha}\mathbf{R}_{\mu\beta\gamma\delta} + \pi^{\mu}_{\beta}\mathbf{R}_{\alpha\mu\gamma\delta} + \pi^{\mu}_{\gamma}\mathbf{R}_{\alpha\beta\mu\delta} + \pi^{\mu}_{\delta}\mathbf{R}_{\alpha\beta\gamma\mu}) + \frac{3}{8}\mathrm{tr}\pi\mathbf{R}_{\alpha\beta\gamma\delta}$, we see that the commutation of $\hat{\mathbf{L}}_X$ with (B) produces error terms depending on the vector X only through the trace-less part of its deformation tensor π. Thus, if the background metric is that of the Minkowski space-time the modified Lie derivative relative to a conformal Killing vectorfield commutes with the Bianchi equations (B). These properties of (B) allow us to define generalized energy norms for $\mathbf{R}$, as we have done previously for F. These norms allow us to control the asymptotic behaviour of $\mathbf{R}$. We illustrate this in Minkowski space-time as follows. Assume that W is a 4-covariant traceless tensor verifying all the symmetry properties of the curvature tensor $\mathbf{R}$. Assume that W is a solution of the Bianchi equations (B) in $\mathbf{M}^{3+1}$. Let $\mathbf{Q}_0(t), \mathbf{Q}_1(t), \mathbf{Q}_2(t)$ be the quantities,

$$\mathbf{Q}_0(t) = \int_{\mathcal{H}_t} \mathbf{T}(W)(\overline{K_0}, \overline{K_0}, T_0 T_0)dx$$

$$\mathbf{Q}_1(t) = \sum_{a=1}^{3} \int_{\mathcal{H}_t} \mathbf{T}(\hat{\mathcal{L}}_{\Omega(a)}W)(\overline{K_0}, \overline{K_0}, T_0 T_0)dx$$

$$+ \int_{\mathcal{H}_t} \mathbf{T}(\hat{\mathcal{L}}_{T_0}W)(\overline{K_0}, \overline{K_0}, \overline{K_0} T_0)dx$$

$$\mathbf{Q}_2(t) = \sum_{a,b=1}^{3} \int_{\mathcal{H}_t} \mathbf{T}(\hat{\mathcal{L}}_{\Omega(a)}\hat{\mathcal{L}}_{\Omega(b)}W)(\overline{K_0}, \overline{K_0}, T_0, T_0)dx$$

$$+ \sum_{a=1}^{3} \int_{\mathcal{H}_t} \mathbf{T}(\hat{\mathcal{L}}_{\Omega(a)}\hat{\mathcal{L}}_{T_0}W)(\overline{K_0}, \overline{K_0}, \overline{K_0}, T_0)dx +$$

[6] These properties hold true as a consequence of the symmetries of $\mathbf{R}$ and, in view of the E-V equations, the condition $\mathbf{R}_{\alpha\beta} = 0$.

$$+\int_{\mathcal{H}_t} \mathbf{T}(\hat{\mathcal{L}}_S\hat{\mathcal{L}}_{T_0}W)(\overline{K_0},\overline{K_0},\overline{K_0},T_0)dx$$

$$+\int_{\mathcal{H}_t} \mathbf{T}(\hat{\mathcal{L}}_{T_0}\hat{\mathcal{L}}_{T_0}W)(\overline{K_0},\overline{K_0},\overline{K_0},T_0)dx \tag{6}$$

with $T_0, \Omega, S, \overline{K_0}$ as in (5).

The null decomposition of W is given by $\alpha_A = W(e_A, e_+, e_B, e_+)$, $\beta_A = \frac{1}{2}W(e_A, e_+, e_-, e_+)$, $\rho = \frac{1}{4}W(e_-, e_+, e_-, e_+)$, $\sigma = \frac{1}{4}{}^*W(e_-, e_+, e_-, e_+)$, $\underline{\beta}_A = \frac{1}{2}W(e_A, e_-, e_-, e_+)$, $\underline{\alpha}_{AB} = W(e_A, e_-, e_B, e_-)$. We can now state the following

Proposition 2 *Let W be a solution of the Bianchi equations in $\mathbf{M}^{3+1}$ and assume that the initial conditions on the hyperplane $t = 0$ are such that the quantities $\mathbf{Q}_i(0)$ are all bounded. Consider the asymptotic behaviour of W for $t > 0$. In the space-time region $r \leq 1 + \frac{t}{2}$ all components of W behave in the same way, $|W(t,x)| \leq c(1+t)^{-\frac{7}{2}}$. Outside that region we have, $|\alpha(t,x)| \leq (1+t+r)^{-\frac{7}{2}}$, $\beta \leq (1+t+r)^{-\frac{7}{2}}$, $|\rho(t,x), \sigma(t,x)| \leq (1+t+r)^{-3}(1+|t-r|)^{-\frac{1}{2}}$, $|\underline{\beta}| \leq (1+t+r)^{-2}(1+|t-r|)^{-\frac{3}{2}}$, $|\underline{\alpha}(t,x)| \leq (1+t+r)^{-1}(1+|t-r|)^{-\frac{5}{2}}$.*

Both Propositions 1,2 have been proved in [11].

Among the classical field theories quoted above only the Maxwell equations and the Einstein field equations are known to have direct physical content. The others, like the Yang-Mills equation, are believed to become physically relevant only after they suffer the painful process of quantization. In view of this fact one can describe the basic goals of the mathematical theory of classical fields as follows:

(1) Investigate the mathematical consequences of the Einstein Field Equations in so far as they describe the physical world.
(2) Investigate the mathematical properties of the other nonlinear field theories from the perspective of:
   (a) In view of the fact that the Einstein-field equations are exceedingly complicated some of the other field theories provide a simplified testing ground for new ideas. From this point of view the final goal remains the understanding of the Einstein field equations but we hope to get important insights from the simpler nonlinear field theories. To this one might add our belief that any new ideas which lead to significant progress made on problems of nonlinear classical fields will prove useful in other areas of Mathematical Physics where nonlinear P.D.E's, in particular hyperbolic, are at the heart of the subject.
   (b) Investigate the mathematical properties of nonlinear field theories in view of their possible relevance to quantum field theory.

This final goal is most vague. One can try to make it slightly more meaningful by pointing out the importance of uncovering the fundamental properties of classical mechanical systems[7], as carried out by mathematicians such as Euler, Lagrange, Hamilton, Poincare and others, to the very formulation of Quantum Mechanics.

Since according to the above scheme, General Relativity and the Einstein field equations play the dominant role in the mathematical theory of classical fields, I will concentrate most of my attention to it and refer to other field theories from the perspective of 2a. The mathematical structure of the Einstein-Vacuum equations, or shortly E-V, is already sufficiently complicated so that we might as well concentrate our attention on them.

In view of the four contracted Bianchi identities mentioned above the E-V equations can be viewed as a system of $10 - 4 = 6$ equations for the 10 components of the metric tensor $\mathbf{g}$. The remaining 4 degrees of freedom correspond to the general covariance of the equations. Thus, if ϕ is a diffeomorphism of $\mathcal{M}$, then the pair $(\mathcal{M}, \mathbf{g})$ and $(\mathcal{M}, \phi_* \mathbf{g})$ represent the same solution of the field equations. Written explicitly in an arbitrary system of coordinates the E-V equations lead to a nonlinear degenerate system of partial differential equations in $\mathbf{g}$. Indeed the principal part of the Ricci curvature $\mathbf{R}_{\mu\nu}$ is $\frac{1}{2}\, \mathbf{g}^{\alpha\beta}\Big(\partial_\mu \partial_\alpha \mathbf{g}_{\beta\nu} + \partial_\nu \partial_\alpha \mathbf{g}_{\beta\mu} - \partial_\mu \partial_\nu \mathbf{g}_{\alpha\beta} - \partial_\alpha \partial_\beta \mathbf{g}_{\mu\nu}\Big)$. Nevertheless the E-V equations are seen to be hyperbolic when one takes into account the geometric equivalence of metrics related by diffeomorphisms. This indicates that it must be possible to choose coordinates relative to which the E-V equations can be written as a system of nonlinear wave equations. One well known such choice is that of "wave coordinates" namely coordinates x^μ which satisfy the wave equation $\mathbf{g}^{\alpha\beta} D_\alpha D_\beta x^\mu = 0$. In this case the principal part of $\mathbf{R}_{\mu\nu}$ becomes $-\frac{1}{2} g^{\alpha\beta} \partial_\alpha \partial_\beta \mathbf{g}_{\mu\nu}$ and the E-V equations take the form:

$$\mathbf{g}^{\alpha\beta} \partial_\alpha \partial_\beta \mathbf{g}_{\mu\nu} = F_{\mu\nu}(\mathbf{g}, \partial \mathbf{g}) \tag{7}$$

with $F_{\mu\nu}$ a polynomial in $(\mathbf{g}, \partial \mathbf{g})$ quadratic in $\partial \mathbf{g}$. The form (7) allowed Y. Choquet-Bruhat [6] to prove the well posedness of the local Cauchy problem in wave coordinates by appealing to the local theory of systems of nonlinear wave equations developed by J. Leray. The reliance on fixed coordinate conditions is however very problematic if one wants to study the E-V equations in the large. Indeed, as pointed by Y. Chocquet-Bruhat [6], the "wave coordinates" are unstable in the large even when one starts with initial conditions close to flat. This problem, often referred to as the problem of coordinates, is the first major difficulty one has to overcome in the construction of global solutions to the E-V equations.

But, before discussing global solutions, let us first shortly recall what we mean by the Cauchy problem for the E-V equations. To do this, assume that $\mathcal{H}$ is a space like hypersurface embedded in a given space-time $(\mathcal{M}, \mathbf{g})$

[7] at least at a formal level

verifying (E-V). Let g, k be the first and second fundamental forms of $\mathcal{H}$ and denote by ∇ the induced covariant differentiation. Since $\mathcal{H}$ is space-like the metric g is Riemannian. Now the Gauss-Codazzi equations of the embedding of $\mathcal{H}$ in $\mathcal{M}$ together with the equations (E-V) lead to the following nonlinear system of equations satisfied by g and k in $\mathcal{H}$.

$$\begin{aligned} \nabla^j k_{ji} - \nabla_i \mathrm{tr}\ k &= 0 \\ R - |k|^2 + (\mathrm{tr}\ k)^2 &= 0 \end{aligned} \tag{8}$$

called the constraint equations. Here R is the scalar curvature of g, tr$k$ is the trace of k relative to g and $|k|^2 = k_{ij}k^{ij}$. In view of this we define an initial data set for (E-V) to be a triplet $(\mathcal{H}, g, k)$ formed by a Riemannian 3-D manifold $(\mathcal{H}, g)$ together with k, a symmetric tensor of rank 2, which verify the constraint equation (8). A development of an initial data set consists of an Einstein-Vacuum space-time $(\mathcal{M}, \mathbf{g})$ together with an embedding i: $\mathcal{H} \to \mathcal{M}$ such that g, k are the induced first and second fundamental forms of the embedding.

One can formulate the central question of the mathematical theory of General Relativity as that of studying the evolution of general initial data sets. Now this goal is too broad since without an appropriate asymptotic restriction the evolution of an arbitrary initial data set can be very bad. A reasonable physical restriction is to consider initial data sets which look flat outside a sufficiently large compact set of $\mathcal{H}$. The evolution of such initial data sets correspond to "isolated physical systems". These systems are particularly important in G.R. since, as was stressed by Geroch [15], "it is only through a suitable notion of an isolated system that one acquires an ability at all to deal individually with various subsystems in the universe; in particular to assign to subsystems such physical attributes as mass, angular momentum, character of emitted radiation, etc." Indeed, as we have mentioned above, the Einstein equations have no energy-momentum tensor and therefore no local concepts of energy, linear and angular momentum. Moreover for any field theory in which the metric is kept fixed these concepts arise from the symmetries of the background space-time. These do not make sense in G.R.; general solutions have no symmetries. Global concepts of energy, linear and angular momentum are nevertheless possible when we restrict our attention to space-times with asymptotic symmetries at infinity. Such is the case of an isolated physical system, at large distances from the source the corresponding space-time becomes flat and thus acquires the symmetries of the Minkowski space-time.

An initial data set $(\mathcal{H}, g, k)$ is said to be asymptotically flat if there exists a system of coordinates x^1, x^2, x^3 globally defined, with a possible exception of a relatively compact set in $\mathcal{H}$ such that $g_{ij} - (1 + \frac{2M}{r})\delta_{ij} \to 0$, $k_{ij} \to 0$ as $r^2 = \Sigma(x^i)^2 \to \infty$. More precisely, we will say that the initial data set is strongly asymptotically flat, or S.A.F., of order[8] k

$$g_{ij} - (1 + \frac{2M}{r})\delta_{ij} = o_k(r^{-3/2}), \quad k_{ij} = o_{k-1}(r^{-3/2}) . \qquad (S.A.F)$$

Given an S.A.F. initial data set one can define its energy E, linear momentum P and angular momentum J by $E = \frac{1}{16\pi} \lim_{r\to\infty} \int_{S_r} \sum_{i,j} \Big(\partial_i g_{ij} - \partial_j g_{ii}\Big) N^j da$, $P_i = 1/8\pi \lim_{r\to\infty} \int_{S_r} (k_{ij} - \mathrm{tr} k\, g_{ij}) N^j da$, $i = 1,2,3$ and, $J_i = 1/8\pi \lim_{r\to\infty} \int_{S_r} \epsilon_{iab} x^a (k^{bj} - g^{bj} \mathrm{tr} k) N_j da$, $i = 1,2,3$ where S_r is the coordinate sphere of radius r, N the exterior unit normal to it and da its area element. In view of (S.A.F) we also have $E = M$, $P = 0$. Thus the S.A.F. condition implies that the initial data set is in a center of mass frame. In view of the positive mass theory, M must be a positive number vanishing only if the initial data set is flat (i.e. its development is flat).

We thus redefine our central theme as being the study of the evolution of asymptotically flat initial data sets. The main features under consideration are the study of local well posedness, regularity and asymptotic behaviour. In what follows I will list a few number of more concrete questions reflecting some of the current activity in General Relativity, pertinent to this theme, and also my personal taste and philosophy.

Question 1. *Under what assumptions on the initial conditions is the Cauchy problem locally well posed?*

Technically this is the question of local in time existence and uniqueness of the development of an initial data set. Our present technology requires too much differentiability on the initial data set. Lowering these differentiability requirements is crucial in understanding the regularity properties of the solutions to (E-V) and also the general Einstein equations.

Question 2. *Under what assumptions on the initial data sets do there exist global, smooth and geodesically complete solutions of the Einstein-Vacuum equations ? What is their asymptotic behaviour ?*

This question is intimately connected to that of stability of the Minkowski space-time. Namely the Minkowsky space-time $\mathbf{M}^{3+1}$ is a special solution of E-V free of singularities. As mentioned above an initial data set is said to be flat if its development is diffeomorphic to $\mathbf{M}^{3+1}$. It is thus natural to ask what happens to the developments of initial data sets which are small perturbations of a flat initial data set. This problem is important from two

[8] A function f is said to be $o_m(r^{-k})$, resp. $O_m(r^{-k})$, as $r \to \infty$ if $\partial^l f(x) = o(r^{-k-l})$, resp. $O(r^{-l-k})$, for any $l = 0, 1 \ldots m$, where ∂^l denote all the partial derivatives of order l relative to the coordinates x^1, x^2, x^3.

points of view. First, it seems[9] that all attempts to find explicit solutions of (E-V) equations, in the A.F. regime have lead to singular space-times. Second, any A.F. initial data set can be interpreted, outside a sufficiently large relatively compact set $\mathcal{K}$, as a small perturbation of a flat initial data set. Thus the study of the global stability of the Minkowski space-time is also the study of the asymptotic properties of the development of any A.F. initial data set outside the future set of a sufficiently large set $\mathcal{K} \subset \mathcal{H}$.

The problem of the stability of the Minkowski space-time has been recently addressed in my joint work with D. Christodoulou [12]. The result which we were able to prove asserts the following,

Theorem ([12]) *Any S.A.F.*[10] *initial data set which satisfies, in addition, a Global Smallness Assumption, leads to a unique, smooth and geodesically complete development, solution of the Einstein-Vacuum Equations. Moreover, this development is globally asymptotically flat, by which we mean that its Riemann curvature tensor approaches zero*[11] *on any causal or space-like geodesic, as the corresponding affine parameter tends to infinity.*

The global smallness assumption requires that an appropriate L^2-norm of up to 2 derivatives of the curvature tensor of g and 3 derivatives of k are small. A more primitive version of our result requires one derivative less, in line with the statement of the local existence theorem. Further improvements of our result will depend crucially on progress made on Question 1.

Question 3. *Under what conditions do initial data sets develop black holes and singularities?*

The famous incompleteness theorem of Penrose asserts that if an initial data set of a space-time verifying the Einstein field equations (with very general assumptions on the energy-momentum tensor **T**) has a trapped sphere[12] then some outgoing null geodesics normal to S must be future-incomplete. It is nevertheless not clear that trapped surfaces form at all.

---

[9] To my knowledge. In particular this is the case of the two parameter family of stationary solutions, called the Kerr family, which are all singular with the exception of the trivial member of the family- the Minkowski space-time.

[10] of order $k = 4$. We note that the precise fall-off conditions of the initial data set are in fact given in L^2-weighted norms, and thus differ slightly from those we have discussed below.

[11] Our result gives precise information on the rate of decay of different components of the curvature tensor.

[12] i.e. a space-like sphere S on $\mathcal{H}$ with a compact filling such that the outgoing null normal to S are everywhere converging

Question 4. *What are the global regularity features of arbitrary developments of A.F. initial data sets ?*

One of the main distant goals of the mathematical theory of General Relativity is to give a correct formulation and solve what is referred to as the "Cosmic Censorship" conjecture. Loosely speaking the conjecture asserts that there are no singularities outside black holes or, more picturesque, there exist no naked singularities.

Question 5. *What are the main, long-time, features of the solutions?*

This question relates to the previous ones, Q2-Q4, but it includes many other aspects of G.R. Among them I could mention the question of the stability of the Kerr Space-Times. At this point we do not have even a satisfactory analysis of the linear stability of these space-times. Moreover even the question of the uniqueness of the Kerr solutions among all stationary solutions of (E-V) is not satisfactory answered[13]. Another fundamental but very difficult question is the study of the asymptotic features of a space-time in which two or more black holes interact.

In what follows I will give a short description on some of the progress made recently on the questions 1,3,4 and give a short description of the proof of Theorem ([12]) mentioned in connection with question 2.

Q1. As we have mentioned below the first solution to this problem was given Y.C. Bruhat under the assumption that the initial metric g has 3 derivatives, and the tensor k has 2 derivatives locally on $L^2(\mathcal{H})$. Her proof is based on two ingredients. The first consists on an auxilliary construction of wave coordinates relative to which the E-V equations take the form (7) of a system of nonlinear wave equations. The special choice of wave coordinates is in fact not important, they can be replaced by other more geometric structures [14]. The second ingredient is more basic as it reflects on our present, limited, techniques in dealing with the local theory for all systems of nonlinear hyperbolic equations in more than 2 space-time dimensions[15] It combines stan-

[13] The classic uniqueness proof of Hawking requires that the space-time is real analytic. This assumption is not a purely technical one as it circumvents the fundamental difficulty of the ill posedness of the Cauchy problem outside the future set of a characteristic hypersurface.

[14] In [12] the local existence theorem is proved relative to a maximal foliation. In other words the space-time is constructed locally together with a "time function" t whose level hypersurfaces are maximal, i.e. $\mathrm{tr}k = 0$.

[15] In 2 space-time dimensions the situation is radically different because of the method of characteristics which allows much better results. This remains true in higher dimensions if one considers only solutions with spherical symmetry.

dard energy estimates for derivatives of solutions of the hyperbolic system together with Sobolev inequalities and an iteration procedure. This procedure has a built in limitation in that it requires too many derivatives on the initial data. For E-V this limitations means that we have to restrict ourselves to initial data sets with one derivative of the curvature tensor and two derivatives of k in $L^2_{loc}(\mathcal{H})$. This limitation is very significant when one passes from the local to the global study of the regularity properties of the system. To overcome the limitation it seems imperative to look first at a much simpler field theory. This is the case of Wave Maps from the Minkowski space-time $\mathbf{M}^{3+1}$ to an arbitrary Riemannian manifold $(\mathcal{N}, h)$. Relative to a local chart in $\mathcal{N}$ the W.M. equations take the following form,

$$\Box\phi^I + \sum_{J,K} \Gamma^I_{JK}(\phi) Q(\mathbf{D}\phi^J, \mathbf{D}\phi^K), \quad I = 1, \dots, N \tag{9}$$

where

$$Q(\mathbf{D}\phi, \mathbf{D}\psi) = \mathbf{D}_\alpha\phi \mathbf{D}^\alpha\psi = \mathbf{m}^{\alpha\beta}\mathbf{D}_\alpha\phi\mathbf{D}_\beta\psi \ . \tag{10}$$

Here $\phi = (\phi^I)_{I=1,\dots,N}$ is a vector valued function defined on $\mathbf{M}^{3+1}$. The functions Γ^I_{JK} are given arbitrary smooth functions of ϕ.
We consider the initial value problem on the hyperplane $t = x^0 = 0$,

$$\phi(0,x) = f_0(x), \quad \partial_t(0,x) = f_1(x) \ . \tag{11}$$

The classical local existence result for equations of type (9) requires $f_0 \in H_3(\mathbb{R}^3)$, $f_1 \in H_2(\mathbb{R}^3)$. Recently, in collaboration with M. Machedon [24] we were able to prove the following,

Theorem ([24]) *Consider systems of nonlinear wave equations in* $\mathbf{M}^{3+1}$ *of the type* (9), (10) *subject to the initial conditions* (11) *under the assumptions* $f_0 \in H_2(\mathbb{R}^3)$, $f_1 \in H_1(\mathbb{R}^3)$. *There exists a* $T_* > 0$ *and a unique solution* ϕ *defined in the slab* $\mathcal{D}_* = [0, T_*] \times \mathbb{R}^3$ *and verifying the following,*

(i) $\sup_{\mathcal{D}_*} |\phi(t,x)| < \infty$,

(ii) $\int\int_{\mathcal{D}_*} (|Q(\phi,\phi)|^2 + |\mathbf{D}Q(\phi,\phi)|^2)\, dtdx < \infty$,

(iii) $\sup_{[0,T_*]} \|\phi(t,\cdot)\|_{H^2(\mathbb{R}^3)} < \infty$.

As we have mentioned above the L^2 estimates, provided by the energy method, are not enough to prove a result like this. On the other hand any attempt to replace L^2 with any other L^p is known to fail (see [25]) even for the linear problem, in $\mathbf{M}^{n+1}$,

$$\Box\phi = F; \ \phi(0,x) = f_0(x), \, \partial_t(0,x) = f_1(x) \ . \tag{12}$$

The only other possibility to get additional information is to consider space-time integrals. The best known result of this type, for the wave equation (12), is due to Strichartz [36],[37] who has proved the following inequality,

$$\|\phi\|_{L^q} \leq C\Big(\|f_0\|_{H^{1/2}} + \|f_1\|_{H^{-1/2}} + \|F\|_{L^q}\Big), \tag{13}$$

where L^p, L^q are space-time norms with exponents $p = 2\frac{n+1}{n+3}$, $q = 2\frac{n+1}{n-1}$.

The novelty of our result consists of estimating, in space-time, the Lorentz invariant bilinear form $Q(\phi, \psi)$ and derive for it an estimate which is stronger than that which could be derived from the Strichartz inequality. This allows us to gain a full derivative over the classical result. The basic new tool is contained in the following

Proposition 3 *Let ϕ, ψ be solutions in $\mathbf{M}^{3+1}$ to the equations $\Box\phi = F$, $\Box\psi = G$ with initial conditions $\phi(0,x) = f_0(x)$, $\partial_t\phi(0,x) = f_1(x)$ and $\psi(0,x) = g_0(x)$, $\partial_t\psi(0,x) = g_1(x)$. Assume that $f_0, g_0 \in H_2(\mathbb{R}^3)$, $f_1, g_1 \in H_2(\mathbb{R}^3)$ and that the integrals $\int_0^T \|\mathbf{D}F(t,\cdot)\|_{L^2(\mathbb{R}^3)}\, dt$, $\int_0^T \|\mathbf{D}F(t,\cdot)\|_{L^2(\mathbb{R}^3)}\, dt$ are both finite. Then the space-time integral $\int_0^T \int_{\mathbb{R}^3} |\mathbf{D}Q(\phi,\psi)|^2\, dtdx$ is also bounded, with a bound depending only on the product of $\Big(\|f_0\|_{H^2(\mathbb{R}^3)} + \|f_1\|_{H^1(\mathbb{R}^3)} + \int_0^T \|\mathbf{D}F(t,\cdot)\|_{L^2(\mathbb{R}^3)}\, dt\Big)$ with $\Big(\|g_0\|_{H^2(\mathbb{R}^3)} + \|g_1\|_{H^1(\mathbb{R}^3)} + \int_0^T \|\mathbf{D}G(t,\cdot)\|_{L^2(\mathbb{R}^3)}\, dt\Big)$.*

Remark.1 The result of the proposition is wrong if we replace the bilinear form Q with an arbitrary bilinear form in the space-time gradients of ϕ, ψ. It is conceivable that the result of Theorem ([24]) is also wrong if one considers in (9) general expressions quadratic in the $\mathbf{D}\phi$. This suggests that any future improvements of the classical local existence and uniqueness results require special features of the equations one is interested in.

Remark.2 The proof of Theorem ([24]) depends heavily on the fact that the nonlinear terms in (9), (10) do not depend on the top derivatives. It is reasonable to believe however that the result of the theorem can be extended to certain classes of quasilinear equations. Most important we believe that the result holds true for the Einstein vacuum equations, in other words the (E-V) equations are well posed for initial data sets $(\mathcal{H}, g, k)$ for which the components curvature tensor $R(g)$ and the first derivatives of k are bounded in L^2.

Q2. In proving the result on the stability of the Minkowski space-time mentioned above we had to overcome the following major obstacles,

a. *The problem of coordinates.*
b. *The strongly nonlinear character of the equations.*

c. *The long range character of the initial data set.*
d. *The nontrivial character of the asymptotic properties of the causal structure of any small perturbation of the Minkowski space-time.*
e. *The conjunction of all of the above.*

The scope of this lecture doesn't allow me to discuss in detail each of the above difficulties and describe our strategy for overcoming them. For those interested I would like to refer to the extensive introduction of [12] or the recent Bourbaki seminar [4]. In what follows I will only give a very short description of the main ideas in [12].

The difficulty (a) has already been alluded to. In short one is faced with the following dilemma. To write the equations in a meaningful way one seems forced to introduce an additional structure, e.g. wave coordinates. Such a structure seems necessary even to allow the formulation of well posed Cauchy problem and a proof of a local in time existence result. Nevertheless, as the particular case of wave coordinates illustrates, the structure, if not carefully chosen, may lead, in the large, to problems of its own making. In [12] we solve this problem by constructing our space-times together with two functions. One is a time function t whose level hypersurfaces are maximal. The second, much more important, is a solution u of the Eikonal equation $\mathbf{g}^{\alpha\beta}\partial_\alpha u \partial_\beta u = 0$, whose level hypersurfaces are outgoing null hypersurfaces with correct asymptotic properties at null future infinity.

The other major obstacle in the study of the Einstein equations consists in their hyperbolic and strongly nonlinear character. As we have already mentioned the only apriori bounds we have available in the study of quasi-linear hyperbolic equations, in the physical space-time dimension, are those based on energy estimates. Yet the classical energy estimates are limited to proving estimates which are local in time. The difficulty has to do with the fact that, in order to control the higher energy norms of the solutions, one has to control the integral in time of their bounds in uniform norm. For this we need to control the decay of the L^∞ norms for which we have no direct information.

This difficulty can be overcome with the help of the extended energy method as illustrated by the Propositions 1,2. The method was first applied to prove global or long time existence results for nonlinear wave equations ([21],[22]). The implementation of a similar strategy for the E-V equations is based on the Bianchi identities (B) and the norms[16] $\mathbf{Q}_1(t), \mathbf{Q}_2(t)$ introduced in (6). The major departure from the linear theory of Prop. 2 is that the vectorfields T_0, Ω, S, K_0 are now themselves unknown and have to be constructed together with our space-time. This is due to the difficulty (d). Indeed, if the asymptotic behaviour of the causal structure of the space-time we construct would have been trivial we could have chosen as vectorfields

[16] Because of the difficulty (c) we have to avoid the use of $\mathbf{Q}_0(t)$ which would be infinite at $t = 0$.

T_0, Ω, S, K_0 the ones given to us in $\mathbf{M}^{3+1}$. As it stands, due to the mass term which appears in the Schwarzschild part of an (S.A.F.) initial data set, has the long range effect of changing the asymptotic position of the null geodesic cones relative to the maximal foliation. They are expected to diverge logarithmically from their corresponding position in flat space-time. In addition to this their asymptotic shear[17] differs drastically from that in the Minkowski space-time. This difference reflects the presence of gravitational radiation in any nontrivial perturbation of the Minkowski space-time. To take this effect into account we rely heavily on our optical function u and construct the vectorfields T_0, Ω, S, K_0 based on its properties[18]. The construction of u itself is very elaborate and requires a systematic study of the geometry of null hypersurfaces.

In linear theory the time derivatives of the $\mathbf{Q}_1, \mathbf{Q}_2$ are zero. In the case of the (E-V) equations they give rise to cubic error[19] terms which depend linearly on the traceless part of the deformation tensors of K_0, T, S, Ω , and quadratic with respect to $\mathbf{R}$ and its covariant and Lie derivatives in the direction of T, S, Ω. The crucial point of our overall strategy is to control the time integral of these error terms. This depends on the one hand on the asymptotic behaviour of all components of $\mathbf{R}$ and its covariant derivatives, which are themselves controlled by the basic quantities $\mathbf{Q}_1, \mathbf{Q}_2$. On the other hand it depends on the asymptotic behavior of the deformation tensors of our vectorfields, and finally, due to the general covariance of the equations, on the cancellations of the "worst possible " cubic terms. It is well known that arbitrary quadratic nonlinear perturbations of the scalar wave equation, even when derivable from a Lagrangean, could lead to formation of singularities unless a certain structural condition, which we have called the Null condition, is satisfied (see [7], [22]). It turns out that the appropriate, tensorial version of this structural condition is satisfied by the Einstein equations. Roughly speaking one could say that the troublesome nonlinear terms, which could have led to formation of singularities, are in fact excluded due to to the covariance and algebraic properties[20] of the Einstein equations.

---

[17] the traceless part of their null second fundamental form

[18] and also those of the maximal foliation induced by the time function t.

[19] generated each time we commute the Bianchi Identities with a one of the vectorfields used in the definition of $\mathbf{Q}_1, \mathbf{Q}_2$

[20] These basic algebraic properties of the Einstein equations, which allow us to prove the above stated global existence result, are in sharp contrast with the nonlinear hyperbolic equations of classical continuum mechanics. Indeed the equation of Nonlinear Elasticity [20] and of Compressible fluids [34], in four space and time dimensions, form singularities even for arbitrary small initial conditions.

Putting together all the elements discussed above requires an elaborate boot-strap argument based on the method of continuity. In the end we show that $\mathbf{Q}_1, \mathbf{Q}_2$ can never become large, if they are sufficiently small at $t = 0$. The local existence theorem then guarantees that we can extend our space-time indefinitely.

Finally it remains to note that our work has many important conclusions. Far from being an abstract proof of existence it provides precise information on the asymptotic nature of gravitational radiation. Some of our conclusions confirm the non-rigorous results obtained by Bondi, Sachs, Penrose etc. Some of them are however new. One of these has lead D.Christodoulou to a real experimental prediction. In [10] he shows that gravitational waves generated by astronomical sources can have a nonlinear effect on laser interferometer detectors on Earth. This effect is shown to be of the same order of magnitude as the linear effects upon which all previous efforts to detect gravitational waves were based. Moreover the nonlinear effect is shown to produce a permanent displacement of test masses after the passage of a wave. It can thus alter significantly the strategy upon which the experimentalists plan to build their detectors.

Q3. At the present time we have no results concerning the formation of black holes for the (E-V) equations. Since the general problem is too hard, one needs to look first at simplified situations. Unfortunately the (E-V) does not allow interesting A.F. solutions with additional symmetries, e.g. spherically symmetric. To overcome this one has to consider the general Einstein field equations coupled with some matterfield. The simplest such field is the linear scalar wave equation (S.W.) with $V = 0$. This coupling allows one to obtain nontrivial dynamics even in the spherically symmetric case. In this case the group $SO(3)$ acts as an isometry group on the space-time $(\mathcal{M}, \mathbf{g})$. The group orbits are space-like metric 2-spheres S of Gauss curvature r^{-2} where r is the area radius of S, i.e. $A(S) = 4\pi r^2$. Due to the $SO(3)$ symmetry the field equations can be reduced to quotient Q of the space-time by the group. This a 2-dimensional manifold with boundary. The boundary corresponds to the set of fixed points of the group action and forms a time-like geodesic Γ. Choosing, on Q, a pair of conjugate optical functions u, v with u constant on any future directed null curve initiating at Γ, v constant along each of the conjugate family of null curves and such that both functions are increasing towards the future, the induced metric on Q has the form $-\Omega^2 du dv$. Remark that u, v are determined up to general transformations of the form $u \mapsto f(u)$, $v \mapsto f(v)$ with f,g arbitrary increasing functions. The reduced field equations form a system on Q for the functions r, Ω, ϕ,

$$r\frac{\partial^2 r}{\partial u \partial v} = -\frac{1}{4}\Omega^2 - \frac{\partial r}{\partial u}\frac{\partial r}{\partial v}$$

$$\frac{\partial^2 r}{\partial u^2} = 2\Omega^{-1}\frac{\partial r}{\partial u}\frac{\partial \Omega}{\partial u} - r\left(\frac{\partial \phi}{\partial u}\right)^2$$
$$\frac{\partial^2 r}{\partial v^2} = 2\Omega^{-1}\frac{\partial r}{\partial v}\frac{\partial \Omega}{\partial v} - r\left(\frac{\partial \phi}{\partial v}\right)^2$$
$$r\frac{\partial^2 \phi}{\partial u \partial v} = -\frac{\partial r}{\partial u}\frac{\partial \phi}{\partial v} - \frac{\partial r}{\partial v}\frac{\partial \phi}{\partial u}\,. \tag{14}$$

The above system is invariant under the transformations $u \mapsto f(u)$, $v \mapsto g(v)$ and $\Omega \mapsto (f'g')^{1/2}\Omega$.

The program of analyzing the spherically symmetric solutions of the coupled Einstein-scalar wave equation was initiated and carried out with remarkable success by D. Christodoulou, see [8], [9]. A lot of his analysis is based on the fundamental monotonicity properties of the mass function m defined by the formula $1 - \frac{2m}{r} = -4\Omega^{-2}\frac{\partial r}{\partial u}\frac{\partial r}{\partial v}$. We have $\frac{\partial m}{\partial v} \geq 0$ and $\frac{\partial m}{\partial u} \leq 0, = 0, \geq 0$ according to whether $1 - \frac{2m}{r} \leq 0, = 0, \geq 0$. In [4] Christodoulou combines this monotonicity properties of m together with the powerful method of characteristics, typical to 2-dimensional hyperbolic problems, to prove a deep result concerning the formation of black holes for the system (14). He considers the evolution of regular initial conditions defined on an outgoing null curve C_0^+, starting at a point on Γ, and shows that under reasonable assumptions of the data trapped spheres must form.

Q4. Once again the understanding of the global regularity properties of the general Einstein equations is beyond our reach. To make progress we have to either restrict our attention to solutions with spherical symmetry or to consider simpler field theories. Concerning the first approach I have to mention once more the pioneering work of Christodoulou on the coupled E-S.W. equations. Following his result on the formation of black holes, which was mentioned above, he has concentrated his efforts on the regularity of solutions outside black holes. His preliminary results suggests that though "naked singularities" are possible they are unstable in the sense that the initial conditions from which these could evolve form an exceptional set of codimension 1 in the set of all allowable initial conditions.

Finally I want to report to you on recent results concerning the global regularity of simpler field[21] theories.

The simplest nonlinear field theory we have is (S.W.) For simplicity we restrict our attention to nonlinear wave equations with

[21] We want to stress once more that our motivation for studying the regularity problem for the simpler non-physical field theories is purely mathematical. We believe that the mathematical difficulties are deep and that to overcome them we need to develop new ideas, ideas which could later be relevant to G.R. and possible other "physical" equations.

$V(\phi) = -\frac{1}{p+1}|\phi|^{p+1}, p > 0$, in Minkowski space-time $\mathbf{M}^{n+1}$. The energy identity (4) applied to the time-like Killing vector field $T_0 = \frac{\partial}{\partial t}$ takes the form, $E(t) = E(t_0)$ where,

$$E(t) = \int \Big(\frac{1}{2}|\mathbf{D}\phi(t,x)|^2 + \frac{1}{p+1}\,|\phi(t,x)|^{p+1}\Big)dx^1 \cdots dx^n \tag{15}$$

where $|\mathbf{D}\phi|^2 = |\partial_t\phi|^2 + |\partial_1\phi|^2 + \dots |\partial_n\phi|^2$.

Prescribing to the space-time variables $t = x^0, x^1, \dots, x^n$ the same scale L the variable ϕ, which verifies the equation $\Box\phi + V'(\phi) = 0$ acquires the scale $L^{-\frac{2}{p-1}}$. Therefore the energy E has the scale L^s where s is the exponent $s = n - 2 - \frac{4}{p-1}$.

The case when the exponent s is strictly negative is quite easy to analyze and has lead to the well-known global regularity result of Jörgens [19]. The case $s = 0$ is called the "critical" while $s > 0$ is called supercritical. In the supercritical regime we have no results, even for spherically symmetric solutions, despite the relatively large attention it has received. The critical case has been recently settled by the combined efforts of Struwe [38], Grillakis [16] and more recently Shatah-Struwe [32].

Theorem ([32]) *Consider the initial value problem $\Box\phi + V'(\phi) = 0$ with initial conditions $\phi(0,x) = f_0(x)$, $\partial_t\phi(0,x) = f_1(x)$ which, for simplicity, we may assume in C_0^∞. Assume that $n \le 7$ and that the exponent $s \le 0$. Then the equations admits unique smooth solutions globally in $\mathbf{M}^{n+1}$.*

The proof of the result turns out to be quite simple and I will try to sketch it below.

Proceeding by contradictions one may assume that there exists a singular point of a solution ϕ of the equation. In view of the local existence theorem the first such point p should be such that its causal past $J^-(p)$ is entirely regular. Let t_* be the time corresponding to p. For all $t < t_*$ we denote by $E(t;p)$ the local energy at time t included in the causal past of p i.e. $E(t;p) = \int_{D(t,p)} \frac{1}{2}|D\phi(t,x)|^2 + \frac{1}{2}|\phi(t,x)|^{p+1}\,dx$ where $D(t,p)$ is the intersection of the t-hypersurface with $J^-(p)$. In view of the energy identity applied to the Killing vector field T_0 one can easily show that $E(t,p)$ is decreasing as $t \uparrow t_*$. Replacing the vector field T_0 with the scaling vector field $S = (x^\alpha - \overline{x}^\alpha)\partial_\alpha$ based at p one can prove that, in fact

$$\int_{D(t,p)} |\phi(t,x)|^{p+1}\,dx \to 0\,. \tag{16}$$

This crucial fact, which was first observed by Struwe [38], lies at the heart of the whole proof.

In the remaining of the argument I restrict my attention to $n = 3$ and $p = 5$. I will present a fake argument[22] based on the following inequality for solutions to the inhomogeneous problem $\Box\phi = F$, with $\phi(t_0, x) = f_0(x)$, $\partial_t\phi(t_0, x) = f_1(x)$,

$$\int_{t_0}^{T} \|\phi(t,x)\|^2_{L^\infty(\mathbb{R}^3)} dt \leq C\Big(\|f_0\|_{H^1(\mathbb{R}^3)} + \|f_1\|_{L^2(\mathbb{R}^3)} + \int_{t_0}^{t} \|F\|_{L^2(\mathbb{R}^3)} dt\Big)\,. \tag{17}$$

The inequality is correct in the spherically symmetric case but diverges logarithmically in the general case [24]. If we assume it to be true however the proof of the theorem follows easily as follows. First it is easy to see, by analyzing the proof of the local existence theorem, that ϕ can only become singular at p if the integral $\int_0^t \|\phi(t',x)\|^2_{L^\infty(D(t',p))} dt'$ becomes infinite as $t \to t_*$. Now, in view of (16), $\int_{D(t,p)} |\phi(t,x)|^6 \leq \epsilon$ if t is larger than some $0 \leq t_0 < t_*$. Applying the inequality (17) to $\Box\phi = \phi^5$, we have in view of the energy identity $I(t) \leq C(E(0) + \epsilon^3 I^2)$, where $I(t)^2 = \int_{t_0}^t \|\phi(t',x)\|^2_{L^\infty(D(t',p))} dt'$. It then immediately follows that $I(t)$ has to remain bounded as $t \to t_*$ and therefore p cannot be a singular point.

As we have mentioned above the estimate (17) holds true only in the spherically symmetric case, where it is an immediate consequence of the Hardy-Littlewood maximal function inequality (see [24]). The correct proof of Theorem ([34]) is similar but uses instead of (17) the deeper result of Strichartz, the inequality (13) mentioned before.

Time does not allow me to talk about other regularity results. I would have liked to mention the current work being done on the regularity of Wave-Maps defined on $\mathbf{M}^{2+1}$, see [13],[33],[17]. I would also have like to mention the older, but beautiful result of Eardley-Moncrief [14] on the global regularity of the Yang-Mills equations, result which despite its originality has received relatively little attention. Finally, in connection with Question 5, one should mention the classic work of Morawetz-Strauss [26] on the asymptotic behaviour of the nonlinear Klein-Gordon equations, work which can be considered in many ways as one of the early pioneering works in the subject which I have tried to present here.

Acknowledgements. This paper was supported by the N.S.F. Grant DMS-9103613.

[22] The argument is correct for spherically symmetric solutions

References

1. Bel, L.: Introduction d'un tensur du quatrieme ordre, Comptes Rendus **247**, pp. 1094, 1297 (1978)
2. Bondi, H.: Nature, **186**, 535 (1960)
3. Bondi, H. van der Burg, M.G.J., Metzner, A.W.K.: Gravitational waves in General Relativity vii. Proc.Roy.Soc.Lond. **A269**, 21-52 (1962)
4. Bourguignon, J.P.: Stabilite par deformation non-lin. de la metrique de Minkowski. Seminaire N. Bourbaki **1990-1991** June 1991.
5. Bruhat C.: Theoreme d'existence pour certain systemes d'equations aux deriveès partielles nonlinaires. Acta Matematica **88**, 141-225 (1952)
6. Bruhat C.: Un theoreme d'instabilite pour certain equations hyperb. non-lineaire. C.R.Acad.Sci.Paris **276A**, 281 (1973)
7. Christodoulou, D.: Global solutions for nonlinear hyperbolic equations for small data. Comm.Pure Appl.Math. **39** 267-282 (1986)
8. Christodoulou, D.: A Mathematical Theory of Gravitational Collapse. Comm.Math.Phys. **109**, 613 (1987).
9. Christodoulou, D.: The Formation of Black Holes and Singularities in Spherically Symmetric Gravitational Collapse. Comm.Pure Appl.Math. **44**, 339-373 (1991).
10. Christodoulou, D.: The Nonlinear Nature of Gravitation and Gravitational Wave Experiments. Preprint, June 1991.
11. Christodoulou, D., Klainerman, S.: Asymptotic Properties of Linear Field Equations in Minkowski Space-Time. Comm.Pure Appl.Math. **43**, 137-199 (1990).
12. Christodoulou, D., Klainerman, S.: The nonlinear stability of the Minkowski space-time. Preprint.
13. Christodoulou, D., Shadi, A.: Tahvildar-Zadeh Regularity of Spherically Symmetric Harmonic Maps of the $2+1$ dim. Minkowski Space. Preprint
14. Eardley, D., Moncrief, V.: The Global Existence of Yang-Mills-Higgs fields in $\mathbf{M}^{3+1}$. Comm.Math.Phys. **83**, 171-212 (1982)
15. Geroch, R.: *Asymptotic structure of space-time.* P. Esposito and L. Witten (eds.), Plenum, New York (1976)
16. Grillakis, M.: Regularity and Asympt. Behaviour of the Wave Equation with as critical nonlinearity. Ann. of Math. **132** , 485-509 (1990).
17. Grillakis, M.: Classical solutions for the Equivariant Wave Maps in 1+2 dimensions. Preprint.
18. Hörmander, L.: On Sobolev spaces associated with some Lie algebras. Report No.4, Inst. Mittag-Leffler (1985)
19. Jörgens, K.: Das Anfangswertproblem im Grossen für eine Klasse nichtlinearer Wellengleichungen. Math.Z. **77**, 295-307 (1961).
20. John, F.: Formation of Singularities in Elastic Waves. Lecture Notes in Phys., Springer-Verlag, 190-214 (1984)
21. Klainerman, S.: Uniform decay estimates and the Lorentz invariance of the classical wave equation. Comm.Pure Appl.Math. **38**, 321-332 (1985)
22. Klainerman, S.: The null condition and global existence to nonlinear wave equations. Lectures in Appl.Math. **23**, 293-326 (1986)

23. Klainerman, S.: Remarks on the global Sobolev inequalities in Minkowski space. Comm.Pure Appl.Math. **40**, 111-117 (1987)
24. Klainerman, S., Machedon, M.: Space-Time estimates for null forms and the local existence theorem. Preprint.
25. Littman, W.: The wave operator and L^p norms. Journal of Math. and Mech. **12**, 55-68 (1963)
26. Morawetz, C., Strauss, W.: Decay and scattering of solutions to a nonlinear relativistic wave equation. Comm.Pure Appl.Math. **25**, 1-31 (1972) and **26**, 47-54 (1973)
27. Newman, E.T., Todd, K.P.: *Asymptotically flat space-times, General Relativity and Gravitation*, vol. 2, A. Held, Plenum, 1980
28. Penrose, R.: *Structure of Space-Time*, Batelle Rencontre, C.M. DeWitt and J.A. Wheeler, (1967)
29. Penrose, R.: Zero rest mass fields including gravitation: asymptotic behaviour. Proc.Roy.Soc.London **A284**, 159-203 (1962)
30. Sacks, R.K.: Gravitational waves in General Relativity viii. Proc.Roy.Soc. London **A270**, 103-126 (1962)
31. Segal, I.: Nonlinear Semigroups. Ann.Math. **78**, 339-364 (1963)
32. Work in progress.
33. Shatah, J., Tahvildar-Zadeh, A.: Regularity of harmonic maps from the Minkowski space-time into rotationally symmetric manifolds. Preprint.
34. Sideris, T.: Formation of Singularities in 3-d Compressible Fluids. Comm. Math.Phys. **101**, 275-485 (1985)
35. Strauss, W.: Decay and asymptotics for $\Box u = F(u)$. J.Funct.Anal. **2**, 409-457 (1968)
36. Strichartz, R.S.: A Priori Estimates for the Wave Equation and some Applications. J.Funct.Anal. **5**, 218-235 (1970)
37. Strichartz, R.S.: Restrictions of Fourier Transforms to Quadratic Surfaces and Decay of Solutions of Wave Equation. Duke Math.J. **44**, No 3, 705-714 (1977)
38. Struwe, M.: Globally Regular Solutions to the Φ^5-Klein-Gordon equation.

Representations of Quantized Differential Forms in Non-Commutative Geometry

Joachim Cuntz

Mathematisches Institut d. Universität Heidelberg,
Im Neuenheimer Feld 288, W-6900 Heidelberg,
Fed. Rep. of Germany

Abstract: Besides giving a survey of some basic structures and ideas in K-theory and cyclic cohomology for non-commutative algebras, we describe a new way to realize algebras of abstract differential forms, over a given algebra A, and their "quantum" deformations. For this we use subalgebras and quotients of an algebra $A[D,F]$ obtained from A by adjoining two additional elements D, F. This is closely related to the notion of a Fredholm module.

1 Introduction

In this article we describe some constructions which serve to define and to compute the basic invariants in non-commutative geometry. In particular, we discuss various forms of the algebra of differential forms over a quantum space (non-commutative algebra) and show how these can be used to define K-theory and Connes' cyclic cohomology. Besides the algebra ΩA of abstract differential forms over A (generated by the non-commutative Kähler differentials) which has been used by Connes, Karoubi, Kastler and others, [20], we consider a certain deformation of ΩA which in fact turns out to be better suited for the purposes of non-commutative geometry.

K-theory and cyclic cohomology (non-commutative de Rham cohomology) can be defined naturally using these "quantum" differential forms. We show in this article that all these algebras of differential forms can be embedded canonically into a very simple algebra $A[D,F]$ generated by A together with two additional elements D, F and that relations between these algebras can be explained by relations in $A[D,F]$. This algebra is also closely related to the important notion of a Fredholm module over A. In fact, such a Fredholm module is simply a representation of $A[D, F]$.

The organization of the article is as follows

First, we recall the construction of involutive algebras given by generators and relations, illustrated by some simple examples encountered in physics. The spirit here is quite similar to algebraic geometry. The next step consists in the $\mathcal{C}^\infty$-completion of these non-commutative "algebraic varieties" giving non-commutative analogues of differential structures. Here already we use an algebra $\mathcal{D}A$ of abstract derivatives of elements of A. If, in this algebra, one suppresses all derivatives of degree ≥ 2 and introduces an abstract $\mathbb{Z}/2$-grading operator F one obtains algebras ΩA, RA, QA of quantum differential forms which are the basis for K-theory and cyclic cohomology.

We realize all the algebras $\mathcal{D}A$, ΩA, RA, QA as subalgebras of the algebra $A[D, F]$ mentioned above. Next we introduce K-theory and cyclic cohomology in terms of RA, QA. Finally, we use the fact that a Fredholm module is simply a representation of $A[D, F]$ to associate cyclic cocycles with finitely summable Fredholm modules. The formulas we obtain are not quite the standard ones even though the resulting cocycles are the same as those of Connes in [7].

2 Non-Commutative Algebras and Non-Commutative Operator Algebras

An involutive algebra is a complex algebra with an antilinear isomorphism $^* : A \to A$ of order 2 such that $(xy)^* = y^* x^*$. All our algebras in the following will be assumed to be involutive, even though in some cases this assumption is not really necessary. Usually we will also assume that the algebras are unital since this simplifies many constructions. In many cases the algebras are given by a set of generators $\{x_i\}_{i \in I}$ subject to certain relations among the x_i and their formal adjoints x_i^*: let $\mathcal{P} = \mathcal{P}(\{x_i\}_{i \in I})$ denote the set of all complex polynomials in x_i and x_i^* divided by the ideal generated by a certain self-adjoint set of relations. Every element of $\mathcal{P}$ is a linear combination of monomials of the form $z = y_1^{k_1} y_2^{k_2} \cdots y_s^{k_s}, y_j \in \{x_i\}_I \cup \{x_i^*\}_I$. Then $\mathcal{P}$ is involutive if we set $z^* = y_s^{*k_s} \cdots y_2^{*k_2} y_1^{*k_1}$ where $(x_i^*)^* \stackrel{\text{def}}{=} x_i$, and extend antilinearly. A norm $\|\cdot\|$ on an involutive algebra $\mathcal{P}$ is called a C^*-norm if $\|x^* x\| = \|x\|^2$ for all $x \in \mathcal{P}$. One has the following two well known theorems (going back to Gelfand-Naimark):

1) if there exists a C^*-norm on $\mathcal{P}$, then $\mathcal{P}$ is isometrically isomorphic to an algebra of bounded operators on a Hilbert space H, the involution corresponding to the adjoint of operators.

2) if $\mathcal{P}$ is commutative and if there is a C^*-norm on $\mathcal{P}$, then there is a compact space X (the spectrum of $\mathcal{P}$) such that $\mathcal{P}$ is isometrically isomorphic to a dense subalgebra of the algebra $\mathcal{C}(X)$ of continuous complex-valued functions on X with the sup-norm, the involution corresponding to taking the complex conjugate.

Definition Let $\mathcal{P} = \mathcal{P}(\{x_i\})$ be an involutive algebra as above and assume that $\|x\|_\infty = \sup\{\|x\| \mid \|\cdot\| \text{ is a } C^*\text{-seminorm on } \mathcal{P}\}$ exists for each $x \in \mathcal{P}$ (which is quite often the case). Then we denote by $C^*(\{x_i\})$ the C^*-algebra obtained as the completion of $\mathcal{P}/N$, $N = \{x \in \mathcal{P} \mid \|x\|_\infty = 0\}$ with respect to $\|\cdot\|_\infty$.

The idea of operator algebras given by generators and relations is quite commonly used in physics and mathematical physics. The first appearance of this idea probably goes back to Heisenberg's commutation relation (which of course can not be realized by bounded operators). We list here some very simple examples of such algebras used by physicists:

1. (a) Take one generator a with relations $a^2 = 0$, $a^*a + aa^* = 1$. Then $C^*(a) \cong M_2(\mathbb{C})$, the complex 2×2-matrices, with a corresponding to $\left(\begin{smallmatrix} 0 & 1 \\ 0 & 0 \end{smallmatrix}\right)$.
   (b) Take one generator b with relation $b^*b - bb^* = 1$. This relation can not be realized by bounded operators. Therefore $C^*(b) = \{0\}$.
2. (a) Take 3 generators $\sigma_1, \sigma_2, \sigma_3$ satisfying

$$\sigma_i\sigma_j + \sigma_j\sigma_i = 0, \quad i \neq j, \quad \sigma_i = \sigma_i^*, \quad \sigma_i^2 = 1, \quad \sigma_1\sigma_2\sigma_3 = 1.$$

   Then $C^*(\sigma_1, \sigma_2, \sigma_3) \cong M_2(\mathbb{C})$.
   (b) Take 2 generators e_1, e_2 satisfying $e_i^2 = e_i = e_i^*$, $i = 1, 2$ and $e_ie_je_i = \tau e_i$, $i \neq j$, for some $0 < \tau < 1$. Then $C^*(e_1, e_2) \cong M_2(\mathbb{C})$.
3. (a) Take one generator x such that $x^*x = xx^* = 1$ (x is unitary). Then $C^*(x)$ is commutative and isomorphic to the algebra $\mathcal{C}(S^1)$ of continuous functions on the unit circle.
   (b) One generator x such that $x^*x = 1$. Here $C^*(x)$ is non-commutative and can be shown to be isomorphic to the algebra of Tolplitz operators on the unit disc.
4. (a) Take two unitary ($u_i^*u_i = u_iu_i^* = 1$) generators u_1, u_2 satisfying $u_1u_2 = u_2u_1$. Then $C^*(u_1, u_2)$ is commutative and isomorphic to $\mathcal{C}(\mathbb{T}^2)$, the continuous functions on the 2-torus.
   (b) Take two unitary generators u_1, u_2 satisfying $u_1u_2 = e^{2\pi i\theta}u_2u_1$, $0 \leq \theta < 1$. Then $A_\theta = C^*(u_1, u_2)$ is a "non-commutative 2-torus" which shares many properties with the algebra A_0 of continuous functions on the ordinary 2-torus (4(a)). The algebra A_θ has been used for an explanation of the quantum Hall effect [2]. Recent results [5] show that for irrational θ, as a quantum space A_θ is in some sense

discrete: every self-adjoint element of A_θ can be approximated arbitrarily closely by self-adjoint elements with finite spectrum. Moreover, a natural subalgebra of A_θ, which contains the almost Mathieu Hamiltonian of [1], is even the completion of a union of matrix algebras [6].

5. Take n generators $s_1, \cdots, s_n$ satisfying $s_i^* s_i = 1, \;\; s_1 s_1^* + \cdots + s_n s_n^* = 1$. Then $C^*(s_1, \cdots, s_n)$ exists and is simple [10]. This algebra carries a canonical action of $SU(n)$ and has been used by Doplicher-Roberts in their work on duality for compact groups and on existence of gauge groups in algebraic quantum field theory, [14],[15].
6. Take $\mu \in [-1, 1]$ and two generators α, γ with relations $\alpha^*\alpha + \gamma^*\gamma = 1$, $\alpha\alpha^* + \mu^2\gamma^*\gamma = 1, \gamma^*\gamma = \gamma\gamma^*, \alpha\gamma = \mu\gamma\alpha, \alpha\gamma^* = \mu\gamma^*\alpha$. Then $C^*(\alpha, \gamma)$ for $\mu = 1$ is isomorphic to $\mathcal{C}(SU(2))$. For $\mu \in [-1, 1]$, $C^*(\alpha, \gamma)$ is denoted by $SU_\mu(2)$ and is the basic example for a compact quantum group in the sense of Woronowicz [21]. One checks easily that the elements $\alpha' = \alpha \otimes \alpha - \mu\gamma^* \otimes \gamma, \gamma' = \gamma \otimes \alpha + \alpha^* \otimes \gamma$ of $SU_\mu(2) \otimes SU_\mu(2)$ satisfy the same relations so that one obtains a homomorphism $SU_\mu(2) \longrightarrow SU_\mu(2) \otimes SU_\mu(2)$ mapping α to α' and γ to γ' – the coproduct – which makes $SU_\mu(2)$ a Hopf algebra.

3 Non-Commutative Differential Forms and Their "Quantum Deformations"

Let M be a compact $\mathcal{C}^\infty$-manifold. With M, one associates classically the algebra $A = \mathcal{C}^\infty(M)$ of $\mathcal{C}^\infty$-functions on M and, in order to analyze the topology and geometry of M one associates classically with A further two important auxiliary algebras: the algebra of differential forms on M, which we denote by $\Omega_{\mathrm{dR}} A$ (the de Rham complex), and the algebra $M_\infty(A)$ of $\mathbb{N} \times \mathbb{N}$-matrices over A with only finitely many non-zero entries. Cohomology classes are represented by closed forms in $\Omega_{\mathrm{dR}} A$, while K-theory classes are represented by vector bundles or equivalently, by elements e of $M_\infty(A)$ satisfying $e - e^2 = 0$. Among the three constructions $M \to \mathcal{C}^\infty M$, $A \to \Omega_{dR} A$ and $A \to M_\infty A$ only the last one carries over naturally to the non-commutative context. The definition of $\mathcal{C}^\infty(M)$ makes use of coordinates on M, while the definition of $\Omega_{\mathrm{dR}} A$ relies heavily upon the commutativity of A.

We first discuss a version of $\mathcal{C}^\infty$-structure in the non-commutative setting. For this, let $\mathcal{D}A$ be the universal algebra generated by elements $p_i(a)$, $i = 0, 1, 2, \cdots$, $a \in A$, which are linear in a and satisfy the relations $p_k(xy) = \sum_{i+j=k} p_i(x) p_j(y)$. $\mathcal{D}A$ is an $\mathbb{N}$-graded algebra if we define the degree of $p_i(x)$ to be i (if A is involutive, then $\mathcal{D}A$ is also involutive with $p_i(x)^* = -p_i(x^*)$, but this is not so important here). Moreover $\mathcal{D}A$ carries a natural derivation δ, defined by

$$\delta\left(p_{i_1}(x_1)\cdots p_{i_n}(x_n)\right) = \sum_{j=1}^{n}(i_j+1)p_{i_1}(x_1)\cdots p_{i_j+1}(x_j)\cdots p_{i_n}(x_n).$$

(To see that δ is well defined, one checks that

$$(k+1)p_{k+1}(xy) = \sum_{i+j=k}(i+1)p_{i+1}(x)p_j(y) + \sum_{i+j=k}(j+1)p_i(x)p_{j+1}(y).)$$

Then $p_j(x) = \frac{1}{j!}\delta^j(x)$ and $\mathcal{D}A$ is generated by $A \cong p_0(A)$ and the images of all δ^j applied to A. It is clear that $\mathcal{D}A$ has the following universal property: whenever $\varphi : A \to B$ is a homomorphism and δ' is a derivation of B, then there exists a unique homomorphism $\psi : \mathcal{D}A \to B$ such that $\psi(a) = \varphi(a)$, $a \in A$, and $\delta' \circ \psi = \psi \circ \delta$.

There is a formal homomorphism $e^\delta : A \to \mathcal{D}A$ defined by $e^\delta(x) = p_0(x) + p_1(x) + p_2(x) + \cdots$. This sum converges and e^δ becomes an actual homomorphism if we divide $\mathcal{D}A$ by the ideal of all elements of degree $> n$ or if we complete $\mathcal{D}A$ with respect to a suitable norm. The homomorphism e^δ has the following universal property: if $\varphi : A \to B$ is a homomorphism from A into an $\mathbb{N}$-graded algebra B, then there is a unique homomorphism $\psi : \mathcal{D}A \to B$ such that $\varphi = \psi \circ e^\delta$.

Assume now that A is equipped with a C^*-norm, i. e. a norm satisfying $\|x^*x\| = \|x\|^2$, $x \in A$, but not complete (e. g. A is the algebra of polynomials in finitely many generators and their adjoints).

In $\mathcal{D}A$, let J_k denote the ideal generated by all elements of degree $\geq k$. Make the quotient $\mathcal{D}A/J_k$ into locally convex algebra with the topology given by all algebra seminorms which are bounded by a multiple of $\|\cdot\|$ on $A \cong p_0(A) \subset \mathcal{D}A/J_k$ and for which the projection onto the degree k part is bounded. Let A_k be the completion of $e^\delta(A)$ in $\mathcal{D}A/J_k$. The $\mathcal{C}^k$-completion $\mathcal{C}^kA$ is then defined as $\mathcal{C}^kA := A_k/(A_k \cap J_1)$. It contains A as a dense subalgebra and we have inclusions $A \subset \cdots \subset \mathcal{C}^{k+1}A \subset \mathcal{C}^kA \subset \cdots \subset \mathcal{C}^0A$, where $\mathcal{C}^0A$ is the completion of A with respect to the C^*-norm $\|\cdot\|$. The $\mathcal{C}^k$-completion has (among others) the following properties, [4]:

(1) For each $x \in A$ the spectrum of x in $\mathcal{C}^kA$ equals its spectrum in $\mathcal{C}^0A$.
(2) Let $x = x^* \in \mathcal{C}^k(A)$ and f a $\mathcal{C}^{k+1}$-function, $\varepsilon > 0$, on $\mathbb{R}$. Then $f(x) \in \mathcal{C}^kA$.
(3) Let δ be a derivation of A. Then δ extends to a continuous derivation $\mathcal{C}^{k+1}A \to \mathcal{C}^kA$.

Suppose now that, on $\mathcal{D}A$, we try to enforce that $\delta^2 = 0$, (as is the case for the de Rham operator d), i. e. we divide by the ideal generated by all $\delta^2(x)$, $x \in A$. Then $2\delta(x)\delta(y) = \delta^2(xy) - x\delta^2(y) - \delta^2(x)y = 0$, so that all that remains in the quotient are elements of the form $x + y\delta(z)$, while forms $x_0\delta(x_1)\cdots\delta(x_n)$ of degree $n>1$ will be 0. Thus, to construct an algebra of differential forms ΩA with an operator d satisfying $d^2=0$ one has to modify δ to a graded derivation. A graded derivation is defined by the property $d(\omega_1\omega_2)=d(\omega_1)\omega_2+(-1)^{|\omega_1|}\omega_1 d\omega_2$. This can be done in the following way: adjoin an element F to $\mathcal{D}A$, i.e. take the algebra generated by $\mathcal{D}A$ and F, satisfying $F^2=1$ and $F\omega=(-1)^{|\omega|}\omega F$ ($|\omega|=$ degree of ω), $\omega\in\mathcal{D}A$, and define $d\omega=F\delta(\omega)$. This gives a graded derivation $d:\mathcal{D}A\to\mathcal{D}A[F]$. It remains to define dF. Before we discuss this in detail we first turn directly to the definition of the algebra ΩA of abstract differential forms over A which carries a graded derivation. By definition, ΩA is the universal algebra generated by $x\in A$ with the relations of A and elements dx which are linear in x and satisfy $d(xy)=xdy+d(x)y$. Elements of ΩA are linear combinations of elements of the form $x_0dx_1\cdots dx_n$ (we use the convention $1dx=d(x)1=dx$ and $d(1)=0$). If we define the degree of such an element to be n, ΩA becomes an $\mathbb{N}$-graded algebra (in fact a differential graded algebra, and ΩA is the universal differential graded algebra over A). There is a graded derivation $d:\Omega A\to\Omega A$ with $d^2=0$ defined by $d(x_0dx_1\cdots dx_n)=dx_0dx_1\cdots dx_n$.

Next, we introduce a ("quantum") deformation of ΩA which is fundamental for K- and KK-theory and also for cyclic homology, the noncommutative version of de Rham cohomology, cf. [7], [8]. For each $t\in\mathbb{C}$, define a new product $\circ_t$ on ΩA by

$$\omega_1\circ_t\omega_2=\omega_1\omega_2-(-1)^{|\omega|}t\,d\omega_1 d\omega_2.$$

One checks that $\circ_t$ is associative.

Denote by Q_tA the algebra ΩA with product $\circ_t$. For $t=0$, Q_0A is isomorphic to ΩA, while for all other $t\neq 0$, Q_tA is isomorphic to Q_1A under the isomorphism $\omega\mapsto t^{-\frac{1}{2}|\omega|}\omega$ for homogeneous elements ω). We denote Q_1A by QA.

The product $\circ_t$ on the algebra of ordinary differential forms had already been considered by Fedosov [16]. The algebra QA has been introduced in [11] and described as a deformation of ΩA in [9], [13].

QA is isomorphic to the free product $A*A$ and can also be described as the universal $\mathbb{Z}/2$-graded algebra generated by A.

We are now going to explain how $\mathcal{D}A$, ΩA and its deformation QA can be canonically embedded into an algebra $A[D,F]$ generated by A together with two additional elements D and F. Since Fredholm modules over A are basically representations (where D is possibly represented by an unbounded operator) of the algebra $A[D,F]$, this is useful in order to associate cyclic cocyles and K-homology classes with Fredholm modules. Moreover, this construction illuminates the connections between $\mathcal{D}A$, ΩA and QA and finally, it points to a duality between QA and M_2A. The relations imposed on the elements of $A[D,F]$ are, besides the relations of A for the elements of A, the following:

$$F^2 = 1; \quad [F,a] = 0,\ a \in A; \quad FD = -DF.$$

We can make $A[D,F]$ into an involutive algebra by setting $F^* = F,\ D^* = D$. This algebra is $\mathbb{N}$-graded, if we define A, F to be of degree 0 and D of degree 1. It carries a natural (inner) derivation δ given by $\delta(\omega) = [D,\omega]$, $\omega \in A[D,F]$. One checks immediately that $D = \frac{1}{2}\delta(F)F$. Furthermore $A[D,F]$ has a graded derivation d (satisfying $d(\omega_1\omega_2) = d(\omega_1)\omega_2 + (-1)^{|\omega_1|} \cdot \omega_1 d\omega_2$) defined by $d(\omega) = F\delta(\omega)$.

Proposition 3.1 *The natural map $\varphi : \mathcal{D}A \to A[D,F]$ mapping $p_j(x)$ to $\frac{1}{j!}\delta^j(x)$ is injective. One has $\varphi\circ\delta = \delta\circ\varphi$ (where we use the same letter δ for the derivations of $\mathcal{D}A$ and $A[D,F]$).*

Proof. Let End$(\mathcal{D}A)$ denote the space of linear operators $\mathcal{D}A \to \mathcal{D}A$ and, given $\omega \in \mathcal{D}A$, $\lambda(\omega) \in$ End$(\mathcal{D}A)$ the operator given by left multiplication. Let $\widehat{D}, \widehat{F} \in$ End$(\mathcal{D}A)$ be the operators given by $\widehat{D}(\xi) = \delta(\xi)$, $\widehat{F}(\xi) = (-1)^{|\xi|}\xi$, $\xi \in \mathcal{D}A$. Then $[\widehat{D}, \lambda(\omega)] = \lambda(\delta(\omega))$ since δ is a derivation. Since $\lambda(A), \widehat{D}, \widehat{F}$ satisfy the relations of $A[D,F]$, there is a map $A[D,F] \to B$ where B is the algebra generated by $\lambda(A), \widehat{D}, \widehat{F}$ and the composition $\mathcal{D}A \to A[D,F] \to B$ is simply λ. Now, λ is injective (note that $1 \in \mathcal{D}A$) whence the proposition. □

Proposition 3.2 *Let $\widehat{A}, \widehat{D}, \widehat{F}$ be a representation of $A[D,F]$ (i. e. $\widehat{A}, \widehat{F}, \widehat{D}$ satisfy the relations of A, D, F), B the algebra generated by $\widehat{A}, \widehat{D}, \widehat{F}$ and $\hat{d}$ the image of d, i. e. $\hat{d}(\omega) = \widehat{F}[\widehat{D},\omega]$. Then $\hat{d}^2 = 0$ if and only if $[\widehat{D}^2, x] = 0$, $\forall x \in \widehat{A}$.*

Proof. One has

$$\hat{d}^2(\omega) = \widehat{F}\left(\widehat{D}\widehat{F}(\widehat{D}\omega - \omega\widehat{D}) - \widehat{F}(\widehat{D}\omega - \omega\widehat{D})\widehat{D}\right) = (-\widehat{D}^2\omega + \omega\widehat{D}^2)$$

whence the only if part.

If $[\widehat{D}^2, x] = 0$ then every element of B can be written as a linear combination of elements of the form $z_1 = F\widehat{D}^k a_0 \hat{d}(a_1)\cdots\hat{d}(a_n)$ (note that a_0 may be 1) or $z_2 = \widehat{D}^l b_0 \hat{d}b_1 \cdots \hat{d}b_m$. One has

$$\begin{aligned}\hat{d}(z_1) &= -2\widehat{D}^{k+1}a_0\hat{d}a_1\cdots\hat{d}a_n + (-1)^k F\widehat{D}^k\hat{d}a_0\hat{d}a_1\cdots\hat{d}a_n,\\ \hat{d}^2(z_1) &= (-1)^{k+1}\left(-2\widehat{D}^{k+1}\right)\hat{d}a_0\hat{d}a_1\cdots\hat{d}a_n\\ &\quad + (-1)^k\left(-2\widehat{D}^{k+1}\right)\hat{d}a_0\cdots\hat{d}a_n = 0,\end{aligned}$$

and $\quad \hat{d}(z_2) = (-1)^l\widehat{D}^l\hat{d}b_0\cdots\hat{d}b_n, \quad \hat{d}^2(z_2) = 0,$

since $\widehat{d}^2 x = 0$ for $x \in \widehat{A}$. □

We denote by $A[\widehat{D}, F]$ the universal algebra generated by A together with two additional elements $\widehat{D}, F$ satisfying the relations of $A[D, F]$ ($[F, a] = 0$, $F\widehat{D} + \widehat{D}F = 0$, $F^2 = 1$) and in addition $[\widehat{D}^2, a] = 0$, $a \in A$.

Then $A[\widehat{D}, F]$ is of course a quotient of $A[D, F]$. We still denote by δ, d the derivation, resp. graded derivation of $A[\widehat{D}, F]$ given by $\delta(x) = [\widehat{D}, x]$, $dx = F\delta x$. One has $\widehat{D}\delta(x) = -\delta(x)\widehat{D}$ and $\delta^n(x) = (2D)^{n-1}\delta(x)$. The natural map $\mathcal{D}A \to A[\widehat{D}, F]$ is no longer injective. Under the composition $A \xrightarrow{e^\delta} \mathcal{D}A \longrightarrow A[\widehat{D}, F]$, an element $x \in A$ is mapped to $x + \frac{1}{2D}\left(e^{2D} - 1\right)\delta(x)$.

Proposition 3.3 *Let $A[F]$ denote the universal algebra generated by A with one additional element F satisfying $F^2 = 1$, $[F, a] = 0$, $a \in A$. Then $A[\widehat{D}, F]$ is isomorphic to $\Omega(A[F])$ divided by the relation $F dx = -dx F$ (note that this relation is compatible with d so that d passes to the quotient). Under this isomorphism $\widehat{D}$ corresponds to $-\frac{1}{2}dF$ and d corresponds to d.*

Proof. Denote the quotient of $\Omega(A[F])$ by the additional relation by $\Omega_F(A)$. In this algebra, one has $0 = d(F^2) = d(F)F + FdF$ and $dF = -\frac{1}{2}F[dF, F]$. Also, $d(x) = d(FxF) = d(F)xF - dx + FxdF$ whence $2dx = -F[dF, x]$. Since $d^2 = 0$, by Proposition 3.2 one has $[(dF)^2, x] = 0$, $\forall x \in A$. Thus $\Omega_F(A)$ is generated by A together with F and dF and these operators satisfy the relations of $A[\widehat{D}, F]$. There is a surjective map $A[\widehat{D}, F]$ to $\Omega_F A$ mapping $\widehat{D}$ to $-\frac{1}{2}dF$, F to F and transforming d into d. The existence of the converse map $\Omega_F A \to A[\widehat{D}, F]$ follows from the fact that $\Omega(A[F])$ is the universal DG-algebra over $A[F]$. □

For each $t \in \mathbb{C}$, define a linear map $\varphi_t : \Omega A \to \Omega_F A$ (where $\Omega_F A$ is defined as in the preceding proof) by

$$\varphi_t(x_o dx_1 \cdots dx_n) = (x_0 - t\delta(x_0))\, dx_1 \cdots dx_n .$$

It is important to note that the image under φ_t of $\Omega^{\text{even}} A$ lies in fact in the subalgebra $A[\widehat{D}]$ generated by A and $\widehat{D}$ in $\Omega_F A$ (this follows from the equality $dxdy = -\delta x \delta y$).

Proposition 3.4 *φ_t is injective for each t and $\varphi_t(\omega_1 \circ_{t^2} \omega_2) = \varphi_t(\omega_1)\varphi_t(\omega_2)$. Thus, for $t = 0$, φ_0 is an embedding of ΩA into $\Omega_F A$ and, for $t \neq 0$, φ_t is an embedding of the deformation $Q_{t^2} A$ into $\Omega_F A$. Moreover, one has $\varphi_0 d = d\varphi_0$.*

Proof. Let λ be the representation by left multiplication of ΩA in $\text{End}(\Omega A)$ and let $\widetilde{F}, \widetilde{dF} \in \text{End}(\Omega A)$ be defined by $\widetilde{F}(\omega) = (-1)^{|\omega|}$, $\widetilde{dF}(\omega) = d\omega$. Since $\widetilde{F}, \widetilde{dF}$ satisfy the relations for $\Omega_F A$ and since λ is injective, it follows that φ_0 is injective. The injectivity of φ_t then follows from the relation $\varphi_t(\omega) = (1 - t\delta)\varphi_0\omega$ and the fact that δ is nilpotent in End(ΩA), so that $(1 - t\delta)$ is invertible.

In fact, with $\widetilde{D} = -\frac{1}{2}d\widetilde{F}$, one has $\widetilde{D}^2 = 0$ and $\delta(x) = [\widetilde{D}, x]$. Thus $\delta^3(x) = (2\widetilde{D})^2\delta(x) = 0$.

Finally,

$$\begin{aligned}
&\varphi_t(\omega_1 \circ_{t^2} \omega_2) = \varphi_t(\omega_1\omega_2 - (-1)^{|\omega_1|}t^2 d\omega_1 d\omega_2) \\
&= \varphi_0(\omega_1\omega_2) - t\delta\varphi_0(\omega_1\omega_2) - (-1)^{|\omega_1|}t^2\varphi_0(d\omega_1 d\omega_2) \\
&= \varphi_0(\omega_1)\varphi_0(\omega_2) - t\delta(\varphi_0\omega_1\varphi_0\omega_2) + t^2\delta(\varphi_0\omega_1)\delta(\varphi_0\omega_2) \\
&= \varphi_t(\omega_1)\varphi_t(\omega_2).
\end{aligned}$$

□

Remark. The preceding proof shows more:

(1) Let $\Omega_F A/(D^2{=}0)$ denote the quotient of $\Omega_F A$ by the additional relation $D^2 = 0$. By the argument above, $\Omega_F A/(D^2{=}0)$ can be represented in End(ΩA) and the composition

$$\Omega A \xrightarrow{\varphi_t} \Omega_F A \longrightarrow \Omega_F A/(D^2{=}0)$$

remains injective.

(2) For each t, φ_t can be lifted to an injective homomorphism $\varphi_t : Q_{t^2}A \to A[D, F]$. The quotient by the relation $[D^2, x] = 0,\ x \in A$, was needed only in order to have $\varphi_0 \circ d = d \circ \varphi_0$.

Another natural relation that can be imposed on $\Omega_F A$, is $D^2 = 1$. This corresponds to representations of $\Omega_F A$ by bounded Fredholm modules.

Proposition 3.5 *The composition*

$$\Omega A \xrightarrow{\varphi_t} \Omega_F A \longrightarrow \Omega_F A/(D^2{=}1)$$

remains injective and defines an embedding of the algebra $Q_{t^2}A$ into $\Omega_F A/(D^2{=}1)$.

Proof. This follows from the description of QA as the free product $A * A$. It can be shown that $\Omega_F A/(D^2{=}1)$ is the second crossed product $(QA \rtimes \mathbb{Z}/2) \rtimes \mathbb{Z}/2$ by the $\mathbb{Z}/2$-grading automorphisms of QA and $QA \rtimes \mathbb{Z}/2$, thus isomorphic to $M_2(QA)$. □

4 K-Theory and Cyclic Cohomology in Terms of Non-Commutative Differential Forms

In section 3 we introduced RA and QA as deformations of $\Omega^{\text{even}}A$ and ΩA respectively. If A is an involutive algebra with a C^*-norm, then RA, QA are involutive algebras as subalgebras of $A[D,F]$ and QA is generated by the images of the two homomorphisms (for Fedosov product) $A \to QA$ given by $x \mapsto x + dx$ and by $x \mapsto x - dx$, $x \in A$. Therefore QA admits a natural universal C^*-completion, see section 2. We can also complete RA naturally as a subalgebra of QA.

The linear combinations of forms of degree $\geq k$ form ideals I^k and J^k in RA and QA, respectively. Let $\rho A = I^1$, $qA = J^1$. There are surjective homomorphisms

$$\begin{aligned} \pi_0 : QA &\longrightarrow A & \qquad \pi_1 : RA &\longrightarrow A \\ x &\longmapsto x & x &\longmapsto x \\ dx &\longmapsto 0 & dxdy &\longmapsto 0 \end{aligned}$$

with $\operatorname{Ker} \pi_0 = qA$, $\operatorname{Ker} \pi_1 = \rho A$.

Given a C^*-algebra A, the K-theory groups K_0, K_1 and the K-homology groups K^0, K^1 can now be defined as follows

$$\begin{aligned} K_0 A &= \{\text{homotopy classes of } ^*\text{-homomorphisms } q\mathbb{C} \longrightarrow \overline{M_\infty A}\} \\ K_1 A &= \{\text{homotopy classes of } ^*\text{-homomorphisms } \rho\mathbb{C} \longrightarrow \overline{M_\infty A}\} \\ K^0 A &= \{\text{homotopy classes of } ^*\text{-homomorphisms } qA \longrightarrow \overline{M_\infty \mathbb{C}}\} \\ K^1 A &= \{\text{homotopy classes of } ^*\text{-homomorphisms } \rho A \longrightarrow \overline{M_\infty \mathbb{C}}\}. \end{aligned}$$

Here $\overline{M_\infty A}$ denotes the natural C^*-completion of $M_\infty A$. The sets $K_0 A$, $K_1 A$, $K^0 A$,$K^1 A$ are in fact abelian groups in a natural way. Due to the effort of many mathematicians and especially due to Kasparov's KK-theory, [18],[19], these groups can be computed in virtually all cases of interest, [3]. In the way of examples let us mention that the non-commutative tori A_θ (cf. 2.4 b) have the same K-theory as the ordinary 2-torus $K_0 A_\theta = \mathbb{Z}^2$, $K_1 A_\theta = \mathbb{Z}^2$. The Woronowicz deformation $SU_\mu(2)$ (cf. 2.6) has, for $\mu \neq 1$, $K_0 = \mathbb{Z}$, and $K_1 = \mathbb{Z}$. For $\mu = 1$, $SU_\mu(2)$ is isomorphic to the continuous functions on $SU(2)$ and $SU(2)$ is topologically a 3-sphere. Thus, in this case, too, $K_0 = \mathbb{Z}$, $K_1 = \mathbb{Z}$ - the K-theory of the 3-sphere.

We turn now to the description of cyclic homology and cohomology in terms of differential forms. Given an algebra R, let $\Omega^1 R \subset \Omega R$ be the R-bimodule of 1-forms: $\Omega^1 R = \{xdy \in \Omega R \mid x, y \in R\}$. Let moreover $\Omega^1 R_\natural$ denote the quotient of $\Omega^1 R$ by the subspace generated by commutators of the form $x\omega - \omega x$, $\omega \in \Omega^1 R$, $x \in R$. There is a very simple $\mathbb{Z}/2$-graded complex, the X-complex [13],

$$R \underset{\beta}{\overset{\delta}{\rightleftarrows}} \Omega^1 R_\natural$$

with operators $\delta x = dx_\natural$ and $\beta((xdy)_\natural) = [x,y]$. One checks easily that β is well defined and that $\beta\delta = \delta\beta = 0$.

To get a slightly more familiar picture consider the dual complex

$$R^* \underset{\beta^*}{\overset{\delta^*}{\leftrightarrows}} (\Omega^1 R_\natural)^*$$

where, given a linear space X, X^* denotes the dual space of linear functionals on X. Since the image of β is the commutator subspace of R, the cohomology of this dual complex is

$$HX^{\text{even}} R = \{\text{traces on } R\}/\{\text{traces of the form } T\circ\delta,\ T \text{ a trace on } \Omega^1 R\}$$

$$HX^{\text{odd}} R = \{\text{closed traces on } \Omega^1 R\}/\{\text{traces of the form } T(xdy) = f(xy - yx),\ \ f \in R^*\}.$$

The cyclic homology and cohomology for A can be defined very naturally by applying the X-complex to RA, QA and their ideals I^k, J^k. Let HC^k denote the cyclic cohomology in dimension k. Cum grano salis, one has

$$HX^{\text{even}} I^{2k} = HC^{2k-1}A, \quad HX^{\text{odd}} I^{2k} = HC^{2k}A$$

$$HX^{\text{even}} J^{2k+1} = HC^{2k}A, \quad HX^{\text{odd}} J^{2k+1} = HC^{2k+1}A$$

$$HX^{\text{even}} RA/I^{2k} = HC^{2k-2}A, \quad HX^{\text{odd}} RA/I^{2k} = HC^{2k-1}A$$

$$HX^{\text{even}} QA/J^{2k+1} = HC^{2k-1}A, \quad HX^{\text{odd}} QA/J^{2k+1} = HC^{2k}A$$

These identities are not literally correct. There are some small correction terms which are too cumbersome to write here. This is related to the fact that the $\mathbb{N}$-grading of cyclic cohomology is not entirely natural. The important point is that cyclic cocycles are represented by *traces* on RA, QA, J^k, J^k and by *closed traces* on $\Omega^1 RA, \Omega^1 QA, \Omega^1 I^k, \Omega^1 J^k$.

The picture is even more natural for periodic and entire cyclic homology and cohomology. In the first case one completes RA, QA algebraically to obtain $\widehat{R}A = \varprojlim_k RA/I^k$, $\widehat{Q}A = \varprojlim_k QA/J^k$. Under the identification $RA \cong \Omega^{\text{even}} A$, $QA \cong \Omega A$ with Fedosov product this means that $\widehat{R}A, \widehat{Q}A$ correspond to forms with infinite support. The periodic cyclic cohomology HP is then given by the dual X-complex with *continuous* functionals

$$HP^{\text{even}} A \cong HX^{\text{even}} \widehat{R}A \cong HX^{\text{odd}} \widehat{Q}A$$

$$HP^{\text{odd}} A \cong HX^{\text{odd}} \widehat{R}A \cong HX^{\text{even}} \widehat{Q}A$$

Finally, entire cyclic cohomology can be defined in the case where A is a normed algebra by appropriately completing RA, QA with a norm derived from the norm on A to obtain $\overline{R}A, \overline{Q}A$, and again taking continuous functionals in the X-complex, by

$$H^{\text{even}}_{\text{ent}} A = HX^{\text{even}}\overline{R}A \cong HX^{\text{odd}}\overline{Q}A$$
$$H^{\text{odd}}_{\text{ent}} A = HX^{\text{odd}}\overline{R}A \cong HX^{\text{even}}\overline{Q}A$$

Some further variants are possible by using the $\mathbb{Z}/2$-grading on QA and supertraces in the place of traces.

One great advantage of the description of cyclic cohomology in terms of the X-complex is that homotopy invariance of periodic and entire cyclic cohomology is immediate [13]. Moreover, it shows that cyclic cohomology is functorial (can be transported) under linear maps with small curvature, i. e. linear maps that differ not too much from a homomorphism. (By definition the curvature of a linear map α between two algebras is the function of 2 variables $\omega(x,y) = \alpha(xy) - \alpha(x)\alpha(y)$.) We sketch a proof of both facts for the case of periodic cyclic cohomology: let $\alpha : A \to B$ be a linear map between two algebras. Then α induces a homomorphism $\tilde{\alpha} : RA \to B$. Assume that $\tilde{\alpha}$ extends to a continuous homomorphism $\hat{\alpha} : \widehat{R}A \to B$ from the completion $\widehat{R}A$ (this means that the curvature of α is nilpotent). Suppose that we have a cocycle on B, represented by a trace T on $\widehat{R}B$. We now use the fact, mentioned above, that RA is the free unital algebra over A, i. e. the algebra generated by $p(x)$, $x \in A$, where $p(x)$ is linear in x and with only further relation $p(1) = 1$. There is a natural homomorphism $\widehat{R}A \xrightarrow{\kappa} \widehat{R}(\widehat{R}A)$ (Kasparov product), mapping px to $p(px)$. Thus we can pull back T to a cyclic cohomology class on B through the composition

$$\widehat{R}A \xrightarrow{\kappa} \widehat{R}\widehat{R}A \xrightarrow{\widehat{R}\hat{\alpha}} \widehat{R}B.$$

If now $\alpha_t : A \to B$, $t \in [0,1]$ is a differentiable homotopy of linear maps with nilpotent curvature, then $\gamma_t = \widehat{R}\hat{\alpha}_t \circ \kappa$ is differentiable. Let $\dot{\gamma}_t$ denote its derivative. We obtain a map of bimodules

$$\sigma_t : \Omega^1 \widehat{R}A \longrightarrow \widehat{R}B$$

mapping xdy to $\gamma_t(x)\dot{\gamma}_t(y)$.

If $T_t = T \circ \gamma_t$ and $\dot{T}_t$ denotes its derivative, then $\dot{T}_t = (T \circ \sigma_t) \circ d$.

Therefore, $\dot{T}_t$ is obtained by composing a trace on $\Omega^1 \widehat{R}A$ with d, so that the corresponding cyclic cocycle is a coboundary. This shows that the cyclic cohomology class associated with T_t does not depend on t.

5 Fredholm Modules

A bounded Fredholm module over A is a representation of $A[D]$ or of $A[D,F]$ by bounded operators on a Hilbert space $\mathcal{H}$ satisfying certain conditions.

Firstly, the commutators $[D,x]$, $x \in A$, should be represented by compact operators. If these elements are represented by operators in the Schatten p-class $l^p(\mathcal{H})$, we say that the Fredholm module is p-summable. Second, it is important to impose a condition on the representation of D, since otherwise, replacing D by tD, $t \in \mathbb{R}$, we get a homotopy with 0. As in K-theory, there are two basic possible conditions:

1) D should be idempotent ($D^2 = D$)
2) D should be invertible (here we don't assume $D = D^*$).

In both cases, one could ask for the condition to be satisfied only modulo compact or l^p-operators:

$$(D - D^2)x \text{ compact or } l^p, \quad \forall x \in A, \text{ resp.}$$
$$\exists\, D',\ DD'x - x, \quad \text{compact or } l^p,\ \forall x \in A.$$

Replacing D by $D' = 2D - 1$ in 1) we could also impose the condition $D'^2 = 1$.

Replacing D by $D' = \begin{pmatrix} 0 & D \\ D^{-1} & 0 \end{pmatrix}$ in 2) and introducing $F = \begin{pmatrix} 1 & 0 \\ 0 & -1 \end{pmatrix}$ we could instead of a representation of $A[D]$ where D is invertible consider representations of $A[D,F]$ where $D^2 = 1$.

In order to associate cyclic cocycles with a Fredholm module, we have to find traces or supertraces on $RA, QA, \Omega^1 RA$ or on $I^k, J^k, \Omega^1 I^k$ where I, J are the natural ideals in RA, QA. The most straightforward way to do this, would be to use the embeddings $RA \hookrightarrow A[D]$, $QA \hookrightarrow A[D,F]$ discussed in section 3 and to evaluate the trace Tr on the image of I^k, J^k under the representation given by the Fredholm module (these images lie in $l^1(\mathcal{H})$ if $k \geq p$ and the module is p-summable). However, these embeddings are homotopic to the trivial maps $RA \to A[D]$, $QA \to A[D,F]$ which vanish on I, J via the homotopy which replaces δx by $t\delta x$.

Therefore one either has to use different embeddings $RA \longrightarrow A[D]$, $QA \longrightarrow A[D,F]$ which take into account the special conditions on D, or one has to use a functional different from Tr. We discuss two typical cases:

1) Let (A, D) be a p-summable Fredholm module where $D^2 = D$. The map

$$\varphi : x_0 dx_1 \cdots dx_{2n} \longrightarrow D\,(x_0 - \delta x_0)\,\delta x_1 \cdots \delta x_{2n}$$

is a homomorphism of $RA = \Omega^{\text{even}} A$ with Fedosov product into $\mathcal{L}(\mathcal{H})$. In fact, if ψ is the map $RA \to A[D]$ of section 3, then $\varphi(\omega) = D\psi(\omega)$ and one checks easily that $D\psi(\omega) = D\psi(\omega)D$ if $D^2 = D$. One obtains a trace on I^p, putting $\tau(\omega) = \mathrm{tr}(\varphi\omega)$. Equivalently $f(\cdot) = \mathrm{tr}(D\cdot)$ defines a trace on $\psi(I^p)$, so that $\tau(\omega) = f(\psi\omega)$ is a (the same) trace on I^p.

2) Let (A, D, F) be a p-summable Fredholm module where $D^2 = 1$. Let $P = \frac{1}{2}(D+1)$ and $P^\perp = 1-P$. One has $P^2 = P$ and $FP = P^\perp F$. Moreover, with $\delta x = [D, x]$, $dx = F\delta x$ as in Section 3, one has $[P, x] = \frac{1}{2}\delta x$, $Pdx = dxP$, $P^\perp \delta x = \delta x P$ and $P\left(x - \frac{1}{2}\delta x\right) P = P\left(x - \frac{1}{2}\delta x\right)$. Thus, if ψ is the representation of $Q_{\frac{1}{4}}A = \Omega A$ with Fedosov product mapping $x_0 dx_1 \cdots dx_n$ to $(x_0 - \frac{1}{2}\delta x_0)dx_1 \cdots dx_n$ (cf. 3.4), then $\mathrm{tr}\left(P\psi(\omega)\right) = \tau(\omega)$ defines a trace on $J^p \subset Q_{\frac{1}{4}}A \cong QA$. In fact $\omega \mapsto P\psi(\omega)$ is a homomorphism from $Q_{\frac{1}{4}}A$ if $D^2 = 1$. We should point out that this construction of the Chern character for the Fredholm module is different from the usual one even though it gives the same cyclic cocycle.

Finally, to illustrate the usefulness of the RA-formalism let us look at the case 1) but with $D^2 = D$ replaced with $\left(D^2 - D\right)x \in l^{p/2}(\mathcal{H})$, $x \in A$. Define a homomorphism $\varphi : RA \to \mathcal{L}(H)$ by $px \mapsto DxD$ or

$$RA = \Omega^{\mathrm{even}} A \ni x_0 dx_1 \cdots dx_{2n} \longrightarrow Dx_0 D\omega(x_1, x_2) \cdots \omega(x_{2n-1}, x_{2n})$$

where

$$\omega(x, y) = DxyD - DxDDyD = -D[D, x][D, y] + \alpha(x, y)$$
$$\text{with} \qquad \alpha(x, y) \in l^{p/2}(\mathcal{H}).$$

Thus $\omega(x, y) \in l^{p/2}(\mathcal{H})$ if the module is p-summable. $\tau(\omega) = \mathrm{tr}(\varphi\omega)$ defines a trace on I^k for $k \geq p$. If $D^2 = D$, then $\alpha(x, y) = 0$ and we obtain the same homomorphism as above: $\varphi(\omega) = D\psi(\omega)$.

References

1. Bellissard, J.: K-theory of C^*-algebras in solid state physics, statistical mechanics and field theory, mathematical aspects. Lecture Notes in Phys. **257**, 99-156 (1986)
2. Bellissard, J.: Ordinary quantum Hall effect and noncommutative cohomology. Proc. of the Bad Schandau Conf. on Localization, Teubner, Leipzig (1987)
3. Blackadar, B.: *K-theory for Operator Algebras*. Springer-Verlag, Heidelberg/Berlin/New York/Tokyo, 1986
4. Blackadar, B., Cuntz, J.: Differential Banach algebra norms and smooth subalgebras of C^*-algebras. Preprint
5. Blackadar, B., Kumjian, A., Rørdam, M.: Approximately central matrix units and the structure of non-commutative tori. Preprint
6. Bratteli, O., Kishimoto, A.: Non-commutative spheres III: Irrational rotations. Preprint
7. Connes, A.: Non-commutative differential geometry. Publ. Math. IHES **62**, 257-360 (1985)

8. Connes, A.: Entire cyclic cohomology of Banach algebras and characters of θ-summable Fredholm modules. K-theory **1**, 519-548 (1988)
9. Connes, A., Cuntz, J.: Quasi-homomorphismes, cohomologie cyclique et positivité. Comm.Math.Phys. **114**, 515-526 (1988)
10. Cuntz, J.: Simple C^*-algebras generated by isometries. Comm.Math.Phys. **57**, 173-185 (1977)
11. Cuntz, J.: A new look at KK-theory. K-theory **1**, 31-51 (1988)
12. Cuntz, J.: Cyclic cohomology and K-homology. Proc.Int.Congr.Math. Kyoto 1990, to appear
13. Cuntz, J., Quillen, D.: Cyclic homology and nonsingularity for noncommutative algebras. In preparation.
14. Doplicher, S., Roberts, J.: A new duality theory for compact groups. Invent.Math. **98**, 157-218 (1989)
15. Doplicher, S., Roberts, J.: Why there is a field algebra with a compact gauge group describing the superselection structure in particle physics. Comm.Math.Phys. **131**, 57-107 (1990)
16. Fedosov, B.V.: Analytic formulas for the index of elliptic operators (translation from Russian). Trans.Moscow Math.Soc. **30**, 159-240 (1974)
17. Jaffe, A., Lesniewski, A., Osterwalder, K.: Quantum K-theory: the Chern character. Comm.Math.Phys. **118**, 1-14 (1988)
18. Kasparov, G.G.: The operator K-functor and extensions of C^*-algebras. Izv.Akad.Nauk SSSR, Ser.Math. **44**, 571-636 (1980)
19. Kasparov, G.G.: Equivariant KK-theory and the Novikov conjecture. Invent.Math. **91**, 147-201 (1988)
20. Kastler, D.: *Cyclic cohomology.* Hermann, Paris, 1988
21. Woronowicz, S.L.: Compact matrix pseudogroups. Comm.Math.Phys. **111**, 613-665 (1987)

Part II

Session Lectures

Section 1

Analysis on Manifolds and Classical Mechanics

Organizer: K.D. Elworthy

Solutions of Nonlinear Wave Equations and Localization Theory

Walter Craig [1], C. Eugene Wayne [2]

[1]Department of Mathematics, Brown University,
Providence, RI 02912
[2]Department of Mathematics, Pennsylvania State University,
University Park, PA 16802

In this note we describe a new method for constructing time periodic solutions of nonlinear wave equations of the form

$$u_{tt} = u_{xx} - g(x,u) \qquad 0 < x < \pi, \tag{1}$$

with either periodic or Dirichlet boundary conditions. For a complete description of the results and methods with detailed proofs, see [2]. This article will attempt only to explain the general ideas behind this method and some of its possible extensions.

Our main result is the following:

Theorem *Consider equation (1) with periodic boundary conditions. There is a set of nonlinearities $g(x,u)$, which is dense in $L^2_{\mathrm{per}}(0,\pi)$ and has open intersection with the set of analytic functions, for which the following statements are true:*
There exists a Cantor set $C \subset \mathbb{R}$, of positive measure, and a C^∞ curve $\rho(\Omega)$ such that for every $\Omega \in C$, equation (1) has a non-trivial, analytic, time periodic solution $u(x,t;\Omega)$, with angular frequency Ω, with $\|u(\cdot,\cdot;\Omega)\|_{L^2} = \rho(\Omega)$.

Similar results hold if periodic boundary conditions are replaced with Dirichlet boundary conditions.

The conditions which determine whether or not the Theorem applies to a given nonlinear term are finite in number, involve only the first three terms in the Taylor series for $g(x,u)$, and can, at least in principle, be explicitly verified for a given $g(x,u)$. As a particular example of the sort of equation to which the Theorem applies, consider the Klein-Gordon equation,

$$u_{tt} = u_{xx} - m^2 u + u^3 \quad .$$

Then in the case of Dirichlet boundary conditions, we are able to show that the Theorem applies to this equation for an open set of values of m of full Lebesgue measure. (In this example, and in any cases in which the nonlinear term $g(x,u) = g(u)$, a simple analysis of the phase plane gives the existence result when periodic boundary conditions are imposed.)

Our method is perturbative in nature and thus starts from an analysis of the linearized problem. We assume that $g(x,u)$ is analytic and has a Taylor series at $u = 0$ of the form $g(x,u) = g_1(x)u + g_2(x)u^2 + \ldots$ so that to first order (1) becomes

$$v_{tt} = (\partial_x^2 - g_u(x,0))v \ , \quad 0 < x < \pi \, . \tag{2}$$

This equation is easily solved. If the eigenvectors and eigenvalues of the Sturm-Liouville operator $\left(-\frac{d^2}{dx^2} + g_u(x,0)\right)$ with the appropriate boundary conditions are $\{\psi_j(x)\}_{j=1}^{\infty}$ and $\{\omega_j^2\}_{j=1}^{\infty}$ (assume for simplicity that all the eigenvalues are positive, though this is not necessary) then $v^{(j)}(x,t) = \psi_j(x)\sin(\omega_j t)$ is a periodic solution of (2), for every $j = 1, 2 \ldots$. We wish to show that there are solutions of the full non-linear equation (1), which are "close" to these solutions of the linearized equations. With only a slight loss of generality we will focus on the case $j = 1$ – i.e. we look for solutions of (1), with frequency close to ω_1. If such a solution exists, we can write it in an eigenfunction-Fourier expansion as

$$u(x,t) = \sum_{j=1}^{\infty} \sum_{k \in \mathbb{Z}} \widehat{u}(j,k)\psi_j(x)e^{i\Omega kt} \tag{3}$$

where $\Omega \approx \omega_1$. If we substitute (3) into (1), we obtain an infinite set of coupled equations for the coefficients $\{\widehat{u}(j,k)\}$, which we write schematically as

$$F(\widehat{u})(j,k) = W(\widehat{u})(j,k) + V(\Omega)\widehat{u}(j,k) = 0, \quad j = 1,2\ldots, \ ; \ k \in \mathbb{Z} \ . \tag{4}$$

Here, $V(\Omega)$ is the diagonal operator $V(\Omega)\widehat{u}(j,k) = (\omega_j^2 - k^2\Omega^2)\widehat{u}(j,k)$, which arises from the linearized equation (2), while $W(\widehat{u})$ is a non-diagonal operator coming from the non-linear part of the function $g(x,u)$, whose form we will comment upon below.

We will construct solutions $\{\widehat{u}(j,k)\}$ of (4) by applying Newton's method, taking as our starting point the function $\varphi(p) = p\delta(1,1) + \overline{p}\delta(1,-1)$, where $p \in \mathbb{C}$. Note that φ is a solution of the linearized problem (i.e. when $W = 0$) if $\Omega = \omega_1$. (Here, $\delta(j,k)$ is the Kronecker δ-function on the lattice $\mathbb{Z}^+ \times \mathbb{Z}$.)

The most difficult part of this process is to control the inverse of the operator

$$D_{\tilde{u}}F = D_{\tilde{u}}W + V(\Omega) \tag{5}$$

which occurs when we linearize about an approximate solution $\widetilde{u}$. We control this operator in two stages. The first problem we encounter is that when $\Omega = \omega_1$, the linear operator $V(\omega_1)$ has a two-dimensional null space $\ell^2(N)$, where $N = \{(1,1),(1,-1)\}$. This is a common problem in bifurcation theory and we treat it using a standard method, the Lyapunov-Schmidt method. Define Q = orthogonal projection onto $\ell^2(N)$ and $P = \mathbf{1} - Q$. Then (4) is equivalent to a pair of equations,

$$QF(\widehat{u}) = QW(\widehat{u})(j,k) + (\omega_1^2 - \Omega^2)\widehat{u}(j,k) = 0, \quad (j,k) \in N \ , \tag{6}$$

and

$$PF(\widehat{u}) = PW(\widehat{u})(j,k) + (\omega_j^2 - k^2\Omega^2)\widehat{u}(j,k) = 0, \quad (j,k) \notin N. \tag{7}$$

Note that the "Q-equation" (*i.e.* (6)) is a finite dimensional problem (in fact two dimensional) and can be handled by ordinary implicit function theorem arguments. Thus, we will concentrate in the remainder of this note on how one can solve the "P-equation" (7).

Equation (7) is solved inductively by finding better and better approximate solutions, on larger and larger subsets of the lattice, $\mathbb{Z}^+ \times \mathbb{Z}$. In the limit, the subset converges to the whole lattice and the solution converges to a true solution of (7). In order to start the induction we make certain assumptions about finitely many of the frequencies $\{\omega_j\}$, which is equivalent to making assumptions about the first order term in the Taylor expansion of the non-linear term $g(x,u)$ in (1). We comment below on just how restrictive these assumptions are.

The initial "box" on which we solve (7) is the set $B_0 = \{(j,k) \in \mathbb{Z}^+ \times \mathbb{Z} \mid j + |k| \leq L_0\}$, with L_0 a large constant. We then assume that for (j,k) in B_0 the frequencies of the linearized problem satisfy (i) $|\omega_j^2 - k^2\omega_1^2| \geq d_0$ and (ii) $|k\omega_1 - j| \geq c_0(j+|k|)^{-\tau}$, where d_0, c_0 and τ are positive constants. Furthermore we assume that the nonlinearity satisfies a "twist" condition. This condition ensures that the bifurcation curve $\rho(\Omega)$ has non-zero curvature at $\Omega = \omega_1$, and it can be explicitly computed from a formula involving the second and third order terms $g_2(x)$ and $g_3(x)$ of the Taylor series of the nonlinearity and the eigenfunctions of the linearized operator. When restricted to B_0, (7) becomes a finite dimensional problem and conditions (i) and (ii) guarantee that it can be solved by the ordinary implicit function theorem. One finds further that the coefficients $\{u^{(0)}(j,k)\}$ of this solution decay exponentially in $(|j|+|k|)$, which means that $u^{(0)}$ fails to be a true solution of (7) (i.e. a solution on the whole lattice $\mathbb{Z}^+ \times \mathbb{Z}$ rather than just on B_0) only by an amount of $\mathcal{O}(e^{-\sigma L_0})$, for some $\sigma > 0$. Thus, if L_0 is large we have a very good approximate solution.

To improve this solution we use Newton's method. If we restrict (7) to a new, larger lattice region $B_1 \supset B_0$, where B_1 is defined analagously to B_0, but with L_0 replaced by $L_1 > L_0$, and try to solve $PF(u^{(0)} + v^{(0)})\,|_{B_1} = 0$ by Newton's method, we find

$$v^{(0)} = -\{P(D_{u^{(0)}}W + V(\Omega))\,|_{B_1}P\}^{-1}\,PF|_{B_1}(u^{(0)}). \tag{8}$$

From the comments above, we know that $PF|_{B_1}$ is small. Thus, the only difficulty we encounter is controlling the inverse of the operator $P(D_{u^{(0)}}W + V(\Omega))|_{B_1}P$. We have an explicit form for the diagonal term $V(\Omega)$, so this causes no problems. To understand the off-diagonal term $D_{u^{(0)}}W$, we note that for the example of the Klein-Gordon equation, $u_{tt} = u_{xx} - m^2u + u^3$, one can compute D_uW exactly if one takes u to be the solution of the linearized problem. One finds that $D_uW + V(\Omega) = \gamma\Delta + V(\Omega)$, where γ is a constant, and Δ is the lattice laplacian. Thus, in this example, the task of inverting $D_{u^{(0)}}W + V(\Omega)$ is the same as controlling the inverse of a lattice Schrödinger operator. This is a problem that has been studied in depth in the physics literature. In particular, Fröhlich and Spencer [3] have shown that if one can insure that the "resonances" in $V(\Omega)$ are not too close together and not too severe, then the matrix elements of $(\gamma\Delta + V(\Omega))^{-1}$ will decay exponentially fast as one moves away from the diagonal. For more general non-linearities than that in the Klein-Gordon equation, or if we evaluate the derivative at a point other than the solution of the linearized equation, one will not obtain $D_{u^{(0)}}W = \gamma\Delta$. However, $D_{u^{(0)}}W$ will still be a bounded operator with matrix elements which decay exponentially fast as one moves away from the diagonal, and the methods of Fröhlich and Spencer are still applicable.

Applying the methods of Fröhlich and Spencer places restrictions on the allowed frequencies of our periodic orbits. In the present problem the "resonances" correspond to points (j,k) at which the potential $V(\Omega)(j,k)$ approaches zero. A simple calculation using the fact that asymptotically the linear frequencies in this problem satisfy $\omega_j \sim j$, shows that if we eliminate frequencies Ω which fail to satisfy the diophantine inequality $|k\Omega - j| \geq c(j + |k|)^{-\tau}$, then for any remaining frequency, Ω, two lattice sites at which $V(\Omega)$ is small, must be widely separated. One must also insure that at any remaining sites where $V(\Omega)$ is small, the linearized operator $(D_uW + V(\Omega))$ does not have points in its spectrum that are too close to zero. This entails a further excision of frequencies Ω, so that at the conclusion of this induction argument we are left with a family of periodic orbits, but the frequencies in this family form a Cantor set of positive measure, not an interval as one typically finds in ordinary finite dimensional bifurcation theory. Once one makes these excisions of frequencies, the inverse of the linearized operator in Newton's method is controllable, and the induction argument converges to give a family of periodic orbits. We remark that methods related to ours have been applied to construct periodic solutions for some nonlinear versions of the Anderson localization model by Albanese, Fröhlich and Spencer [1].

There are several related problems which we think can be treated by this method. First of all, there is the existence of quasi-periodic (in time) orbits for (1). One can again make an eigenfunction-Fourier expansion similar to (3), but with n-frequencies in time – i.e. the exponential becomes $\exp\{i(\sum_{\ell=1}^{n} \Omega_\ell k_\ell)t\}$. In this case the original partial differential equation is reduced to a problem on the lattice $\mathbb{Z}^+ \times \mathbb{Z}^n$. One can then apply Newton's method just as in the periodic case. While formally the same, the problems one encounters when studying the resonances of the linearized problem are more complicated than in the periodic case, and we have not yet finished this analysis. However, we believe that this construction will also quasi-periodic solutions of (1). We note that if successful this method would also give a new construction of quasi-periodic orbits of finite dimensional hamiltonian systems of the form $H(p,q) = \sum_{j=1}^{N} \frac{1}{2} p_j^2 + V(q)$, in the neighborhood of an elliptic equilibrium point which is different from the usual construction using the KAM Theorem.

Another case which seems difficult to treat with KAM methods, but which we believe the present methods can treat, is the case of resonances in the linear problem. Suppose, for instance, that the linear frequencies ω_1 and ω_2 satisfy $\omega_2 = 2\omega_1$. In this case when we convert the problem to one on the lattice we find the null space of the diagonal operator $V(\Omega)$ is four dimensional rather than two dimensional. We can still proceed with the Lyapunov Schmidt procedure, however, and the "P-equation" (7) is solved just as in the non-resonant case. This leaves us with the "Q-equation", (6). This is now a finite (in this example, four) dimensional problem. Although the resonance makes the ordinary implicit function theorem inapplicable, we believe other methods will allow us to solve this problem — either perturbative techniques like those of Schmidt [6], or variational methods like those used by Weinstein [8] and Moser [5] to extend Lyapunov's center theorem to the case in which the linear problem contains resonances. The advantage that the present method offers, in our opinion, is that it allows one to treat quite independently the problems arising from the linear resonance ("Q-equation") and the infinite dimensionality of the problem ("P-equation").

We conclude with a short discussion of how these methods compare with KAM approaches to the problem which have been used by Kuksin [4] and Wayne [7]. The principle difference is in the sort of non-resonance conditions that are imposed on the frequencies of the linear problem. The conditions imposed by the present method require essentially that no multiple of the frequency of the periodic orbit we are constructing resonate with any of the other linearized frequencies. This condition is an infinite dimensional generalization of the hypothesis of the Lyapunov center theorem. In contrast, the KAM type methods require in addition that certain non-resonance conditions hold among those linear frequencies that we are not perturbing. These additional conditions arise because in the KAM method one transforms the hamiltonian of the problem to some normal form. The present method is not a transformation theory in this sense, but rather a more direct implicit function theorem. In contrast to previous approaches like that of Zehnder [9], which used the "group structure" given by the transformation theory to invert the linearized operator in Newton's method explicitly, in the present method we obtain bounds on this inverse without using a transformation theory. Indeed, even the hamiltonian nature of the problem plays little explicit role in this approach, so that we hope it may be applicable to a wide variety of problems.

Acknowledgements: The authors would like to thank the Dèpartement de Physique Thèorique, Universitè de Genève, and Oxford University for their hospitality, and the National Science Foundation (grants #DMS-8858218 and #DMS-9002059) and the Alfred P. Sloan Foundation for their support of our research.

References

1. Albanese, C., Fröhlich, J.: Periodic solutions of some infinite-dimensional hamiltonian systems associated with non-linear partial difference equations: Part I. Comm.Math.Phys. **116**, 475-502 (1988)
   Albanese, C., Fröhlich, J., Spencer, T.: Periodic solutions of some infinite-dimensional hamiltonian systems associated with non-linear partial difference equations: Part II. Comm.Math.Phys. **119**, 677-699 (1988)
2. Craig, W., Wayne, C.E.: Newton's method and periodic solutions of nonlinear wave equations. Preprint June 1991
3. Fröhlich, J., Spencer, T.: Absence of diffusion in the Anderson tight binding model for large disorder or low energy. Comm.Math.Phys. **88**, 151-184 (1983)
4. Kuksin, S.: Perturbation of quasiperiodic solutions of infinite-dimensional linear systems with an imaginary spectrum. Functional Anal.Appl. **21**, 192-205 (1987)
   Kuksin, S.: Perturbation theroy for quasiperiodic solutions of infinite-dimensional hamiltonian systems; Parts I-III. Preprint of Max-Plank-Institut, Bonn 1990
5. Moser, J.: Periodic orbits near an equilibrium and a theorem by Alan Weinstein. Comm.Pure Appl.Math. **29**, 727-747 (1976)
6. Schmidt, D. S.: Periodic solutions near a resonant equilibrium of a hamiltonian system. Celestial Mech. **9**, 81-103 (1974)
7. Wayne, C. E.: Periodic and quasi-periodic solutions of nonlinear wave equations via KAM theory. Comm.Math.Phys. **127**, 479-528 (1990)
8. Weinstein, A.: Normal modes for non-linear hamiltonian systems. Invent. Math. **20**, 47-57 (1973)
9. Zehnder, E.: Generalized implicit function theorems with applications to some small divisor problems I. Comm.Pure Appl.Math. **28**, 91-140 (1975)

Connections on Symplectic Manifolds and Characteristic Classes of Hamiltonian Systems

Valerii V. Trofimov

Department of Geometry and Topology
Faculty of Mathematics and Mechanics
Moscow State University, Moscow 119899 U.S.S.R.

Abstract: We present a method for constructing the generalization of Maslow-Arnold classes to a certain class of symplectic manifolds.

1 Classical Constructions

Let us consider Hamiltonian system of differential equations

$$\dot{q}_i = \frac{\partial H}{\partial p_i}, \quad \dot{p}_i = \frac{\partial H}{\partial q_i}, \qquad i = 1, \ldots, n,$$

which is integrable in the Liouville sense. This means that there exists a family $F = \{ f_1, \ldots, f_n \}$ of smooth functions on $\mathbb{R}^{2n}$ such that $f_1, \ldots, f_n$ are functionally independent on $\mathbb{R}^{2n}$ and $\{ f_i, f_j \} = 0$ relative to the standard Poisson bracket on $\mathbb{R}^{2n}$. In this situation we shall say, than F is completely involutive. Such families appear in various problems of geometry and its applications in mechanics. In this situation the compact connected component of the level surface $N = \{ f_1 = c_1, \ldots, f_n = c_n \}$ is a lagrangian submanifold diffeomorphic torus T^n. This is well-known Liouville theorem.

A lagrangian submanifold of the symplectic space $\mathbb{R}^{2n}$ possesses a natural topological invariant, that is the Maslov index, and more generally the Maslov-Arnold characteristic classes.

If M is a Lagrangian submanifold of the standard symplectic space $\mathbb{R}^{2n}$, the the correspondence which sends a point $x \in M$ to the tangent space T_xM translated to the origin determinates a map $f : M \to \Lambda(n)$ of the manifold M to the Lagrangian Grassmannian $\Lambda(n)$, which allows one to define the characteristic classes of the submanifold M. More precisely, if $a \in H^*(\Lambda(n))$, then $f^*(a) \in H^*(\Lambda(n))$ is the requisite invariant.

2 Construction of the Generalization of the Maslov-Arnold Classes

This classical construction may be generalized, using unexpected idea. Such way, new characteristic classes for Lagrangian submanifolds in arbitrary phase spaces was discovered [1]. In this case of the standard symplectic space $\mathbb{R}^{2n}$, these new classes are equal to classical classes of Arnold-Maslov.

Let us give the main construction. To do this, I need the notion of the connection compatible with symplectic structure. If on manifold M the differential form ω_{ij} is given, then we say that a symmetric connection Γ^i_{jk} is compatible with ω_i if $\nabla_k \omega_{ij} = 0$. We use the standard notations of the tensor analysis. Such connection is called the *symplectic connection*. It is easy to see that if for a form ω there exists a connection compatible with ω then $d\omega = 0$ There exists infinitely many connections compatible with a symplectic structure.

Let $N^n \subset M^{2n}$ be a Lagrangian submanifold of the symplectic manifold M^{2n}. Consider on M^{2n} a symplectic connection Γ^i_{jk} compatible with the symplectic structure ω_{ij}. Denote by $C(x_0)$ the set of paths starting and ending at $x_0 \in M^{2n}$. Parallel translation along paths $\gamma \in C(x_0)$ generates a group $H_{x_0}(M^{2n})$ of linear transformations of the space $T_{x_0}M^{2n}$. Since the connection is compatible with the symplectic structure ω_{ij}, the group maps the manifold of Lagrangian subspaces of $T_{x_0}M^{2n}$ into itself. Hence, there exists an action of $H_{x_0}(M^{2n})$ on the Lagrangian-Grassmannian $\Lambda(T_{x_0}M^{2n}) = \Lambda(T_{x_0}M^{2n})/H_{x_0}(M^{2n})$. Let $\gamma(0) = x \in N$ and $\gamma(1) = x_0$. Using the symplectic connection, parallel translate the tangent space T_xN along the path $\gamma(t), 0 \le t \le 1$, to the point $x_0 \in N$. We get a subspace $\tau(T_xN) \in \Lambda(T_{x_0}M^{2n})$. This subspace is Lagrangian since the connection is compatible with the symplectic structure. This gives rise to a mapping $f_\gamma : N \to \Lambda(T_{x_0}M^{2n})$. The mapping, of course, depends on the choice of the path $\gamma(t)$, connecting x and x_0. Consequently, we get a well defined mapping of Lagrangian submanifold N into the reduced Grassmannian $H\Lambda(T_{x_0}M^{2n})$, that is $f : N \to H\Lambda(T_{x_0}M^{2n})$. This mapping induces a map in cohomology $f^* : H^*(H\Lambda(T_{x_0}M^{2n})) \to H^*(N)$. If $a \in H^*(H\Lambda(T_{x_0}M^{2n}))$, then it defines a characteristic class $a(N) = f^*(a) \in H^*(N)$ of the Lagrangian submanifold $N \subset M^{2n}$. So we have the following assertion.

Theorem 1 *Let M^{2n} be a symplectic manifold, and N^n a Lagrangian submanifold of M^{2n}. Then every cohomology class $a \in H^*(H\Lambda(T_{x_0}M^{2n}))$ of the reduced Grassmannian $H\Lambda(T_{x_0}M^{2n})$ defines a natural characteristic Maslov class $a(N) \in H^*(N)$ of the Lagrangian submanifold $N^n \subset M^{2n}$.*

Now we want to compare the characteristic classes of the distinct submanifolds of a given manifold M^{2n} with connection Γ^i_{jk}. We have to consider the oriented submanifolds $N^n \subset M^{2n}$. In this case, there is a fundamental homological class $[N^n] \in H_n(N^n, \mathbb{R})$. For any cohomological class $f^*(a) \in H^n(N^n, \mathbb{R})$ is defined Kronecker's index $a[N^n] = < f^*(a), [N^n] > \in \mathbb{R}$ which is called the *index of a submanifold* $N^n \subset M^{2n}$. The properties of this index we can find in [2].

Theorem 2 *Let M^{2n} be a symplectic manifold. There are natural characteristic classes*

$$a_c(F) \in \Lambda(w_1, \ldots, w_s), \qquad c \in \mathbb{R}^n, \quad s \leq n,$$

of a completely commutative set F of functions on the manifold M^{2n}. If F_1 and F_2 are two functionally equivalent complete commutative sets, then $a_c(F_1) = a_{\varphi(c)}(F_2)$ for the mapping $\varphi : \mathbb{R}^n \to \mathbb{R}^n$. Here $\Lambda = \Lambda(w_1, \ldots, w_n)$ is an exterior algebra, $\Lambda = H^*(TS^n, \mathbb{R}), m \leq n$.

References

1. Trofimov, V.V. : Symplectic connections, Maslov index and Fomenko's conjecture. Dokl. Akad. Nauk SSSR, **304** (6), 1302–1305 (1989) (Russian). English transl. in Soviet Math.Dokl. **39** (1), 214–217 (1989)
2. Trofimov, V.V.: Connection on manifolds and new characteristic classes. Acta Appl.Math. **22**, 283–312 (1991)

Symplectic Twist Maps and the Theorem of Conley-Zehnder for General Cotangent Bundles

Christophe Golé

SUNY at Stony Brook

Abstract: We announce the existence and multiplicity of periodic orbits for time dependent Hamiltonian systems on cotangent bundle of arbitrary compact manifolds. We discretize the variational problem by decomposing the time 1 map into a product of symplectic twist maps.

0 Introduction

The celebrated theorem of Poincaré-Birkhoff states the existence of at least two fixed points for an area preserving map of the annulus $\mathbb{S}^1 \times [0,1]$ which "twists" the boundaries in opposite directions.

In the 60's, Arnold proposed a generalisation of this theorem for a time 1 map F of a time dependent Hamiltonian of $\mathbb{T}^n \times \mathbb{B}^n$ (where $\mathbb{B}^n$ is the closed ball in $\mathbb{R}^n$). While the Hamiltonian condition naturally generalises the area preservedness, a linking of the boundary of each fiber (a sphere in $\mathbb{R}^n$) with its image by F in the boundary of $\mathbb{T}^n \times \mathbb{B}^n$ was to replace the twist condition. Arnold [1] conjectured that such a map would have at least as many fixed points as a real valued function on $\mathbb{T}^n$ has critical points. The philosophy was that fixed points for symplectic maps should arise from Morse theory and not, say, from Lefshetz theory.

In 1982, Conley and Zehnder gave a first proof of this conjecture. However, they were not able to use the linking of spheres in its full generality. The result remains crucial since it was the first non perturbational one in this direction. The boundary condition that they used is expressed on the Hamiltonian in the following way. Letting $(\mathbf{q}, \mathbf{p})$ be the coordinates on $\mathbb{T}^n \times \mathbb{B}^n$ which is endowed with the canonical symplectic structure $d\mathbf{p} \wedge d\mathbf{q}$, they set:

$$H(\mathbf{q},\mathbf{p},t) = < A\mathbf{p},\mathbf{p} > \text{ for } \|\mathbf{p}\| \geq K ,$$

where $A = A^t$ is a non degenerate $n \times n$ matrix. This condition implies the linking of spheres at the boundary.

We propose here a version of this theorem on the cotangent bundle of an arbitrary compact manifold, with a quite different method of proof than theirs: whereas they use cut-offs on Fourier expansion, we decompose the time 1 map into symplectic wist maps to get a finite dimensional variational problem. On the other hand, our method is directly related to the so-called method of broken geodesics reintroduced in symplectic geometry by Chaperon [2] . The idea of decomposition comes from the work of LeCalvez [5] on twist maps of the annulus.

1 The Result

Let (M,g) be a compact Riemannian manifold. Define

$$\mathbb{B}^*M = \{(\mathbf{q},\mathbf{p}) \in T^*M \mid g(\mathbf{q})(\mathbf{p},\mathbf{p}) = \|\mathbf{p}\|^2 \leq R^2\} ,$$

where R is the radius of injectivity of (M,g). Let π denote the canonical projection $\pi : \mathbb{B}^*M \to M$.

Theorem 1 *Let $F : \mathbb{B}^*M \to \mathbb{B}^*M$ be the time 1 map of a Hamiltonian H on $\mathbb{B}^*M$, where H satisfies the boundary condition:*

$$H(\mathbf{q},\mathbf{p},t) = g(\mathbf{q})(\mathbf{p},\mathbf{p}) \text{ for } \|\mathbf{p}\| = R .$$

Then F has $cl(M)$ distinct fixed points and $sb(M)$ if they are all non degenerate.

We remind the reader that $cl(M)$ is the cup length of M, which is known to be a lower bound for the number of critical points of any real valued function on M. Non degenerate means that no floquet multiplier is equal to one and $sb(M)$ is a lower bound for the number of critical points for a Morse function on M. The cuplength estimate might not be optimal since in general $cat(M) \geq cl(M)$.

The key tool is the following:

Definition A diffeomorphism $F : \mathbb{B}^*M \to \mathbb{B}^*M$ is a **symplectic twist map** if and only if:

1) $F^*\mathbf{p}d\mathbf{q} - \mathbf{p}d\mathbf{q} = dS$ for some $S : \mathbb{B}^*M \to \mathbb{R}$,
2) $\pi \circ F|_{\pi^{-1}(\mathbf{q})}$ is a diffeomorphism onto its image.

Example Let T be the time $R^{1/2}$ map of the hamiltonian $H(\mathbf{q},\mathbf{p},t) = \frac{1}{2}g(\mathbf{q})(\mathbf{p},\mathbf{p})$. Then T is a symplectic twist map: the first condition is satisfied due to its Hamiltonian nature, the second one derives from the fact that the flow associated to H is, on each energy level, the geodesic flow (up to reparametrization). The second condition is given in this case, by the classical fact that the exponential map is a local diffeomorphism.

The following simple lemma was first proven by Le Calvez [5] for diffeomorphisms of the bounded annulus that are isotopic to the identity.

Lemma (LeCalvez, Banyaga, Golé): *Let F be the time 1 map of a (time dependent) Hamiltonian on $\mathbb{B}^*M$. If F preserves the boundary of $\mathbb{B}^*M$, then F can be decomposed into a finite product of symplectic twist maps:*

$$F = F_{2N} \circ \ldots \circ F_1 \quad .$$

Proof. Let G_t be the time t map of the Hamiltonian, starting at $t = 0$. We can write :

$$F = G_1 \circ G^{-1}_{\frac{N-1}{N}} \circ \ldots \circ G_{\frac{k}{N}} \circ G^{-1}_{\frac{k-1}{N}} \circ \ldots \circ G_{\frac{1}{N}} \circ Id$$

and each of the $G_{\frac{k}{N}} \circ G^{-1}_{\frac{k-1}{N}}$ is an exact symplectic map, which we can make as close as we want to the Id by increasing N. The twist condition 2) in the definition of symplectic twist is an open condition. Hence,if T is the map introduced in the example, the map $F_{2k-1} := T^{-1} \circ G_{\frac{k}{N}} \circ G_{\frac{k-1}{N}}$ must satisfy 2) for big enough N (here, the compactness of $\mathbb{B}^*M$ is needed). We then set $F_{2k} = T$ for all k to get the decomposition advertised. □

We sketch the proof of the theorem.

It is not hard to see that the fixed points of F correspond to the critical points of

$$W(\underline{\mathbf{q}}) = \sum_{k=1}^{2N} S_k(\mathbf{q}_k, \mathbf{q}_{k+1}) \quad ,$$

where $\mathbf{q}_{2N+1} = \mathbf{q}_1$, and S_k is the generating function corresponding to the symplectic twist map F_k. We can then use Conley's theory on the finite dimensional configuration space of $\underline{\mathbf{q}} \in M \times U$, where U is an open set in $\mathbb{R}^{2n-1}$: the boundary condition enables one to construct an isolating neiborhood for the gradiant flow of W that contains the topology of M, and conclude to the existence of at least cl(M) critical points.

We conclude by a theorem of a more local nature:

Theorem 2 (Banyaga, Golé): *If F is a symplectic twist map on $\mathbb{B}^*M$ which satisfies the linking of spheres condition, then F has as many fixed points as a real function on M has critical points.*

Proof. (Sketch) One is looking for critical points of $S(\mathbf{q}, \mathbf{q})$ in this case, where S is the generating function for F. The linking condition insures that the diagonal in $M \times M$ is fully included in the image of $\mathbb{B}^*M$ under the embedding $(\mathbf{q}, \mathbf{p}) \to (\mathbf{q}, \mathbf{Q})$, where $F(\mathbf{q}, \mathbf{p}) = (\mathbf{Q}, \mathbf{P})$.

The Arnold conjecture in its full generality is still open, to our knowledge.

References

1. Arnold, V.I.: Sur une propriété topologique des applications globalement canoniques de la mécanique classique. C.R.Acad.Sci.Paris, **261**, 3719-3722 (1965)
2. Chaperon, M.: Une idée du type "géodésiques brisées" pour les systèmes hamiltoniens. C.R.Acad.Sci.Paris Sér I.**298**, no 13, 293-296 (1984)
3. Golé, C.: Periodic points for monotone symplectomorphisms of $\mathbb{T}^n \times \mathbb{R}^n$. Ph.D. Thesis, Boston University (1989)
4. Golé, C.: Ghost circles for twist maps. IMA preprint (1990). To appear in J.Differential Equations
5. LeCalvez, P.: Existence d'orbites de Birkhoff généralisées pour les difféomorphismes conservatifs de l'anneau. Preprint, Université Paris Sud, Orsay

Floer Homology for Mapping Cylinders

Stamatis Dostoglou, Dietmar Salamon

Mathematics Institute, University of Warwick,
Coventry CV4 7AL, Great Britain

Abstract: The symplectic and instanton Floer homologies are presented. When applied to a symplectomorphism on the space of flat connections on a Riemann surface and and for a 3-dimensional mapping cylinder of the Riemann surface, respectively, the resulting homologies are naturally isomorphic.

1 Floer Homology for Symplectic Fixed Points

In [1] Floer invented a new approach to infinite dimensional Morse theory to prove the Arnold conjecture for an exact symplectomorphism ψ of a symplectic manifold $(\mathcal{M}, \omega)$ where ω is proportional to the first Chern class of $\mathcal{M}$ on spheres: The number of fixed points is bounded below by the sum of the Betti numbers of $\mathcal{M}$.

This approach leads to a new invariant called Floer homology. It is the homology HF_* of a chain complex generated by the fixed points of ψ. This is analogous to finite dimensional Morse theory: In [2] Witten constructed a chain complex C_k generated by the critical points of index k of a Morse function f on a Riemannian manifold M. The boundary operator $\partial : C_k \to C_{k-1}$ is determined by the number $n(x,y)$ of gradient orbits connecting x and y (counted modulo 2):

$$C_k = \bigoplus_{index(x)=k} \mathbb{Z}_2 < x >, \quad \partial < x >= \sum_{index(y)=k-1} n(x,y) < y > .$$

Then $\partial^2 = 0$ and the homology of the complex is the homology of M. This construction generalizes to arbitrary coefficient rings if the connecting orbits are counted with appropriate orientations.

For the symplectic fixed point problem there is a variational formulation on the infinite dimensional space of paths $\Omega_\psi = \{\gamma : \mathbb{R} \to \mathcal{M} : \gamma(s+1) = \psi(\gamma(s))\}$, see [1]. The 1-form

$$da : T_\gamma \Omega_\psi \to \mathbb{R} \quad ; \quad \xi \mapsto \int_0^1 \omega(\dot{\gamma}, \xi) ds$$

represents the differential of the symplectic action functional. The critical points (where $da = 0$) are the fixed points of ψ. Let J be an almost complex structure on $\mathcal{M}$ such that $g(\xi,\eta) = \omega(\xi, J\eta)$ is a Riemannian metric. A gradient flow line connecting the fixed points x and y is a solution of the non-linear Cauchy-Riemann equations

$$\frac{\partial u}{\partial s} + J\frac{\partial u}{\partial t} = 0, \qquad \lim_{t\to-\infty} u(s,t) = x, \qquad \lim_{t\to+\infty} u(s,t) = y.$$

The dimension of the space of solutions is the *relative Morse index* $\mu_{symp}(x,y)$. It is given by the spectral flow of the self adjoint operator family $\xi \mapsto J\nabla\xi$ along u as t varies. The index depends on the homotopy class of u and is defined modulo $2N$ where N is given by $c_1(\pi_2(\mathcal{M})) = N\mathbb{Z}$. Here c_1 denotes the first Chern class of the almost complex manifold $\mathcal{M}$. The chain complex is then defined by

$$C_k = \bigoplus_{\mu(x,x_0)=k \bmod 2N} \mathbb{Z}_2 < x >, \quad \partial < x >= \sum_{\mu(y)=(k-1) \bmod 2N} n(y,x) < y >,$$

where $n(x,y)$ is the (finite) number of connecting orbits in the 1-dimensional part of the space of orbits joining y with x. An argument involving the glueing of connecting orbits shows that $\partial^2 = 0$. The Floer homology groups for the symplectomorphism ψ are defined by

$$HF_k^{symp}(\mathcal{M};\psi) = \frac{Ker(\partial : C_k \to C_{k-1})}{Im(\partial : C_{k+1} \to C_k)}.$$

They are independent of the almost complex structure J and depend on ψ only up to exact symplectic isotopy. If ψ is an exact symplectomorphism then $HF_*^{symp}(\mathcal{M};\psi)$ is isomorphic to the homology of $\mathcal{M}$.

2 Instanton Homology

This approach to Morse Theory can be applied to the Chern-Simons functional

$$\mathcal{CS}(a_0 + a) = \frac{1}{2}\int_M tr\left(d_{a_0}a \wedge a + \frac{1}{3}[a\wedge a]\wedge a\right)$$

defined on the space of connections $\mathcal{A}(M)$ on a bundle Q over a 3-manifold M, see [3]. Now the homology groups are *not* isomorphic to the homology of the manifold but give rise to new invariants.

The gradient of $\mathcal{CS}$ at a is given by $*F_a$. Hence critical points of the Chern-Simons functional on Q are flat connections and the gradient flow equation

$$\frac{\partial a}{\partial t} + *F_a = 0$$

is the self-dual Yang-Mills equation for the connection $a(t)$ on $M \times \mathbb{R}$ with respect to the product metric. The relative Morse index μ_{inst} of two flat connections is given by the spectral flow of the operator family $*d_a$ along a path $a(t)$ joining the flat connections. The chain complex is defined by

$$C_k = \bigoplus_{\mu_{inst}(a)=k} \mathbb{Z}_2 < a >, \qquad \partial < a >= \sum_{\mu_{inst}(b)=(k-1)} n(a,b) < a >$$

where $n(a,b)$ is the number of gradient flow lines joining a and b mod 2. The homology groups $HF_k^{inst}(M;Q)$ of this complex are called *instanton homology*. They are independent of the metric used to construct them.

3 A Comparison Theorem

In [6] we consider the special case where $\mathcal{M} = \mathcal{M}_F(P)$ is the symplectic manifold of gauge classes of flat connections of the non-trivial $SO(3)$ bundle P over a Riemann surface Σ. The symplectomorphism $\psi = \psi_f$ is the pull back action on connections of an automorphism $f : P \to P$. We also consider the bundle $Q = P_f$ over the 3-dimensional mapping cylinder $M = \Sigma_h$ where $h : \Sigma \to \Sigma$ is the diffeomorphism induced by f. Note that flat connections on P_f correspond to fixed points of the symplectomorphism ψ_f.

Theorem $HF_*^{symp}(\mathcal{M}_F(P);\psi_f) = HF_*^{inst}(\Sigma_h;P_f)$.

The proof consists of showing that the relative Morse indices of both theories agree and the number of orbits connecting the common critical points are the same. This requires a comparison of pseudoholomorphic curves and self-dual Yang-Mills connections and involves rescaling the metric on the mapping cylinder.

The above Theorem is a special case of a conjecture in [4]. The case of the symplectic Floer homology for Lagrangian intersections arising from a Heegard splitting of a homology 3-sphere is being worked out in [6].

References

1. Floer, A.: Comm.Math.Phys **120**, 575 (1989). For the Lagrangian intersections point of view: Floer,A.: J.Differential Geom. **28**, 513 (1988) and Floer's papers connected to this. See also Golé's article in this volume.
2. Witten, E.: J.Differential Geom. **17**, 661 (1982)
3. Floer, A.: Comm.Math.Phys. **118**, 215 (1988)
4. Atiyah, M.: Proc.Sympos.Pure Math. **48** (1988)
5. Dostoglou, S., Salamon, D.: Warwick Preprint, 1990
6. Dostoglou,S., Robbin, J., Salamon, D.: In progress.

Section 2

Quantum Groups and Non-Commutative Differential Geometry

Organizer: S.L. Woronowicz

κ-Deformation of (Super)Poincaré Algebra

J. Lukierski

Institute of Theoretical Physics, University of Wrocław,
Cybulskiego 36, 50-205 Wrocław, Poland

1 Introduction

The notion of quantum groups and quantum algebras (see e.g. ref. [1]-[6]) can be used in order to study the deformations of space-time symmetries as well as their supersymmetric extensions. In order to obtain the quantum deformation of semisimple Lie algebras describing Minkowski or Euclidean group of motions mostly the contraction techniques have been used. In particular there were obtained:

a) quantum deformation of $D = 2$ and $D = 3$ Euclidean and Minkowski geometries, described as quantum Lie algebra or quantum Lie group [7],[8]
b) quantum deformation of $D = 4$ Poincaré algebra [9],[10]
c) quantum deformations of $D = 2$ supersymmetry algebra in its Minkowski as well as its Euclidean version [11]-[13] .

We should also note here that the inhomogeneous quantum groups were studied by contracting the Gelfand - Tzetlin basis of simple quantum Lie algebra [14] as well as by introducing the inhomogeneous quantum groups as matrix quasitriangular Hopf algebras with noncommuting entries [15].

Below we shall describe briefly the results of [9], where the quantum deformation of $D = 4$ Poincaré algebra is obtained by contracting the Cartan-Weyl basis of a real form $U_q(O(3,2))$ of the quantum complex $D = 4$ de-Sitter algebra $U_q(Sp(4;C))$. We shall present also few basic relations describing $D = 4$ superPoincaré algebra. The contraction limit (denoted by $\left\{ {R\to\infty \atop q\to 1} \right\}$) which provides finite quantum inhomogeneous algebras is defined as follows:

$$\left\{ \begin{matrix} R\to\infty \\ q\to 1 \end{matrix} \right\} : \qquad R\to\infty; \qquad R\,\log\, q \to i^{\epsilon}\kappa^{-1} \qquad (0<\kappa<\infty)\,, \tag{1}$$

where q is the quantum deformation parameter in Drinfeld-Jimbo scheme [3],[4], $\epsilon = 0$ for q real, and $\epsilon = 1$ for $|q| = 1$. [1] We see that the prescription (1) provides a particular way of approaching the limit $q \to 1$ in such a way that it appears a new mass- like parameter κ. In this report we shall show that the limit (1) provides for finite κ a nonlinear modification of the Poincaré algebra("κ - deformation") accompanied by the consistent modification of coproduct and antipode formulae.

It should be stressed that for the quantum algebras of rank ≥ 2 physically more important seems to be the Cartan-Weyl basis, which has the following two advantages over the Cartan-Chevalley basis:

i) The q-deformed Cartan-Weyl basis describes the quantum deformation of Lie algebra generators (i.e. in the limit $q \to 1$ one obtains the ordinary Lie algebra relations).

ii) The nonlinear (three-linear or fourlinear for classical Lie algebras) Serre relations are replaced by bilinear relations.

It appears that not every involution of Cartan-Chevalley basis can be extended to the full Cartan-Weyl basis. Besides it was shown (see [10])that one can introduce 16 inner involutions of Cartan-Weyl basis of $U_q(Sp(4;C))$, but only two seem to be suitable as candidates providing after contraction (1) the q-deformation of Poincaré algebra. These two contractions - one with $|q| = 1$ [9] and second with q real lead in the case $|q| = 1$ to the deformation of Klein -Gordon equation with trigonometric nonlinearity, and for q real provides *sinh* and *cosh* nonlinearities. At the end of this presentation we shall present few formulae describing $D = 4$ quantum superPoincaré algebra.

2 $D = 4$ Deformation κ-of Poincaré Algebra

In reference [9] the commutation relations, comultiplication, antipode and one of the Casimir operators were given for the quantum κ-Poincaré algebra, where the parameter κ is introduced via contraction (1). One gets the following formulae a) Lorentz sector M_i, L_i; $i = 1,2,3$ $(M_\pm = M_1 \pm iM_2$, $L_\pm = L_1 \pm iL_2)$

$$[M_+, M_-] = 2M_3 \qquad [M_3, M_\pm] = \pm M_\pm \tag{2}$$

$$\begin{aligned} [L_-, L_+] &= 2M_3 \cos\frac{P_0}{\kappa} - \frac{1}{\kappa}\{P_3, L_3\} + \frac{1}{2\kappa^2}P_3^2 \\ [L_3, L_\pm] &= \mp e^{\mp iP_0/\kappa} M_\pm \pm \frac{1}{2\kappa} L_\pm P_3 + \frac{i}{2\kappa} L_3 P_\mp \end{aligned} \tag{3}$$

[1] For $\epsilon = 0$ such limit was firstly introduced by Firenze group [7],[8] for studying the contractions of rank one quantum algebras $(U_q(Sl(2;C))$ and its real forms $U_q(SU(2))$ and $U_q(SU(1,1)))$.

$$[M_3, L_3] = 0 \qquad [M_3, L_\pm] = \pm L_\pm$$
$$[M_+, L_3] = -L_+ - \frac{i}{2\kappa} M_3 P_- \qquad [M_-, L_3] = L_- - \frac{i}{2\kappa} P_+ M_3$$
$$[M_+, L_-] = 2L_3 + \frac{i}{2i\kappa} P_+ M_+ + \frac{1}{\kappa} P_3 M_3 \tag{4}$$
$$[M_-, L_+] = -2L_2 + \frac{1}{2i\kappa} M_- P_- - \frac{1}{\kappa} P_3 M_3$$
$$[M_\pm, L_\pm] = \frac{1}{2i\kappa} M_\pm P_\mp$$

b) Translation sector ($P_\pm = P_2 \pm iP_1$)

$$\begin{aligned} &[P_\mu, P_\nu] = 0 && (\mu, \nu = 1,2,3,0) \\ [M_i, P_0] = 0 \qquad &[M_i, P_k] = i\epsilon_{ikl} P_l && (i,k,l = 1,2,3) \end{aligned} \tag{5}$$

$$\begin{aligned} [L_3, P_0] &= iP_3 & [L_3, P_3] &= i\kappa \sin \frac{P_0}{\kappa} \\ [L_3, P_2] &= \frac{1}{2i\kappa} P_1 P_3 & [L_3, P_1] &= -\frac{1}{2i\kappa} P_2 P_3 \\ [L_\pm, P_0] &= iP_1 \mp P_2 & [L_\pm, P_3] &= \frac{1}{2i\kappa} P_\mp P_3 \\ [L_\pm, P_2] &= \mp\kappa \sin \frac{P_0}{\kappa} - \frac{1}{2i\kappa} P_3^2 & [L_\pm, P_1] &= i\kappa \sin \frac{P_0}{\kappa} \pm \frac{1}{2\kappa} P_3^2 \,. \end{aligned} \tag{6}$$

The coproduct is

$$\Delta M_i = M_i \otimes I + I \otimes M_i$$

$$\Delta P_i = P_i \otimes e^{\frac{iP_0}{2\kappa}} + e^{\frac{-iP_0}{2\kappa}} \otimes P_i \qquad \Delta P_0 = P_0 \otimes I + I \otimes P_0$$

$$\Delta L_3 = L_3 \otimes e^{\frac{iP_0}{2\kappa}} + e^{\frac{-iP_0}{2\kappa}} \otimes L_3 + \frac{i}{2\kappa} e^{\frac{-iP_0}{2\kappa}} M_+ \otimes P_+ - \frac{i}{2\kappa} P_- \otimes M_- e^{\frac{iP_0}{2\kappa}}$$

$$\begin{aligned} \Delta L_+ = {} & L_+ \otimes e^{\frac{iP_0}{2\kappa}} + e^{\frac{-iP_0}{2\kappa}} \otimes L_+ + \frac{i}{2\kappa} P_- \otimes M_3 e^{\frac{iP_0}{2\kappa}} \\ & - \frac{i}{2\kappa} e^{\frac{-iP_0}{2\kappa}} M_3 \otimes P_- + \frac{1}{\kappa} e^{\frac{-iP_0}{2\kappa}} M_+ \otimes P_3 \end{aligned}$$

$$\begin{aligned} \Delta L_- = {} & L_- \otimes e^{\frac{iP_0}{2\kappa}} + e^{\frac{-iP_0}{2\kappa}} \otimes L_- + \frac{i}{2\kappa} P_+ \otimes M_3 e^{\frac{iP_0}{2\kappa}} \\ & - \frac{i}{2\kappa} e^{\frac{-iP_0}{2\kappa}} M_3 \otimes P_+ + \frac{1}{\kappa} P_3 \otimes M_- e^{\frac{iP_0}{2\kappa}} \,. \end{aligned} \tag{7}$$

The antipode is

$$\begin{aligned} S(P_0) &= -P_\mu \qquad S(M_i) = -M_i \\ S(L_3) &= -L_3 - \frac{1}{2\kappa}P_3 + \frac{i}{2\kappa}(M_+P_+ - P_-M_-) \\ S(L_+) &= -L_+ - \frac{i}{\kappa}P_- + \frac{1}{\kappa}M_+P_3 \\ S(L_-) &= -L_- + \frac{i}{\kappa}P_+ + \frac{1}{\kappa}P_3M_- \,. \end{aligned} \tag{8}$$

The κ-deformation of the first Casimir operator describing the relativistic mass operator is

$$\begin{aligned} I_2 &= P_1^2 + P_2^2 + P_3^2 + 2\kappa^2\left(cos\frac{P_0}{\kappa} - 1\right) \\ &= P_1^2 + P_2^2 + P_3^2 - P_0^2 + \frac{1}{12\kappa^2}P_0^4 + O(\kappa^{-4})\,. \end{aligned} \tag{9}$$

The formula (9) gives a dispersion relation for light in vacuum: $dE/dp = f(E)$, and a corresponding modification of the velocity of massive particles. It remains to be seen what is the value of κ compatible with the experimental limits.

There exists also the choice of involution of $U_q(Sp(4))$providing after the contraction (1) the κ-deformation of Poincaré algebra with q real. In such a case one gets analogous formulae to the relations (2-9), obtained by replacing $cos(P_0/\kappa)$ by $ch(P_0/\kappa)$ and $e^{\pm\frac{iP_0}{2\kappa}}$ by $e^{\pm\frac{P_0}{2\kappa}}$.

3 $D = 4$ κ-Deformation of SuperPoincaré Algebra

We would like here to announce some results concerning the description of $D = 4$ quantum Poincaré superalgebra, obtained from the contraction (1) of the Cartan-Weyl basis of $U_q(OSp(1|4))$. We shall describe here only the most important results concerning the algebra sector.

Let us introduce firstly the Cartan-Weyl basis for $U_q(OSp(1|4))$. The root space of $OSp(1|4) \simeq B(0,2)$ is spanned by two simple bosonic roots $\pm\alpha \equiv \alpha_{\pm 1}$ and two simple fermionic roots $\pm\beta \equiv \alpha_{\pm 2}$, with the following Cartan matrix [16]-[18]

$$\alpha^{sym} = \begin{pmatrix} 2 & -1 \\ -1 & 1 \end{pmatrix}\,. \tag{10}$$

The Cartan-Weyl basis is obtained by considering all nonsimple roots. The corresponding generators ($\alpha_i \Rightarrow e_\alpha$ etc.) are

$$\begin{aligned} \text{bosonic}: &\; e_\alpha, e_{\pm(\alpha+2\beta)}, e_{\pm(2\alpha+2\beta)}, e_{\pm 2\beta}, h_\alpha, h_\beta \\ \text{fermionic}: &\; e_{\pm(\alpha+\beta)}, e_\beta\,, \end{aligned} \tag{11}$$

where h_α, h_β describe two bosonic Cartan generators.

Using Drinfeld-Jimbo q-deformation of Cartan-Chevalley basis $(e_{\pm\alpha}, e_{\pm\beta}, h_\alpha, h_\beta)$ one can calculate the q-deformation of Cartan-Weyl basis of $U_q(OSp(1;4))$. We introduce the fourspinor of supercharges

$$Q_a = \begin{pmatrix} e_{-(\alpha+\beta)} \\ e_{-\beta} \\ e_\beta \\ e_{\alpha+\beta} \end{pmatrix} \tag{12}$$

and identify the reality condition for the fermionic $U_q(OSp(1;4))$ generators with the Majorana condition for the fourspinor Q. It can be shown that after rescaling the supercharges (12) by $\sqrt{R}$ $(S_a = \sqrt{R}\, Q_A)$ one obtains in the limit $R \to \infty$ the following κ-deformation of the supersymmetry algebra

$$\begin{aligned} &[P_\mu, P_\nu] = 0 \\ &\{S_a, \bar{S}_b\} = \kappa\gamma^0_{ab} sin\left(\frac{P_0}{\kappa}\right) + \gamma^k_{ab} P_k \\ &[P_\mu, S_a] = 0 \ , \end{aligned} \tag{13}$$

where $\{\gamma_\mu, \gamma_\nu\} = 2\eta_{\mu\nu}$ and $\gamma_0 = \begin{pmatrix} 1 & 0 \\ 0 & -1 \end{pmatrix}$, $\gamma_k = \begin{pmatrix} 0 & \sigma_k \\ -\sigma_k & 0 \end{pmatrix}$. The full q-deformed $N = 1$ Poincaré superalgebra is obtained if we add to (13) the covariance relations for supercharges and the q-deformed Poincaré algebra. We get the following covariance relations:

$$\begin{aligned} &[M_{kl}, S_a] = i(\sigma_{kl})_{ab} S_b \\ &[M_{k4}, S_a] = i(\sigma_{0k})_{ab} S_b - \frac{i}{\kappa} P_3 (\gamma_5 \sigma_{0k})_{ab} S_b \end{aligned} \tag{14}$$

where $\sigma_{\mu\nu} = \frac{1}{4}[\gamma_\mu, \gamma_\nu]$.

The Poincaré sector of q-deformed superPoincaré algebra *differs* from the one obtained by the contraction presented in Sect. 3. We obtain that

i) The Poincaré generators belonging to the quantum $N = 1$ superalgebra do not form a quantum subalgebra.
ii) The commutators of Lorentz generators depend on fermionic generators of $U_q(OSp(1|4))$.

More details of quantum $D = 4$ κ-deformed superPoincaré algebra are given in [19].

4 Discussion

The limit $\left\{ {R\to\infty \atop q\to 1} \right\}$ is the quantum generalization of the (super) - de-Sitter contraction. In our approach we underline strongly the appearance in the contraction limit of mass-like parameter κ and the resulting quantum algebras we call as "κ-deformed". It appears as a very interesting question how to interpret physically the fundamental mass parameter κ. We feel that it would be desirable to find for example its relation with the Planck mass.

Let us briefly summarize the main features of the κ-deformations in four dimensions:

i) The subalgebra E_3^κ describing three-dimensional group of motions (space rotations + space translations) remains unchanged. The quantum nature of the algebra E_3^κ shows up only in the coalgebra sector with the comultiplication rules for the three-momenta modified by the energy-depending factor. Because the energy operator P_0 commutes with all the generators of E_3^κ, one can say that the quantum Hopf algebra E_3^κ is obtained from the classical one ($E_3 \equiv E_3^\infty$), by the central extension in the coalgebra sector. We see therefore that contrary to the proposals presented in [20]-[24] the bialgebra of three - rotations remains unchanged. The only modification of the nonrelativistic kinematics for $\kappa < \infty$ appears in the addition law for the three-momenta.

ii) The four momenta operators $P_\mu = (P_i, P_0)$ form Abelian subalgebra for κ-deformed $D = 4$ Poincaré as well as for $D = 4$ superPoincaré algebras. It appears therefore that one can introduce the relativistic Schrödinger realization of the generators P_μ

$$P_\mu \psi(x) = \frac{1}{i} \frac{\partial}{\partial x^\mu} \psi(x) \tag{15}$$

The formula (15) introduces classical Minkowski space by a Fourier transformation of the fourmomentum eigenfunctions. It should be mentioned that in other proposals for the q-deformed translation generators (see e.g. [14],[15]) the commutators of the fourmomentum operators are q-deformed.

iii) The κ-deformed Lorentz algebra generators do not form quantum subalgebra. Moreover, the relations satisfied by κ-deformed Lorentz generators are different for κ-deformed Poincaré and κ-deformed superPoincaré algebras. We would like here to recall that these features are related with the structure of quantum algebras. In general the subalgebras (e.g. bosonic subalgebra of a superalgebra) after quantum deformation cease to be further quantum subalgebras.

References

1. Faddeev, L., Reshetikhin, N., Takhtajan, L.: Algebra i Analiz **1**, 178 (1989)
2. Woronowicz, L.: Comm.Math.Phys. **111**, 613 (1987); ibid. **122**, 125 (1989)
3. Drinfeld, V.G.: Quantum groups. In: Proceedings of the International Congress of Math. Berkeley, USA (1986). p. 793
4. Jimbo, M.: Lett.Math.Phys. **10**, 63 (1985); ibid. **11**, 247 (1986)
5. Manin, Yu.T.: Quantum groups and non-commutative geometry. Centre de Recherches Math., Univ. de Montreal, 1989
6. Proceedings of First EIMI Workshop on Quantum Groups, October-December 1990, ed. P. Kulish, Springer Verlag, 1991
7. Celeghini, E., Giacchetti, R., Sorace, E., Tarlini, M.: J.Math.Phys. **32**, (1991) 1155; ibid. 1159
8. Celeghini, E., Giacchetti, R., Sorace, E., Tarlini, M.: Contraction of quantum groups. Contribution to [6]
9. Lukierski, J., Nowicki, A., Ruegg, H., Tolstoy, V.N.: Phys.Lett.B **264**, 331 (1991)
10. Lukierski, J., Nowicki, A., Ruegg, H.: Geneve Univ. preprint UGVA/DPT-08-740
11. Schmidke, W.P., Vokos,S.P., Zumino, B.: Z.Phys.C **48**, 249 (1990)
12. Lukierski, J., Nowicki, A.: Wrocław Univ. preprint UWr 777/91, June 1991
13. Celeghini, E., Giacchetti, R., Kulish, P.P., Sorace, E., Tarlini, M.: Firenze Univ. preprint DFF 139/6/91
14. Chakrabarti, A.: J.Math.Phys. **32**, 1227 (1991)
15. Schlieker, M., Weich, W., Weixler, R.: Munich University preprint LMU-TPW 1991-3, April 1991
16. Tolstoy, V.N.: Proceedings of Summer Workshop "Quantum Groups", Clausthal, Germany, July 1989. Springer Verlag
17. Kulish, P.P., Chaichian, M.: Phys.Lett.B **234**, 72 (1990)
18. Koroshkin, S.M., Tolstoy, V.N.: Nuclear Physics Institute, Moscov State University, preprint 91-13/217
19. Lukierski, J., Nowicki, A., Ruegg, H.: Quantum Deformations of Poincaré Algebra and the Supersymmetric Extensions. Bordeaux Univ. preprint, October 1991; to be published in Proceedings of the International Symposium on Topological and Geometrical Methods in Field Theory, Turku, May 1991
20. Podleś, P., Woronowicz, S.L.: Comm.Math.Phys. **130**, 381 (1990)
21. Carow-Watamura, V., Schlieker, M., Scholl, M., Watamura, S.: Z.Phys.C **48**, 159 (1990)
22. Carow-Watamura, V., Schlieker, M., Scholl, M., Watamura, S.: Internat.J.Modern Phys.A **6**, 3081 (1991)
23. Woronowicz, S.L.:New Quantum Deformation of $SL(2;C)$ - Hopf Algebra Level. Warsaw University preprint, 1990
24. Schmidke, W.P., Wess, L., Zumino, B.: Max Planck Institute preprint MPI-Ph/91-15, March 1991

Quantum Physics as Non-Commutative Geometry

Arthur Jaffe

Harvard University, Cambridge, MA 02138, U.S.A.

1 Non-Commutative Geometry

For a person in mathematical physics, notions of non-commutative geometry (NCG) seem very natural. Related ideas to those in NCG occur in quantum theory – especially supersymmetric quantum theory – and also in statistical mechanics. One can interpret NCG as a quantization of geometry, in the sense that quantum theory is a quantization of classical physics. Many basic notions of non-commutative geometry can be understood by thinking of NCG as a way to define and to integrate differentials $a_0 da_1 \cdots da_n$ in a framework more general than that of differential forms on manifolds. The quantum functions $a_0, \ldots, a_n$ are operators; their integrals can be thought of as quantum mechanical expectation values. What results is a theory in which classical notions of geometry carry over. In particular, there is a natural interpretation of NCG in terms of a cohomology theory. This cohomology reduces to de Rham theory in the usual commutative case.

NCG has another unusual aspect; NCG has a natural and straightforward extension to an infinite dimensional setting. The concept of dimension of the underlying space of interest M arises indirectly, and can be studied in terms of the spectrum of an associated Laplace operator. In the infinite dimensional setting, the study of the cohomology problem has profited from recent ideas with their origins in the mathematical theory of quantum fields. Of course, this approach can also be used in the usual, finite dimensional theory.

In de Rham theory, one studies a manifolds M by means of functions on M and the associated tangent, cotangent vector and spinor bundles. In NCG, the algebra $C_0^\infty(M)$ is replaced by an operator algebra $\mathcal{A}$ acting on a Hilbert space. In the special case that $\mathcal{A}$ is the algebra of smooth functions, i.e. $\mathcal{A} = C_0^\infty(M)$, acting as an algebra of multiplication operators, then Connes has shown that NCG reduces to de Rham theory [2].

In de Rham theory, the space M can naturally be recovered from the algebra $\mathcal{A}$ of functions on M; it is given by the maximal ideals of $\mathcal{A}$. Likewise, in NCG the underlying space M is determined by $\mathcal{A}$. Granted this does not give a very intuitive notion of how M looks. However there are advantages to working in terms of $\mathcal{A}$, for a natural theory of invariants arises. By invariants, we mean integrals of forms which are invariant under smooth changes of the differential d. In special cases, these invariants can be interpreted as the index (in the sense of Atiyah and Singer) of the differential d, and in this case these invariants of NCG reduce to standard integer invariants of manifolds which are unchanged by homotopies of M. The invariants of NCG are more general and, they are not in general integers. (Non-integer indices are common, for example, in von Neumann algebra theory.) The invariants are stable under small deformations of d.

The connection between NCG and ideas from quantum physics centers about the fact that the NCG differential da is given by a graded commutator. Here the algebra $\mathcal{A} = \mathcal{A}_+ \oplus \mathcal{A}_-$ can be decomposed into its even and odd parts under a $\mathbb{Z}_2$ grading, $\gamma(\mathcal{A}_\pm) = \pm\mathcal{A}_\pm$. Generally, the physics interpretation is that $\mathcal{A}_+$ and $\mathcal{A}_-$ denote bosonic and fermionic operators respectively. The graded commutator, given by an odd operator D, is

$$da = \begin{cases} Da - aD\,, & a \in \mathcal{A}_+ \\ Da + aD\,, & a \in \mathcal{A}_-\,. \end{cases}$$

Thus for all a,

$$d^2a = D^2a - aD^2\,.$$

We therefor infer that if $D^2 = \lambda I$ is a multiple of the identity, then

$$d^2 = 0\,.$$

In physics examples, D will often be linear in fermionic annihilation operators, so $D^2 = 0$.

A second natural differential given by the graded commutator is the adjoint differential

$$d^*a = \begin{cases} D^*a - aD^* & a \in \mathcal{A}_+\,, \\ D^*a + aD^* & a \in \mathcal{A}_-\,. \end{cases}$$

So if $D^2 = \lambda I$ then also $D^{*2} = \bar{\lambda} I$ and $d^{*2} = 0$.

The signature-type operator

$$\delta = d + d^*$$

yields the differential

$$\delta a = [Q, a]\,, \quad Q = D + D^*\,,$$

where by $[Q, a]$ we denote the graded commutator. Then

$$\delta^2 a = Q^2 a - aQ^2 \ ,$$

so defining $H = Q^2$, we have

$$\delta^2 a = [H, a] \ .$$

If $D^2 = 0$, then the Hamiltonian H of super-symmetric quantum theory can be written in the form of a Laplacian, $H = DD^* + D^*D$.

The differential δa can be thought of as an infinitesimal symmetry – namely supersymmetry – generated by the called the supercharge Q .

The properties of $H = Q^2$ are reflected in three cases:

(i) *Compact, Finite Dimensional Space:* For p sufficiently large,

$$\mathrm{Tr}(\mathrm{H} + \mathrm{I})^{-\mathrm{p}/2} < \infty \ . \tag{1}$$

Then $\dim = \inf p$ for which (1) is finite. This is the p-summable case of Connes [2].

(ii) *Compact, Infinite Dimensional Space:* This is the Θ-summable case [3],[10], for which

$$\mathrm{Tr}(\mathrm{e}^{-\mathrm{tH}}) < \infty \ , \quad \text{all } \mathrm{t} > 0 \ . \tag{2}$$

(iii) *Non-Compact Space, Finite or Infinite Dimensional:* In this case there is a functional ω on A which satisfies the super-KMS constraints [11],[15]:

$$\omega(\delta a) = 0 \qquad \text{infinitesimal invariance,}$$

and

$$\omega(ab) = \omega(\gamma(b)a(1)) \qquad \text{super-KMS condition.}$$

Here γ is the $\mathbb{Z}_2$ grading and $a(t) = e^{-tH} a e^{tH}$.

One fundamental result is the existence of field theories corresponding to (ii) and (iii).

Theorem 1 *Polynomial, $N = 2$ Supersymmetric quantum fields on a cylindrical space-time exist and are Θ-summable. (See* [7].*) The $\mathbb{R}^2$ limit of space-time yields a super-KMS functional* [14].

2 NCG as Cohomology

The appropriate cohomology theory is cyclic cohomology. In the infinite-dimensional case, it is natural to work with a version called entire cyclic cohomology, which I describe here. It is striking how much the mathematical framework of entire cyclic cohomology looks like the mathematical framework of quantum field theory and statistical mechanics, when they are studied through expectation values. This has the consequence on the side of mathematics that representations and computational tools of mathematical physics, such as functional integrals, become natural tools in understanding cohomology. On the side of physics, this connection reveals a cohomological (geometric) content to quantum field theory.

The actual similarity between NCG and physics can be illustrated through the study of cohomology through the study of dual space to $\mathcal{A}$, or rather to the dual space to the tensor algebra over $\mathcal{A}$,

$$\mathcal{A} = \mathbb{C} \oplus \mathcal{A} \oplus \mathcal{A}^2 \oplus \cdots ,$$

where $\mathcal{A}^n$ is the n-fold tensor product $\mathcal{A} \otimes \mathcal{A} \otimes \cdots \otimes \mathcal{A}$. The space C_n of $(n+1)$-multilinear functionals $f_n(a_0, a_1, \ldots, a_n)$ is the natural space of cochains of NCG. A cochain, evaluated on $a_0, da, \ldots, da_n$ can be interpreted as an integral of the differential $a_0 da_1 \cdots da_n$,

$$f(a_0, da, \ldots, da_n) = \int a_0 da_1 \cdots da_n \ . \qquad (NCG)$$

We are familiar in the context of mathematical physics of studying the algebra $\mathcal{A}$ of obeservables (e.g. in a Haag-Kastler framework, or in the Wightman field framework) in terms of expectation values

$$f(a_0, da, \ldots, da_n) = \langle a_0 da_1 \cdots da_n \rangle \ . \qquad (QFT)$$

From the expectation values in a cycle state $\langle \ \ \rangle$, we can reconstruct the algebra $\mathcal{A}$ (or the field operators) through a GNS construction.

In entire cyclic cohomology, one restricts the cochains (expectations) in the direct one $C = \oplus_n C_n$ to satisfy a growth condition in n. In fact, C_n has a natural norm, given by a norm $\|\cdot\|$ on $\mathcal{A}$, with

$$\|f_n\| = \sup_{\|a_j\|=1} |f_n(a_0, \ldots, a_n)| \ .$$

The entire growth condition of Connes is the assumption that every sequence $f = (f_0, f_1, f_2, \ldots, f_j, \ldots) \in C$ satisfies

$$n^{1/2} \|f_n\|^{1/n} \longrightarrow 0 \ .$$

Another way to describe a cochain f, in using language of quantum field theorists, is as an element of the dual to the algebra of fields. This approach was pursued in the 1960's by Borchers, and the algebra is often given his name. A cochain f is an element of the Borchers algebra restricted by a growth condition, namely: $\Sigma_n \|f_n\| |z|^n$ is bounded by a function of exponential order 2. Just as in the theory of analytic functions a growth condition ensures certain uniqueness of interpolations, this growth condition ensures a certain type of uniqueness of a normalized representation of cohomology classes described below.

On the space $\mathcal{C}$ there are two natural coboundary operators, the Hochschild "creation" operator

$$b : \mathcal{C} \longrightarrow \mathcal{C} , \quad b : \mathcal{C}_n \longrightarrow \mathcal{C}_{n+1} ,$$

and Connes' "annihilation" operator

$$B : \mathcal{C} \longrightarrow \mathcal{C} , \quad B : \mathcal{C}_{n+1} \longrightarrow \mathcal{C}_n .$$

These operators satisfy

$$b^2 = B^2 = 0 , \quad \text{as well as} \quad Bb + bB = 0 .$$

These commutation conditions mean that the double complex described mean that the double complex described by b and B has a total cohomology, determined by the coboundary operator

$$\partial = b - B , \quad \text{which also satisfies} \quad \partial^2 = 0 .$$

Explicitly

$$\begin{aligned}(bf_n)(a_0, a_1, \ldots, a_{n+1}) &= \sum_{j=0}^{n} (-1)^j f_n(a_0, \ldots, a_j a_{j+1}, \ldots, a_{n+1}) \\ &\quad + (-1)^{n+1} f_n(a_{n+1} a_0, a_1, \ldots, a_n) ,\end{aligned}$$

and

$$\begin{aligned}(Bf_n)(a_0, \ldots, a_{n-1}) &= f_n(I, a_0, \ldots, a_{n-1}) - (-1)^n f_n(a_0, \ldots, a_{n-1}, I) \\ &\quad + \text{cyclic antisymmetrization} .\end{aligned}$$

3 The Cocycle Equation

The fundamental cohomology classes described by ∂ are given by solutions of the cocycle equation

$$\partial\tau = 0 , \quad \tau \in \mathcal{C} .$$

This could also be written $b\tau = B\tau$. A solution is said to be trivial if $\tau = \partial G$ for some $G \in \mathcal{C}$. Thus the classes of cocycles τ are determined by the equivalence relation

$$\tau \cong \tau + \partial G .$$

The relation between solutions to the cocycle equation and geometric invariants are provided by an abstract result of Connes [3] which pairs projections in A with cocycle classes τ:

Theorem 2 *Given a projection $e \in A$ and a solution $\tau \in C$ to the cocycle equation, the pairing*

$$\langle \tau, e\rangle = \sum_{n=0}^{\infty}(-1)^n \frac{(2n)!}{n!}\tau_{2n}(e, e, \cdots, e) \qquad (*)$$

depends only on the cohomology class of τ. In particular, $\langle \partial G, e\rangle = 0$.

The pairing $(*)$ provides invariants $\langle \tau, e\rangle$ under a homotopy τ^λ of τ, for a family indexed by $\lambda \in [0,1]$. The problem of finding invariants of $\mathcal{A}$ (and hence of the underlying commutative or non-commutative space M) can be understood in terms of finding solutions to the cocycle equation. Thus what nontrivial τ's satisfy $\partial\tau = 0$?

At the present time only one type of solution for τ is known. Given a differential $\delta a = [Q, a]$ on $\mathcal{A}$, there is a solution τ^Q to the cocycle equation. While two different formulas have been proposed as solutions, they have recently been shown to be equivariant [4].

Here I describe the formula τ_{JLO} for a solution which was discovered in [10]. There is an especially simple expression for a generating function τ^Q for τ_n^Q,

$$\tau_{JLO}^Q(a) = \sum_{n=0}^{\infty}\frac{1}{n!}\tau_{JLO,n}^Q(a, a, \ldots, a)\,.$$

Here Str denotes the graded (super-) trace, and

$$\tau_{JLO,n}^Q(a, a, \ldots, a) = \frac{d^n}{ds^n}\,\mathrm{Str}(a \exp(-Q^2 + s\delta a))\Big|_{s=0}\,. \qquad (JLO)$$

Then $\tau_{JLO,n}^Q(a_0, \ldots, a_n)$ is obtained by polarization.

Theorem 3 *Let e^{-tQ^2} be trace class for $t > 0$. Then the sequence (JLO) is a solution to the cocycle equation $(*)$.*

In fact, the explicit form of τ_n yields a bound $n\|\tau_n\|^{1/n} \leq e$, so τ_n has exponential growth of order 1.

The pairing $\langle \tau^Q, e\rangle$ actually provides invariants which are stable under small deformations of Q:

Theorem 4 *Let q be odd. Let q be bounded, or more generally let q be relatively bounded with respect to Q^2 as described in [6]. Then τ_{JLO}^Q and τ_{JLO}^{Q+q} are cohomologous, namely for some $G \in C$,*

$$\tau_{JLO}^Q = \tau_{JLO}^{Q+q} + \partial G\,.$$

What we have shown, is that given a graded (super-) derivation Q, there exists a functional τ^Q_{JLO} on projections e which yields a cohomology invariant $\langle\tau^Q, e\rangle$, given by the pairing formula

$$\langle\tau^Q_{JLO}, e\rangle = \sum_{n=0}^{\infty}(-1)^n \frac{(2n)!}{n!}\frac{d^{2n}}{ds^{2n}}\mathrm{Str}(e\exp(-Q^2 + s[Q,e]))\Big|_{s=0} . \qquad (**)$$

Since (**) is expressed as a graded trace of a heat kernel, it may have a functional integral representation. This can be extremely useful from a computational point of view. The examples of Theorem 1 do have the functional integral interpretation of $\langle\tau^Q_{JLO}, e\rangle$.

The cocycle τ^Q_{JLO} has a very simple geometric interpretation. If Q_+ denotes the part of Q which maps $\mathcal{H}_+$ to $\mathcal{H}_-$ (even to odd subspaces of $\mathcal{H}$ with respect to the grading γ) then

$$\langle\tau^Q_{JLO}, e\rangle = \mathrm{Index}((eQe)_+) .$$

The formula (**) is a generalized McKean-Singer type of representation for the index. The functional τ^Q is called the Chern character of Q. We pose the general question: Is there a second, noncohomologous solution to the cocycle equation? Such a solution might yield torsion or other invariants.

4 Quantum Space

We have seen that the underlying geometric space M arising in a non-commutative geometry may not be a classical manifold. In fact, the notion of a quantum space M arises naturally in this context. We should then ask whether quantum spaces play (or will play) an important role in theoretical physics? If so, then in what way do they? It is best to approach these questions in terms of examples. Is quantum physics described by quantum field theory on a quantum space?

There are not so many concrete examples of quantum spaces. This is understandable, since the connection between the algebra $\mathcal{A}$ and the space M is not straightforward. Clearly it would be good to have intuitive pictures of these and of other examples. Let me mention several examples which do exist:

1) Connes' 2-sheeted modification of space-time which leads to a beautiful picture (at the classical level) of the standard electroweak model [5], and which is described by him at this meeting.
2) The quantum torus, studied by many authors. In this case the algebra $\mathcal{A}$ is generated by p, q where

$$pq = \gamma qp$$

and $\gamma \in \mathbb{C}\,,\quad |\gamma| = 1.$

3) The quantum unit disk, QD. This algebra $\mathcal{A}$ is generated by $z, \bar{z}$, and one can imagine functions given by power series

$$f(z,\bar{z}) = \sum a_{nm} z^n \bar{z}^m .$$

Classically z and $\bar{z}$ commute. In the quantum disk one has $\|z\| \leq 1$ and for the real parameter h, $0 < h < 1$,

$$[z,\bar{z}] = h(1-\bar{z}z)(1-z\bar{z}) . \qquad (QD \text{ algebra})$$

The QD was introduced by Berezin [1] and studied by Klimek and Lesniewski [16]. It follows that $z\bar{z}$ and $\bar{z}z$ commute, and that $[z,\bar{z}] \geq 0$. In fact, it was shown in [16] that $SU(1,1)$ acts on this algebra $\mathcal{A}$, so that $z' = (az+b)(\bar{b}z+\bar{a})^{-1}$ and the corresponding $\bar{z}' = (\bar{a}\bar{z}+\bar{b})(b\bar{z}+a)^{-1}$, where $|a|^2 - |b|^2 = 1$, also satisfy the QD algebra.

Given the invariance of the QD algebra under $SU(1,1)$, one can use the QD to define a quantum Riemann surface, QRS. In particular ,

$$QRS = A/\text{Discrete Subgroup of } SU(1,1) .$$

The QRS is specified by the deformation parameter h, and goes over continuously to the classical RS when $h \to 0$.

Clearly the QRS as a family of deformations of a Poisson structure has many aspects in common with geometric quantization. But geometric quantization seems to yield the QRS only for a discrete subset of values of h. This is a question to understand better in the future.

5 Quantization Map

The QD algebra can also be represented by a quantization in terms of Toeplitz operators. Let $dw_h(\zeta)$ represent the measure on the classical unit disc, CD, given by

$$dw_h(\zeta) = \frac{i}{2}(1-|\zeta|^2)^{-1+\frac{1}{h}} d s d\bar{\zeta} .$$

Let us denote the L_2 Hilbert space

$$\mathcal{E} = L_2(CD, dw_h) ,$$

with the inner product

$$\langle f, g\rangle_h = \int_{CD} \bar{f} g dw_h .$$

Let $\mathcal{H} \subset \mathcal{E}$ denote the Hilbert subspace of holomorphic functions on CD, and let P be the orthogonal projection from $\mathcal{E}$ to $\mathcal{H}$. (The projection P is given by the Bergmann kernel.) Then we can consider the operators on $\mathcal{E}$ to operators on $\mathcal{H}$ given by the commutative diagrams,

$$\begin{array}{ccc} \mathcal{E} & \xrightarrow{\zeta} & \mathcal{E} \\ {\scriptstyle P}\downarrow & & \downarrow{\scriptstyle P} \\ \mathcal{H} & \xrightarrow[z]{} & \mathcal{H} \end{array} \quad \text{and} \quad \begin{array}{ccc} \mathcal{E} & \xrightarrow{\bar{\zeta}} & \mathcal{E} \\ {\scriptstyle P}\downarrow & & \downarrow{\scriptstyle P} \\ \mathcal{H} & \xrightarrow[\bar{z}]{} & \mathcal{H} \end{array} . \qquad \text{(Quantization Map)}$$

Such operators given by multiplication followed by projection are called Toeplitz operators in the mathematics literature.

Theorem 5 [16] *The Toeplitz operators $z, \bar{z}$ defined by the quantization map $z = P\zeta$ and $\bar{z} = P\bar{\zeta}$, satisfy $\bar{z} = z^*$ and provide a representation of the QD algebra.*

In fact, this Toeplitz operator quantization classical space is very reminiscent to the quantization of field operators (on classical space) which is standard in the theory of quantized fields. This quantization relates the Hilbert space of quantum theory $\mathcal{H}$, and field operators on $\mathcal{H}$, to the projection of classical coordinate functions on a "Euclidean" Hilbert space of classical fields. Consider $\mathcal{E} = L_2(\mathcal{S}', d\mu)$, where $d\mu$ is a measure on $\mathcal{S}'(\mathbb{R}^d)$ which satisfies reflection positively. Also let $\mathcal{H} \subset \mathcal{E}$ be the subspace of $\mathcal{E}$ obtained by projecting onto positive times and the reflection positive inner product. The Toeplitz quantization of the coordinate function φ on $\mathcal{E}$ yields the Osterwalder-Schrader quantization of a Euclidean quantum field, namely $\varphi_{\mathrm{quantum}}$ which acts on $\mathcal{H}$. The main difference between this Osterwalder-Schrader construction and the Toeplitz quantization of the classical unit disc described above, is the use of a different inner product on $\mathcal{H}$ from that on $\mathcal{E}$.

$$\begin{array}{ccc} \mathcal{E} & \xrightarrow{\varphi} & \mathcal{E} \\ {\scriptstyle P}\downarrow & & \downarrow{\scriptstyle P} \\ \mathcal{H} & \xrightarrow[\varphi_{\mathrm{quantum}}]{} & \mathcal{H} \end{array}$$

Having remarked that similar quantization maps appear naturally both in the quantization of space and the quantization of fields, we can ask whether quantum fields on quantum space provide an appropriate physical theory? Here lie many open questions, and we may dream of the proper formulation for the next revolution in quantum physics. For 65 years quantum theory has remained central to our thinking, and there is no indication that this will change. It is tempting, however, when seeing the close connection between the ideas of quantum theory and those of NCG, to speculate that NCG will play a key role in the physics of the future.

References

1. Berezin, F.: General Concepts of Quantization. Comm.Math.Phys. **40**, 153-174 (1975)
2. Connes, A.: Noncommutative differential geometry. Publ.Math.IHES **62**, 257-360 (1986)
3. Connes, A.: Entire cyclic cohomology of Banach algebras and characters of Θ-summable Fredholm modules. K-Theory **1**, 519-548 (1988)
4. Connes, A.: On the Chern Character of Θ- summable Fredholm Modules. Comm.Math.Phys. **139**, 171-181 (1991)
5. Connes, A., Lott, J.: Non-Commutative Geometry and Particle Physics. Nuclear Phys. B (1990) and Comm.Math.Phys., to appear
6. Ernst, K., Feng, P., Jaffe, A., Lesniewski, A.: Quantum K-Theory, II. Homotopy Invariance of the Chern Character. J.Funct.Anal. **90**, 355-368 (1990)
7. Jaffe, A., Lesniewski, A.: Supersymmetric Field Theory and Infinite Dimensional Analysis. In: Nonperturbative Quantum Field Theory. Proceedings of the 1987 Cargèse Summer School. G. 't Hooft *et. al.*, Editors, Plenum Press 1988
8. Jaffe, A., Lesniewski, A.: Geometry of Supersymemtry. 1988 Erice Lectures, G. Velo and A. Wightman, Editors, by Plenum Press, New York, (1990)
9. Jaffe, A., Lesniewski, A.: 1990-1991 Harvard University Lectures on 'Quantum Fields, Geometry, and Analysis in Infinite Dimesnions', to appear
10. Jaffe, A., Lesniewski, A., Osterwalder, K.: Quantum K-Theory, I. The Chern Character. Comm.Math.Phys. **118**, 1-14 (1989)
11. Jaffe, A., Lesniewski, A., Osterwalder, K..: On Super-KMS Functionals and Entire Cyclic Cohomology. K-Theory **2**, 675-682 (1989)
12. Jaffe, A., Lesniewski, A., Wisniowski, M.: Deformations of Super-KMS Functional. Comm.Math.Phys. **121**, 527-540 (1989)
13. Jaffe, A., Osterwalder, K.: Ward Identities for Non-Commutative Geometry. Comm.Math.Phys. **132**, 119-130 (1990)
14. Janowsky, S., Weitsman, J.: A Vanishing Theorem for Supersymmetric Quantum Field Theory and Finite Size Effects in Multiphase Cluster Expansion. Comm.Math.Phys., to appear
15. Kastler, D.: Cyclic Cocycles from Graded KMS Functionals. Comm.Math. Phys. **121**, 345-350 (1989)
16. Klimek, S., Lesniewski, A.: Quantum Riemann Surfaces. Comm.Math.Phys., to appear

An Operator Algebraic Framework for the Duality of Quantum Groups

Tetsuya Masuda [1], Yoshiomi Nakagami [2]

[1]Institute of Mathematics, University of Tsukuba
Tsukuba, Ibaraki, 305 Japan
[2]Department of Mathematics, Yokohama City University
Seto, Kanazawa-ku, Yokohama, 236 Japan

Dedicated to Professor Jun Tomiyana on his sixtieth anniversary

The duality for a locally compact group established by Pontrjagin, Tannaka, Krein, Steinspring, Eymard and Tatsuuma is an important theoretical basis for the harmonic analysis. At the formal level of pure algebras, the notion of Hopf algebras is used to deal with the algebraic groups, discrete groups, or their dual objects at the same time. In order to control the infinite dimensional unitary representations, functional analysis is necessarily combined with the algebraic framework of Hopf algebras. This consideration suggests us to introduce the notion of Kac-algebras in the language of the Neumann algebras, in with the above mentioned duality is generalized by Kac [5], Takesaki [9] and Enock and Schwartz [2].

In the recent study of the quantum inverse scattering method, non-commutative and non-cocommutative Hopf algebras were introduced in the articles by Drinfeld [1] and Jimbo [4], as a natural "q-deformation" of the universal enveloping algebra of semisimple Lie algebras and called "quantum groups". Around the same time, Woronowicz [11] also discovered independently a C*-algebra which looks like a non-commutative deformation of the Hopf algebra of complex valued continuous functions on the compact Lie group SU(2). It then turned out that these objects are essentially the same (see [7], for example).

At the present stage we have two kinds of quantized algebras mentioned in the above for the coordinate rings of locally compact groups. The purpose of this paper is to unify these two non-commutative algebras by introducing the deformation automorphism in the framework of the Kac algebra.

Definition 1 The family (M, δ, R, τ, h) with the following conditions is called *Woronowicz algebra*:

(1) M is a von Neumann algebra with the coassociative **comultiplication** $\delta : M \to M \otimes M$ which is a $*$-homomorphism with $(\delta \otimes \mathrm{id}) \circ \delta = (\mathrm{id} \otimes \delta) \circ \delta$.
(2) **Unitary antipode** is the mapping $R : M \to M$ which is a $*$- anti-homomorphism such that $R^2 = \mathrm{id}$ and $(R \otimes R) \circ \delta = \sigma \circ \delta \circ R$, where $\sigma : M \otimes M \to M \otimes M, x \otimes y \mapsto y \otimes x$ is the flip mapping.
(3) **Deformation automorphism** is the continuous one parameter group of $*$-automorphisms $\tau : \mathbb{R} \to \mathrm{Aut}(M)$ such that the mappings δ and R are $\{\tau_t\}$-invariant i.e. $(\tau_t \otimes \tau_t) \circ \delta = \delta \circ \tau_t$ and $\tau_t \circ R = R \circ \tau_t$ for $t \in \mathbb{R}$.
(4) (left invariant) **Haar weight** h is a $\{\tau_t\}$-invariant semifinite faithful normal weight on the von Neumann algebra M such that the following conditions hold:
   (i) **Left invariance:** For any positive $\phi \in M_*$, the weight $(\phi \otimes h) \circ \delta$ coincides with the weight $\phi(1)h$.
   (ii) **Symmetry condition:** For any $x, y, \in \mathfrak{n}_h$, and $\phi \in M_*$ which is entire analytic with respect to the action of the deformation automorphism $\{\tau_t\}$ on the predual M_*, the following equality holds:

$$(\phi \otimes h)((1 \otimes y^*)\delta(x)) = (\phi \circ \tau_{-i/2} \circ R \otimes h)(\delta(y^*)(1 \otimes x)).$$

   (iii) **Commutativity:** The two weights h and $h \circ R$ commute (i.e. the corresponding modular automorphisms commute).

It is worth while noticing that the antipode is formally of the form $R \circ \tau_{-i/2}$ and if the deformation automorphism is trivial in the above definition, our setting is the same as the Kac algebra.

Let $\{(Dh \circ R : Dh)_t : t \in \mathbb{R}\}$ be the Radon-Nikodym cocycle of the weight $h \circ R$ and h. Since $h \circ R$ commutes with h, there exists a positive self-adjoint operator ρ affiliated with M such that $(Dh \circ R : Dh)_t = \rho^{-it}$ for $t \in \mathbb{R}$. Let $\{L_t\}$ be a one parameter family of mappings of M into itself defined by $L_t(x) = \rho^{it}x, t \in \mathbb{R}$ and $x \in M$. Let $\mathfrak{a}_0$ be the set of entire analytic elements in $\mathfrak{n}_h \cap (\mathfrak{n}_h)^*$ with respect to $\{\sigma_t^h\}, \{\sigma_t^{h \circ R}\}, \{\tau_t\}$ and $\{L_t\}$, where $(\mathfrak{n}_h)^* = \{x : x^* \in \mathfrak{n}_h\}$. Let $(M_*)_\tau$ be the set of entire analytic elements in M_* with respect to $\{\tau_t\}$. Then $(M_*)_\tau$ is an involutive algebra with respect to the product $\phi * \psi$ and the involution $\phi^\#$ of ϕ defined by

$$(\phi * \psi)(x) = (\phi \otimes \psi)(\delta(x)) \text{ and } \phi^\#(x) = \overline{\phi(\tau_{i/2}(R(x^*)))},\ x \in M.$$

Let $\{\pi_h, \mathfrak{H}_h, \eta_h\}$ be the GNS construction of $\{M, h\}$. If $x, y \in \mathfrak{a}_0$, then the linear functional $\omega_{\eta_h(x), \eta_h(y)}$ restricted to M is an element in $(M_*)_\tau$, where $\omega_{\xi,\eta}(x) = (x\xi \mid \eta)$ for $x \in M$.

Definition 2 Let $\mathfrak{b}$ be the involutive subalgebra of $(M_*)_\tau$ generated by the set $\{\omega_{\eta_h(x), \eta_h(y)} : x, y \in \mathfrak{a}_0\}$.

Throughout this note the von Neumann algebra M is assumed to be standard: $\{M, \mathfrak{H}\} = \{\pi_h(M), \mathfrak{H}_h\}$.

Let W be the Kac-Takesaki operator, i.e. the operator on $\mathfrak{H} \otimes \mathfrak{H}$ defined by

$$W\eta_{h\otimes h}(x \otimes y) = \eta_{h\otimes h}(\delta(y)(x \otimes 1)),\ x, y \in \mathfrak{n}_h\,.$$

Then W is an isometry on $\mathcal{L}(\mathfrak{H}\otimes\mathfrak{H})$ by the left invariance of the Haar weight. It is easy to see that W belongs to $M \otimes \mathcal{L}(\mathfrak{H})$ and satisfies the pentagonal relation $W_{12}W_{23} = W_{23}W_{13}W_{12}$ by the coassociativity of coproduct, where W_{ij} is the tensor product of W acting on the i-th and the j-th spaces of $\mathfrak{H} \otimes \mathfrak{H} \otimes \mathfrak{H}$ and the identity on the remaining space. The Kac-Takesaki operator is interpreted as the left regular representation of the Woronowicz algebra or the "quantum group".

An element $\phi \in M_*$ is said to be L^2 – *bounded* if there exists a constant $C \geq 0$ such that

$$|\ \phi(y^*)\ | \leq C \parallel \eta_h(y) \parallel \text{ for } y \in \mathfrak{n}_h.$$

We denote by $\widehat{\eta}(\phi)$ the unique element in $\mathfrak{H}$ satisfying $\phi(y^*) = (\widehat{\eta}(\phi) \mid \eta_h(y))$ for $y \in \mathfrak{n}_h$. Then $\widehat{\eta}$ is a linear mapping from the linear space of L^2-bounded elements to a dense subspace of $\mathfrak{H}$. It is shown that if $\phi * \psi$ is L^2-bounded, then $\phi * \psi$ is also L^2-bounded and $\parallel \widehat{\eta}(\phi * \psi) \parallel \leq \parallel \phi \parallel \parallel \widehat{\eta}(\psi) \parallel$. There exists then a unique bounded operator $\widehat{\pi}(\phi)$ on $\mathfrak{H}$ defined by $\widehat{\pi}(\phi)\widehat{\eta}(\psi) = \widehat{\eta}(\phi * \psi)$. Then $\widehat{\pi}$ is a multiplicative linear mapping of M_* to $\mathcal{L}(\mathfrak{H})$. Denote by $\widehat{M}$ the von Neumann algebra generated by the range $\{\widehat{\pi}(\phi) : \phi \in M_*\}$ of $\widehat{\pi}$. Since $\widehat{\pi}(\phi^{\#}) = \widehat{\pi}(\phi)^*$ by the symmetry condition, the mapping $\widehat{\pi}$ is a $*$-homomorphism, interpreted as the Fourier transform of $(M_*)_\tau$ to $\widehat{M}$.

Let S and $\widehat{S}$ denote the closure of conjugate linear operators defined by $S\eta_h(x) = \eta_h(x^*)$ and $\widehat{S}\widehat{\eta}(\phi) = \widehat{\eta}(\phi^{\#})$. By polar decompositions we have two kinds of the modular operators $\Delta, \widehat{\Delta}$ and the modular conjugations $J, \widehat{J}$ with $S = J\Delta^{1/2}$ and $\widehat{S} = \widehat{J}\widehat{\Delta}^{1/2}$. Then we have

Theorem 1

(i) *The Kac-Takesaki operator* W *is a unitary in* $M \otimes \widehat{M}$.
(ii) $W(\widehat{\Delta} \otimes \Delta)W^* = \widehat{\Delta} \otimes \Delta$ *and* $(\widehat{J} \otimes J)W(\widehat{J} \otimes J) = W^*$.

Using the unitarity of the Kac-Takesaki operator, we obtain the following theorem.

Theorem 2

(i) *An element in the subalgebra* $\mathfrak{b}$ *of* $(M_*)_\tau$ *is* L^2*-bounded.*
(ii) *Let* $\widehat{\mathfrak{A}} = \widehat{\eta}(\mathfrak{b})$. *Then* $\widehat{\mathfrak{A}}$ *is a left Hilbert algebra.*

Since the deformation automorphism $\{\tau_t\}$ preserves the Haar weight h, there exists a positive self-adjoint operator H with $H^{it}\eta_h(x) = \eta_h(\tau_t(x))$, $t \in \mathbb{R}, x \in \mathfrak{n}_h$.

Proposition 3

(i) *The four operators Δ, H, ρ and $J\rho J$ are positive self-adjoint and commute to each other.*
(ii) $JHJ = \hat{J}H\hat{J} = H^{-1}$ *and the modular operator $\widehat{\Delta}$ is the closure of $J\rho JH$.*

We define on the von Neumann algebra $\widehat{M}$ the coproduct $\widehat{\delta}$, the unitary antipode $\widehat{R}$, the deformation automorphism $\{\widehat{\tau}_t\}$ and the Haar weight $\widehat{h}$ by formulas $\widehat{\delta}(x) = \widehat{W}(1 \otimes x)\widehat{W}^*$, $\widehat{R}(x) = Jx^*J$, $\widehat{\tau}_t(x) = H^{it}xH^{-it}$ and

$$\widehat{h}(\widehat{\pi}_\ell(\xi)^*\widehat{\pi}_\ell(\xi)) = \begin{cases} (\xi \mid \xi) & \text{if } \xi \in \widehat{\mathfrak{A}}'' \\ \infty & \text{otherwise} \end{cases},$$

where $\widehat{W}$ denotes the unitary $\sigma(W^*)$ on $\mathfrak{H} \otimes \mathfrak{H}$ and $\widehat{\mathfrak{A}}''$ is the achieved left Hilbert algebra for $\widehat{\mathfrak{A}}$. Then $(\widehat{M}, \widehat{\delta}, \widehat{R}, \widehat{\tau}, \widehat{h})$ is a Woronowicz algebra, which is called the *dual* Woronowicz algebra of (M, δ, R, τ, h). In this case, $\widehat{R}(\widehat{\pi}(\phi)) = \widehat{\pi}(\phi \circ R)$ and $\widehat{\tau}_t(\widehat{\pi}(\phi)) = \widehat{\pi}(\phi \circ \tau_{-t})$ for all $\phi \in M_*$. Further, the dual Woronowicz algebra $(\widehat{\widehat{M}}, \widehat{\widehat{\delta}}, \widehat{\widehat{R}}, \widehat{\widehat{\tau}}, \widehat{\widehat{h}})$ of $(\widehat{M}, \widehat{\delta}, \widehat{R}, \widehat{\tau}, \widehat{h})$ is called the *bidual* Woronowicz algebra of (M, δ, R, τ, h).

Let $\widehat{\widehat{\mathfrak{A}}}$ be the left Hilbert algebra obtained in Theorem 2 corresponding to the dual Woronowicz algebra and $\widehat{\widehat{S}} = \widehat{\widehat{J}}\widehat{\widehat{\Delta}}^{1/2}$ the polar decomposition of the closed conjugate linear operator corresponding to $\widehat{S}$.

Lemma 4

(i) *The left Hilbert algebra $\widehat{\widehat{\mathfrak{A}}}$ is a subalgebra of the achieved left Hilbert algebra $\mathfrak{A}_h = \eta_h(\mathfrak{n}_h \cap (\mathfrak{n}_h)^*)$ and $\widehat{\widehat{S}} \subset S$.*

(ii) *The modular operator $\widehat{\widehat{\Delta}}$ commutes with the modular operator Δ.*

Theorem 5 (Duality) *The Woronowicz algebra (M, δ, R, τ, h) is $*$-isomorphic to the bidual Woronowicz algebra $(\widehat{\widehat{M}}, \widehat{\widehat{\delta}}, \widehat{\widehat{R}}, \widehat{\widehat{\tau}}, \widehat{\widehat{h}})$.*

If a Woronowicz algebra is Abelian, then M is isomorphic to a group von Neumann algebra $L^\infty(G)$ for some locally compact group G.

The quantum matrix pseudogroup $\mathrm{SU}_q(2)$, $0 < q < 1$, is $*$-isomorphic to the C*-algebra $\mathfrak{A}$ generated by the operators x and u on $\ell^2(\mathbb{Z}_+) \otimes \ell^2(\mathbb{Z})$ defined by

$$x = \sum_{n \geq 0} (1 - q^{2n+2})^{1/2} e_{n+1,n} \otimes 1, \quad u = \sum_{n \geq 0} q^n e_{n,n} \otimes T$$

and the Haar measure h_0 is given by $h_0(a) = (1-q^2)(\mathrm{Tr}\otimes\mathrm{tr})(u^*ua), a \in \mathfrak{A}$, where T is the shift operator on $\ell^2(\mathbb{Z})$ and $\mathrm{tr}(\sum_{n\in\mathbf{Z}} \lambda_n T^n) = \lambda_0$. Let h'' be the weight on $\mathfrak{A}$ obtained as to be restriction of the right hand side of h_0 to $\mathfrak{A}$. Now we consider the standard representation $\{M, h\}$ of $\{\mathfrak{A}'', h''\}$. The coproduct δ on M is induced from that of $\mathfrak{A}$. The unitary antipode R and the deformation automorphism $\{\tau_t\}$ in M are defined by $R(x) = x^*$, $R(u) = -u$, $\tau_t(x) = x$, $\tau_t(u) = q^{2it}u$. Then (M, δ, R, τ, h) is a Woronowicz algebra.

References

1. Drinfeld, V.G.: Quantum groups. Proc. ICM, Berkeley, Vol.**1**, 798-820 (1986)
2. Enock, M., Schwartz, J.M.: Une dualite dans les algèbre de von Neumann. Bull.Soc.Math.France, Suppl.Mèmoir **44**, 1-144 (1975)
3. Enock, M., Schwartz, J.M.: Sur la structure des algèbres de Kac, II. Proc. London Math.Soc. (3) **41**, 465-480 (1980)
4. Jimbo, M.: A q-difference analogue of U($\mathfrak{g}$) and the Yang-Baxter equation. Lett.Math.Phys. **10**, 63-69 (1985)
5. Kac, G.I.: Generalization of the group principle of duality. Soviet Math.Dokl. **2**, 581-584 (1961)
6. Masuda, T., Mimachi, K., Nakagami, Y., Noumi, M., Ueno, K.: Representation of quantum groups, "Mapping of operator algebras". Proc. Japan-U.S. Joint Seminar, Univ.Pennsylvania, 119-128 (1988), edited by H.Araki and R.Kadison, Birkhäuser, Boston-Basel-Berlin
7. Rosso, M.: Comparison des groupes SU(2) quantiques et de Woronowicz. C.R.Acad.Sci.Paris **304**, 323-326 (1987)
8. Schwartz, J.M.: Sur la structure des algèbres de Kac, I. J.Funct.Anal. **34**, 370-406 (1979)
9. Takesaki, M.: Duality and von Neumann algebras. Lecture Notes in Math. **247**, 665-786 (1972), Springer-Verlag, Berlin-Heidelberg-New York
10. Vainerman, L.I., Kac, G.I.: Non unimodular ring groups and Hopf-von Neumann algebras. Math. USSR Sbornik **23**, 185-214 (1974)
11. Woronowicz, S.L.: Twisted SU(2) group. An example of a non-commutative differential calculus. Publ.RIMS, Kyoto Univ. **23**, 117-181 (1987)
12. Woronowicz, S.L.: Tannaka-Krein duality for compact matrix pseudogroups. Twisted SU(N) group. Invent.Math. **93**, 35-76 (1988)

Section 3

Chaotic Quantum Systems

Organizer: M. Combescure

Distribution of Energy Levels in Quantum Systems with Integrable Classical Counterpart. Rigorous Results.

P.M.Bleher

Raymond and Beverly Sackler Faculty of Exact Sciences
School of Mathematical Sciences, Tel Aviv University
Ramat Aviv 69978, Israel

Let $E_0 \le E_1 \le E_2 \le \dots$ be the energy levels (eigenvalues) of the Schrödinger operator $H = -\frac{1}{2}\Delta + U(q)$ on a closed d-dimensional Riemannian manifold M^d. Here

$$-\Delta = -\frac{1}{\sqrt{g}}\frac{\partial}{\partial q^i}(\sqrt{g}\, g^{ij}\frac{\partial}{\partial g^j}) \tag{1}$$

is the Laplace-Beltrami operator and to ensure the discreteness of the spectrum of H we assume, in the case of a non-compact M^d, that $\lim_{q\to\infty} U(q) = \infty$. For simplicity we assume also that M^d has no boundary. Otherwise it is neccessary to supply H with Dirichlet (or some other) boundary conditions.

The main problem we are interested in is the statistics of the energy levels $\{E_k, k \ge 0\}$ in large spectral intervals. To be more precise we describe, following works [4], [7], two concrete problems. Let $N(E) = \#\{k \mid E_k \le E\}$ be the counting function of the energy levels. By $\#A$ we denote the number of elements in a set A.

1 Distribution of Level Spacings. Let $\Delta E_k = E_k - E_{k-1}, k \ge 0$, be spacings between neighboring energy levels. We are interested in the limit distribution of ΔE_k in the spectral interval $E_0 \le E_k \le E$, when $E \to \infty$. Let us define the distribution function of the spacings,

$$P(x;E) = \frac{\#\{k \mid E_k \le E,\, \Delta E_k \le x\Delta E\}}{N(E)} \quad ,$$

where

$$\Delta E = \frac{E - E_0}{N(E)} \tag{2}$$

is the mean spacing. The problem is: Does the limit

$$P(x) = \lim_{E\to\infty} P(x;E) \tag{3}$$

exist and if so, what is $P(x)$?

2 Distribution of the Number of Energy Levels in Intervals of a Prescribed Length. Let $\lambda > 0$ be a fixed constant. Let us define

$$A_n(E) = \{E_0 \le E' \le E \mid \#\{k \mid E' \le E_k \le E' + \lambda\Delta E\} = n\},\ n \ge 0\,, \tag{4}$$

and

$$\pi(n;E) = (E - E_0)^{-1} \int_{A_n(E)} dE' \quad .$$

One can see that $\pi(n;E)$ is the probability that a spectral interval $E' \le E_k \le E' + \lambda\Delta E$, with E' uniformly distributed in $[E_0, E]$, contains exactly n energy levels. The problem is the existence and the calculation of

$$\lim_{E\to\infty} \pi(n;E) = \pi(n)\,, \quad n \ge 0\,. \tag{5}$$

A general conjecture is the following. Assume that the flow in the cotangent bundle T^*M^d over M^d generated by the Hamilton equations

$$\dot{q} = \frac{\partial H(p,q)}{\partial p} \qquad \dot{p} = -\frac{\partial H(p,q)}{\partial q}$$

with $H(p,q) = \frac{1}{2}p^2 + U(q)$ is a completely integrable system, i.e. there exist d Poisson-commuting independent integrals of the motion $I_1(p,q), \ldots, I_d(p,q)$. Then in 'typical case' the limit distribution $P(x)$ in (3) is exponential, i.e.

$$p(x) \equiv -\frac{dP(x)}{dx} = \exp(-x) \tag{6}$$

and the limit distribution $\pi(n)$ in (5) is Poisson, i.e.

$$\pi(n) = \frac{\lambda^n}{n!}\exp(-\lambda) \quad . \tag{7}$$

It is noteworthy, that the limit distributions (6), (7) are the same as if the energy levels E_k were independent random variables, uniformly distributed in the interval $E_0 \le E_k \le E$. Therefore (6), (7) can be interpreted as the absence of an interaction between energy levels.

A proof of (6) was given in the pioneer work [4] of Berry and Tabor, but it was not completely rigorous. The ideas of [4] were used by Sinai in [7] to prove (7) for the Laplace-Beltrami operator on two-dimensional revolution surfaces. In that case the geodesic flow is integrable because of the Clairaut integral. The strategy of Sinai was the following. Let M^2 be a periodic revolution surface in $\mathbf{R}^3$ which is defined in cylindric coordinates (r, φ, z) by the equation $r = f(z)$, where $f(z) > 0$ is a smooth periodic function, $f(z+a) = f(z)$. Let ds^2 be the Euclidean metric, restricted to M^2. Because of the periodicity of $f(z)$ we may consider z as a cyclic coordinate, so that M^2 is a compact Riemannian manifold diffeomorphic to a torus. For the sake of simplicity we will assume that $f(z)$ has only one maximum and one minimum, both non-degenerate, in a period. One can choose a new cyclic coordinate $s = s(z)$ in such a way that the eigenvalue problem $-\Delta u = Eu$ on M^2 is written as

$$-\frac{\partial^2 u}{\partial s^2} - \frac{\partial^2 u}{\partial \varphi^2} = Ef^2(s)u \quad . \tag{8}$$

We look for u in the form $u(s,\varphi) = v(s)e^{in\varphi}$, and for $v(s)$ we get the equation

$$-v''(s) + \left(Ef^2(s) - n^2\right) v(s) = 0 \tag{9}$$

with periodic boundary condition. The integer n is a parameter of the equation. For every n we have, obviously, a sequence of eigenvalues E_{mn}, $m \geq 0$, of (9).

Now the problem is decomposed into two stages. At the first stage we solve (9) in the quasi-classical approximation (QCA) and at the second one we deal with a geometric probability problem. The main difficulty of the first stage is to obtain uniform estimates of the remainder term of the QCA and to show that almost all eigenvalues of (9) can be found in the QCA. This problem was solved in [1]. It is important to stress here an aspect of multiplicity of the eigenvalues. Let us note first that actually every eigenvalue E_{mn} is twice degenerate, when $n \neq 0$, since $E_{m,-n} = E_{mn}$. Besides, an analysis of the QCA shows that for every $n \geq 0$ there are two series of eigenvalues,

$$\Sigma_n^{(1)} = \{E_{mn} \mid E_{mn} f_{min}^2 - n^2 > 0\}, \qquad f_{min} = \min_s f(s),$$
$$\Sigma_n^{(2)} = \{E_{mn} \mid E_{mn} f_{min}^2 - n^2 < 0\}\,,$$

such that the eigenvalues of the first series are asymptotically twice degenerate, namely for all $N > 0$,

$$\mid E_{2m'-1,n} - E_{2m',n} \mid = O\left((m'^2 + n^2)^{-N}\right) \quad ,$$

while the eigenvalues of the second series are strongly separated each from other. The Bohr-Sommerfeld quantization condition is written as

$$\frac{1}{\pi}\int_0^{s(a)} \sqrt{E_{mn} f^2(s) - n^2}\, ds = 2m', \qquad m = 2m'-1,\ 2m'$$

for the first series of eigenvalues and as

$$\frac{1}{\pi}\int_{s1}^{s2} \sqrt{E_{mn} f^2(s) - n^2}\, ds = m + \frac{1}{2} \quad ,$$

where s_1, s_2 are the turning points, $E_{mn}f^2(s_1) - n^2 = E_{mn}f^2(s_2) - n^2 = 0$, for the second one. Because of different asymptotic multiplicity of eigenvalues in the two series, it is neccessary to establish the Poisson law (7) for each of them separately. With the help of quasi-classical quantization conditions this problem is reduced to a geometric probability problem. Simplifying it somewhat, we can formulate it in the following way (see [7]; for a full formulation see [1]).

Let γ be a curve in the plane which is defined in polar coordinates (r, α) by the equation $r = G(\alpha), \alpha_0 \leq \alpha \leq \alpha_1$, where $G(\alpha)$ is a positive continuous function. Let γ_R with $R > 0$ be the curve which is defined by the equation $r = RG(\alpha)$, $\alpha_0 \leq \alpha \leq \alpha_1$, so that γ_R is just the homothety of γ with the coefficient R. Let Π_R be the strip in the sector $\alpha_0 \leq \alpha \leq \alpha_1$ between γ_R and $\gamma_{R+\frac{\lambda}{2R}}$, where $\lambda > 0$ is the same as in (4). Let $\pi_k(E)$ be the probability that Π_R contains exactly k integer points (m, n) assuming that R^2 is uniformly distributed in $[0, E]$. The problem is to prove that

$$\lim_{k \to \infty} \pi_k(E) = \frac{\lambda^k}{k!} \exp(-\lambda) \quad . \tag{10}$$

A 'stochastic' version of this problem was solved by Sinai and Major (see [8], [5]). They introduced a class of probability distributions on the space $\text{Lip}^+([\alpha_0, \alpha_1])$ of positive continuous Lipschitz functions on $[\alpha_0, \alpha_1]$, for which they proved the following theorem: For every probability distribution P from this class, (10) is true for P-almost every function $G(\alpha) \in \text{Lip}^+([\alpha_0, \alpha_1])$. In particular, it proves an existence theorem of curves $\gamma = \{r = G(\alpha)\}$, for which (10) is true. In this respect it resembles the famous Cantor's proof of the existence of transcendental numbers. The situation here is however even more surprising: No example of γ for which (10) is true is known now. On the contrary, there exist examples for which (10) is known to be false, e.g. ellipses with rational principal axes (it corresponds to a plain torus with rational periods as M^2) and some others.

The Sinai - Major result combined with the estimates of QCA in [1] supports strongly the following 'stochastic' version of the Poisson conjecture (7): For almost every revolution surface with respect to a probability distribution P, $\lim_{E \to \infty} \pi(n, E) = \pi(n)$ exists and is a mixture of Poisson distributions with multiplicities 2 and 4. This problem remains open.

Let us turn now to linear quantum systems. In that case the Schrödinger operator is written as

$$H = \frac{1}{2} \sum_{j=1}^{d} \left(-\frac{\partial^2}{\partial q_j^2} + \omega_j^2 q_j^2 \right)$$

and its energy levels are $E_m = E_0 + (m, \omega)$, where $m = (m_1, \ldots, m_d)$, $m_1, \ldots, m_d \geq 0$, $\omega = (\omega_1, \ldots, \omega_d)$, $(m, \omega) = m_1\omega_1 + \ldots + m_d\omega_d$, $E_0 = 1/2(\omega_1 + \ldots + \omega_d)$, so we are dealing here with the statistics of numbers of the form (m,ω). Let $d = 2$. Since $m_1\omega_1 + m_2\omega_2 = \omega_1(m_1 + m_2\alpha)$, $\alpha = \omega_2/\omega_1$, it is enough to consider the numbers $\{\lambda_{mn} = m+n\alpha; m,n \geq 0\}$. Let $\Delta\lambda_{mn} = \lambda_{mn} - \lambda_{m'n'}$ be the spacing between λ_{mn} and the previous level, and $P_\alpha(x;E)$ be the distribution of $\Delta\lambda_{mn}/\Delta\lambda$ in the spectral interval $0 \leq \lambda_{mn} \leq E$. Here $\Delta\lambda = E/n(E)$ is the mean spacing. Again the problem is: Does $\lim_{E\to\infty} P_\alpha(x;E) = P_\alpha(x)$ exist and what is $P_\alpha(x)$? This problem was studied in works [4], [6], [2,3] and others. We present here the main result of [2,3].

Let $0 < \alpha < 1$ be an irrational number, represented by a continued fraction $[a_1, a_2, \ldots]$. Let $\frac{p_k}{q_k} = [a_1, a_2, \ldots, a_k], k \geq 1$, be the approximants of α. We are interested in

$$\lim_{k\to\infty} P_\alpha(x;p_k) = P_\alpha(x)$$

and we formulate two results, negative and positive.

Theorem 1 (negative result) *For almost every α, $\lim_{k\to\infty} P_\alpha(x;p_k)$ does not exist.*

Theorem 2 (positive result) *For almost every α, $\lim_{k\to\infty} \frac{1}{k} \sum_{j=1}^{k} P_\alpha(x;p_j) = P(x)$ exists and does not depend on α.*

Actually in [2,3] a more detailed description of $p_\alpha(x;E)$ is given, but we have no place here to discuss it.

References

1. Bleher, P.M.: Quasi-classical expansion and the problem of quantum chaos. Preprint CARR Rept. in Math.Phys., 1990, No 9/90 (to appear in Lecture Notes in Mathematics)
2. Bleher, P.M.: J.Statist.Phys. **61**, 869-876 (1990)
3. Bleher, P.M.: J.Statist.Phys. **63**, 261-283 (1991)
4. Berry, M.V., Tabor, M.: Proc.Roy.Soc.London Ser. A **356**, 375-394 (1977)
5. Major, P.: Poisson law for the number of lattice points in a random strip with finite area. Preprint Mathematical Institute of the Hungarian Academy of Sciences, 1991
6. Pandey, A., Bohigas, O., Giannoni, M.J.: J.Phys. A: Math.Gen.**22**, 4083-4088 (1989)
7. Sinai, Ya.G.: Mathematical problems in the theory of quantum chaos. Distinguished Raimond and Beverly Sackler Lectures. Tel Aviv University, 1990 (to appear in Lecture Notes in Mathematics)
8. Sinai, Ya.G.: Poisson distribution in a geometrical problem. Advances in Soviet Mathematics, Publications of AMS (1991)

Entropy-Invariants of Dynamical Systems and Perturbations of Operators

Dan Voiculescu*

Department of Mathematics, University of California,
Berkeley, California 94720

Abstract: In [12] we proved a sharp extension to the case of commuting n-tuples, of the well-known invariance of absolutely continuous spectra in trace-class scattering theory. Our proof relied on invariants $k_{\mathcal{J}}(\tau)$ ($\mathcal{J}$ a normed ideal, τ an n-tuple of operators), playing the role of a "size $\mathcal{J}$"-dimensional measure of τ.

We recently found the entropy of dynamical systems is related to $k_{\mathcal{J}}(\tau)$ with $\mathcal{J} = \mathcal{C}_\infty^-$ the Macaev ideal.

This report is about $k_{\mathcal{J}}(\tau)$ and its use in constructing entropy-like invariants. We also explain the connection between $k_1(\tau)$ and the entropy of Bogoliubov automorphisms ([11]).

1 The Normed Ideals $\mathcal{C}_p^-$

General normed ideals in the ring of bounded operators $B(\mathcal{H})$ on the separable complex Hilbert space $\mathcal{H}$ will be denoted by $(\mathcal{J}, |\,|_{\mathcal{J}})$ (see [7],[10]). $(\mathcal{C}_p, |\,|_p)$ is the Schatten-von Neumann p-class.

The ideal $(\mathcal{C}_p^-, |\,|_p^-)$ is defined by the norm $|T|_p^- = \sum_j \lambda_j j^{-1+1/p}$ where $\lambda_1 \geq \lambda_2 \geq \ldots$ are the eigenvalues of $(T^*T)^{\frac{1}{2}}$. We have $\mathcal{C}_1^- = \mathcal{C}_1$ while for $p > 1$, $\mathcal{C}_p^- \subset \mathcal{C}_p$ without being equal. Also $\mathcal{C}_r \subset \mathcal{C}_s^-$ if $r < s$. $\mathcal{C}_p^-$ corresponds to the $(p,1)$ Lorentz norm. The ideal $\mathcal{C}_\infty^-$ is also called the Macaev ideal.

The dual of $(\mathcal{J}, |\,|_{\mathcal{J}})$ is the normed ideal $(\mathcal{J}^d, |\,|_{\mathcal{J}^d})$ (the duality given by $\mathrm{Tr}XY$). Then $(\mathcal{C}_p^-)^d = \mathcal{C}_q^+$ $(p^{-1} + q^{-1} = 1)$ an ideal slightly larger than $\mathcal{C}_q$.

The finite rank operators are dense in $\mathcal{C}_p$ and $\mathcal{C}_p^-$, but not in $\mathcal{C}_q^+$.

*Supported in part by National Science Foundation grant DMS 89-12362.

2 The Invariant $k_{\mathcal{J}}(\tau)$

Throughout we assume the finite-rank operators are dense in $\mathcal{J}$.

If $\tau = (T_1, \ldots T_n)$ with $T_k \in B(\mathcal{H})$, $(1 \leq k \leq n)$ we define

$$k_{\mathcal{J}}(\tau) = \liminf_{a \in R_1^+(\mathcal{H})} \|[A, \tau]\|_{\mathcal{J}} ,$$

where $R_1^+(\mathcal{H}) = \{A \in B(\mathcal{H}) | 0 \leq A \leq I\}$, the lim inf being w.r.t. the natural order on $R_1^+(\mathcal{H})$ and we write $\|[A,\tau]\|_{\mathcal{J}}$ for $\max\{\|[A, T_k]\|_{\mathcal{J}} | 1 \geq k \geq n\}$.

If $\mathcal{I} \supset \mathcal{J}$ then $k_{\mathcal{I}}(\tau) \geq k_{\mathcal{J}}(\tau)$ and if $\tau \equiv \tau'$ (mod$\mathcal{J}$) then $k_{\mathcal{J}}(\tau) = k_{\mathcal{J}}(\tau')$.

One evaluates $k_{\mathcal{J}}(\tau)$ using the following facts ([12],[13]).

Fact 1: If $A_m \in \mathcal{R}_1^+(\mathcal{H})$ and $A_m \overset{w}{\rightarrow} I$ as $m \to \infty$ then $\liminf_{m \to \infty} \|[A_m, \tau]\|_{\mathcal{J}} \geq k_{\mathcal{J}}(\tau)$.

Fact 2: Let τ be an n-tuple of self-adjoint operators. We have $0 < C < k_{\mathcal{J}}(\tau)$ iff there are $X_m = X_m^* \in \mathcal{J}^d$ $(1 \leq m \leq n)$ so that $\sum |X_m|_{\mathcal{J}^d} \leq 1, Y = \sum i[T_m, X_m] \in (B(\mathcal{H}))^+ + \mathcal{C}_1$ and Tr$Y > C > 0$. ($(B(\mathcal{H}))^+$ denotes the positive part of $B(\mathcal{H})$.)

If $\mathcal{J} = \mathcal{C}_p^-$ or $\mathcal{C}_p$ we denote $k_{\mathcal{J}}(\tau)$ by $k_p^-(\tau)$ and $k_p(\tau)$ respectively.

If $1 < p \leq \infty$, $k_p(\tau)$ takes only values 0 and ∞ ([12]).

For sharp results one uses $k_p^-(\tau)$ which is a kind of p-dimensional measure of τ.

Example (a) If τ is an n-tuple of commuting self-adjoint operators, then

$$(k_n^-(\tau))^n = \gamma_n \int_{\mathbb{R}^n} m(s) d\lambda(s) \quad ,$$

where $m(s)$ is the multiplicity function of the Lebesgue absolutely continuous part of τ, $d\lambda$ is Lebesgue measure and γ_n a universal constant ([12]).

If τ_Ω is multiplication by the coordinate functions in $L^2(\Omega, d\lambda)$, $(\Omega \subset \mathbb{R}^n)$ then $(k_n^-(\tau_\Omega))^n = \gamma_n \lambda(\Omega)$, which is why k_n^- is a kind of n-dimensional measure.

Example (b) Let $1 < p < n$ (p not necessarily an integer) and μ a compactly supported probability measure so that

$$\mu(B(x,r)) \leq C r^p \quad .$$

This implies $k_r^-(\tau) = \infty$ if $r < p$ and $k_s^-(\tau) = 0$ if $s > p$, where τ is multiplication by the coordinate functions in $L^2(\mu)$.

Connes' dimension $d(\mathcal{A})$ for a set of operators $\mathcal{A}$ ([2]) is the inf over $p \in [1, \infty]$ so that there are unbounded operators $D \geq 0$ with $[D,T]$ bounded for all $T \in \mathcal{A}$ and $(I + D)^{-1} \in \mathcal{C}_p$. In [13] we showed $k_p(\tau) > 0$ for some $\tau \subset \mathcal{A}$, implies $p \leq d(\mathcal{A})$ (this is like positive p-measure implies dimension

$\geq p$). A.Connes refined this to a sharp upper estimate of $k_p^-(\tau)$, if there is D as above, with $\mathcal{C}_p$ replaced by $\mathcal{C}_p^+$ ([3]).

3 The Invariant $k_\infty^-(\tau)$

If $p = \infty$ the corresponding invariant $k_\infty^-(\tau)$ has remarkable properties:

(1) For n-tuples τ, $k_\infty^-(\tau) \leq 2\|\tau\| \log(2n+1)$. Unlike other $k_{\mathcal{J}}$, k_∞^- is always finite.
(2) If $\mathcal{J} \supset \mathcal{C}_\infty^-, \mathcal{J} \neq \mathcal{C}_\infty^-$ then $k_{\mathcal{J}}(\tau) = 0$.
(3) $k_\infty^-(\tau \otimes I_{\mathcal{H}'}) = k_\infty^-(\tau)$ (k_∞^- does not depend on multiplicity).
(4) If $S_1, \ldots, S_n$ are isometries with orthogonal ranges, then $\log n \leq k_\infty^-(S_1, \ldots, S_n) \leq \log(2n+1)$.

4 The Perturbation-entropy $H_P(\Omega, M)$

Let $M \subset B(\mathcal{H})$ be a von Neumann algebra and let $\Omega \subset B(\mathcal{H})$ be unitaries such that $u \in \Omega \Rightarrow uMu^* = M$. In [14] we defined

$$H_P(\Omega, M) = \sup\{k_\infty^-(\tau \cup \omega | \tau \subset M, \dim C^*(\tau) < \infty, \omega \subset \Omega, \omega, \tau \, finite\} \quad ,$$

where $C^*(\tau)$ is the C^*-algebra generated by τ.

If α is a set of automorphisms of a W^*-algebra M there is a unique set of unitaries Ω on the Hilbert space of the standard form implementing α and preserving a certain cone (see [9]). We define $H_P(\alpha, M)$ by $H_P(\alpha, M) = H_P(\Omega, M)$.

Let T be an ergodic measure-preserving automorphism of a finite measure-space X and let α_T be the corresponding automorphism of $L^\infty(X)$.

Theorem [14] *Let $h(T)$ be the entropy of T. Then*

(a) $h(T) \leq 2H_P(\alpha_T, L^\infty(X)) \leq 12h(T)$.
(b) If T is Bernoulli then $\gamma h(T) = H_P(\alpha_T, L^\infty(X))$ for some universal constant γ.

We conjecture (b) holds in general. Note that H_P is defined without averaging over a group.

Without a Shannon-McMillan-Breiman-type theorem for automorphisms of von Neumann algebras, an analogue of the above theorem for the entropies in [3],[4] seems out of reach.

H_P is well defined for automorphisms with quasi-invariant measure and depends only on the class of the measure. However, we don't know whether H_P can be finite and $\neq 0$ in absence of an invariant measure.

5 Smooth Perturbation-Entropy

Let $\mathcal{A} \subset B(\mathcal{H})$ be some algebra, Ω a set of unitaries normalizing $\mathcal{A}$ and $\mathcal{J}$ a normed ideal. Then

$$H_P^{\mathcal{J}}(\Omega|\mathcal{A}) = \sup\{k_{\mathcal{J}}(\omega \cup \tau) | \omega \subset \Omega, \tau \subset \mathcal{A} \text{ finite sets}\}$$

is called the smooth $\mathcal{J}$-perturbation entropy ([15]) of $(\Omega, \mathcal{A})$. It is ∞ unless $k_{\mathcal{J}}(\tau) = 0$ for all finite subsets $\tau \subset \mathcal{A}$.

The typical example is $\mathcal{A} = C^\infty(M)$, M a compact C^∞-manifold and Ω the unitaries induced in $L^2(M, \mu)$ by a set of diffeomorphisms of M, where μ is a smooth measure (with density > 0 in any coordinate patch). Then the smooth $\mathcal{J}$-entropy does not depend on the choice of μ, i.e. it is an invariant of the family of diffeomorphisms. Note also that we should assume $\mathcal{J} \subset \mathcal{C}_n^-$, $\mathcal{J} \neq \mathcal{C}_n^-$, where $n = \dim M$.

Very little is known at present about the case $\mathcal{J} = \mathcal{C}_p^-$, $p < \infty$ which seems related to "polynomial growth rates".

Example ([15]) Let A_θ be the rotation C^*-algebra generated by unitaries u, v satisfying $uv = e^{i\theta} vu$. Let $\mathcal{A}$ be the algebra of C^*-elements w.r.t. the natural $\mathbb{T}^2$-action. Let α be the cat-map automorphism studied in [1]. We have shown [15] that the smooth entropy ($\mathcal{J} = \mathcal{C}_\infty^-$) for the unitary implemented by α in the representation of the canonical trace-state of A_θ is independent of θ and of the same order as the usual entropy for the automorphism of $\mathbb{T}^2$.

6 Bogoliubov Automorphisms and k_1

Let $U, A \in B(\mathcal{H})$, U unitary, $0 \leq A \leq I$ and $[U, A] = 0$. Let α_U be the Bogoliubov automorphism of the CAR-algebra $\mathcal{A}(\mathcal{H})$ and ω_A the quasifree state. In joint work with E.Störmer ([11]) we showed that the Connes-Narnhofer-Thirring entropy, assuming A has pure point spectrum, is

$$h(\alpha_U) = \frac{1}{2\pi} \int_0^{2\pi} \mathrm{Tr}(\eta(A(\theta) + \eta(I - A(\theta)))d\theta$$

($A(\theta)$ is the disintegration of A w.r.t. to $e^{i\theta}$ in the spectrum of U and $\eta(x) = -x \log x$.) This formula agrees with the entropy-density of a related statistical mechanics model [8]).

If $A = \frac{1}{2}I$, ω_A is the trace-state and $H(\alpha_U)$ is proportional to $k_1(U)$. Thus $U \to \alpha_U$ relates $\mathcal{C}_1$-properties of U to entropy.

References

1. Benatti, F., Narnhofer, H., Sewell, G.L.: Preprint
2. Connes, A.: Ergodic Theory Dynamical Systems **9**, 207-220 (1989)
3. Connes, A.: Nuclear Phys. B (Prac. Suppl.) **5B**, 65-70 (1988)
4. Connes, A., Narnhofer, H., Thirring, W.: Comm.Math.Phys. **112**, 691-719 (1987)
5. Connes, A., Störmer, E.: Acta Math. **134**, 289-306 (1975)
6. David, G., Voiculescu, D.: J.Funct.Anal. **94**,no.1, 14-26 (1990)
7. Gohberg, I.C., Krein, M.G.: "Introduction to the theory of non-self-adjoint operators", Moscow, 1965
8. Fannes, M.: Comm.Math.Phys. **31**, 279-290 (1973)
9. Haagerup, U.: Math.Scand. **37**, 271-283 (1975)
10. Simon, B.: *London Mathematical Society Lecture Notes 35*, Cambridge University Press
11. Störmer, E., Voiculescu, D.: Comm.Math.Phys. **133**, 521-542 (1990)
12. Voiculescu, D.: J.Operator Theory **1**, 3-37 (1979);
    Voiculescu, D.: J.Operator Theory **5**, 77-100 (1981)
13. Voiculescu, D.: J.Funct.Anal. **91**, 1-36 (1990)
14. Voiculescu, D.: Entropy of dynamical systems and perturbations of operators, I, II. Preprints IHES and Berkeley (1990)
15. Voiculescu, D.: In preparation

Chaotic Motion in Coulombic Potentials

Markus Klein, Andreas Knauf

Technische Universität, Fachbereich 3 - Mathematik, MA 7–2,
Strasse des 17. Juni 135, D-1000 Berlin 12

1 Introduction

The classical motion of a particle in an attracting Coulomb $(-1/r)$ potential is described by conic sections, whereas the equations of motion in the field of two fixed centres have been solved by Euler and Jacobi.

It is natural to study the classical and quantum dynamics of a particle in the field of n fixed centres, or in a periodic potential with Coulombic singularities. This question has been adressed in [1]-[4].

So we start by considering the classical motion of a particle in the plane whose Hamiltonian has the form

$$\hat{H}(\boldsymbol{q},\boldsymbol{p}) = \tfrac{1}{2}\boldsymbol{p}^2 + V(\boldsymbol{q})\,. \tag{1}$$

In the case of n centres,

$$V(\boldsymbol{q}) = \sum_{l=1}^{n} \frac{-Z_l}{|\boldsymbol{q}-\boldsymbol{s}_l|} + W(\boldsymbol{q})\,, \qquad Z_l > 0\,, \tag{2}$$

where $W : \mathbb{R}^2 \to \mathbb{R}$ is a smooth potential, and we assume that, for $|\boldsymbol{q}|$ large, V (and some of its derivatives) has a fall-off like a Coulomb potential: $V(\boldsymbol{q}) \sim -Z_\infty/|\boldsymbol{q}|$. In the periodic case, we have

$$V(\boldsymbol{q}+\boldsymbol{l}_1) = V(\boldsymbol{q}+\boldsymbol{l}_2) = V(\boldsymbol{q})\,,$$

where $\boldsymbol{l}_1, \boldsymbol{l}_2$ is a basis of the lattice, and V has Coulombic singularities at the points $\boldsymbol{s}_1, \ldots, \boldsymbol{s}_n$ in a unit cell.

Figure 1 shows trajectories of positive energy particles in the configuration plane.

The corresponding quantum problem consists in the analysis of the quadratic form given by

$$H_\hbar := -\frac{\hbar^2}{2}\Delta + V(\boldsymbol{q}) \qquad \text{on } L^2(\mathbb{R}^2)\,.$$

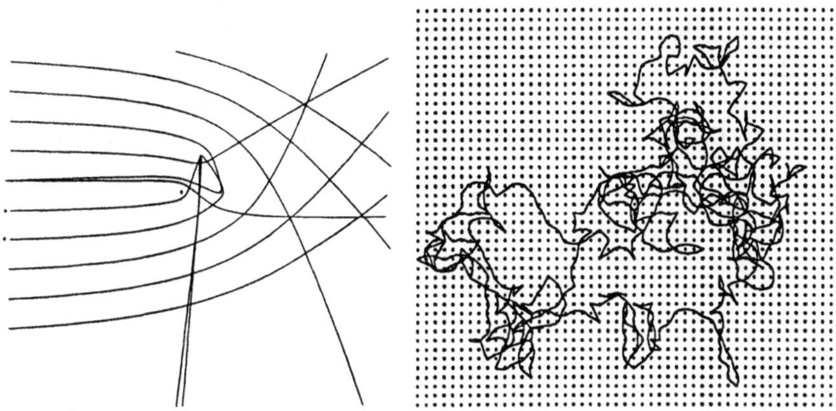

a b

Fig. 1. a) Scattering trajectories for fixed initial angle and energy in the field of $n = 3$ centers. b) Motion in the periodic potential

2 Regularization

$\hat{H}$ is defined on the incomplete phase space $T^*\hat{M}$, where $\hat{M}$ equals the configuration space $\mathbb{R}^2$ minus the positions $\boldsymbol{s}_l$ (resp. $\boldsymbol{s}_l + k_1\boldsymbol{l}_1 + k_2\boldsymbol{l}_2$) of the Coulombic singularities.

We show first the existence of a *smooth* extension (P, ω, H) of the Hamiltonian system $(T^*\hat{M}, d\boldsymbol{q} \wedge d\boldsymbol{p}, \hat{H})$, so that the smooth Hamiltonian flow $\Phi^t : P \to P$ on the four-dimensional phase space P is complete [3]. So the Coulomb singularity leads to a smooth knot in phase space.

To gain a better understanding of the non-trivial topology induced by the singularities, one applies a second regularization scheme which globalizes the method of Levi-Civita. [1],[3]

The geodesics in the *Jacobi metric* $(1 - V(\boldsymbol{q})/E)\delta_{ik}$ coincide with the energy E solution curves for (1), if $E > V$. Using the complex coordinate q in the configuration plane, we define a branched covering surface $\mathbf{M}$ of that plane, which in the case of n centres is given by

$$\mathbf{M} := \left\{ (q, Q) \in \mathbb{C} \times \mathbb{C} \,\middle|\, Q^2 = \prod_{l=1}^{n}(q - s_l) \right\}\,. \tag{3}$$

In the case of a periodic potential, one compactifies the problem by considering the configuration torus $\mathbb{T}^2 := \mathbb{R}^2/(\boldsymbol{l}_1\mathbb{Z}\oplus\boldsymbol{l}_2\mathbb{Z})$ and defining a compact branched covering surface of $\mathbb{T}^2$ using elliptic functions. For energies $E > V$ one finds a unique smooth completion $\mathbf{g}_E$ of the lifted Jacobi metric, so that $(\mathbf{M}, \mathbf{g}_E)$ is a smooth complete Riemann surface (see Figure 2).

In many cases (say, for the potential (3) with $W \equiv 0$) the Gaussian curvature of that surface is strictly negative, so that the classical motion is unstable.

The corresponding quantum problem is regularized by a similar method, leading to the study of the Laplace-Beltrami operator on $(\mathbf{M}, \mathbf{g}_E)$ [2].

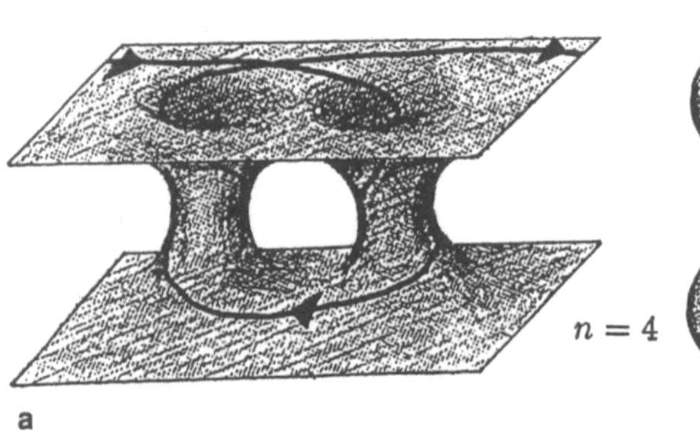

Fig. 2. The surfaces $(\mathbf{M}, \mathbf{g}_E)$ for a) $n = 4$ centres, b) the periodic case

3 The Periodic Case

It is well-known that the geodesic motion on a closed surface of negative curvature is mixing. Moreover, the Central Limit Theorem applies to a large class of phase space functions.

As a consequence, the high-energy motion in the periodic potential is diffusive in the sense that

$$\lim_{t\to\infty} \frac{\langle \boldsymbol{q}^2(t)\rangle}{t} = D > 0 ,$$

if one averages $\boldsymbol{q}^2(t)$ over a probability measure of initial conditions $(\boldsymbol{q}_0, \boldsymbol{p}_0)$ which is absolutely continuous w.r.t. Liouville measure. More precisely, the density of $\boldsymbol{q}(t)/\sqrt{t}$ becomes asymptotically Gaussian [1].

On the other hand, it is clear from Bloch's Theorem that the quantum motion is ballistic, *i.e.*

$$\lim_{t\to\infty} \frac{\langle q^2(t)\rangle}{t^2} = v^2 > 0 \, .$$

Nevertheless, the correspondence principle of quantum mechanics is true in the sense that the group velocity v goes to zero in the semiclassical limit $\hbar \searrow 0$. This is a consequence of the fact (first observed by Shnirelman) that observables of typical eigenfunctions converge to their classical mean values if the classical motion is ergodic [2].

So quantum chaos slows down the high-energy motion in the crystal.

4 Scattering by n Centres

To understand scattering, it turns out to be important to first analyse the *bounded* orbits of positive energies. For *all* potentials of the form (3) and large energies E, there are no bounded orbits for $n = 1$ centre, one closed orbit for $n = 2$, and for $n \geq 3$ a Cantor set of bounded orbits which is of measure zero. All these orbits are unstable and can be fully analysed using symbolic dynamics.

We obtain upper and lower bounds for the following quantities:

1. The *topological entropy* of the motion for energy E is bounded by $C \cdot \ln|2n-3|\sqrt{E}$.
2. For $n \geq 3$ the *Hausdorff dimension* of the set of bounded orbits is bounded by $1 + C/\ln E$.
3. *Escape rate:* The measure of the scattering orbits with time delay larger than T goes like $\exp(-C\sqrt{E}\ln E \cdot T)$.

We classify the scattering orbits of given energy, incoming and outgoing direction by the elements of a semidirect product $\pi_1(\mathbf{M})\star\mathbb{Z}_2$, the fundamental group $\pi_1(\mathbf{M})$ being a free group on $n-1$ generators. The classical cross section is shown to be smooth and to have the same forward asymptotics as the Rutherford cross section [3].

The quantum problem can be separated for the case of two centres. Then semiclassically the resonances in the complex energy plane come in strings and are indexed by two integers, the coefficients being determined by the Jacobi distance between the centres and the Lyapunov exponent of the closed orbit.

For $n \geq 3$, we expect that methods developed by Helffer and Sjöstrand lead to an upper bound

$$N(\hbar) \leq C\hbar^{-\frac{1}{2}(1+\dim(b_E))}$$

for the number of resonances in a rectangle $[E_1, E_2] \times [-c\hbar, 0]$. Here $\dim(b_E)$ denotes the (fractal) dimension of the bounded orbits of energy E [4].

References

1. Knauf, A.: Ergodic and Topological Properties of Coulombic Periodic Potentials. Comm.Math.Phys. **110**, 89-112 (1987)
2. Knauf, A.: Coulombic Periodic Potentials: The Quantum Case. Ann.Physics **191**, 205-240 (1989)
3. Klein, M., Knauf, A.: Classical Planar Scattering by Coulombic Potentials. Preprint, TU Berlin, 1991
4. Klein, M., Knauf, A.: In preparation

Generalized Floquet Operator for Quasiperiodically Driven Quantum Systems

H.R. Jauslin, J.L.Lebowitz

Rutgers University, New Brunswick, USA

The behaviour of systems with time dependent perturbations can be qualitatively different from the one of isolated quantum systems, e.g. in a bounded spatial domain, or confined by a potential going rapidly to infinity. Since the energy levels of such a system are discrete, the wave function and thus all expectation values and correlation functions will of necessity be almost period in time [1]. This rules out any good ergodic properties or other forms of classical chaoticity, which require decay of correlation functions. Of course as the size of the system becomes large and the spacing between levels becomes very small, the ensemble averages of same classes of quantum variables may exhibit good decay properties over long time periods. These can become, in suitable limits, indistinguishable from those given by chaotic dynamics.
The situation is different for systems described by time dependent Hamiltonians where the relevant spectrum is the one of the quasi-energy operator, which can be continuous, leading to decaying correlations and even strong ergodic properties.
The case of periodic time dependence has been extensively studied [2]. There is strong evidence that such systems usually have pure point quasi-energy spectrum. In models like the periodically kicked rotator this property manifests as a "quantum limitation of diffusion". The case of non-periodic perturbations is quite different. Existing studies indicate that in this case the quantum limitation of diffusion is weaker or non-present. This has been observed numerically for the quasi-periodically kicked rotator, with two or three incommensurate frequencies [3],[4]. In [5] it was shown that for a class of two-level systems with an almost-periodic force generated by the Fibonacci sequence the response generically is not almost-periodic. Other examples of two-level systems were studied numerically in [6]-[8]. Kicked systems modulated by substitution sequences (e.g. Morse-Thue) of different degrees of randomness were studied in [9]. In these examples the correlation functions have singularly continuous spectrum. For the randomly kicked rotator, it has been shown that the energy grows unbounded [10]. This is in

fact the case for a general class of quantum systems, in which the time-dependence is given by a Markov process [11],[12].
Here we will discuss some results for the case of quasiperiod time dependence, concerning the relation between the quasi-energy spectrum and stabillity formulated in terms of the asymptotic growth of the energy.

1 Quasi-Energy Operator

We consider the time dependent Hamiltonian $H = H_0(x) + V(x; \underline{\theta}(t))$ on a Hilbert space $\mathcal{H}$, where x are the variables of the system and the force $\underline{\theta}(t) = g_t\underline{\theta} \in \Omega$ is described by an invertible flow with an ergodic invariant measure μ. Let $U(t, s; \underline{\theta})$ be the corresponding propagator. The generalized quasi-energy operator (QEO) K is defined [13] on the enlarged space $\mathcal{K} = \mathcal{H} \otimes L_2(\Omega, d\mu)$ by

$$\left[e^{-iKt}\Psi\right](\underline{\theta}) \equiv U(0, -t; \underline{\theta})\mathcal{T}_{-t}\Psi(\underline{\theta}) \equiv \mathcal{T}_{-t}U(t, 0; \underline{\theta})\Psi(\underline{\theta}),$$

where $[\mathcal{T}_{-t}\Psi](\underline{\theta}) = \Psi(g_{-t}\underline{\theta})$. K acts as $[K\Psi](\underline{\theta}) = -i\frac{d}{dt}\Psi(g_t\underline{\theta})_{|t=0} + H(\underline{\theta})\Psi$.

Example 1 Periodic force: $\Omega = S^1$ is a circle, $g_t\theta = \theta + \omega t, d\mu = d\theta$, and $K = -i\frac{\partial}{\partial\theta} + H(\theta)$.

Example 2 Quasi-periodic force with two frequencies ω_1, ω_2 : Ω is a torus $S^1_{T_1} \times S^1_{T_2}$, $T_j = 1/\omega_j$, the flow $g_t(\theta_1, \theta_2) = (\theta_1 + t, \theta_2 + t)$, $d\mu = d\theta_1 d\theta_2$, and $K = -i\frac{\partial}{\partial\theta_1} - i\frac{\partial}{\partial\theta_2} + H(\theta_1, \theta_2)$.

2 Generalized Floquet Operator

In the periodic case there is a one to one relation between the spectrum of the QEO and the one of the Floquet operator defined by $U(T, 0; 0)$. In the quasiperiodic case there is the following natural generalization: Define the generalized Floquet operator [14] $U_F : \mathcal{K}_1 \to \mathcal{K}_1$; $\mathcal{K}_1 = \mathcal{H} \otimes L_2(S^1_{T_1}, d\theta_1)$ as the unitary operator $U_F(\theta_1) = \mathcal{T}^1_{-T_2}U(T_2, 0; \theta_1, 0) \equiv U(0, -T_2; \theta_1, 0)\mathcal{T}^1_{-T_2}$, where $[\mathcal{T}^1_{-T_2}\phi](\theta_1) = \phi(\theta_1 - T_2)$.

Theorem 2.1

(i) If $\phi \in \mathcal{K}_1$ is an eigenfunction of $U_F\phi = e^{-i\lambda T_2}\phi$ then the function ψ defined by

$$\psi(\theta_1, \theta_2) = e^{i\theta_2\lambda}U(0, -\theta_2; \theta_1, \theta_2)\phi(\theta_1 - \theta_2) \qquad (2.1.)$$

belongs to $\mathcal{K}$ and is an eigenfunction of the QEO, with eigenvalue λ: $e^{-iKt}\psi = e^{-i\lambda t}\psi, \quad \forall t \in \mathbb{R}$.

(ii) *Conversely, if $\psi \in \mathcal{K}$ is an eigenfunction of the QEO, then there is a function $\phi \in \mathcal{K}_1$ such that ψ can be represented by (2.1), and ϕ is an eigenfunction of the generalized Floquet operator.*

3 Point Spectrum and Stability

Theorem 3.1 *For initial states $\varphi \in \mathcal{H}$ such that $\varphi \otimes 1$ is in the pure point subspace $\mathcal{K}_{pp} \subset \mathcal{K}$ associated with the QEO the time evolution $\varphi(t)$ is almost-periodic (for almost all $\underline{\theta}$).*

Corollary 3.2 *If $\varphi \otimes 1 \in \mathcal{K}_{pp}$ then φ has bounded energy for almost all $\underline{\theta}$:*

$$\varphi \in \mathcal{H}_{be} = \{\varphi \in \mathcal{H} \mid \lim_{E\to\infty} \sup_{t\geq 0} \|F(H_0 > E)U(t,0;\theta_1\theta_2)\varphi\|_{\mathcal{H}} = 0\} ,$$

where $F(H_0 > E) : \mathcal{H} \to \mathcal{H}$ is the projector into the subspace generated by the eigenfunctions of H_0 with energy than E.

Remark. As in the case of periodic force, one cannot conclude that the expectation value of the energy is finite, since the decay of the tail might be too slow.

4 Continuous Spectrum, RAGE Theorem

As oposed to the periodic case, [15],[16], continuous spectrum of the QEO does not necessarily imply unbounded growth of the energy of the system. One can construct counterexamples, e.g. for a spin 1/2 in a quasi-periodic magnetic field, for which the energy is trivialy bounded but the spectrum of the QEO is continuous [17]. Here we consider some energy-like operators that play a role analogous to that of the unperturbed energy operator H_0 in the periodic case.

Let $\mathcal{K}_{1,pp}$ and $\mathcal{K}_{1,c}$ be the subspaces of $\mathcal{K}_1$ corresponding respectively to the point and continuous spectrum of the generalized Floquet operator U_F.

(i) Let $\Pi_0(E_0) : \mathcal{H} \to \mathcal{H}$ be the projector into the subspace of $\mathcal{H}$ generated by the eigenfunctions of $H_0\varphi^{(0)} = E\varphi^{(0)}$ with $E < E_0$.

(ii) Let $\Pi_1(E_1)$ be the projector of $L_2(S^1_{T_1}, d\theta_1)$ into the subspace generated by the eigenfunctions of $H_1 \equiv -\partial^2/\partial\theta_1^2$ with energy smaller than E_1.

(iii) $\Pi(E_0, E_1) = \Pi_0(E_0) \otimes \Pi_1(E_1)$ is then a projector in $\mathcal{K}_{1,c}$ into the subspace in which both energies are bounded $H_0 < E_0 < E_1$. Continuous spectrum implies that one or both of the energies H_0, H_1 grows unbounded:

Theorem 4.2 *Let $\varphi \in \mathcal{H}$ be such that $\varphi \otimes 1 \in \mathcal{K}_{1,c}$. Then for any fixed energies E_0, E_1,* $\sup_{t \geq 0} \|[\mathbf{1}_1 - \Pi(E_0, E_1)]U(t,0;\, \theta_1, 0)\varphi\|^2_{\mathcal{K}_1} = 1$.

Corollary 4.5 *Let $\varphi(t) = U(t,0;\theta_1,0)\varphi$. The expectation value $< H_0 + H_1 > (t) \equiv < \varphi(t), [H_0 + H_1]\varphi(t) >_{\mathcal{K}_1}$ is unbounded. More precisely, $\forall E > 0$ there is a time t_1 such that either $< H_0 + H_1 > (t_1) > E$ or $\varphi(t_1)$ is not in the domain of $H_0 + H_1$ (i.e. $< H_0 + H_1 > (t_1)$ is infinite).*

Acknowledgements:
This work was supported in part by AFOSR Grant 87-0010. H.R. Jauslin was supported by a grant of the Swiss National Science Foundation.

References

1. Reed, M., Simon, B.: Methods of Modern Mathematical Physics, Vol IV, Thm XIII.76; Academic Press, New York 1979
2. Casati, G., Molinari, L.: Progr.Theoret.Phys.Suppl. **98**, (1989) 287
3. Shepelyansky, D.L.: Physica **8D**, 208 (1983)
4. Samuelides, M., Fleckinger, R., Touzillier, L., Bellissard, J.: Europhys. Lett. **1**, 203 (1986)
5. Luck, J.M., Orland, H., Smilansky, U.: J.Statist.Phys. **53**, 551 (1988)
6. Pomeau, Y., Dorizzi, B., Grammaticos, B.: Phys.Rev.Lett. **56**, 681 (1986)
7. Badii, R., Meier, P.F.: Phys.Rev.Lett. **58**, 1045 (1987)
8. Miloni, P.W., Ackerhalt, J.R., Goggin, M.E.: Phys.Rev. **A 35**, 1714 (1987)
9. Combescure, M.: J.Statist.Phys. **62**, 779 (1991) ; and preprint Orsay 1991
10. Guarneri, I.: Lett.Nuovo Cimento **40**, 171 (1984)
11. Pillet, C.-A.: Comm.Math.Phys. **102**, 237 (1985) and **105**, 259 (1986)
12. Pillet, C.-A.: In: *Stochastic Processes in Classical and Quantum Systems.* Edited by S. Albeverio, G. Casati and D. Merlini, Springer Verlag, Berlin 1986
13. Bellissard, J.: In *Stochastic Processes in Classical and Quantum Systems.* Edited by S. Albeverio, G. Casati and D. Merlini, Springer Verlag, Berlin 1986
14. Jauslin, H.R., Lebowitz, J.L.: *Chaos* **1** (1991)
15. Bunimovich, L., Jauslin, H.R., Lebowitz, J.L., Pellegrinotti, A., Nielaba,P.: J.Statist.Phys. **62**, 793 (1991)
16. Enss, V., Veselic, K.: Ann.Inst.H.Poincaré **A 39**, 159 (1983)
17. Bleher, P., Jauslin, H.R., Lebowitz, J.L.: In preparation

Section 4

Equilibrium Statistical Mechanics

Organizer: S.B. Shlosman

Magnetization and Slow Decay of Correlations in Continuum $1/r^2$ Ising Models

Luiz R. G. Fontes*

Courant Institute of Mathematical Sciences†
251 Mercer Street, New York, NY 10012

1 Introduction

The continuum Ising model is a (one dimensional) spin system defined on the real line $\mathbb{R}$ instead of on the discrete lattice $\mathbb{Z}$. It arises in the study of a quantum mechanical model of the motion of a particle in a double well potential, subjected to a field. Localization in this model is equivalent to magnetization in the continuum Ising model (see [1]).

Its configurations are piecewise constant functions in $\mathbb{R}$ taking the values $+1$ or -1. An arbitrary configuration will be denoted by $\sigma \equiv \sigma(t)$. Consider now the space of all configurations restricted to a finite interval, say $[-T, T]$, denoted by Ω_T. Let $P_{\epsilon,T}$ denote the (a priori) measure on Ω_T such that the the flips of the configurations form a Poisson process of positive rate ϵ and $\sigma(-T)$ is ± 1 with equal probabilities. The finite volume continuum Ising measure corresponding to $[-T, T]$ is defined as follows.

$$d\rho_{T,\epsilon,\alpha}(\sigma) = \frac{1}{Z_{T,\epsilon,\alpha}} e^{-H(\sigma)} dP_{\epsilon,T}(\sigma) , \tag{1.1}$$

where H is the Hamiltonian

$$H(\sigma) = -\frac{\alpha}{4} \int_{-T}^{T} \int_{-T}^{T} W(|t-s|)\sigma(t)\sigma(s)dtds , \tag{1.2}$$

with W a nonnegative function, say continuous, on $[0, \infty]$ such that

$$r^2 W(r) \to 1, \text{ as } r \to \infty,$$

and α is a positive constant. $Z_{T,\epsilon,\alpha}$ is a normalizing factor.

---

*Supported by FAPESP, proc. no. 87/2629-7.

†On leave from Instituto de Matemática e Estatística, USP, Caixa Postal 20570, São Paulo, SP, 01498, Brazil.

The infinite volume limit of $\rho_{T,\epsilon,\alpha}$ as $T \to \infty$ exists by standard arguments and is denoted $\rho_{\epsilon,\alpha}$. It corresponds to free boundary conditions. The notation $\langle\cdot\rangle_{\epsilon,\alpha}$ is used for its expectations. We can define a + b.c. continuum Ising measure by starting with an a priori measure such that the flips form a Poisson process *conditioned to having an even number of flips in* $[-T,T]$ and such that $\sigma(-T) = +1$. We then follow the same steps as above, but with Hamiltonian

$$H'(\sigma) = -\frac{\alpha}{4}\int_{-\infty}^{\infty}\int_{-T}^{T} W(|t-s|)\sigma(t)\sigma(s)dtds \; , \tag{1.3}$$

with $\sigma(t) = +1$ in $\mathbb{R}\backslash[-T,T]$ and obtain an infinite volume measure $\rho^+_{\epsilon,\alpha}$ (with expectations denoted $\langle\cdot\rangle^+_{\epsilon,\alpha}$).

The measures $\rho_{\epsilon,\alpha}$ and $\rho^+_{\epsilon,\alpha}$ can be obtained as weak limits of ordinary discrete Ising measures (with appropriate interactions) in finer and finer lattices, as explained in [1].

Below, we state previous and new results concerning this model together with a sketch of arguments.

2 Magnetization

The previous results concerning the phase diagram of the continuum Ising model state the following (see [1]). For $\alpha \leq 1$ and any $\epsilon > 0$, the model exhibits no magnetization (meaning that $\langle\sigma(0)\rangle^+_{\epsilon,\alpha} = 0$). For any $\alpha > 2$ and ϵ close enough to 0, there is magnetization (i.e., $\langle\sigma(0)\rangle^+_{\epsilon,\alpha} > 0$).

The gap in the α-space was due to use of the technique of representing the continuum Ising model as a random cluster model (FK representation). This allows a comparison to independent continuum percolation and the use of arguments derived for the discrete case (see [2] and [3]), but introduces a factor of 2 in the inverse temperature (see [1] and [4]).

The new result is the closing of this gap:

Theorem 1 *The continuum Ising model exhibits magnetization for any* $\alpha > 1$ *and* ϵ *close enough to* 0.

This was expected, since it is what happens in the discrete $1/r^2$ Ising model (see [2] and [5]).

The proof is obtained by comparing the continuum to a discrete Ising model. For $A = \{t_1,\ldots,t_n\}$, a set of points in $[-T,T]$, let σ_A denote the product $\sigma(t_1)\ldots\sigma(t_n)$. We need the following inequality:

$$\langle\sigma_A\rangle_{T,\epsilon,\alpha} \geq \langle S_{A'}\rangle_{K,J,\alpha'} \; . \tag{2.1}$$

Here, $\langle\cdot\rangle_{K,J,\alpha'}$ is the expectation of a (translation invariant) discrete $1/\,|\,i-j\,|^2$ Ising model in $\{-K\,,\,-K+1,\ldots,\,K-1\}$ $\;(K = K(T) \to \infty$

as $T \to \infty$), with nearest neighbor coupling J_1 large and α' close to α when ϵ is small. S is a discrete configuration, $A' = \{\lfloor t_1/L \rfloor, \ldots, \lfloor t_n/L \rfloor\}$, for some large L (the t_i's should be so separated that A' does not have repetitions), and $S_{A'} = \prod_{i \in A'} S(i)$. The boundary conditions are free on both sides of (2.1).

By (2.1), magnetization in the continuum model follows from *long long range order* in the discrete one (by which we mean that $\langle S(0)S(l)\rangle_{K,J,\alpha'}$ is uniformly positive in K and l), which was established in [5] for $\alpha > 1$ and J_1 large.

Previous attempts at proving the above theorem were based on either using the discrete Ising model approximating the continuum one, which met the difficulty of conflicting limits of interactions in short and long distances (the short range ones going to infinity while the long distance ones go to zero), or by discretizing the a priori measure together with the interaction function W, in which process too much correlation at short distances is lost, so that inequality (2.1) can be obtained, but not with J_1 as large as needed.

Our approach is the following. We first discretize the model by partitioning $[-T, T]$ into intervals I_i, $i = -K, \ldots, K-1$, of length L. We then consider the model obtained similarly as the continuum Ising model, but with $P_{\epsilon,T}(\sigma)$ replaced by $\prod_i P_\epsilon^{(i)}(\sigma_i)$, where σ_i is a continuum configuration in I_i and $P_\epsilon^{(i)}$ is a measure similar to $P_{\epsilon,T}$, but on configurations restricted to I_i. As is intuitively clear, the correlations of this modified model are smaller than the original ones.

We next notice that the modified model is a sort of annealed site diluted Ising model, the dilution corresponding to ocurrence of flips in σ_i.

We then derive a correlation inequality between this model and an ordinary discrete quenched site diluted Ising model, with the help of GKS and Harris-FKG inequalities (see [6]).

Jensen's inequality is the main ingredient in the last step of the proof, which is a correlation inequality between the quenched and an ordinary (non-diluted) discrete Ising model. We get (2.1) with

$$J_{ij} = \int_{I_i} \int_{I_j} W(|t-s|)dtds$$

and $\alpha' = \eta^2 \alpha$, where

$$\eta = \frac{1}{\Gamma} \log[1 - (1 - e^{-\Gamma})(1 - O(\epsilon))]^{-1},$$

with Γ a constant.

Notice that $|i-j|^2 J_{ij} \to 1$, as $|i-j| \to \infty$, $J_1 \to \infty$, as $L \to \infty$ and $\eta \to 1$, as $\epsilon \to 0$.

3 Slow Decay of Correlations

As in the discrete $1/r^2$ Ising model, the physical relevance of magnetization (more precisely, long long range order) for $\alpha > 1$ is that it implies the existence of an *intermediate phase* with slow decay of correlations.

Theorem 2 *For $\alpha > 1$, $\delta > 0$ and ϵ close enough to zero, we have*

$$\langle \sigma(0)\sigma(t)\rangle^{+}_{\epsilon,\alpha} - (\langle \sigma(0)\rangle^{+}_{\epsilon,\alpha})^2 \geq const/t^{2\gamma}, \tag{3.1}$$

where $\gamma = \min(1, \alpha - 1 + \delta)$.

The proof follows the corresponding proof for the discrete case (see [5]), using the FK representation of the continuum Ising model and deriving lower bounds for corresponding quantities in the corresponding continuum random cluster model (which one does by resorting to FKG domination to compare this to continuum independent percolation). Minor modifications are needed to account for the extra randomness of the continuum model.

Upper bounds for the truncated two-point function, which exist for the discrete model, were not derived in the continuum case.

Acknowledgements. This work is part of the author's PhD research. Thanks are due to Charles M. Newman, who supervised it, Herbert Spohn, for pointing out the problem and to the National Science Foundation, for travel support.

References

1. Spohn, H.: Ground state(s) of the spin-boson Hamiltonian. Comm.Math. Phys. **123**, 277 (1989)
2. Aizenman, M., Newman, C.M.: Discontinuity of the percolation density in one dimensional $1/|x-y|^2$ percolation models. Comm.Math.Phys. **107**, 611 (1986)
3. Newman, C.M., Schulman, L.S.: One dimensional $1/|i-j|^s$ percolation models: The existence of a transition for $s \leq 2$. Comm.Math.Phys. **104**, 547 (1986)
4. Aizenman, M., Chayes, J.T., Chayes, L., Newman, C.M.: Discontinuity of the Magnetization in one dimensional $1/|x-y|^2$ Ising and Potts models. J.Statist.Phys. **50**, 1 (1988)
5. Imbrie, J.Z., Newman, C.M.: An intermediate phase with slow decay of correlations in one dimensional $1/|x-y|^2$ percolation, Ising and Potts models. Comm.Math.Phys. **118**, 303 (1988)
6. Fontes, L.R.G.: An Ordered Phase with Slow Decay of Correlations in Continuum $1/r^2$ Ising Models. N.Y.U. preprint (1991)

Existence of the Kosterlitz-Thouless Phase for Two-Dimensional Coulomb Gases at Inverse Temperatures above 8π

Abel Klein*

Department of Mathematics, University of California, Irvine, CA 92717, USA

A two-dimensional (lattice) Coulomb gas is a system of classical particles with electric charges ± 1, whose possible positions range over a finite array of sites $\Lambda \subset \mathbb{Z}^2$, interacting via a two-body Coulomb potential.

At high temperature the Coulomb gas is in a plasma phase exhibiting Debye screening and exponential decay of truncated correlations (Brydges [1], Brydges and Federbush [2], Yang [3]).

In two dimensions the Coulomb potential is logarithmic, behaving like $\frac{1}{2\pi} \log r$ for large distances r. This potential tends to bind oppositely charged particles into dipoles (or more generally into neutral multipoles), which cannot screen. Kosterlitz and Thouless [4] (see also Berezinskii [5]) predicted the existence of a dipole phase at low temperature and low density, characterized by power law decay of correlations. The existence of such a phase was proved by Frohlich and Spencer [6].

A heuristic description of the mechanism behind the dipole phase is given by the following energy-entropy argument [4],[6]: a neutral dipole of length r has a Gibbs weight $\approx e^{-\frac{\beta}{2\pi} \log r}$. There is a entropy factor of the order of r^3(a factor of r^2 is the area over which one charge may range, the remaining factor of r being the possible positions of the other charge). Thus the probability of a dipole of length r is $\approx r^3 e^{-\frac{\beta}{2\pi} \log r}$. If $\beta > 8\pi$, this probability is summable in r, so long dipoles are suppressed. Thus, if $\beta > 8\pi$ and the density is low, the Coulomb gas looks like a dilute gas of not too long dipoles which cannot screen so that test particles feel the logarithmic potential.

Such result has been proven by Marchetti and myself [7]. Our proof makes rigorous the energy-entropy argument. Marchetti, Perez and myself

* Partially supported by the NSF under grants DMS-8905627 and INT-8703059

[8] had already proven the result for $\beta > 24\pi$, improved by Marchetti [9] to $\beta > \frac{8(1+\sqrt{3})}{3-\sqrt{3}}\pi \approx 17.2\pi$ and by Braga [14] to $\beta > 16\pi$.

In the sine-Gordon representation the partition function of the gas is given by

$$Z_\Lambda = \int \prod_{x\in\Lambda} \hat{\lambda}_z(\phi(x))d\mu_\beta(\phi) \quad , \tag{1}$$

where $d\mu_\beta$ is the Gaussian measure with covariance $\beta(-\Delta)^{-1}$. $\hat{\lambda}_z$ is the Fourier transform of the "a priori" charge weight at particle activity $z \geq 0$. We have $\hat{\lambda}_z(\phi) = 1 + z\cos\phi$ for the hard-core gas, $\hat{\lambda}_z(\phi) = e^{z\cos\phi}$ for the standard gas.

The external charges correlation function is given by $G_{\xi,\Lambda}(x) = Z_{\xi,\Lambda}(x)/Z_\Lambda$, where

$$Z_{\xi,\Lambda}(x) = \int e^{i\xi(\phi(0)-\phi(x))} \prod_{x\in\Lambda} \hat{\lambda}_z(\phi(x))d\mu_\beta(\phi) \tag{2}$$

is the external charges partition function. Here $\xi \in \mathbb{R}$. $G_{\xi,\Lambda}(x)$ measures the average energy required to pump two opposite charges of strength ξ, located at 0 and x, into the system. Using Jensen's inequality (and making explicit the dependence on z) we have [10]

$$G_{\Lambda,\xi,z}(x) \geq G_{\Lambda,\xi,0}(x) \approx \frac{1}{|x|^{\beta\xi^2/2\pi}} \quad .$$

By $G_\xi(x)$ we will denote a thermodynamic limit. When screening occurs (e.g., β small) we have $G_\xi(x) \to C > 0$ exponentially fast as $|x| \to \infty$ ([1],[2],[3], see comments in [10]). In the dipole phase $G_\xi(x)$ should decay polynomially, at least for fractional charges.

Frohlich and Spencer [6] proved the existence of the Kosterlitz-Thouless phase by showning that there is no Debye screening for β large, in the sense that a certain sum rule characteristic of Debye screening is violated. They also showed that $G_\xi(x) \leq C|x|^{-c\beta\eta^2}$ for $\beta > b/c\eta^2$, where C, b, c are constants, $0 < \xi < 1, \eta = \min(\xi, 1-\xi)$ (see also Braga [11]). Marchetti, Perez and myself [8] proved that if $\beta > 24\pi$, then for z small enough $G_\xi(x)$ decays polynomially for all fractional charges.

But the fact that the critical β should be 8π is supported not only by heuristic arguments, but also by the hierarchical model studied by Marchetti and Perez [12] and by the proof of Dimock and Hurd [13] that the equilibrium measure of a standard Coulomb gas on $\mathbb{R}^2$ is driven, under a renormalization group transformation, to a Gaussian measure if $\beta > 8\pi$ and z is sufficiently small.

Marchetti and I [7] proved:

Theorem *Let $\beta > 8\pi$. Then there exist $z(\beta) > 0, \theta = \theta(\beta) > 0$, such that for $0 < z < z(\beta)$ we have*

$$G_\xi(x) \leq \frac{C_\xi}{|x|^{\theta\eta^2}}$$

for all $\xi \in \mathbb{R}$, *where* $\eta = dist(\xi, \mathbb{Z})$, *and* C_ξ *is bounded on compact subsets of* $\mathbb{R} \setminus (\mathbb{Z} \setminus \{0\})$.

Our proof uses the main ingredients of the Frohlich-Spencer proof: expectations in the Coulomb gas are written as convex combinations of expectations in diluted gases of neutral multipoles of various sizes by a simple trigonometric identity and falloff is extracted from charged multipoles by an analytic continuation argument (the energy estimate).

There are two main technical improvements in our proof:

i) our proof is organized to optimize the entropy estimates. The geometric picture of the isolated neutral multipoles is fixed "a priori". This was done in [8].
ii) we control the effect of the imaginary shifts, needed for the energy estimates, on the background of neutral multipoles. To do that we consider a more general form for the partition function.

We will now sketch the main ideas in the proof. Let us fix $\beta > 8\pi$, pick $\alpha > 1$, and an initial scale L_1. We set $L_{k+1} \approx L_k^\alpha, k = 1, 2, \cdots$. We write $B(x, L) = \{y \in \mathbb{Z}^2; |y - x|_\infty < \frac{L}{2}\}$. We take $\Lambda = B(0, R)$, and pick N such that $L_{N-1} < R \leq L_N$. We let $\Lambda^{(k)} = \Lambda \cap L_k\mathbb{Z}^2, B_k(y) = B(y, L_k)$. Notice $\Lambda^{(N)} = \{0\}$.

Our optimized trigonometric identity is:

Lemma 1 [8] *Let* $z_i > 0$, $\phi_i \in \mathbb{R}$, $i = 1, 2, \cdots, n$. *Then*

$$\prod_{i=1}^{n}(1 + z_i \cos\phi_i) = \sum_{\sigma \in g_n} c_\sigma(1 + z_\sigma \cos\phi(\sigma)) \quad ,$$

where $g_n = \{\sigma : \{1, 2, \cdots, n\} \to \{0, 1, -1\}, \sigma \not\equiv 0\}$, $\phi(\sigma) = \sum_{i=1}^n \phi_i\sigma_i$, $0 < c_\sigma, \sum_{\sigma\in g_n} c_\sigma = 1$, *and*

$$z_\sigma = \prod_{i=1}^{n}(b_n z_i)^{|\sigma_i|}$$

with $b_n = (2^{1/n} - 1)^{-1} \leq n/\log 2$.

Using Lemma 1 inside each $B_1(y), y \in \Lambda^{(1)}$, we rewrite (1) as the convex combination of partition functions of the form

$$\int \prod_{y\in\Lambda^{(1)}} (1 + z(\rho_y^{(1)}) \cos\phi(\rho_y^{(1)}))d\mu_\beta(\phi) \quad , \tag{3}$$

where $\rho_y^{(1)}$ is a charge density with support in $B_1(y)$ and renormalized activity $z_{\rho_y}^{(1)} \leq \bar{z}^{(1)} \equiv \frac{L_1^2}{\log 2} z$, which can be made to satisfy the desired "a priori" bound on the 1st scale by picking z sufficiently small.

We could now repeat the precedure and express each partition function of the form given by (3) as a convex combination of partition functions of the same form but on the 2nd scale. But now we do not have a free parameter (previously z) to control our renormalized activities. If $\rho_y^{(2)}, y \in \Lambda^{(2)}$, comes from two or more charge densities from the 1st scale, we get $z_{\rho_y}^{(2)} \leq ((L_2/L_1)^2(1/\log 2)\bar{z}^{(1)})^2$ which, because of the square, will satisfy the "a priori" bounds in the 2nd scale (actually we take "a priori" bounds of the form $L_k^{-(r+2(\alpha-1))}$ where r is suitably chosen so the above procedure reproduces the bounds). But if $\rho_y^{(2)}$ is isolated, i.e., as a charge density $\rho_y^{(2)} = \rho_y^{(1)}$, this procedure does not work. If $\rho_y^{(2)}$ is neutral, we put it into the background of neutral multipoles. If it is charged, we need the energy estimate to extract falloff to renormalize the activity. This is done as follows:

Lemma 2 [6],[8] *Let ρ be a charge density with support in $B(0,\ell), q = \sum_x \rho(x) \neq 0$, and let $F(\phi)$ be a function depending only on $\{\phi(x), x \notin B(0,L)\}$, where $L > \ell$. Then*

$$\int e^{i\phi(\rho)} F(\phi) d\mu_\beta(\phi) = Y \int e^{i\phi(\bar{\rho})} F(\phi) d\mu_\beta(\phi)$$

where

$$\left(\frac{L}{\sqrt{2}\ell}\right)^{-\frac{\beta q^2}{2\pi(1+a)}} \leq Y \leq \left(\frac{L}{\sqrt{2}\ell}\right)^{-\frac{\beta q^2}{4\pi(1+a)}},$$

where $\bar{\rho}$ is a charge density with charge q with support in $B(0,L)$, and $a = a(\ell) \to 0$ as $\ell \to \infty$.

This is proved by making the imaginary shift $\phi \to \phi + i\mathcal{E}$ for a suitable function $\mathcal{E}$.

We now use Lemma 2(we ignore details) for an isolated charge density $\rho_y^{(2)}$. By Lemma 2, we can replace $\rho_y^{(2)}$ by $\bar{\rho}_y^{(2)}$ with support in $B_2(y)$, and $z_{\rho_y}^{(2)}$ in replaced by

$$\bar{z}_{\rho_y}^{(2)} = Y z_{\rho_y}^{(2)} \leq z_{\rho_y}^{(1)} \frac{1}{\log 2} \left(\frac{L_2}{L_1}\right)^2 \left(\frac{L_2}{\sqrt{2}L_1}\right)^{-\beta/4\pi(1+a)}.$$

This renormalizes the activity (i.e., makes it smaller) if $\beta > 8\pi(1+a)$, which we can obtain by taking L_1 sufficiently large. This is the same energy-entropy competition as before, but in a form that can be used for a rigorous proof.

More generally, we introduce the notion of a k-regular partition function, and prove that a k-regular partition function can be written as a convex combination of $(k+1)$-regular partition functions, for $k = 1, 2, \cdots, N-1$.

But a k-regular partition function cannot be of the same form as (3), because of the background of neutral multipoles we accumulated during the inductive procedure. In [8] a k-regular partition function was given to be of the form

$$\int \prod_{y \in \Lambda^{(k)}} [(1 + z^{(k)}_{\rho^{(k)}_y} \cos \phi(\rho^{(k)}_y)) \prod_{\rho \in N_y} (1 + z(\rho) \cos \phi(\rho))] d\mu_\beta(\phi) \quad ,$$

where $\rho^{(k)}_y$ is a charge density localized in $B_k(y)$, with renormalized activity $\rho^{(k)}_y$ satisfying "a priori" bounds, and N_y is a "sparse" ensemble of neutral multipoles with support in $B_k(y)$.

The sparse background of neutral multipoles interferes with the imaginary shift in the proof of the energy estimate (Lemma 2). Previously [6],[8], the shift had been made constant on the support of neutral multipoles. The falloff factor obtained at scale L is then of the form $Y \approx L^{-(\alpha-1)\beta/4\pi(1+a)}$, with a $\approx L^{3-2\alpha}$. To get $a \to 0$ as $L \to \infty$ we need $\alpha > 3/2$. It turns out that the inverse temperature $\bar{\beta}$ above which the proof in [8] gives power-law falloff is $\bar{\beta} = (8\pi + 2a)\alpha/2 - \alpha$. By taking the initial scale to infinity we get $\bar{\beta} = 8\pi\alpha/2 - \alpha$. If we let $\alpha \to 3/2$, we get $\bar{\beta} = 24\pi$.

Marchetti and myself [7] improved the energy estimate to obtain $a \approx (\log L)^p L^{(1-\alpha)}$ with $2 < p < \infty$ fixed. We can thus let $\alpha \searrow 1$ to get $\bar{\beta} = 8\pi$. To do so we had to consider the effect of the imaginary shift on the neutral charges background; so we had to introduce a more general form of a regular partition. We replace terms of the form $(1 + z_\rho \cos \phi(\rho))$ by

$$(1 + e^{w(\rho,\phi)} \cos(\phi(\rho) + \theta(\rho, \phi))) \quad ,$$

where $w(\rho, \phi), \theta(\rho, \phi)$ are real-valued, real-analytic functions of $\{\phi(u), u$ in the support of $\rho\}, w(\rho, -\phi) = w(\rho, \phi), \theta(\rho, -\rho) = -\theta(\rho, \phi)$, with certain properties and "a priori" bounds. We prove a more general version of Lemma 2 and control w and θ as we change scales by an inductive procedure. A new and important ingredient is that we only add to the neutral charges background multipoles whose total "absolute charge" $|\rho| = \sum |\rho(x)|$ is not too big by scale (say, $|\rho| \leq (\log L_k)^p$).

A similar procedure can be carried out for $Z_{\xi,\Lambda}(x)$, and the quotient $Z_{\xi,\Lambda}(x)/Z_\Lambda$ can be written as a convex combination of similar quotients of N-regular partition functions, which can be uniformly estimated yielding polynomial decay [8],[7].

References

1. Brydges, D.: Comm.Math.Phys. **58**, 313-350 (1978)
2. Brydges, D., Federbush, P.: Comm.Math.Phys. **73**, 197-246 (1980)
3. Yang, W.S.: J.Statist.Phys. **49**, 1-32 (1987)
4. Kosterlitz, J.M., Thouless, D.J.: J.Phys. C **6**, 1181-1203 (1973)
5. Berezinskii, V.: Soviet Phys.JETP **32**, 493 (1971)
6. Frohlich, J., Spencer, T.: Comm.Math.Phys. **81**, 527-602 (1981)
7. Marchetti, D.H.U., Klein, A.: J.Statist.Phys. **64**, 135-162 (1991)
8. Marchetti, D.H.U., Klein, A., Perez, J.F.: J.Statist.Phys. **60**, 137-166 (1990)
9. Marchetti, D.H.U.: J.Statist.Phys. **61**, 909-911 (1990)
10. Frohlich, J., Spencer, T.: J.Statist.Phys. **24**, 617 (1981)
11. Braga, G.A., Thesis, Ph.D.: New York University (1989)
12. Marchetti, D.H.U., Perez, J.F.: Phys.Lett. A **778**, 74 (1986); Marchetti, D.H.U., Perez, J.F.: J.Statist.Phys. **55**, 141 (1989)
13. Dimock, J, Hurd, T.: Comm.Math.Phys. **137**, 263-287 (1991)
14. Braga, G.A.: J.Statist.Phys. **63**, 285-289 (1991)

Large Deviation Behavior of Statistical Mechanics Models in the Multiphase Regime

R. L. Dobrushin [1], S. B. Shlosman [2]

[1]Institute for Problems of Information Transmission, Moscow;
Visiting Professor, Massachusetts Institute of Technology
[2]Institute for Problems of Information Transmission, Moscow

1. The theory of large deviations was initiated by the study of the asymptotical behavior of probabilities of large deviations of sums of n identically distributed independent random variables. In this case the logarithms of the probabilities are asymptotically equal to $-nI$, where I is a constant which can be defined by a minimization of the so-called action functional. The theory of large deviations is now a well-developed branch of the probability theory (see the books [1], [2] for example) devoted mainly to generalization of the mentioned asymptotics to wide classes of random processes. Also these types of results were recently generalized on a wide class of Gibbsian random fields (see [3], [4]).

The main special feature of multidimensional Gibbsian fields is a possibility of phase transition. It was noted recently by Schonman [5] and Föllmer and Ort [6] that in the case of phase transitions of the first order the constant I vanishes typically and so the usual formulations of the theory of large deviations became insipid. They showed that in such situations the logarithm of the probability of large deviations really is proportional to $n^{d-1/d}$ where $d > 1$ is the dimension of the field and gave some estimates for the corresponding proportionality coefficient. Recently (see [7], [8], [9]) a close connection between the theory of large deviation and a well known physical theory describing shapes of droplets of phases was established. It was found that in the case of $d = 2$ and a field on a torus with n sites (periodic boundary conditions) the logarithms of the probabilities of large deviations are asymptotically equal to $-\hat{I}n^{1/2}$ where $\hat{I}$ can be found as a solution of the so-called Wulff variational problem for the surface tension. The main aim of this report is a discussion of a more general problem connected with a description of large deviations of heterogeneities of spatial behavior of densities of the field. It turns out that the answer of the problem can be described by the help of a natural generalization of the Wulff construction.

For simplicity we restrict ourselves to the case of the symmetric two-dimensional Ising model with the periodic boundary conditions and large enough value of parameter β. The generalizations to other two-dimensional Gibbs fields with first-order phase transitions are direct enough. It seems also that similar results would be found for more than two dimensions but there are serious mathematical difficulties in proving this (compare discussions in [7]).

2. Let T_N be a discrete two-dimensional torus of the size $N \times N$ and $\Omega_N = \{-1,1\}^{T_N}$ be the space of realizations of Ising model in T_N. Let the Hamiltonian of the Ising model with a magnetic field $h \in \mathbb{R}^1$

$$H(\sigma) = - \sum_{s,t \in T_N : |s-t|=1} \sigma_s \sigma_t - h \sum_{t \in T_N} \sigma_t, \ \sigma = (\sigma_t, \ t \in T_N), \tag{2.1}$$

where the sum is taken over all nearest neighboring pairs of sites in T_N. The Gibbs distribution P_N^β in T_N with an inverse temperature $\beta \geq 0$ is given by

$$P_N^{\beta,h}(\sigma) = \exp\{-\beta H(\sigma)\} / Z(\beta, h, N) \ , \tag{2.2}$$

where $Z(\beta, h, N)$ is the normalization factor called also a partition function. In the following we shall also use the free energy function defined as

$$f(\beta, h) = \beta^{-1} \lim_{N \to \infty} N^{-2} \ln Z(\beta, h, N) . \tag{2.3}$$

We shall suppose in the following that $\beta > \bar{\beta}$ where $\bar{\beta}$ is a large enough constant. Let

$$S_N(\sigma) = N^{-2} \sum_{t \in T_N} \sigma_t . \tag{2.4}$$

It is well known that for $h \neq 0$ the probability distribution of S_N induced by $P_N^{\beta,h}$ converges if $N \to \infty$ to the probability distribution concentrated at the point

$$m(\beta, h) = \frac{\partial f(\beta, h)}{\partial h} . \tag{2.5}$$

If $h = 0$ it converges to the probability distribution concentrated in two points $\pm m(\beta)$ having probabilities 1/2, where

$$m(\beta) = \lim_{h \to 0+} m(\beta, h) = - \lim_{h \to 0-} m(\beta, h) . \tag{2.6}$$

3. Let T be a continuous 1×1 two-dimensional torus and $M(T)$ be a set of all signed measures on T endowed with the weak topology. Let $\mathcal{P}(M(T))$ be a space of all probabilistic measures on the Borel σ-algebra of the space $M(T)$, i.e., a space of probability distributions of random measures on T.

To compare the probabilities $P_N^{\beta,0}$ for different N we interpret them as elements of $\mathcal{P}(M(T))$. For it we let

$$\mu_\sigma = N^{-2} \sum_{t \in T_N} \sigma_t \delta_{t/N}, \quad \sigma \in \Omega_N, \tag{3.1}$$

where $\delta_x \in M(T)$ is a δ-measure at a point $x \in T$. Let $\hat{P}_N^\beta \in \mathcal{P}(M(T))$ be the probability distribution induced from $P_N^{\beta,0}$ by the help of the transformation $\sigma \to \mu_\sigma$.

Theorem 1 *There exists an action function $I(\mu), \mu \in M(T)$ such that for any Borel set $B \subset M(T)$*

$$-\inf_{\mu \in B^0} I(\mu) \le \liminf_{N\to\infty} N^{-2} \ln \hat{P}_N^\beta(B) \le \limsup_{N\to\infty} N^{-2} \ln \hat{P}_N^\beta(B)$$
$$\le -\inf_{\mu \in \bar{B}} I(\mu), \tag{3.2}$$

where $\bar{B}$ is the closure of B and B^0 is the interior of B. The action function $I(\mu)$ can be defined by the following construction. Let a measure μ have a density $p(x), x \in T$ with respect to the Lebeque measure on T, such that $|p(x)| < 1$. Then

$$I(\mu) = \int_T H(p(x))\, dx, \tag{3.3}$$

where for $|p| > m(\beta)$

$$H(P) = \beta(f(\beta, h(p)) - f(\beta, 0)) + ph(p) \tag{3.4}$$

and $h(p)$ is a (unique) solution of the equation

$$m(\beta, h(p)) = p \tag{3.5}$$

and where if $|p| \le m(\beta)$ then

$$H(p) = 0. \tag{3.6}$$

At last

$$I(\mu) = \infty \tag{3.7}$$

for all other measures $\mu \in M(T)$.

Roughly speaking the functional (3.3) describes asymtotically the probability of a disposition of particles on T with a density $p(x)$. This result is not surprising because it can be treated as a discrete-argument variant of the results which are well known for continuous-time processes (see [2] for example).

In the case when

$$B = B_{q_1,q_2} = \{\mu \in M(T) : q_1 \le \mu(T) \le q_2\}, \quad 0 \le q_1 < q_2, \tag{3.8}$$

the infinums in (3.2) achieve if $p(x) \equiv q_1$ and so (3.2) is reduced to results proven in [3], [4].

4. Because of (3.6) the infinums in (3.2) vanish for a wide class of sets B and in these cases the result of the theorem 1 is not very informative.

Theorem 2 *There exists a function $\tilde{I}(\mu), \mu \in M(T)$ such that if for a Borel set $B \subset M(T)$ the value*

$$\inf_{\mu \in \bar{B}} I(\mu) = 0\,, \tag{4.1}$$

then

$$-\operatorname*{Inf}_{\mu \in B^0} \tilde{I}(\mu) \leq \liminf_{N\to\infty} N^{-1} \ln \hat{P}_N^\beta(B) \leq \limsup_{N\to\infty} N^{-1} \ln \hat{P}_N^B(B) \leq -\operatorname*{Inf}_{\mu \in \bar{B}} \tilde{I}(\mu)\,. \tag{4.2}$$

The function $\tilde{I}(\mu)$ can be defined by the following construction. Let a measure μ have a density $p(x)$, $x \in T$ with respect to the Lebeque measure on T such that

$$|p(x)| = m(\beta), \qquad x \in T. \tag{4.3}$$

Let the boundary Γ between the sets $\{x \in T\colon p(x) = m(\beta)\}$ and $\{x \in T\colon p(x) = -m(\beta)\}$ consists of a finite set of rectifiable curves $\Gamma = \{\gamma_1, \ldots, \gamma_k\}$. Then

$$\tilde{I}(\mu) = \beta \sum_{i=1}^{k} \int_{\gamma_i} \tau(\beta, n_l)\, dl\,, \tag{4.4}$$

where the integrals are taken along the lines γ_i and n_l is a normal vector to the curve γ_i at a point l. Here $\tau(\beta, n)$, $n \in \mathbb{S}^1$ is a positive function which is called the surface tension and is defined by the relation

$$\tau(\beta, n) = -\beta^{-1} \lim_{N\to\infty} (N \cos(\widehat{n, e}))^{-1} \ln \frac{Z(\beta, 0, N, n)}{Z(\beta, 0, N, +)}\,. \tag{4.5}$$

Here $Z(\beta, 0, N, n)$ is a partition function in the square box $\left[-\frac{N}{2}, \frac{N}{2}\right]^2 \bigcap \mathbb{Z}^2$ for the Ising model with parameters β and $h = 0$ and boundary condition

$$\sigma_t^n = \begin{cases} +1 & \text{if} \quad tn \geq 0 \\ +1 & \text{if} \quad tn < 0 \end{cases}\,, \qquad t \in \mathbb{Z}^2, \tag{4.6}$$

$Z(\beta, 0, N, +)$ is the similar partition function with boundary condition $\sigma_t^+ \equiv +1$ and $(\widehat{n, e})$ is the angle between the vector n and the coordinate ort e in $\mathbb{Z}^2$ (we supposed here that $|(\widehat{n, y})| \leq \pi/4$ but $\tau(\beta, n)$ can be extended to other n by periodicity.) For all other $\mu \in M(T)$ we let

$$\tilde{I}(\mu) = \infty\,. \tag{4.7}$$

As a comment to the previous construction we note that the set of measures for which $\tilde{I}(\mu)$ is defined by (4.4) is dense in the set of all $\mu \in M(T)$ such that

$$|\mu(B)| \leq m(\beta)|B| \tag{4.8}$$

for any Borel set $B \subset T$ (here $|B|$ is the Lebeque measure of B). For example the measure having a constant density $p(x) = q$, where $|q| < m(\beta)$ can be approximated if we consider a periodic partition of T on small squares in which a density takes values $\pm m(\beta)$ such that the portion π of cubes with the value $+m(\beta)$ satisfies the equation

$$\pi m(\beta) + (1-\pi)(-m(\beta)) = q\,. \tag{4.9}$$

It is clear that the corresponding value $\tilde{I}(\mu) \to \infty$ when the diameter of this partition goes to 0. So roughly speaking the infinums in (4.2) are reached for the coarsest partition consisting with the condition B.

In the case of the set $B = B_{q_1,q_2}$ (see (3.8)) the condition (4.1) is true if $|q_1| < m(\beta)$. In this case the statement of Theorem 2 reduces to a result proved in [7]. If $\frac{1}{2} < q_2 < m(\beta)$ and β is large enough the infinums (4.2) are reached for a density p taking the value $m(\beta)$ on a convex set which is well-known in statistical mechanics as the Wulff shape of a droplet. If $O < q_2 < 1/2$ and β is large enough the infinums (4.2) are reached for a density p taking the value $m(\beta)$ on a strip winding around the torus.

References

1. Freidlin, M.I., Ventcel, A.D.: *Random Perturbations of Dynamical Systems.* Springer, Berlin, 1984
2. Deuschel, J.-D., Strook, D.W.: *Large Deviations.* Acad. Press, Boston, 1989
3. Olla, S.: Probab.Theory Related Fields **77**, 343 (1988)
4. Föllmer, H., Orey, S.: Ann.Probab. **16**, 961 (1988)
5. Shonman, R.H.: Comm.Math.Phys. **112**, 409 (1987)
6. Föllmer, H., Ort, M.: Astérisque **157-158**, 173 (1988)
7. Dobrushin, R., Kotecky, R., Shlosman, S.: *Wulff Construction: Global Form.* To be published by AMS
8. Dobrushin, R., Shlosman, S.: Large and Moderate Deviations in the Ising Model and Droplet Condensation. In preparation
9. Shlosman, S.: Comm.Math.Phys. **125**, 91 (1989)

Section 5

Classical Dynamical Systems and Random Perturbations

Organizer: Y. Kifer

A Variational Approach to the Random Diffeomorphisms Type Perturbations of a Hyperbolic Diffeomorphism

Yuri Kifer

Institute of Mathematics
The Hebrew University, Jerusalem, Israel

1. Let f be a diffeomorphism of a compact Riemannian manifold M. The standard model of random perturbations of f is generated by a particle which jumps from x to fx and smears with some distribution close to the δ - function at fx. This model has the continuous time version where a flow is perturbed by a small diffusion. This approach leads to a Markov chain X_n^ε (or diffusion X_t^ε, in the continuous time case) with a small parameter $\varepsilon > 0$ and one is interested whether invariant measures of X_n^ε converge as $\varepsilon \to 0$ to a particular invariant measure of the diffeomorphism f. To describe limiting measures I employed in [5] the Donsker-Varadhan variational formula

$$\lambda^\varepsilon(V) = \sup_{\mu \in \mathcal{P}(M)} \left(\int V d\mu - I^\varepsilon(\mu)\right) \tag{1}$$

for the principal eigenvalues $\lambda^\varepsilon(V)$ of the operators $P_V^\varepsilon g = P^\varepsilon(e^V g)$ where P^ε is the transition operator of the Markov chain X_n^ε, V is a continuous function, and $I(\mu)$ is certain lower semicontinuous convex functional on the space $\mathcal{P}(M)$ of probability measures on M. It turns out that if f is a hyperbolic diffeomorphism then $\lambda^\varepsilon(V)$ converges as $\varepsilon \to 0$ to the topological pressure $Q(V + \varphi^u)$ of f corresponding to the function $V + \varphi^u$ where $\varphi^u = -\log J^u(x)$ and $J^u(x)$ is the Jacobian of the differential Df restricted to the unstable subbundle. It is known that the topological pressure satisfies the variational principle which is convenient to write here in the form

$$Q(V + \varphi^u) = \sup_{\mu \in \mathcal{P}(M)} \left(\int V d\mu - I(\mu)\right), \tag{2}$$

where $I(\mu) = \begin{cases} -\int \varphi^u d\mu - h_\mu(f) & \text{if } \mu \text{ is } f \text{ - invariant} \\ \infty & \text{otherwise} \end{cases}$ and $h_\mu(f)$ is the Kolmogorov-Sinai entropy of f corresponding to μ. Let μ_V^ε and μ_V be the measures at which the supremums in (1) and (2), respectively, are attained.

Both measures are unique under natural circumstances. Then from the convergence as $\varepsilon \to 0$ of $\lambda^\varepsilon(V)$ to $Q(V+\varphi^u)$ for all continuous V it follows by the general convex analysis arguments that μ_V^ε weakly converges to μ_V as $\varepsilon \to 0$.

When $V \equiv 0$ then μ_V^ε turns out to be an invariant measure of X_n^ε and μ_V turns out to be the, so called, Sinai-Bowen -Ruelle measure μ^u which solves the problem we are interested in.

2. In order to prove the convergence of $\lambda^\varepsilon(V)$ as $\varepsilon \to 0$ one has to impose certain homogeneuity assumptions on the transition operator P^ε of X_n^ε and to rely heavily on probabilistic arguments. Following [9] I consider here the more special model of perturbations which enables one to relax conditions on P^ε and to rely only on dynamical systems methods. Still the variational approach will work in this situation, as well, The Donsker-Varadhan formula and the variational principle will be replaced here by the relativized variational principle.

The set up is the following. Let f_0 be a diffeomorphism of a Riemannian manifold M having a compact set $\Lambda \subset M$ as its hyperbolic attractor, i.e. Λ is an Axiom A basic hyperbolic set of f_0 (see, for instance, [1]) and there exists an open set $W \supset \Lambda$ with the compact closure $\overline{W}$ such that $f_0\overline{W} \subset W$ and $\bigcap_{n\geq 0} f_0^n \overline{W} = \Lambda$. Let $\mathcal{D}_\alpha(f_0)$ denotes the set of diffeomorphisms α close to f_0 in thc C^2 topology. Let $\alpha_0 > 0$ be small enough so that any $f \in \mathcal{D}_{\alpha_0}(f_0)$ has a hyperbolic attractor close to Λ and $f\overline{W} \subset W$ for each $f \in \mathcal{D}_{\alpha_0}(f_0)$. Let η^ε be a probability measure on $\mathcal{D}_{\alpha_0}(f_0)$ such that η^ε converges weakly as $\varepsilon \to 0$ to the δ - function concentrated in f_0 and any map $\psi_x : \mathcal{D}_\alpha(f_0) \to M, \psi_x(f) = f_x$ maps η^ε to a measure ν_x^ε which is absolutely continuous with respect to the Riemannian volume. This set up generates a Markov chain X_n^ε such that

$$P\{X_{n+1}^\varepsilon \in \Gamma | X_n^\varepsilon = x\} = \eta^\varepsilon\{f : fx \in \Gamma\} = \nu_x^\varepsilon(\Gamma). \tag{3}$$

A measure μ^ε is called an invariant measure of X_n^ε if $\int \nu_x^\varepsilon d\mu^\varepsilon(x) = \mu^\varepsilon$. In view of the assumption on μ_x^ε any invariant measure μ^ε of X_n^ε is absolutely continuous with respect to the Riemannian volume.

Set $\Omega = (\mathcal{D}_{\alpha_0}(f_0))^{\mathbf{Z}}$, $\Omega_+ = (\mathcal{D}_{\alpha_0}(f_0))^{\mathbb{N}}$.

Theorem *Suppose that the above conditions hold true and α_0 is small enough. Then all invariant measures μ^ε of X_n^ε weakly converge as $\varepsilon \to 0$ to the Sinai-Bowen-Ruelle measure μ^u.*

3. To prove the above Theorem I employ the relativized variational principle for the skew-product transformations (see [2]) which I am going to describe now. Set $\Omega = (\mathcal{D}_{\alpha_0}(f_0))^{\mathbf{Z}}$, $\Omega_+ = (\mathcal{D}_{\alpha_0}(f_0))^{\mathbb{N}}$, $P^\varepsilon = (\eta^\varepsilon)^{\mathbf{Z}}$, $P_+^\varepsilon = (\eta^\varepsilon)^{\mathbb{N}}$, and $F : \Omega \to \mathcal{D}_{\alpha_0}(f_0)$, $F(\omega) = \omega_0$ for $\omega = (\ldots, \omega_{-1}, w_0, w_1, \ldots) \in \Omega$. Let $\theta : \Omega \to \Omega$, $(\theta : \Omega_+ \to \Omega_+)$ be the shift and $\tau : M \times \Omega \to$

$M \times \Omega, \tau(x,\omega) = (F(\omega)x, \theta\omega)$ be the corresponding skew product transformation. If μ^ε is an invariant measure of X_n^ε then $\mu^\varepsilon \times P_+^\varepsilon$ is an invariant measure of τ restricted to $M \times \Omega_+$ and there exists a unique τ-invariant measure $\mu^\varepsilon P^\varepsilon$ on $M \times \Omega$ whose projection on $M \times \Omega_+$ is $\mu^\varepsilon \times P_+^\varepsilon$. Let $\Lambda(\omega)$ be the hyperbolic attractor of $F(\omega), \omega \in \Omega$ and $\Gamma^u(\omega) \oplus \Gamma^s(\omega)$ be the corresponding $DF(\omega)$-invariant hyperbolic splitting of the tangent bundle over Λ so that $\Gamma^u(\omega)$ is the unstable and $\Gamma^s(\omega)$ is the stable subbundle. Set $\varphi^u(x,\omega) = \varphi_\omega(x) = -\log J_\omega^u(x)$ where $J_\omega^u(x)$ is the Jacobian of the restriction of $DF(\omega)$ to $\Gamma_x^u(\omega)$. I extend the function $\varphi_\omega^u(x)$ from $\Lambda \times \Omega$ to $\overline{W} \times \Omega$ so that the extended function $\varphi(x,\omega) = \varphi_\omega(x)$ is a continuous function on $\overline{W} \times \Omega$ and the family $\{\varphi_\omega, \omega \in \Omega\}$ is equi-Hölder continuous in some neighborhood of Λ. This is possible since each φ_ω is Hölder continuous with constants close to ones for the corresponding function φ_0 of f_0.

A subset $E \subset \overline{W}$ is called (ω,δ,n)-separated set if for each pair $x, y \in E, x \neq y$ one has $\max_{0\le\ell\le n-1} \operatorname{dist}(F^\ell(\omega)x, F^\ell(\omega)y) > \delta$ where $F^\ell(\omega) = F(\theta^\ell\omega) \circ F(\theta^{\ell-1}\omega) \circ \ldots \circ F(\theta\omega)F(\omega)$. For any bounded measurable on $M \times \Omega$ and continuous on M function $V_\omega(x) = V(\omega, x)$ one can define the topological pressure $Q^\varepsilon(V)$ by (see, for instance, [2])

$$Q^\varepsilon(V) = \overline{\lim_{\delta\to 0}}\ \overline{\lim_{n\to\infty}} \frac{1}{n} \int \log\sup\Big\{\sum_{x\in E} \exp\sum_{\ell=0}^{n-1} V \circ \tau^\ell(\omega, x) : \\ E \text{ is a } (\omega,\delta,n)-\text{separated set}\Big\} dP^\varepsilon(\omega) \tag{4}$$

and the following variational principle holds

$$Q^\varepsilon(V) = \sup\left\{h_{\mu P^\varepsilon}^{(r)}(\tau) + \int V d\mu P^\varepsilon : \mu P^\varepsilon \text{ is } \tau \text{ invariant}\right\} \tag{5}$$

where $\mu P^\varepsilon(dx \times d\omega) = d\mu_\omega(x) dP^\varepsilon(\omega)$, μ_ω is a family of probability measures on M, and $h^{(r)}$ is the relativized entropy. It is easy to see that μP^ε is τ-invariant if and only if $F(\omega)\mu_\omega = \mu_{\theta\omega}$ P^ε-almost surely (a.s.). The main step in the proof of Theorem is the following result.

Proposition *For all functions V as above,*

$$\lim_{\varepsilon\to 0} Q^\varepsilon(V) = Q^0(V). \tag{6}$$

4. It is convenient to estimate the relativized topological pressure via the following integral

$$I_n^\omega(V) = \int_{\overline{W}} \exp\Big(\sum_{\ell=0}^{n-1} V \circ \tau^\ell(x, w)\Big) dm(x)$$

where m is the Riemannian volume.

Lemma 1 *Suppose that $\{V_\omega, \omega \in \Omega\}$ is an equicontinuous family of functions on $\overline{W}$ then for all $\varepsilon \geq 0$,*

$$\lim_{n\to\infty} \frac{1}{n} \int \log I_n^\omega(V) dP^\varepsilon(\omega) = Q^\varepsilon(V + \varphi). \tag{7}$$

Proof. Let E be a maximal (ω, δ, n)-separated set in $\overline{W}$. Then

$$\bigcup_{x\in E} U_{\delta/2}(\omega, n, x) \subset \overline{W} \subset \bigcup_{x\in E} U_\delta(\omega, n, x), \tag{8}$$

where $U_\delta(\omega, n, x) = \{y \in \overline{W} : \max_{0\leq \ell \leq n-1} \mathrm{dist}(F^\ell(\omega)x, F^\ell(\omega)y) \leq \delta\}$ and the left hand side of (8) is the disjoint union. Employing the same arguments as in the volume lemma from [3] modified similarly to [7] one can show that there exist $c = c_\delta > 0$ independent of $\omega \in \Omega, x \in \overline{W}$, and an integer $n > 0$ such that

$$c_\delta^{-1} \leq m(U_\delta(\omega, n, x)) \exp(-\sum_{\ell=0}^{n-1} \varphi \circ \tau^\ell(x, \omega)) \leq c_\delta. \tag{9}$$

Since the left hand side of (8) is the disjoint union then (8) and (9) together with (4) and the equicontinuity of the family $\{V_\omega, \omega \in \Omega\}$ yield (7) (cf. [6], Proposition 3.2). □

We will prove (6) showing that $Q_n^\varepsilon(V + \varphi) = \frac{1}{n} \int \log I_n^\omega(V) dP^\varepsilon(\omega)$ is uniformly in n close for small ε to $Q_n^0(V + \varphi)$.

Lemma 2 *Suppose that α_0 is small enough. Then there exists $C > 0$ such that for any $\gamma > 0, \omega \in \Omega$, an integer $n > 0$, and points $x_0, x_1, \ldots, x_n \in \overline{W}$ satisfying*

$$\mathrm{dist}(F(\theta^i\omega)x_i, x_{i+1}) \leq \gamma \tag{10}$$

for all $i = 0, \ldots, n-1$ one can find a point $y \in \overline{W}$ such that

$$\mathrm{dist}(F^i(\omega)y, x_i) \leq C\gamma \tag{11}$$

for all $i = 0, \ldots, n-1$.

Proof. This "random" shadowing property can be proved in the same way as the usual one since for small α_0 the hyperbolic splittings of all $F(\omega)$ are close to the corresponding splitting of the hyperbolic diffeomorphism f_0. □

By the same reasons one has also the following version of the specification property from [1].

Lemma 3 *If α_0 is small enough then for each $\gamma > 0$ there is an integer $N(\gamma)$ such that for any integers $i_\ell < j_\ell \leq n-1, \ell = 1, \ldots, k$ satisfying $i_{\ell+1} \geq j_\ell + N(\gamma), i_1 \geq 0$ and any points $x_1, \ldots, x_k \in \overline{W}$ one can find a point $y \in \overline{W}$ such that for all $q \in [i_\ell, j_\ell)$ and $\ell = 1, \ldots, k$,*

$$\operatorname{dist}(F^q(\omega)y, F^{q-i_\ell}(\theta^{i_\ell}\omega)x_\ell) \leq \gamma. \tag{12}$$

For $\omega \in \Omega$ define $i_\ell = i_\ell(\omega)$ and $j_\ell = j_\ell(\omega)$ such that $0 \leq i_\ell < j_\ell \leq n-1$, $j_\ell \leq i_{\ell+1}$, $\ell = 1, \ldots, k(n, \omega)$ and $F(\theta^q\omega) \in \mathcal{D}_\gamma(f_0)$ if and only if $q \in [i_\ell, j_\ell)$. By the law of large numbers

$$\lim_{n\to\infty} n^{-1} \sum_{\ell=1}^{k(n,\omega)} (j_\ell - i_\ell - 1) = \beta_\gamma(\varepsilon) \quad P^\varepsilon - a.s. \tag{13}$$

and

$$\lim_{\varepsilon\to 0} \beta_\gamma(\varepsilon) = 1. \tag{14}$$

Moreover, (13) and (14) imply

$$\lim_{n\to\infty} n^{-1} k(n,\omega) \leq 1 - \beta_\gamma(\varepsilon) \to 0 \text{ as } \varepsilon \to 0. \tag{15}$$

Denote by Ω_0 the set of $\omega \in \Omega$ for which (13) holds true. Let $E^\omega, \omega \in \Omega_0$ be a maximal (ω, γ, n) - separated set in $\overline{W}$ and E be a maximal (γ, n) - separated set for f_0 in $\overline{W}$. By Lemma 2, for each $y \in E^\omega$ there exist points $x_1, \ldots, x_k \in \overline{W}$, $k = k(n, \omega)$ such that

$$\operatorname{dist}(F^q(\omega)y, f_0^{q-i_\ell} x_\ell) \leq C\gamma \tag{16}$$

for all $q \in [i_\ell, j_\ell), \ell = 1, \ldots, k(n, \omega)$. By Lemma 3 applied to $\omega = (\ldots, w_{-1}\omega_0, w_1, \ldots), \omega_i = f_0$ for all i one finds a point $a \in \overline{W}$ such that

$$\operatorname{dist}(F^q(\omega)y, f_0^q z) \leq (C+1)\gamma \tag{17}$$

for each $q \in [i_\ell, j_\ell - N(\gamma))$ and all $\ell \leq k(n, \omega)$ such that $j_\ell - i_\ell > N(\gamma)$. Next, we can pick up $v \in E$ such that

$$\operatorname{dist}(F^q(\omega)y, f_0^q v) \leq (C+2)\gamma \tag{18}$$

for the same q and ℓ as in (17). Let $\{V_\omega, \omega \in \Omega_0\}$ be an equicontinuous and uniformly bounded family of functions on $\overline{W}$ then

$$\begin{aligned} &|\sum_{\ell=0}^{n-1} V \circ \tau^\ell(y,\omega) - \sum_{\ell=0}^{n-1} V_{\theta^\ell\omega}(f_0^\ell \varphi)| \leq \\ &\leq \|V\|(k(n,\omega)N(\gamma) + n - \sum_{\ell=1}^{k(n,\omega)} (j_\ell - i_\ell)) \\ &+ n\zeta_V((C+2)\gamma), \end{aligned} \tag{19}$$

where $\zeta_V(\delta) = \sup\{|V_\omega(x_1) - V_\omega(x_2)| : \omega \in \Omega_0, \operatorname{dist}(x_1, x_2) \le \delta\}$ and $\|V\| = \sup\{|V_\omega(x)| : \omega \in \Omega_0, x \in \overline{W}\}$. Since $\gamma > 0$ is arbitrary then Lemma 1 together with (9), (13) - (15) and (19) imply

$$\limsup_{\varepsilon \to 0} Q^\varepsilon(V) \le Q^0(V). \tag{20}$$

In order to obtain a lower bound take $x \in E$. Then by Lemma 2 there exist points $x_1, \ldots, x_k \in \overline{W}$ such that

$$\operatorname{dist}(f_0^q x, F^{q-i_\ell}(\theta^{i_\ell}\omega)x_\ell) \le C\gamma \tag{21}$$

for each $q \in [i_\ell, j_\ell)$ and all $\ell = 1, \ldots, k(n, \omega)$. By Lemma 3, there exists $y \in \overline{W}$ such that $\operatorname{dist}(f_0^q x, F^q(\omega)y) \le (C+1)\gamma$ for each $q \in [i_\ell, j_\ell - N(\gamma))$ and all $\ell \le k(n, \omega)$ for which $j_\ell - i_\ell > N(\gamma)$. By the maximality of E^ω there exists $z \in E^\omega$ such that

$$\operatorname{dist}(f_0^q x, F^q(\omega)z) \le (C+2)\gamma \tag{22}$$

for the same q as above. Again employing (9), (13) - (15), and (19) we derive

$$\liminf_{\varepsilon \to 0} Q^\varepsilon(V) \ge Q^0(V)$$

concluding the proof of (6). □

Now set

$$I^\varepsilon(\nu) = \begin{cases} -h_\nu^{(r)}(\tau) - \int \varphi d\nu & : \nu \in \mathcal{P}(M \times \Omega), \nu \text{ is } \tau\text{-invariant} \\ & \quad \text{and } proj_\Omega \nu = P^\varepsilon; \\ \infty & : \text{otherwise.} \end{cases}$$

Then for any $\varepsilon \ge 0$,

$$Q^\varepsilon(V + \varphi) = \sup\{\int V d\nu - I^\varepsilon(\nu) : \nu \in \mathcal{P}(M \times \Omega)\} \tag{23}$$

and $Q^\varepsilon(V + \varphi) \to Q^0(V + \varphi)$ as $\varepsilon \to 0$. By the convex analysis arguments, similarly to [5], these already imply that all weak limits as $\varepsilon \to 0$ of measures where the supremum in (23) is attained are equilibrium states on Λ of the diffeomorphism f_0 corresponding to $V + \varphi$. In particular, take $V \equiv 0$. By the random diffeomorphisms version of the Pesin formula (see [8]) and of the Ruelle inequality (see [4]) if μ^ε is an invariant measure of the Markov chain X_n^ε then $\mu^\varepsilon \times P^\varepsilon$ is τ-invariant and it maximizes (23) for $V \equiv 0$, i.e. $0 = Q^\varepsilon(\varphi) = \sup\{-I^\varepsilon(\nu) : \nu \in \mathcal{P}(M \times \Omega)\}$, since $h^{(r)}_{\mu^\varepsilon \times P^\varepsilon}(\tau) = -\int_{\overline{W}} \int_{\Omega} \varphi d\mu^\varepsilon dP^\varepsilon$. For $\varepsilon = 0$, there exists a unique measure μ^0 on Λ, called the Sinai-Bowen-Ruelle measure, such that (see [1]), $0 = Q^0(\varphi) = \sup\{\int \varphi d\nu + h_\nu(f_0) : \nu \in \mathcal{P}(M)$ and ν is f_0 - invariant $\} = \int \varphi d\mu^0 + h_{\mu^0}(f_0)$.

Summarizing these arguments we derive the assertion of Theorem.

Remark. The proof in [4] of the Ruelle inequality which I used above should be corrected in the following way. The mistake is in the formula (1.23) on p.164 of [4] where the integration should be taken over ${}^{n}f^{-1}\Omega_k$ in place of Ω_k. To correct this take $\Omega = \mathcal{D}(M)^{\mathbb{N}}$ and the corresponding two sided shift θ (in place of the one sided shift) and take the partition (since θ is one-to-one now) $\alpha_\varepsilon = \{\alpha_\varepsilon^k \times \theta^n \Omega_k, k = 1, 2, \ldots\}$. The rest of the proof remains the same.

References

1. Bowen, R.: Equilibrium states and the ergodic theory of Anosov diffeomorphisms. Lecture Notes in Math., **470**, Springer-Verlag (1975)
2. Bogenschütz, T.: Entropy, pressure, and a variational principle for random dynamical systems. Preprint
3. Bowen, R., Ruelle, D.: The ergodic theory of Axiom A flows. Invent.Math. **29**, 181–202 (1975)
4. Kifer, Y.: *Ergodic Theory of Random Transformations.* Birkhäuser, Boston, 1986
5. Kifer, Y.: Principal eigenvalues, topological pressure, and stochastic stability of equilibrium states. Israel J.Math. **70**, 1–47 (1990)
6. Kifer, Y.: Large deviations in dynamical systems and stochastic processes. Trans.Amer.Math.Soc. **321**, 505–524 (1990)
7. Kifer, Y., Newhouse, S.: A global volume lemma and applications. Israel J.Math. **74**, 209–223 (1991)
8. Ledrappier, F., Young, L.-S.: Entropy formula for random transformations. Probab.Theory Related Fields **80**, 217–240 (1988)
9. Young, L.-S.: Stochastic stability of hyperbolic attractors. Ergodic Theory Dynamical Systems **6**, 311–319 (1986)

Chaotic Mappings and Stochastic Markov Chains

M.L.Blank

N.I.Vavilov Institute of General Genetics USSR Ac. Sci.,
Gubkin St. 3, 117809 Moscow B-333, USSR

The main aim of this paper is to investigate some unusual properties of such well known examples of chaotic deterministic dynamical systems as piecewise expanding maps and to discuss the peculiarities of the so called variational approach proposed in the one-dimensional case by A.Lasota and J.A.Yorke in [1].

Let $X = [0,1]^d \subset \mathbb{R}^d$ (or d-dimensional unit torus) with the uniform norm $| x |= \max\{| x_i |: x \in \mathbb{R}^d, i = 1,\ldots,d\}$; $m = m_d$ - the d-dimensional Lebesgue measure, ∂Y - the boundary of some domain $Y \subset X$, $h : X \to \mathbb{R}^1$. Let us introduce functionals:

$$\mu(Y,h) = \int_Y h(x)\, d\mu(x) \quad ; \quad \mu(Y) = \mu(Y,1) \quad ;$$

$$\mathrm{Osc}(h,Y,x) = \sup\{| h(x) - h(y) |: y \in Y\} \quad ;$$

$$\mathrm{Osc}(h,Y) = \sup\{\mathrm{Osc}(h,Y,x) : x \in Y\} \quad ;$$

$$W(h,Y,t) = m(Y, \mathrm{Osc}(h, B_t(x) \cap Y, x)) \quad ; \quad B_t(x) = \{y \in X :| x - y |\leq t\} \, ;$$

$$\mathrm{var}(h,Y) = \limsup \left\{\inf\left\{t^{-1}W(\hat{h},Y,t) : \hat{h} \in h \subset L^1\right\} : t \to 0\right\} /(2d) \quad ;$$

$$\mathrm{var}(h) = \mathrm{var}(h,X) \quad , \quad \|h\|_v = \mathrm{var}(h) + \|h\| \quad , \quad \|h\| = m(X, | h(x) |) \, .$$

The functional $\mathrm{var}(\cdot,\cdot)$ we shall call a generalized variation. Remark, that this definition differs from previous definitions in [2],[3]. The Banach space of integrable functions with bounded generalized variation, equipped with the norm $\|\cdot\|_v$, we shall denote by $BV(X)$.

The map $f : X \to X$ is called piecewise expanding (PE) if there is a finite or countable partition of the compact X into disjoint connected regular domains $\{X_i\}$, such that for all i the restriction of this map $f_{|\,\mathrm{Int}(X_i)} = f_i$ is a $\mathbb{C}^2$- diffeomorphism from $\mathrm{Int}(X_i)$ onto its image and for all $x \in \mathrm{Int}(X_i)$ the Jacobean $\mathbf{D}f(x)$ of this map is r-expanding with $r > 1$ in a sense that $| \mathbf{D}f(x) * z |\geq r \mid z \mid$ for all $z \in \mathbb{R}^d$ and $\limsup_{t\to 0}\{t^{-1}\mathrm{Osc}(| \det(Df) |, B_t(x) \cap X_i \, , x)/ \mid \det(Df) \mid\} \leq C < \infty$. The partition mentioned above we shall call special partition. A domain $Y \subset X$ is said to be regular if there exists two finite functionals $\Omega(Y)$ and $\Theta(Y)$ such that for all $h \in BV(X)$

$$\mathrm{var}(h * 1(Y), X) \leq \Omega(Y) * \mathrm{var}(h, Y) + \Theta(Y) * m(Y, h).$$

Remark. For the rectangular domain Y these functionals do not depend on the dimension d, and for the cubic domain $\Omega(Y) = 2$ and $\Theta(Y) = 2/l$, where l is the length of its side.

It is well known that in the 1-dimensional case if the special partition $\{X_i\}$ is finite and $\mathrm{var}(1/ \mid f' \mid) < \infty$, then the map f has very strong stochastic properties such as the existence and the intrinsic stability of smooth (absolutely continuous with respect to the Lebesgue measure) probabilistic invariant measure μ, exponential correlations decay with respect to this measure, the Central Limit theorem holds for it and so on. The multidimensional statement is the follows.

Theorem 1 *Let f be a PE map with the finite special regular partition. Then there exists a number r_0, depending only on this partition, such that if the expanding constant $r > r_0$, then for $a = \max_i\{\Omega(X_i)\}/r < 1$ and $b = C + \max_i\{\Theta(X_i)\} < \infty$ we have for the Perron-Frobenius operator P of the map f, that*

$$\|Ph\|_v \leq a * \|h\|_v + b * \|h\| \qquad (*)$$

for all $h \in BV(X)$, and, as a consequence of this inequality, the map f has a smooth invariant measure μ with the same ergodic properties as in the one-dimensional case.

The inequality $(*)$ is the key-inequality of the variational approach because it gives a possibility to apply abstract ergodic theorem of Ionescu-Tulchea and Marinescu (see the suitable for this approach variant in [2]) to obtain the spectral decomposition of the Perron-Frobenius operator P.

There are two problems, which appear both in the one- dimensional case and in the general multidimensional one. The first problem is a generalization of these results to the case of countable special partitions, which appear naturally under the consideration of the PE first return map of the initially nonexpanding map. We shall show that without some strong restrictions such a generalization does not valid even for $d = 1$.

Theorem 2 *For any $r > 1$ there exists a PE map f with a countable Markov partition and $\mid Df(x) \mid \geq r$ on each of its elements, such that under the action of the map each smooth probabilistic measure converges to the delta-measure in the unique point.*

It is sufficient to prove one-dimensional variant of this statement. The idea of the proof is a construction of a Markov map with a countable special partition, which is linear on each of its elements and maps it into itself and its two nearest neighbours. The dynamics of densities under the action of such map may be completely described in terms of the discrete one- dimensional random walk with nonconstant transition probabilities to neighbouring states. From this model it is easily to obtain the statement of Theorem, by constructing a suitable sequence of transition probabilities.

By means of this construction we also may show the existence of smooth nonfinite invariant measures of such maps and the existence of PE maps with nonexponential rates of the correlation decay [2].

The results above demonstrate that the Lyapunov exponent may be positive but the map is not chaotic in any sense, because almost all trajectories converge to the unique point.

The second problem is due to the assumption, that the expanding constant r has to be much greater than 1. In the one-dimensional case this problem is solved by going to some degree of the map f. However in the multidimensional case the functionals $\Omega(X_i^n)$, $\Theta(X_i^n)$ may be not finite after this procedure. This problem is especially interesting in the case of so called coupled mappings, one of the most widespread model of turbulence in hydrodynamics, described as a system of infinite number of coupled oscillators. More formally, let $F_\epsilon^{(d)}$ be a system of d ϵ-coupled oscillators, that is for $\bar{x} \in X^d$

$$\left(F_\epsilon^{(d)}(\bar{x})\right)_i = \epsilon f_{i-1}(x_{i-1}) + (1-2\epsilon) f_i(x_i) + \epsilon f_{i+1}(x_{i+1}) .$$

In [4] for the case of equal smooth expanding Markov maps $f_i = f$, mapping each element of the Markov partition to the whole interval X, and a nonconstant coupling function depending on the point $\bar{x}$, such that the Markov partition remains "invariant" for $F_\epsilon^{(d)}$, it was introduced and proved the existence of so called spacetime chaos, that is a weak ergodic property under the action of time and space dynamics of this system. We shall suppose here only that f_i is a PE map with finite special partitions (with at most N elements) and that the diameters of elements are greater than some constant $C > 0$.

Theorem 3 *Suppose that* $\inf_x |f_i'(x)| > r > 2$ *for all* i *and* $\epsilon > 0$ *is small enough. Then there are constants* $0 < a < 1 < b < \infty$ *(independing on* d*) such that for any* $d > 1$ *and* $h \in BV(X^d)$ *the inequality* $(*)$ *holds for the Perron-Frobenius operator* $P = P\left[F_\epsilon^{(d)}\right]$ *of the map* $F_\epsilon^{(d)}$.

Obviously for each $\epsilon > 0$ small enough the map $F_\epsilon^{(d)}$ is a PE map and it is sufficient to verify the assumptions of Theorem 1 to prove this statement. The main difficulty here is to prove the existence of finite constants a and

b in the inequality $(*)$, because all previous (see [2],[3]) definitions give here the expanding constant r of order d (see Remark above).

Now we shall show that the condition $|f_i'(x)| > 1$ is not sufficient here. To do it we prove even a more strong statement, considering a system of only two coupled equal maps. Let us consider for $1 < r < 2$ and $c = 1/(3 + 1/r)$ a one-dimensional map $W[r] : [0,1] \to [0,1]$:

$$\begin{array}{ll} W[r]\,(x) = 1 - x/(c - c/r) & \text{for} \quad 0 \le x \le c - c/r \\ W[r]\,(x) = r(x - c) + c & \text{for} \quad c - c/r \le x \le c \\ W[r]\,(x) = -r(x - c) + c & \text{for} \quad c \le x \le c + c/r \\ W[r]\,(x) = (x - c - c/r)/(c - c/r) & \text{for} \quad c + c/r \le x \le 1/2 \\ W[r]\,(x) = W[r]\,(1 - x) & \text{for} \quad 1/2 \le x \le 1 \quad . \end{array}$$

This is a PE ergodic one-dimensional map with smooth invariant measure.

Lemma 1 *Let* $F_\epsilon^{(2)} : X^2 \to X^2 = [0,1]^2$ *be such that*

$$F_\epsilon^{(2)}(x,y) = \Big((1-\epsilon)W[r]\,(x) + \epsilon W[r]\,(y),\ \epsilon W[r]\,(x) + (1-\epsilon)W[r]\,(y)\Big).$$

Then for all sufficiently small $\epsilon > 0$ *the map* $F_\epsilon^{(2)}$ *has two smooth invariant measures, converging as* $\epsilon \to> 0$ *to delta- measures, concentrated in points* $(c, 1-c)$ *and* $(1-c, c)$.

The proof of this statement bases on the fact that for $r < 2$ there exist two squares around these points, mapping into themselves under the action of $F_\epsilon^{(2)}$. Remark, that here the stabilization of unstable fixed points occurs as the result of the interaction of two equal strongly unstable dynamical systems.

Now let f be a one-dimensional PE map with a neutral singularity at point $y \in X$ $(|\ f'(y)\ | = 1)$. In the typical case such maps have smooth nonfinite invariant measures. However in the computer modeling it appears a "trap" around the point y, such that if a trajectory hits into its neighbourhood, it never comes out. So to describe this phenomenon we must study not invariant measures but conditionally (with respect to the complement of this "trap") invariant ones. Remind that the measure μ is called conditionally f-invariant with respect to the set $Y \subset X$ if for all Borel sets $B \subset Y$ we have $\mu(B) = \mu(f^{-1}B \cap Y)/\mu(f^{-1}Y \cap Y)$. This measure corresponds to the limit as $n \to \infty$ probability that $f^{n+1}x \in Y$ if $f^k x \in Y$ for all $0 \le k \le n$ and $x \in Y$ be chosen at random.

Theorem 4 [5] *Let* f *be as above,* $y = \bigcap_{\epsilon>0} Y_\epsilon$ *and the domains* Y_ϵ *with* $m(Y_\epsilon) \le \epsilon$ *satisfy some technical assumptions. Then for each* $\epsilon > 0$ *there exists a smooth finite conditionally invariant measure, converging as* $\epsilon \to 0$ *to some nontrivial (nonfinite in the general case)* f*-invariant measure.*

It is interesting to remark that for general PE maps in distinction to Markov ones there exist domains Y without any nontrivial conditionally f-invariant measures.

In the final part of the paper we shall consider stochastically ϵ-perturbed PE maps , that is the superposition of the map and the stochastic Markov process, defined by the transition probability densities $q_\epsilon(\cdot,\cdot)$, such that $q_\epsilon(x,y) = 0$ for all $| x - y | > 2\epsilon$.

Theorem 5 *Let the map f satisfies the same assumptions as in the Theorem 1, and for all $x \in X$ and $\epsilon > 0$*

$$\int \mathrm{var}(q_\epsilon(x+z, y+z))dy \leq R < \infty ,$$

where $\mathrm{var}(q_\epsilon(x+z, y+z))$ is considered on the variable $z \in X$ with fixed $x, y \in X$. Then there exists a number r_0, such that if the expanding constant $r > r_0$, then for all $\epsilon > 0$ the ϵ-perturbed maps have smooth invariant measures, converging as $\epsilon \to 0$ to some smooth f-invariant measure.

In this case it appears the same question about the constant r_0 as in Lemma 1. This question was discussed in a long line of papers and there were a lot of wrong proofs that $r_0 = 1$. Let $d = 1$, $10 < k < \infty$ and let us consider the densities $q_\epsilon(\cdot,\cdot)$ of the form: $q_\epsilon(x,y) = k/(2\epsilon)$ if $(1-1/k)\epsilon \leq -(x-y) \leq (1+1/k)\epsilon$ and $q_\epsilon(x,y) = 0$ in other cases. Remark, that such random perturbations have an exponential rate of the correlations decay.

Lemma 2 *Let $d = 1$, $f(x) = W[r](x)$, $1 < r < 2 - 1/k$. Then for all $\epsilon > 0$ small enough the stochastically ϵ-perturbed map f_ϵ has only two invariant measures, that are smooth and converging to delta-measures as $\epsilon \to 0$.*

The proof bases on the fact that the trajectory of the stochastically ϵ-perturbed map, started from the interval $[\, c - \epsilon(r(k+1) - k + 1)/k \, , c + \epsilon(k+1)/k \,]$ remains in it with the probability 1.

References

1. Lasota, A., Yorke, J.A.: On the existence of invariant measures for piecewise monotone transformations. Trans. Amer.Math.Soc., **186**, 481-488 (1973)
2. Blank, M.L.: Small perturbations of chaotic dynamical systems. Russian Math.Surveys, **44**:6, 3-28 (1989)
3. Blank, M.L.: Stochastic properties of deterministic dynamical systems. Soviet Sci.Rev.Sect.C: Math.Phys.Rev., Vol. **6**, 243-271 (1987)
4. Bunimovich, L.A.: Sinai, Ya.G.: Spacetime chaos in coupled map lattices. Nonlinearity, **1**, 491-516 (1988)
5. Blank, M.L.: Marginal singularities, almost invariant sets and small perturbations of chaotic dynamical systems. Chaos (to appear)

Section 6

Disordered Systems

Organizer: A. Kupiainen

Quantum Spin Systems in a Random Environment

Abel Klein*

Department of Mathematics, University of California,
Irvine, CA 92717, USA

Quantum spin systems with random parameters have been introduced to study the effect of impurities in several physical systems (e.g., [1],[2], where models related to superfluidity and superconductivity are discussed). The finite volume Hamiltonian is typically of the form $H_\Lambda = JH_{0,\Lambda} + V_\Lambda$ where $H_{0,\Lambda}$ favors long-range order and V_Λ has discrete spectrum with a unique ground state in which the elementary excitations of the system are perfectly localized, so for $J = 0$ the ground state correlation function is a δ-function. One or both of $H_{0,\Lambda}$ and V_Λ may contain random parameters. J is a coupling constant.

In the non-random homogeneous case the correlation functions typically decay exponentially for J small and exhibit long range order for J large. In the random case such behavior should persist, albeit with probability one.

The small J regime corresponds to large quantum fluctuations. In fact, some of these models were introduced to study the effect of quantum fluctuations on granular superconductors [3] and on superfluidity [1]. The large quantum fluctuations make the elementary excitations of the system localized in the ground state, so the corresponding correlation functions decay exponentially.

The ground state behavior of d-dimensional quantum systems corresponds to the behavior of $(d+1)$-dimensional classical systems. The novel feature in the random case is that the randomness in the corresponding classical system is only d-dimensional, i.e., it is completely correlated in the extra dimension [1]. This makes the study of these systems much harder than in the usual case (isotropic or $(d+1)$-dimensional disorder), since the singular sets responsable for Griffiths singularities are now cylinders infinitely extended in the extra dimension [4],[5],[6].

We will discuss three models:

---

* Partially supported by the NSF under grants DMS-8905627 and INT-8703059

I Ising Model with a Transverse Field in a Random Environment

$$H_\Lambda = -J \sum_{<x,y>\subset\Lambda} J_{<x,y>}\sigma_3(x)\sigma_3(y) - \sum_{x\in\Lambda} h(x)\sigma_1(x) \quad ,$$

where $\Lambda \subset \mathbb{Z}^d$ finite, $< x, y >$ denotes pairs of nearest neighbors sites, σ_1, σ_3 are the usual Pauli spin 1/2 matrices, $J > 0$, and $\mathbf{J} = \{J_{<x,y>}; < x, y >\subset \mathbb{Z}^d\}$ and $\mathbf{h} = \{h(x); x \in \mathbb{Z}^d\}$ are independent families of independent identically distributed strictly positive random variables (see [7] for physical applications).

H_Λ has a unique ground state Ω_Λ. The imaginary time correlation function

$$G_\Lambda((x,t),(y,s)) = \frac{(\Omega_\Lambda, \sigma_3(x)e^{-|t-s|H_\Lambda}\sigma_3(y)\Omega_\Lambda)}{(\Omega_\Lambda, e^{-|t-s|H_\Lambda}\Omega_\Lambda)}$$

is monotone increasing in Λ and in each $J_{<x,y>}$, and monotone decreasing in each $h(x)$. The infinite volume correlation function $G((x,t),(y,s))$ is thus well defined with the monotonicity in each $J_{<x,y>}$ and $h(x)$ being preserved [5].

The non-random homogeneous case, $h(x) \equiv h > 0$ and $J_{<x,y>} \equiv 1$, is one of the simplest quantum spin systems with a non-trivial phase diagram. The relevant parameter is $\zeta = h/J$. In any dimension d there exists $0 < \zeta_1 \leq \zeta_2 < \infty$ such that if $\zeta > \zeta_2$ the correlation function decays exponentially and, if $\zeta < \zeta_1$ there is long range order [8],[9],[5]. If ζ is large enough, there is Ornstein-Zernike decay [10]. If $d = 1$, we have $\zeta_1 = \zeta_2 = 2$ [11].

It follows from the monotonicity in each $h(x)$ and $J_{<x,y>}$ that, if $h(x)/_{JJ_{<x,y>}} \geq \zeta_2 + \epsilon$ for all $x, < x, y >$ and some $\epsilon > 0$, then $G((x,t),(y,s))$ decays exponentially. Similarly, if $h(x)/_{JJ_{<x,y>}} \leq \xi_1 - \epsilon$ for all $x, < x, y >$ and some $\epsilon > 0$, we have long range order. In particular, if $0 < h_1 \leq h(x) \leq h_2 < \infty$ and $0 < J_1 \leq J_{<x,y>} \leq J_2 < \infty$, we have exponential decay for small J and long range order for large J. Thus the interesting non-trivial case is when $h(x)/_{JJ_{<x,y>}}$ can take values both above ζ_2 and below ζ_1 with non-zero probability, so the system exhibits Griffiths singularities.

We can now give a complete picture for this model:

Theorem 1 [5],[12] *For any $d = 1, 2, \cdots$, there exists $\delta(d) = 4d^2 + 0(d)$, such that, if*

$$\mathbf{E}([\log(1 + 1/h(x)]^\delta) < \infty \quad \textit{and} \quad \mathbf{E}([\log(1 + J_{<x,y>})]^\delta) < \infty$$

for some $\delta > \delta(d)$, then there exists $q > 1$ such that for any $m > 0$ we can find $J_1 > 0$ such that if $0 < J < J_1$, we have, with probability one, that for all $x \in \mathbb{Z}^d$

$$G((x,t),(y,s)) \leq C_{x,\mathbf{h},\mathbf{J}} e^{-m(|x-y|+(\log(1+|t-s|))^q)}$$

for all $y \in \mathbb{Z}^d, t, s \in \mathbb{R}$, *with* $C_{x,\mathbf{h},\mathbf{J}} < \infty$.

Theorem 2 [5] *Let* $d \geq 2, h(x), J_{<x,y>}$ *with arbitrary distributions. There exists* $J_2 < \infty$ *such that if* $J > J_2$ *we have, with probability one,*

$$\overline{\lim}_{|y|\to\infty} G((x,t),(y,s)) > 0$$

for any $x \in \mathbb{Z}^d, t, s \in \mathbb{R}$.

Theorem 3 [13] *Let* $d = 1$. *There exists* $\delta_1 > 0$ *such that, if* $\mathbf{E}([\log(1+1/J_{<x,y>})]^\delta) < \infty$ *and* $\mathbf{E}([\log(1+h(x))]^\delta) < \infty$ *for some* $\delta > \delta_1$, *we can find* $J_3 < \infty$ *such that, if* $J > J_3$ *we have, with probability one,*

$$\overline{\lim}_{y\to\pm\infty} \inf_{s,t\in\mathbb{R}} G((x,t),(y,s)) > 0$$

for all $x \in \mathbb{Z}$.

An important feature of this model is that $G((x,t),(y,s))$ can be written as the limit of two-point functions of $(d+1)$-dimensional classical Ising models with d-dimensional disorder, of the type studied in [4],[14]. It follows that $G((x,t),(y,s)) \leq \tilde{G}((x,t),(y,s))$, where $\tilde{G}$ is the connectivity function of a certain independent percolation process (with random parameters) in $\mathbf{Z}^d \times \mathbf{R}$, closely related to the contact process (see [4],[5],[12],[13]). Theorem 1 is proven by a multiscale analysis in this percolation process. The probabilistic estimates are as in [15]. But we must control the decay in the t-direction inside the singular sets, which are now infinitely extended in the t-direction. This is given by the estimate [4],[5],[12]:

$$-\frac{1}{|t-s|} \log \tilde{G}_\Lambda((x,t),(y,s)) \geq$$

$$\geq \frac{1}{2\rho} e^{-\left[2\rho \sum_{<x,y>\subset\Lambda} J_{<x,y>} - \sum_{x\in\Lambda} \log(1-e^{-\rho h(x)})\right]} \tag{1}$$

for any $\rho > 0, |t-s| >> \rho$, where $\tilde{G}_\Lambda$ is the connectivity function in the cylinder with base $\Lambda \subset \mathbb{Z}^d$.

Since the contact process is the oriented percolation process associated with the above discussed percolation process, Theroem 1 also gives extinction of the contact process in a random environment [12].

Theorem 2 is proven by a simple percolation argument that shows that we can find a one-dimensional model with long range order inside our d-dimensional model, monotonicity in the couplings then imply long range order [4],[5].

Theorem 3 uses a duality argument in the percolation representation. Up to boundary conditions the dual process is of the same form except that $\{JJ_{<x,y>}\}$ and $\{h(x)\}$ are interchanged. Theorem 1 then gives summability of the connecting function of the dual process, and from that one gets long range order [13].

II Disordered Array of Quantum Rotators

$$H_\Lambda = -J \sum_{<x,y>\subset\Lambda} J_{<x,y>} \cos(\varphi(x)-\varphi(y)) - \sum_{x\in\Lambda} \frac{h(x)}{2} \frac{\partial^2}{\partial\varphi(x)^2}$$

with $J, \mathbf{J}, \mathbf{h}$ as in the previous model, $\frac{\partial^2}{\partial\varphi^2}$ acting on $L^2([-\pi,\pi], d\varphi)$ with periodic boundary condition. Such ferromagnetically coupled quantum rotators (Josephson junctions) have been used in the physics literature to describe the effect of quantum fluctuations in granular superconductors [3],[16],[1],[2].

As in I, H_Λ has a unique ground state Ω_Λ and the imaginary time ground state correlation function

$$G_\Lambda((x,t),(y,s)) = \frac{(\Omega_\Lambda, e^{i\varphi(x)}e^{-|t-s|H_\Lambda}e^{-i\varphi(y)}\Omega_\Lambda)}{(\Omega_\Lambda, e^{-|t-s|H_\Lambda}\Omega_\Lambda)}$$

has the same monotonicity properties in Λ, $J_{<x,y>}$ and $h(x)$ so $G((x,t),(y,s))$ is similarly defined with the same monotonicity properties (see [6]).

The non-random homogeneous case is similar to the one for the previous model, except that we do not have long range order in one dimension, just a Kosterlitz-Thouless phase with polynomial decay of correlations [9],[17]. The random (or at least non-homogeneous) case when $h(x)/J_{<x,y>}$ is bounded from above and/or below is again similar to the one discussed in I, except that for $d=1$ we can only guarantee a polynomial lower bound for $h(x)/J_{<x,y>}$ bounded from above and J large. Again the interesting case is when we have Griffiths singularities. We have:

Theorem 4 [6] *Let* $d=1,2,\cdots, \mathbf{E}([\log(1+J_{<x,y>})]^\eta) < \infty$ *with* $\eta > 2d$. *Suppose either*

(i) $\mathbf{E}(h(x)^{-\kappa}) < \infty$ *for some* $\kappa > 0$

or

(ii) $\mathbf{E}([\log(1+1/h(x))]^\delta) < \infty$ *for some* $\delta > 2d$.

Then, given $m > 0$ *we can find* $J_1 > 0$ *such that if* $0 < J < J_1$ *we have, with probability one, that for all* $x \in \mathbb{Z}^d$

$$G((x,t),(y,x)) \leq C_{x,\mathbf{h},\mathbf{J}} e^{-m(|x-y|+g(t-s))}$$

for all $y \in \mathbb{Z}^d, t, s \in \mathbb{R}$, *where either*

$$g(t) = |t|^{1/v} \textit{ for some } v > 1 \textit{ if (i) holds,}$$

or

$$g(t) = (\log(1 + |t|))^q \textit{ for some } q > 1 \textit{ if (ii) holds,}$$

and $C_{x,\mathbf{h},\mathbf{J}} < \infty$.

The proof uses a representation of $G((x,t),(y,s))$ as the limit of two-point function of $(d+1)$-dimentional classical rotators with d-dimensional disorder. Theorem 4 is proven by a multiscale analysis in these classical systems, using the Simon-Lieb-Rivasseau inequality. The singular regions are infinitely extended in the t-direction. The new feature is the presence of a continuous symmety which allows the control of the decay in the t-direction in the singular regions through the use of Spencer-McBryan bounds. This is given by the estimate [6]:

$$-\frac{1}{|t-s|} \log G_\Lambda((x,t),(y,s)) \geq \left[\sum_{x \in \Lambda} 1/h(x)\right]^{-1} . \tag{2}$$

Notice that (2) is better than (1).

Long range order is not yet proven for this model. A version of Theorem 2 still applies for $d \geq 2$, but instead of long range order it only guarantees that the decay is not faster than polynomial, since the non-random homogeneous model has only a Kosterlitz-Thouless transition for $d = 1$.

III Quantum $X - Y$ Model with a Random Transverse Field

$$H_\Lambda = -J \sum_{<x,y> \in \Lambda} (\sigma_+(x)\sigma_-(y) + \sigma_+(y)\sigma_-(x)) + \sum_{x \in \Lambda} h(x)\sigma_3(x)$$

where $J \in \mathbb{R}, \mathbf{h} = \{h(x); x \in \mathbb{Z}^d\}$ are i.i.d.r.v.'s, σ_+, σ_- are Pauli spin 1/2 matrices.

This model was used to study the effect of disorders in superfluidity and superconductivity [1].

If H_Λ has a unique ground state Ω_Λ, we define

$$G_\Lambda(x,y) = (\Omega_\Lambda, \sigma_+(x)\sigma_-(y)\Omega_\Lambda) .$$

$G(x,y)$ will denote a thermodynamic limit.

For this model there is still a lot to be proven, even in the non-random homogeneous case, $h(x) \equiv h$. If $h \neq 0, |J/h|$ small, H_Λ has a unique ground state and $G(x,y)$ decays exponentially [8],[18]. If $d \geq 2$, there is long range order for $h = 0$ [19], nothing is proven for $|J/h|$ large, $h \neq 0$. If $d = 1, h = 0$, we have $G(x,y) \approx C|x-y|^{-2/\pi}$ [20]. The methods of [20],[21] should also give polynomial decay for $|J/h|$ large, $h \neq 0$, establishing a Kosterlitz-Thouless transition.

In the random case there are no results for $d \geq 2$ (see, however, [22]). But for $d = 1$ we have a complete picture (we set $G_L(x,y) = G_\Lambda(x,y)$ for $\Lambda = [-L, L] \cap \mathbb{Z}$):

Theorem 5 [21] *Let $d = 1$, and suppose the distribution of $h(x)$ is not concentrated in a single point and $\mathbf{E}(|h(x)|^\eta) < \infty$ for some $\eta > 0$. Then for any J there exists $m > 0$ such that we have, with probability one, that for all $x \in \mathbb{Z}$*

$$\sup_L |G_L(x,y)| \leq C_{x,\mathbf{h}} e^{-m|x-y|}$$

for all $y \in \mathbb{Z}$ with $C_{x,\mathbf{h}} < \infty$.

This result shows that, in this model, a minimum amount of disorder destroys the Kosterlitz-Thouless transition. But notice that this is not true for the disordered quantum rotator.

The theorem is proven by mapping the model into a free Fermi gas, following [20], whose one particle Hamiltonian turns out to be the one-dimensional Anderson Hamiltonian $-J\Delta + h$. The exponential decay with probability one then follows from one-dimensional Anderson localization.

References

1. Ma, M., Halperin, B.I., Lee, P.A.: Phys.Rev. B **34**, 3136-3143 (1986)
2. Fisher, M.P.A., Weichman, P.B., Grinstein G., Fisher, D.S.: Phys.Rev. B **40**, 546-570 (1989)
3. Doniach, S.: Phys. Rev. B **24**, 5063 (1981)
4. Campanino, M., Klein, A.: Comm.Math.Phys. **135**, 483-497 (1991)
5. Campanino, M., Klein, A., Perez, J.F.: Comm.Math.Phys. **135**, 499-515 (1991)
6. Klein, A., Perez, J.F.: Localization in the ground state of a disordered array of quantum rotators. Preprint (1991)
7. de Gennes, P.G.: Solid State Commun. **1**, 132 (1963); Wang, Y.L., Cooper, B.: Phys.Rev. **172**, 539 (1968)
8. Ginibre, J.: Comm.Math.Phys. **14**, 205-234 (1969)
9. Driessler, W., Landau L.J., Perez, J.F.: J.Stat.Phys **20**, 123-162 (1979)
10. Kennedy, T.: Comm.Math.Phys. **137**, 599-615 (1991)
11. Pfeuty, P.: Ann.Physics **57**, 79-90 (1970)
12. Klein, A.: Extinction of Contact and Percolation Processes in a Random Environment. In preparation
13. Aizenman, M., Klein, A., Newman, C.: In preparation

14. McCoy, B., Wu, T.T.: Phys Rev. B **76**, 631 (1968);
Shankar, R., Murthy, G.: Phys.Rev. B **36**, 536-547 (1987)
15. von Dreifus, H.: Ph.D. Thesis, New York University (1987);
von Dreifus, H., Klein, A.: Comm.Math.Phys. **124**, 283-299 (1989)
16. Abeles, B.: Phys.Rev. B **15**, 2828 (1977);
Jacobs, L., José, J.V.: Physica B **152**, 148 (1988);
Fisher, M.P.A., Grinstein, G.: Phys.Rev.Lett. **60**, 208 (1988);
Kim, S., Choi, M.Y.: Phys.Rev. B **41**, 111 (1990)
17. Bonato, C.A., Perez, J.F.: J.Stat.Phys. **56**, 13-22 (1989)
18. Albanese, C.: J.Stat.Phys. **55**, 279 (1989);
Albanese, C.: J.Stat.Phys.. **55**, 297 (1989)
19. Kennedy, T., Lieb, E.H., Shastry, B.S.: Phys.Rev.Lett. **61**, 2582 (1988)
20. Lieb, E.H., Schultz, T.D., Mattis, D.C.: Ann.Phys. **16**, 407-466 (1961)
21. Klein A., Perez, J.F.: Comm.Math.Phys. **128**, 99-108 (1990)
22. Albanese, C.: Comm.Math.Phys. **134**, 1-27 (1990);
Albanese, C.: Comm.Math.Phys. **134**, 237-272 (1990)

Rounding Effects in Systems with Static Disorder

Jan Wehr

Institute for Advanced Study
Princeton, NJ 08540, USA

Condensed matter physics often has to consider systems with *static disorder* [6],[7], i.e. with impurities, dislocations, substitutions etc. which vary from sample to sample (thus introducing *disorder*) but which do not exhibit thermal fluctuations on relevant time scales (hence the word *static*). To account for such disorder mathematically one often uses lattice spin systems with random parameters in the interaction (e.g. random magnetic fields or random coupling constants). For each fixed realization of these parameters one then obtains a spin system in which the usual quantities of physical interest - magnetization, free energy etc. - can be calculated. Random parameters of this type are often called quenched, to stress the fact that they remain constant during the calculation of spin averages - corresponding to the static nature of the disorder in the modelled physical system.

The present note discusses absence of some first order phase transitions in spin systems related to presence in these systems of quenched randomness. The results presented here are special cases of those obtained in a joint work with Michael Aizenman and published in [1] and [2].

Among the most widely studied spin systems with quenched randomness are Random Field (RF) models, i.e. models with the interaction:

$$H = -\sum_{|x-y|=1} \sigma_x \cdot \sigma_y - \sum_x (h + \varepsilon\eta_x) \cdot \sigma_x \,.$$

Here σ_x are spin variables indexed by elements of the d-dimensional lattice $\mathbb{Z}^d$, and $h + \varepsilon\eta_x$ represents a random field variable at site x. It consists of two terms - deterministic (h) and random ($\varepsilon\eta_x$). In the RF Ising Model σ_x takes values 1 and -1 and $h, \eta_x \in \mathbb{R}$. In the vector case $\sigma_x \in S^N$ (the N-dimensional unit sphere; $N = 2$ for the Heisenberg case) and $h, \eta_x \in \mathbb{R}^{N+1}$. We assume that

$$E(\eta_x) = 0\,; \quad E(\eta_x^2) = 1\,,$$

so that ε sets the effective scale of the random field.

It is known that for a general class of systems, including those introduced above, the free energy per unit volume becomes a nonrandom quantity in the infinite volume limit, i.e.

$$\lim_{|\Lambda|\to\infty}[\frac{1}{\beta|\Lambda|}\log Z_\Lambda] = f(\beta, h, \varepsilon),$$

with f independent of $\{\eta_x\}$.

Theorem 1 *In the RF Ising Model on the two-dimensional lattice* $\mathbb{Z}^d$, f *is a differentiable function of* h *for all* β *and* h, *provided* ε *is positive.*

Remark. For $\varepsilon = 0$ the resulting system is the standard Ising model, which does exhibit spontaneous magnetization at low temperatures. It can be shown that in any dimension the discontinuity of the derivative,

$$\left(\frac{\partial f}{\partial h}\right)_+ - \left(\frac{\partial f}{\partial h}\right)_-$$

equals the spontaneous magnetization averaged over the field variables η_x. The theorem thus says that in the two-dimensional case the average spontaneous magnetization vanishes in the presence of arbitrarily weak randomness. As a consequence, spontaneous magnetization, which is nonnegative, vanishes for almost all configurations $\{\eta_x\}$. The behavior in higher dimensions is drastically different as has been shown by Bricmont and Kupiainen [3] and earlier by Imbrie [4] in the zero temperature case (not discussed here).

Theorem 2 *In the RF Heisenberg Model (or, more generally,* N*-vector model,* $N \geq 2$*) in* $d \leq 4$, f *is a differentiable function of* h *at* $h = 0$ *for any* β, *provided* $\varepsilon > 0$ *and the distribution of* η_x *is spherically symmetric and absolutely continuous.*

Remark. The standard (nonrandom)N-vector model for $d \geq 3$ has long range order at low temperatures, manifested in particular by nondifferentiability of f at 0. The theorem thus says that this effect is wiped out by the presence of random fields with the required properties. Unlike in the Ising case, there are no results known about the behavior of the system for $d > 4$ for large β and small ε.

Theorems 1 and 2 can be summarized, somewhat imprecisely, by saying that the random fields "round off" the first-order phase transitions present in the $\varepsilon = 0$ versions of the considered models. Hence the name "rounding effect" used in [1],[2] for this type of behavior. It should be added that the above rigorous results were preceded by (and based on) an appealing heuristic scaling argument due to Imry and Ma [8].

Random Field models are not the only ones where rounding occurs. As another interesting example let us consider the Random Bond Potts Model, in which the "spin" variables σ_x take values $1, \ldots, q$ and the interaction is defined as

$$H = - \sum_{|x-y|=1} (J + \varepsilon J_{x,y}) \delta_{\sigma_x, \sigma_y} ,$$

where the random couplings $J + \varepsilon J_{x,y}$ have the structure similar to that of the random fields in the models considered above. In particular

$$E(J_{x,y}) = 0 ; \quad E(J_{x,y}^2) = 1 .$$

It is known [5] that for $\varepsilon = 0$, $d \geq 2$ and q large, the free energy of this system (in the infinite volume limit) is a nondifferentiable function of J at a certain value J_c.

Theorem 3 *If $d = 2$ and $\varepsilon > 0$, then the free energy in the above model is everywhere differentiable in J.*

The above theorem can also be formulated in terms of the corresponding order parameter, in this case the latent heat of the system at the critical value of J (see [2]).

The reader is referred to [2] for the general formulation and proofs of theorems about rounding effects. Reference [9] contains a simplified (but complete) proof in the special case of the Random Field Ising Model.

References

1. Aizenman, M., Wehr, J.: Rounding of First-Order Phase Transitions in Systems with Quenched Disorder. Phys.Rev.Lett. **62**, 2503 (1989)
2. Aizenman, M., Wehr, J.: Rounding Effects of Quenched Randomness on First-Order Phase Transitions. Comm.Math.Phys. **130**, 489 (1990)
3. Bricmont, J., Kupiainen, A.: Lower critical Dimension for the Random-Field Ising Model. Phys.Rev.Lett. **59**, 1829 (1987);
   Bricmont, J., Kupiainen, A.: Phase Transition in the Random-Field Ising Model. Comm.Math.Phys. **116**, 539 (1988)
4. Imbrie, J.Z.: Lower Critical Dimension of the Random-Field Ising Model. Phys.Rev.Lett. **53**, 1747 (1984);
   Imbrie, J.Z.: The Ground State of the Three-Dimensional Random-Field Ising Model. Comm.Math.Phys. **98**, 145 (1985)
5. Kotecky, R., Shlosman, R.: First-Order Phase Transitions in Large Entropy Lattice Models. Comm.Math.Phys. **83**, 493 (1982)
6. Balian, R., Manyard, R., Toulouse,G. (eds.): *Ill-Condensed Matter.* North-Holland, Amsterdam (1979)

7. Osterwalder, K., Stora, R. (eds.): *Critical Phenomena, Random Systems, Gauge Theories.* Vol.II. North-Holland, Amsterdam (1986)

8. Imry, Y, Ma, S.-k: Random-Field Instability of the Ordered State of Continuous Symmetry. Phys.Rev.Lett. **35**, 1399 (1975)

9. Wehr, J.: Effects of Static Disorder in Statistical Mechanics: Absence of a Phase Transition in the Two-Dimensional Random-Field Ising Model. In: Morrow, G.J., Yang, W.S. (eds.): Probability Models in Mathematical Physics. World Scientific (1991)

Section 7

Nonequilibrium Statistical Mechanics

Organizer: J. Fritz

Derivation of the Euler Equation from Hamiltonian Systems with Negligible Random Noise

Horng-Tzer Yau

Courant Institute of Mathematical Sciences,
New York University, New York, NY 10012

The Euler equation of conservation law has been one of the fundamental equations in fluid dynamics since its discovery two centuries ago. Although there are disputes regarding its maximum range of applicability, it has been firmly established in suitable region. In principle, it is a logical consequence of Newtonian mechanics and a rigorous derivation of it from Newton's equation should be possible. Certainly some scaling has to be chosen and Euler equation is exact only in the scaling limit. This problem indeed is much harder than it appears, as Euler equation involves thermodynamical quantities such as pressure and temperature while classical Hamiltonian systems are characterized by the pair potential. So there is a link via classical statistical mechanics which does not enter explicitly in the classical Hamiltonain systems. In other words, to prove Euler equation from classical Hamiltonian systems one in a certain sense justifies Boltzmann's principle from classical mechanics. So far no one knows how to achieve this except in some artificial systems. The known results in this direction include: a heuristic derivation of Euler equation by assuming some strong ergodic property and all equilibrium states being Gibbs; one dimensional hard rod systems; Lanford's theorem in the small time and low density limit. (For a review see [3].) In this note, I shall report some recent progress in this direction done in collaboration with S. Olla and S. R. S. Varadhan [2].

To fix the notation let $\{x_\alpha^i, p_\alpha^i\}$, $i = 1, 2, 3$, $\alpha = 1, \ldots, N$ be the coordinates and momenta of N particles in $\mathbb{R}^3$. Let $H(x, p)$ be the Hamiltonian with pair potential V, namely (with all masses set equal to 1)

$$H = \frac{1}{2} \sum_\alpha \sum_i (p_\alpha^i)^2 + \frac{1}{2} \sum_{\alpha \neq \beta} V(x_\alpha - x_\beta) \, .$$

We shall assume that V is short range and stable so that the usual thermodynamic formalism is valid. The classical Hamiltonian systems of these N particles are characterized by the Hamiltonian equation

$$dx_\alpha/dt = \partial H/\partial p_\alpha$$
$$dp_\alpha/dt = -\partial H/\partial x_\alpha .$$

In addition to the total number of particles which is clearly conserved there are four additional conserved quantities of interest. They are the three components of the total momenta and the total energy given by the Hamiltonian itself.

Let $\epsilon > 0$ be a small scale parameter. We shall replace the potential $V(x)$ by $V(x/\epsilon)$ in order that the potential has effective range ϵ. The number of particles is assumed to be of order ϵ^{-3} so that each particle typically interacts with only finite number of particles. Consider the following empirical measures representing the spatial distribution of density, momenta and energy, as functions of time.

$$\begin{aligned}
\xi^0_\epsilon(dx,t) &= N^{-1}\sum_\alpha \delta(x - x_\alpha(t))\,dx \\
\xi^i_\epsilon(dx,t) &= N^{-1}\sum_\alpha \delta(x - x_\alpha(t))\,p^i_\alpha(t)\,dx\ ,\ i = 1,2,3 \\
\xi^4_\epsilon(dx,t) &= N^{-1}\sum_\alpha \delta(x - x_\alpha(t))\,h_\alpha(t)\,dx
\end{aligned} \tag{1}$$

Here h_α is defined by

$$h_\alpha(t) = \frac{1}{2}\sum_i (p^i_\alpha(t))^2 + \frac{1}{2}\sum_{\alpha\neq\beta} V(x_\alpha(t) - x_\beta(t)) .$$

The task to derive Euler equation is to prove that as $\epsilon \to 0$ these empirical measures have limits and the limiting measures have densities $\rho(x,t)$, $\rho(x,t)v(x,t)$ and $\rho(x,t)\theta(x,t)$ which satisfy the Euler equation

$$\begin{aligned}
\left[\frac{\partial}{\partial t} + \sum_{i=1}^3 v_i \frac{\partial}{\partial x_i}\right]\rho(x) + \rho(x)\sum_{i=1}^3 \frac{\partial}{\partial x_i} v_i &= 0 \\
\rho\left[\frac{\partial}{\partial t} + \sum_{i=1}^3 v_i \frac{\partial}{\partial x_i}\right] v_j + \frac{\partial}{\partial x_j} P &= 0 \\
\rho\left[\frac{\partial}{\partial t} + \sum_{i=1}^3 v_i \frac{\partial}{\partial x_i}\right]\theta + \sum_{i=1}^3 \frac{\partial}{\partial x_i}[P v_i] &= 0 .
\end{aligned} \tag{2}$$

Here P is a function of p, v and θ determined via classical statistical mechanics.

We are unable to solve this conjecture as it stands, but with some modifications we have the following results.

Let us start by reviewing our modifications.

A. We shall replace $\mathbb{R}^3$ by T^3, torus of three dimensions.
B. The classical kinetic energy is replaced by relativistic kinetic energy $c\sqrt{p^2+c^2}-c^2$. In the classical limit $c \to \infty$ we recover the classical kinetic energy $p^2/2$.
C. The addition of random noise in the velocity exchanges. Roughly speaking, for two particles of distance of order ϵ apart, they exchange velocity randomly but in a way to preserve the total energy and total momenta. The strength of noise is chosen so that no viscosity term will appear in the final macroscopic equation while it is strong enough to conclude that any extremal invariant measure is constant Lebesgue in any connected component of constant total energy and constant total momenta.

The last modification is intrinsically much more serious than the first two. Without this assumption, we are unable to prove invariant measures of Hamiltonian systems being Gibbs. The precise form of the noise is not important, see discussion in [2].

Main result [2]

Assume the modifications (A), (B) and (C). Assume also that the Euler equation has a smooth solution in $[0, t_0]$ *with the energy, momenta and density strictly lying in the single phase region of the Gibbs states with potential* V. *If the initial distribution is a local Gibbs state characterized by the initial energy, momenta and density then the empirical measures defined in* (1) *have unique limits as* $\epsilon \to 0$. *Furthermore, their densities with respect to the Lebesgue measure satisfy the Euler equation* (2).

The assumption that the initial density is local Gibbs can be relaxed somewhat but the smoothness assumption on the solutions to Euler equations is crucial. We are unable to handle any type of singularities in hyperbolic systems. Our method based on the approach of relative entropy in [5]. This method applies to several other models including the interacting Brownian motions and interacting Ornstein-Uhlenbeck processes in dimension $d > 1$. (The case $d = 1$ was solved in [1],[4].) We are unable to review this method here but refer the interested readers to [5], where a simpler model on lattices was solved by this approach.

Acknowledgments

Research partially supported by U. S. National Science Foundation grant DMS-9101196, A.P. Sloan Foundation fellowship and David and Lucile Packard fellowship.

References

1. Olla, S., Varadhan, S.R.S.: Scaling limit for intereacting Ornstein-Uhlenbeck processes. Comm.Math.Phys. **135**, 355–378 (1991)
2. Olla, S., Varadhan, S. R. S., Yau, H.-T.: Derivation of Euler equation from Hamiltonian systems with negligible random noise. In preparation
3. Spohn, H.: Large scale dynamics of interacting particles. Preprint, 1989. To appear in Lecture Notes in Phys., Springer-Verlag
4. Varadhan, S. R. S.: Scaling limits for interacting diffusions. Comm.Math. Phys. **135**, 313–353 (1991)
5. Yau, H.-T.: Relative entropy and hydrodynamics of Ginzburg-Landau models. Lett.Math.Phys. **22**, 63–80 (1991)

The Spatial Structure in Diffusion Limited Two-Particle Reactions

Maury Bramson [1] and Joel L. Lebowitz [2]

[1]Department of Mathematics, University of Wisconsin, Madison, WI 53706

[2]Departments of Mathematics and Physics, Rutgers University, New Brunswick, NJ 08903

Abstract: We analyze the limiting behavior of the densities $\rho_A(t)$ and $\rho_B(t)$, and the random spatial structure $\xi(t) = (\xi_A(t), \xi_B(t))$ for the diffusion controlled chemical reaction $A + B \to$ inert with equal initial densities. There is a change in behavior from $d \leq 4$, where $\rho_A(t) = \rho_B(t) \approx c_d/t^{d/4}$, to $d \geq 4$, where $\rho_A(t) = \rho_B(t) \approx c_d/t$ as $t \to \infty$. There is a corresponding change in the spatial structure. In $d < 4$, the particle types separate with only one type present locally, and ξ, after suitable rescaling, tends to a random Gaussian process. In $d > 4$, both particle types are, after large times, present locally in concentrations not depending on type or location. In $d = 4$, both particle types are present locally but with random concentrations, and the process tends to a limit.

1 Introduction

Consider a system of particles of two types on $\mathbb{Z}^d$, A and B, which execute simple random walks in continuous time. That is, the motion of different particles is independent and a particle at site x will jump to a given one of its $2d$ nearest neighbors at rate $1/2d$. Particles are assumed not to interact with their own type – multiple A particles or multiple B particles can occupy a given site. However, when a particle meets a particle of the opposite type, both disappear. (When a particle simultaneously meets more than one particle of the opposite type, it will cause only one of these particles to disappear.)

To study the time evolution of this system one needs to specify an initial measure for the process. We consider here the case in which one independently throws down A and B particles according to the homogeneous

Poisson random measures with densities r_A and r_B; if there are initially both A and B particles at x, they immediately cancel each other out as much as possible. We denote by $\xi(t) = (\xi_A(t), \xi_B(t))$ the random state of the system at time t; $\xi(G;t)$ denotes the number of particles of each type in the set $G \subset \mathbb{Z}^d$.

This process can serve as a model for the irreversible chemical reaction $A + B \to$ inert, where both particle types A and B are mobile. A and B can also represent matter and antimatter. There has been much interest in this model over the last several years following papers by Ovchinikov and Zeldovich [1], and Toussaint and Wilczek [2]; see Bramson and Lebowitz [3] for a more complete set of references. One question concerns the behavior of the densities in spatially homogeneous systems, i.e., the expected number of A and B particles per site, $\rho_A(t)$ and $\rho_B(t)$, as $t \to \infty$. Here, we are interested in the case with equal initial densities, that is $0 < \rho_A(0) = \rho_B(0)$; this corresponds to $0 < r_A = r_B$. Since $\rho_B(t) - \rho_A(t)$ must clearly remain constant for all t, one has $\rho_A(t) = \rho_B(t)$. The following is a slight improvement of Theorem 1 in [4].

Theorem 1 *Assume that the initial measure is Poisson with $r_A = r_B > 0$. There exist positive constants $C_d > 0$ such that*

$$\begin{aligned} \rho_A(t) = \rho_B(t) &\approx C_d\sqrt{r_A}/t^{d/4}, \quad d < 4, \\ &\approx C_4(\sqrt{r_A} \vee 1)/t, \quad d = 4, \\ &\approx C_d/t, \quad d > 4. \end{aligned} \tag{1}$$

(By $a(t) \approx b(t)$ we mean that $a(t)/b(t) \to 1$ as $t \to \infty$.)

In addition to the long term behavior of the density, one is also interested in the limiting spatial structure of the model (see, e.g., [2]). Simulations indicate that in $d \leq 3$, the particle types separate as time increases, so that most regions are dominated by one type. This behavior is in accordance with Theorem 1. One does not expect this separation in $d \geq 4$. In Sections 2 and 3, we discuss how after suitable rescaling, $\xi(t)$ converges to a limiting process. Its local densities are given by a parabolic equation with random initial data, and the process itself is locally a Poisson random field. The results are qualitatively different in $d < 4$ and $d > 4$. The behavior in $d = 4$ is a hybrid of the two other cases. A detailed treatment is given in ref. [5]. In ref. [3], a more detailed version of the present article is given.

2 Spatial Structure – Macroscopic Behavior

We wish to analyze the asymptotic behavior of the process $\xi(t) = (\xi_A(t), \xi_B(t))$ as $t \to \infty$. As would be expected, we rescale both space and time to pass to a limiting process. Individual particles execute random walks (until annihilation). Since rescaling time by t and space by $\sqrt{t}$ produces Brownian motion in the limit for such particles, it makes sense to undertake the same rescaling here. We therefore set

$$\xi^t(G; s) = \xi(\sqrt{t}\, G; ts), \tag{2}$$

where $G \subset \mathbb{R}^d$, $\sqrt{t}\, G$ is the set G multiplied by $\sqrt{t}$ in each direction, and $s, t \geq 0$. After rescaling by $\sqrt{t}$, we need to compensate for the increase in particle density in this new scale. Applying Theorem 1, we see that on this scale one has density of magnitude $t^{-d/4} \cdot t^{d/2} = t^{d/4}$ for $d \leq 4$ and $t^{-1} \cdot t^{d/2} = t^{d/2-1}$ for $d \geq 4$ as $t \to \infty$. In either case, there are hopefully enough particles present so that the local behavior is determined by some law of large numbers, and the evolution of the process is on this macroscopic, or hydrodynamic, scale asymptotically deterministic.

A limiting process η attained in such a manner will depend on its initial measure and a rule prescribing its evolution. Since $\xi(0)$ is defined by Poisson random measures which are independent at different sites, rescaling $\xi(0)$ gives white noise, which we denote by Φ. (That is, Φ is the generalized Gaussian random field with covariance structure $E[\Phi(x)\Phi(y)] = \delta(x - y)$.) One might hope that the evolution of η is connected with normal distributions. Here N_s denotes a normal distribution in d dimensions with mean 0 and variance s.

We now state Theorem 2, which gives the behavior of η in $d < 4$. Here $*$ denotes convolution and $[x]^- = -(x \wedge 0)$, $[x]^+ = x \vee 0$.

Theorem 2 *For $d < 4$,*

$$t^{-d/4} \xi^t \xrightarrow{\text{w}} \eta \quad as \quad t \to \infty, \tag{3}$$

where $\eta = (\eta_A, \eta_B)$ and

$$\eta_A(G; s) = [(N_s * \Phi)(G)]^-, \quad \eta_B(G; s) = [(N_s * \Phi)(G)]^+. \tag{4}$$

In (11), $\xrightarrow{\text{w}}$ denotes weak convergence with a slight abuse of notation, since the convergence is not uniform in s close to 0.

Theorem 2 states that for large t, $t^{-d/4}\, \xi^t$ is approximated by η, whose components are specified by $\Psi_s = N_s * \Phi$. This convolution is a smooth function of x in $\mathbb{R}^d$, on account of N_s. Note that whereas Φ is random, N_s is not. So the evolution of the system on this macroscopic scale is specified (in the limit) completely by its initial data. As indicated by (4), there is also only one type of particle locally.

Some motivation for the behavior in (3)–(4) is provided by the following observation. Set

$$W(t) = E[\xi_B(t) - \xi_A(t) \mid \xi(0)], \quad W^t(G;s) = W(\sqrt{t}\, G; ts). \qquad (5)$$

That is, W is the mean signed measure of ξ conditioned on knowledge of the initial configuration, and W^t is the corresponding rescaled measure. The analog of (3)–(4) is then

$$t^{-d/4} W^t \xrightarrow{\text{w}} \eta_B - \eta_A = N * \Phi. \qquad (6)$$

It is easy to see that the density $\overline{u}(x,s)$ of $N * \Phi$ is given by $\overline{u}_s = \frac{1}{2}\Delta \overline{u}$ with $\overline{u}(\,\cdot\,,0) = \Phi$. One might therefore expect that (6) is obtainable in some simpler manner. One can in fact do so by replacing the system $A + B \to$ inert by the corresponding system of independent random walks. That is, particles of each type move as before, but there is no interaction. Scaling space and time as before, random walks are replaced by Brownian motion as $t \to \infty$, and one obtains the density $\overline{u}$ in the limit. One can verify, on the other hand, that the expectations for the interactive and noninteractive systems are the same. Therefore, (6) must also hold for the system $A+B \to$ inert (for all d). The point in Theorem 2 is that the system η in $d < 4$ in fact evolves deterministically for a prescribed initial configuration, with only one type of particle ever present locally. The result is proved in [5].

The asymptotic behavior of ξ^t for the case $d > 4$ is simpler – in this case, the limiting process η does not have any spatial structure, nor does it depend on the particular initial configuration. So one has the following result.

Theorem 3 *For $d > 4$,*

$$t^{1-d/2} \xi^t \xrightarrow{\text{w}} \eta \quad \textit{as } t \to \infty, \qquad (7)$$

where $\eta = (\eta_A, \eta_B)$ is nonrandom and has densities C_d/s, $C_d > 0$.

Note that unlike the case $d < 4$, both types of particles are always present locally, in fact in the same concentration.

The asymptotic behavior of ξ^t for the case $d = 4$ is the most difficult to work with. Both types of particles are present locally, but in different concentrations. So one can neither interpret the limit η in terms of differences of densities as in $d < 4$ nor as a scalar as in $d > 4$. Instead, η has a density which satisfies the system of equations given below.

Theorem 4 *For $d = 4$,*

$$t^{-1} \xi^t \xrightarrow{\text{w}} \eta \quad \textit{as } t \to \infty, \qquad (8)$$

where η *has density* $u(x,s) = (u_A(x,s), u_B(x,s))$ *given by*

$$\frac{\partial u_A}{\partial s} = \tfrac{1}{2}\Delta u_A - \Gamma u_A u_B\,, \quad \frac{\partial u_B}{\partial s} = \tfrac{1}{2}\Delta u_B - \Gamma u_A u_B\,, \tag{9}$$

$\Gamma > 0$, *with*

$$u_A(\,\cdot\,,0) = \Phi^-, \quad u_B(\,\cdot\,,0) = \Phi^+\,. \tag{10}$$

(In (10), Φ^- and Φ^+ denote the negative and positive parts of white noise. Since Φ has infinite variation everywhere, one really needs to be more precise by what one means by these quantities. This is explained in [5].)

3 Spatial Structure – Microscopic Behavior

We wish to investigate the asymptotic behavior of ξ more carefully. Theorems 2–4 gave its behavior when space is rescaled by $\sqrt{t}$, which is the proper scaling to follow the evolution of the process in terms of the local densities of its particles. One may instead wish to view the process on the microscopic level, so that the evolution of ξ maintains its random character. The natural scaling to use in this case is t^β, where $\beta = 1/4$ for $d \le 4$ and $\beta = 1/d$ for $d \ge 4$. Referring to Theorem 1, we see that this is what one needs to compensate for the decay in densities of the A and B particles. We therefore set

$$\hat{\xi}(G;x,t) = \xi(t^\beta G + x; t), \quad E \subset \mathbb{R}^d\,. \tag{11}$$

One can consider, for instance, cubes G which are centered at 0. Then $t^\beta G + x$ provides a "window" around x which is large enough to view the local behavior of ξ. Note that since $\beta < 1/2$ in all cases, the densities calculated in Theorems 2–4 are all asymptotically constant on this scale.

Recall that the limiting random measure η in Theorems 2–4 always has a random density $u(x,s) = (u_A(x,s), u_B(x,s))$, with $x \in \mathbb{R}^d$, $s > 0$. Since u is homogeneous in space and scales in time, we restrict ourselves to $x = 0$, $s = 1$, and denote by

$$F(a,b) = P[u_A(0,1) \le a,\ u_B(0,1) \le b] \tag{12}$$

the corresponding distribution function. Note that in $d < 4$, all of the mass is located at $a = 0$ or $b = 0$, and in $d > 4$, everything is at one point. Also, let $\mathcal{P}_{a,b}$ denote the homogeneous Poisson random field with means a and b for the A and B type particles, and set

$$\mathcal{P}_F = \int \mathcal{P}_{a,b}\, dF(a,b)\,, \tag{13}$$

that is, $\mathcal{P}_F$ is the convex combination of these fields weighted according to F.

Theorem 5 *For all d,*

$$\hat{\xi}(x,t) \xrightarrow{\text{w}} \mathcal{P}_F \quad \textit{as } t \to \infty \,. \tag{14}$$

References

1. Ovchinikov, A.A., Zeldovich, Ya.B.: Chem.Phys. **28**, 214 (1978)
2. Toussaint, D., Wilczek, F.: J.Chem.Phys. **78**, 2642 (1983)
3. Bramson, M., Lebowitz, J.L.: Spacial structure in diffusion limited two-particle reactions. J.Statist.Phys., to appear
4. Bramson, M., Lebowitz, J.L.: J.Statist.Phys. 62, 297 (1991)
5. Bramson, M., Lebowitz, J.L.: The asymptotic spatial structure in diffusion limited two-particle reactions. In preparation

Multilevel Models of Interacting Diffusions and Large Deviations

Donald A. Dawson [1], Jürgen Gärtner [2]

[1]Carleton University Ottawa, Dept. of Mathematics and Statistics, Canada K1S 5B6
[2]Technische Universität Berlin, Fachbereich 3 – Mathematik, Straße des 17. Juni 136, D-W 1000 Berlin 12, Fed. Rep. of Germany

As compared with short range models, hierarchical models of interaction show a more realistic behavior than mean field models concerning both equilibrium and nonequilibrium phenomena. One of the remarkable features of such models is the observation that for large, but finite, system size the cooperative effects are organized in multiple time scales and in fact this provides a caricature of the behavior of short range systems at successively larger spatial scales, cf. [5].

The aim of our present work is to show that some of the relevant features such as *metastability* and *nucleation* are already present in models with a two-level (weak) mean field structure and to develop appropriate large deviation techniques for the study of multi-scale phenomena.

Consider M cells each of which contains N particles (or spins) moving on the real axis according to diffusions in a double-well potential with mean field interaction inside of each cell and (weak) mean field interaction between different cells. Let $x_{ij} = x_{ij}(t)$ denote the position of the j-th particle in the i-th cell. The dynamics of $\{x_{ij}(t);\, 1 \le i \le M, 1 \le j \le N\}$ is governed by the following system of stochastic differential equations:

$$\begin{aligned} dx_{ij} = &- \left\{ V_h'(x_{ij}) + J_{\mathrm{I}} \left(x_{ij} - m_{\mathrm{I}}(X_i^N) \right) \right\} dt + \sigma \, dw_{ij} \\ &- \frac{J_{\mathrm{II}}}{N} \left(m_{\mathrm{I}}(X_i^N) - m_{\mathrm{II}}(X^{MN}) \right) dt. \end{aligned} \tag{1}$$

Here w_{ij} denote independent Brownian motions and

$$V_h(x) = \frac{x^4}{4} - \frac{x^2}{2} + hx$$

is the double-well potential with external field h. The level I and level II interactions (of strength $J_{\mathrm{I}} > 0$ and $J_{\mathrm{II}} > 0$) are given by

$$m_{\rm I}(X_i^N) = \frac{1}{N}\sum_{j=1}^{N} x_{ij} \qquad \text{and} \qquad m_{\rm II}(X^{MN}) = \frac{1}{M}\sum_{i=1}^{M} m_{\rm I}(X_i^N),$$

respectively. An adequate description of this exchangeable two level system is given by the level II empirical process

$$X^{MN}(t) = \frac{1}{M}\sum_{i=1}^{M} \delta_{X_i^N(t)}, \tag{2}$$

where

$$X_i^N(t) = \frac{1}{N}\sum_{j=1}^{N} \delta_{x_{ij}(t)} \tag{3}$$

is the level I empirical process in the i-th cell. The processes X_i^N, $1 \le i \le M$, and X^{MN} live in the spaces $\mathcal{M}_{\rm I} = \mathcal{M}(\mathbb{R})$ and $\mathcal{M}_{\rm II} = \mathcal{M}(\mathcal{M}_{\rm I})$ of probability measures on $\mathbb{R}$ and $\mathcal{M}_{\rm I}$, respectively.

To study large deviations for the law of the level II process X^{MN} (and its equilibrium distribution), let us forget for the moment about the level II interaction. Then X^{MN} may be regarded as the empirical measure of M *independent* level I processes $X_1^N, \ldots, X_M^N$. From our analysis of the mean field model we know that each of the processes X_i^N (and its equilibrium distribution) satisfies the large deviation principle, see [1], [3]. It turns out that this already allows to derive the level II large deviations. This multilevel large deviation result may be formulated in the following abstract setting [4]. Let X and Y be Polish spaces, and let (X_N) be a sequence of subsets of X. Call a sequence (x_N) an X_N-sequence if $x_N \in X_N$ for all N. Assume that each point of X is the limit of an X_N-sequence. Let $\{P_x^N;\, x \in X_N, N \in \mathbb{N}\}$ be a family of probabilities on Y satisfying the *large deviation principle as* $N \to \infty$ *with scale* $\gamma_N \to \infty$ *and rate function* $I : X \times Y \to [0,\infty]$, i.e.

(a) for each X_N-sequence $x_N \to x$ and each open set $G \subseteq Y$:

$$\liminf_{N\to\infty} \gamma_N^{-1} \log P_{x_N}^N(G) \ge -\inf\{I(x;y) : y \in G\};$$

(b) for each X_N-sequence $x_N \to x$ and each closed set $F \subseteq Y$:

$$\limsup_{N\to\infty} \gamma_N^{-1} \log P_{x_N}^N(F) \le -\inf\{I(x;y) : y \in F\};$$

(c) the level sets $\Phi(K;s) = \{y \in Y : I(x;y) \le s \text{ for some } x \in K\}$ are compact for all compact sets $K \subseteq X$ and all $s \ge 0$.

Let $\mathcal{M}(X)$ and $\mathcal{M}(Y)$ denote the spaces of probability measures on X and Y, respectively, furnished with the topology of weak convergence. Denote by $\mathcal{M}^{M,N}$ the subset of $\mathcal{M}(X)$ consisting of M-point empirical measures from X_N, i.e., measures μ of the form

$$\mu = \frac{1}{M}\sum_{i=1}^{M}\delta_{x_i} \qquad \text{with } x_1,\ldots,x_M \in X_N.$$

Given such a measure μ, let us denote by $\mathcal{P}_{\mu}^{M,N}$ the law on $\mathcal{M}(Y)$ of the empirical measure $M^{-1}\sum_{i=1}^{M}\delta_{\xi_i}$ of M independent random variables $\xi_1,\ldots,\xi_M$ having law $P_{x_1}^N,\ldots,P_{x_M}^N$, respectively. Note, that each measure in $\mathcal{M}(X)$ is the limit of an $(\mathcal{M}^{M,N})$-sequence as $M,N\to\infty$.

Theorem 1 *Under the above assumptions,* $\{\mathcal{P}_{\mu}^{M,N};\ \mu\in\mathcal{M}^{M,N},\ M,N\in\mathbb{N}\}$ *satisfies the large deviation principle as* $M,N\to\infty$ *with scale* $M\gamma_N$ *and rate function*

$$S(\mu;\nu) = \inf_{Q\in\mathcal{M}(\mu,\nu)}\int I(x;y)\,Q(dx,dy), \qquad \mu\in\mathcal{M}(X),\ \nu\in\mathcal{M}(Y).$$

Here $\mathcal{M}(\mu,\nu)$ *denotes the set of probability laws on* $X\times Y$ *having marginals* μ *and* ν.

Let us assume in addition that there is a continuous surjective map $\pi: Y\to X$ such that $P_x^N(\pi^{-1}\{x\}) = 1$ for all N and x and that the rate function I has the special form $I(x;y) = \hat{I}(y)$ for $\pi(y) = x$ and $I(x;y) = \infty$ otherwise. We will call $\hat{I}$ *special rate function* for $\{P_x^N\}$. Under these circumstances, it turns out that the rate function S may also be written in the special form $S(\mu;\nu) = \hat{S}(\nu)$ for $\nu\circ\pi^{-1} = \mu$ and $S(\mu;\nu) = \infty$ otherwise. The special rate function $\hat{S}$ has the form

$$\hat{S}(\nu) = \int \hat{I}(y)\,\nu(dy), \qquad \nu\in\mathcal{M}(Y).$$

Let us now examine the large deviation behavior of the equilibrium distribution Π^{MN} of the level II empirical process (2) under the assumption that the external field h_N depends on N and $h_N\to h$. We introduce a level II free energy functional $\mathcal{F}_{\mathrm{II}}$ (with respect to scale MN) by

$$\mathcal{F}_{\mathrm{II}}(\nu) = \int \mathcal{F}_{\mathrm{I}}(\mu)\,\nu(d\mu), \qquad \nu\in\mathcal{M}_{\mathrm{II}}.$$

Here

$$\mathcal{F}_{\mathrm{I}}(\mu) = \mathcal{H}_{\mathrm{I}}(\mu) + \sigma^2\mathcal{K}(\mu;\mu^*)$$

is the free energy functional of the level I mean field model, where $\mathcal{H}_{\mathrm{I}}(\mu) = (J_{\mathrm{I}}/2)\iint|x-\tilde{x}|^2\,\mu(dx)\mu(d\tilde{x})$ denotes the level I interaction energy and $\mathcal{K}(\mu;\mu^*) = \int\log(d\mu/d\mu^*)\,d\mu$ is the Kullback-Leibler information of μ with respect to the equilibrium measure μ^* of a single noninteracting particle. If

$J_{\mathrm{II}} = 0$, then the equilibrium measures of the empirical processes X_i^N satisfy the large deviation principle as $N \to \infty$ with scale N and rate function $\sigma^{-2}\mathcal{F}_{\mathrm{I}} + \mathrm{const}$, see [3]. (Thereby the constant must be chosen so that the minimum of the rate function equals zero.)

Because of this, we may apply Theorem 1 to obtain the following result.

Theorem 2 *$\{\Pi^{MN}\}$ satisfies the large deviation principle as $M, N \to \infty$ with scale MN and rate function $\sigma^{-2}\mathcal{F}_{\mathrm{II}} + \mathrm{const}$.*

The local minima of $\mathcal{F}_{\mathrm{I}}$ and $\mathcal{F}_{\mathrm{II}}$ correspond to level I and level II mean field equilibria, respectively. The level I free energy functional $\mathcal{F}_{\mathrm{I}}$ is not convex. But, as in the lattice case, $\mathcal{F}_{\mathrm{II}}$ is convex.

We know from [2] that there exists a critical temperature σ_{cr}^2 such that the following holds true. If $h \neq 0$ or $\sigma \geq \sigma_{\mathrm{cr}}$, then $\mathcal{F}_{\mathrm{I}}$ has a unique global minimum μ_0. If $h = 0$ and $0 < \sigma < \sigma_{\mathrm{cr}}$, then $\mathcal{F}_{\mathrm{I}}$ has two global minima μ_- and μ_+. In the first case, $\mathcal{F}_{\mathrm{II}}$ has also a unique minimum, namely δ_{μ_0}. In the second case, the set of minima of $\mathcal{F}_{\mathrm{II}}$ is

$$\mathcal{G} = \left\{\lambda\delta_{\mu_-} + (1-\lambda)\delta_{\mu_+} : 0 \leq \lambda \leq 1\right\}.$$

We consider the last case in detail. Theorem 2 is too weak to discover metastability and nucleation effects. To do this, we need more subtle large deviation estimates on $\mathcal{G}$ which may be obtained in the M-scale.

Theorem 3 *Suppose that $0 < \sigma < \sigma_{\mathrm{cr}}$ and $h_N = h^*/N$ for some $h^* \in \mathbb{R}$. Then $\{\Pi^{MN}\}$ satisfies the large deviation principle as $M, N \to \infty$ with scale M and rate function $\sigma^{-2}\mathcal{F}_{\mathrm{II}}^* + \mathrm{const}$. The M-scale free energy functional $\mathcal{F}_{\mathrm{II}}^*$ has the form*

$$\mathcal{F}_{\mathrm{II}}^*(\nu) = \mathcal{H}_{\mathrm{II}}(\nu) + \sigma^2\mathcal{K}(\nu;\nu^*),$$

where $\mathcal{H}_{\mathrm{II}}(\nu) = (J_{\mathrm{II}}/2)\iint |m_{\mathrm{I}}(\mu) - m_{\mathrm{I}}(\tilde{\mu})|^2\,\nu(d\mu)\nu(d\tilde{\mu})$ is the level II interaction energy, $m_{\mathrm{I}}(\mu) = \int x\,\mu(dx)$, and $\mathcal{K}(\nu;\nu^*)$ denotes the Kullback-Leibler information of ν with respect to a certain measure $\nu^* \in \mathcal{G}$ depending on σ and h^*.

A simple analysis leads to the following picture. $\mathcal{F}_{\mathrm{II}}^*$ is finite on the set $\mathcal{G}$ only. If J_{II} does not exceed $J_{\mathrm{II}}^{\mathrm{cr}} = 2\sigma^2|m_{\mathrm{I}}(\mu_+) - m_{\mathrm{I}}(\mu_-)|^{-2}$, then $\mathcal{F}_{\mathrm{II}}^*$ has a single local minimum for every h^*. If $J_{\mathrm{II}} > J_{\mathrm{II}}^{\mathrm{cr}}$, then there exists $h_{\mathrm{cr}} > 0$ such that $\mathcal{F}_{\mathrm{II}}^*$ has exactly two local minima for $|h^*| < h_{\mathrm{cr}}$ and only one for $|h^*| \geq h_{\mathrm{cr}}$. This may be interpreted as a *nucleation effect* in the following sense. If $|h| \ll h_{\mathrm{cr}}$, then this supports the conjecture that there exists a *metastable phase* (a local minimum of the analogue of $\mathcal{F}_{\mathrm{II}}$ when $M \to \infty$ but N remains bounded) if and only if the size N of the cells is smaller than $N_{\mathrm{cr}} = h_{\mathrm{cr}}/|h|$. (Recall that the external field h is assumed to have the form h^*/N.) This picture extrapolates to the multilevel hierarchy in the following way. For each K, level K of the hierarchy consists of N level $(K-1)$ subsystems. Then for any $h > 0$ there exists a minimal level K^* at which $\mathcal{F}_{K^*}$ has a unique minimum and, consequently, the corresponding metastable phase is unstable with respect to a 'nucleus' of size N^{K-1} or larger (recall that this is the analogue of a sufficiently large spatial scale).

The level I empirical processes (3) may be regarded as randomly perturbed dynamical systems in $\mathcal{M}_{\mathrm{I}}$ with perturbation strength $N^{-1/2}$. Concerning the dynamical large deviation behavior, we study a simple caricature of the two-level model (1)–(3). The idea is to replace (3) by a randomly perturbed dynamical system in $\mathbb{R}^d$:

$$dX_i^N = b(X_i^N; X^{MN})\, dt + N^{-1/2}\, dW_i, \qquad i = 1, \dots, M,$$

where W_i are independent d-dimensional Brownian motions and X^{MN} is defined by (2). Relying on Theorem 1, we showed in [4] that under weak restrictions on the vector field b the law of $X^{MN}(\cdot)$ on $\mathcal{C}_{0,T} = C([0,T]; \mathcal{M}(\mathbb{R}^d))$ satisfies the large deviation principle with scale MN. The associated special rate function is given by

$$S(\mu(\cdot)) = \frac{1}{2} \int_0^T \|\dot{\mu}(t) - \mathcal{L}(\mu(t))^* \|_{\mu(t)}^2 \, dt\ ,$$

if $\mu(\cdot) \in \mathcal{C}_{0,T}$ is absolutely continuous and $S(\mu(\cdot)) = \infty$ otherwise. Here $\mathcal{L}(\mu)^*$ is the formal adjoint of the operator

$$\mathcal{L}(\mu) = (b(\cdot\,; \mu), \nabla)$$

and

$$\|\theta\|_\mu^2 = \sup_{f \in \mathcal{D}} \frac{|\langle \theta, f \rangle|^2}{\langle \mu, |\nabla f|^2 \rangle}, \qquad \theta \in \mathcal{D}', \ \mu \in \mathcal{M}(\mathbb{R}^d),$$

where $\mathcal{D}$ and $\mathcal{D}'$ denote the space of C^∞-functions on $\mathbb{R}^d$ with compact support and the associated Schwartz space of distributions, respectively.

References

1. Dawson, D.A., Gärtner,J.: Stochastics **20**, 247 (1987)
2. Dawson, D.A., Gärtner,J.: In: Stochastic Calculus in Application. Proc. Cambridge Symp. Ed. by J. R. Norris, p. 29, Longman (1988)
3. Dawson, D.A., Gärtner,J.: Mem.Amer.Math.Soc. **78** No. 398 (1989)
4. Dawson, D.A., Gärtner,J.: Preprint (1991)
5. Dawson, D.A., Greven, A.: Preprint (1990)

Section 8

General Theory of Quantized Fields

Organizer: D. Iagolnitzer

Fields, Particles and Analyticity: Recent Results

or

30 Goldberg*...(er)** Variations on B.A.C.H.

where

B=Bethe Salpeter Equation, A=Axiomatic Field Theory
C=Complex Variables and H=Harmonic Analysis

J. Bros

Service de Physique Théorique† de Saclay
CE-Saclay, F-91191 Gif-sur-Yvette Cedex, France

ARIA

As it is known, Axiomatic Field Theory (A) implies "double analyticity" of the n-point functions in space-time and energy-momentum Complex Variables (C), with various interconnections by Fourier-Laplace analysis. When the latter is replaced by Harmonic Analysis (H) on spheres and hyperboloids, a new kind of double analyticity results from (A) (i.e. from locality, spectral condition, temperateness and invariance): complex angular momentum is thereby introduced (a missing chapter in (A)). Exploitation of Asymptotic Completeness via Bethe-Salpeter-type equations (B) leads to new developments of the previous theme on (A,C,H) (complex angular momentum) and of other themes on (A,C) (crossing, Haag-Swieca property etc...). Various aspects of (A) + (B) have been implemented in Constructive Field Theory (composite spectrum, asymptotic properties etc...) by a combination of specific techniques and of model-independent methods.

VARIATIONS

Harmonic Analysis with Double Analyticity and the Corresponding "Propagator interpretation"

(1) Composition of analytic kernels and Fourier-Laplace transformation on the complex hyperboloid

---

* for the 250th year of J.S. Bach's Goldberg Variations (B.W.V. 998) composed in Leipzig in 1742.

** for the 25th year of M.L. Goldberger's theorem: "$\forall\varepsilon$, the predictive power of Axiomatic Field Theory is less than ε".

† Laboratoire de la Direction des Sciences de la Matière du Commissariat à l'Energie Atomique

$X_d^{(c)} = \left\{z = \left(z^{(0)}, ..., z^{(d)}\right); z^{(0)^2} - z^{(1)^2} - ... - z^{(d)^2} = -\alpha^2\right\}$ will be outlined in our variations (2-5) and illustrated by applications to two-point functions in more general frameworks of (A). In $X_d^{(c)}$, the real one-sheeted hyperboloid X_d *plays the role of Minkowski space* $\mathbb{R}^d$*, while the sphere* S_d ($z^{(0)} = iy^{(0)}$; $z^{(1)}, ..., z^{(d)}$ real) replaces the euclidean space $\mathrm{E}_d = i\mathbb{R}\times\mathbb{R}^{d-1}$ of the complexified Minkowski space $\mathbb{C}^d$. Correspondingly, a "continuous" Fourier transform on E_d becomes a "discrete" Fourier-Legendre series on S_d; analytic continuation of the former is now turned into a carlsonian analytic interpolation of the latter; the analytic function thus obtained remains a Laplace-type transform $\tilde{F}(\lambda)$ (on X_d instead of $\mathbb{R}^d$) of a jump of the original analytic function.

(2) The one-variable case (S_1 and $X_1^{(c)}$) is illustrated by the propagator $\mathcal{W}\left(z^{(0)}, \boldsymbol{x}\right)$ of a local field at temperature β^{-1}. Locality and K.M.S. (instead of spectral) condition imply that $\mathcal{W}$ is analytic in an $i\beta$-periodic (instead of aperiodic) cut-plane of the complex time variable $z^{(0)}$. The role of $\tilde{F}(\lambda)$ (in (1)) is played by the (retarded) Green function $\tilde{R}(\omega, \boldsymbol{x})$, with $\omega = \beta^{-1}\lambda$.

(3) In Minkowski space $\mathbb{R}^d$, the matrix elements of retarded commutators between bounded energy states are discontinuities of two-point functions analytic in a cut-domain of $\mathbb{C}^d \times \mathbb{C}^d$. Similarly, one can define on the hyperboloid X_d, interpreted as a de Sitter (d-dimensional) space-time, "causal kernels" (with support properties) which are discontinuities $\Delta\mathcal{K}$ of "perikernels" $\mathcal{K}$, analytic in a cut-domain of $X_d^{(c)} \times X_d^{(c)}$. The composition law $\mathcal{K} = \mathcal{K}_1 * \mathcal{K}_2$ of kernels on S_d yields by analytic continuation (see [Bros, Viano 1990]) a "Volterra" composition law [Faraut 1981] $\Delta\mathcal{K} = \Delta\mathcal{K}_1 \diamond \Delta\mathcal{K}_2$ on X_d involving integration on a "double cone" (a de Sitter-space-version of the "Wick rotation" property).

(4) Perikernels $\mathcal{K}(z, z')$ on $X_d^{(c)}$ which are invariant under the complex Lorentz (or de Sitter) group on $\mathbb{C}^{(d+1)}$ are functions $\mathcal{F}(\cos\theta)$ (with $\cos\theta = -z \cdot z'/\alpha^2$) analytic in a cut-plane whose jump $\Delta\mathcal{F}(\equiv \Delta\mathcal{K})$ is carried by $[1, +\infty)$. Such kernels $\Delta\mathcal{K}$ admit a Laplace transform $\tilde{F}(\lambda)$ introduced via "horocyclic sections" of X_d and expressed as an integral transform of $\Delta\mathcal{F}(\cosh v)$ whose kernel is a second-kind Legendre function $Q_\lambda^{(d+1)}(\cosh v)$. The analyticity and boundedness properties of $\cos\theta^{-N} \times \mathcal{F}(\cos\theta)$ in the cut-plane are equivalent to the analyticity and boundedness of $\tilde{F}(\lambda)$ in the half-plane $\Pi_N = \{\lambda;\ \mathrm{Re}\ \lambda > N\}$, while $\left\{\tilde{F}(\ell);\ \ell \text{ integer} > N\right\}$ reproduces the Fourier-Legendre series of $\mathcal{F}$ on S_d. Moreover $\Delta\mathcal{K}_1 \diamond \Delta\mathcal{K}_2$ is turned into $\tilde{F}_1(\lambda) \cdot \tilde{F}_2(\lambda)$ [Faraut 1981, Faraut, Viano 1986, Bros, Viano 1990].

(5) Invariant perikernels $\mathcal{K}$ on $X_d^{(c)}$ are interpreted as propagators on de Sitter space; $\mathcal{K}$ admits an integral representation in terms of $\tilde{F}(\lambda)$ which extrapolates the Källen-Lehmann representation, $\alpha^{-1}[\lambda + (d-1)/2]$ playing the role of the mass [Bros, 1990].

Recent Developments and Results Concerning the Four-Point Function

(6) Several topics concerning the four-point function and the two-particle scattering amplitudes (in the usual framework of complexified Minkowski space) will be reviewed in the following variations (7-21). The introduction of complex angular momentum in the "linear program" (exclusive use of (A)) and its implications in the framework of the "non-linear program" (use of (A)+(B)) will be presented. Various developments of the non-linear program concerning in particular the composite spectrum and the asymptotic properties at large momenta will be described and related results of constructive field theory will be given.

Linear program: introduction of complex angular momentum

(7) In the t-channel ($t = K^2$, $K = k_1 + k_2 = k_1' + k_2'$) and at $t < 0$, the four-point function $F(\underline{\zeta},\, \underline{\zeta}';\, t,\, \cos\,\theta)$ is analytic in a cut-plane of the "off-shell scattering angle" θ, when the mass variables $\underline{\zeta} = (\zeta_1, \zeta_2)$, $\underline{\zeta}' = (\zeta_1', \zeta_2')$ ($\zeta_i = k_i^2$, $\zeta_i' = k_i'^2$) belong to a region Δ_t corresponding to the "euclidean subspace": at $(t, \underline{\zeta}, \underline{\zeta}')$ fixed, the section of this subspace yields a sphere embedded in a complex hyperboloid which is contained in the primitive domain of (A), except for the s-cut and u-cut of the crossed channels; the analytic framework ((C)+(H)) outlined in (1,4) is therefore entirely implemented (except that two cuts instead of one now occur).

(8) A "Laplace-Froissart-Gribov" transform $\tilde{F}\left(\underline{\zeta}, \underline{\zeta}'; t, \lambda\right)$ of F, analytic in a half-plane $\Pi_N = \{\lambda;\ \mathrm{Re}\ \lambda \geq N\}$ of the complex angular momentum variable λ can then be introduced; moreover $\tilde{F} = e^{i\frac{\pi\lambda}{2}}\left[\tilde{F}_s \cos\frac{\pi\lambda}{2} - i\tilde{F}_a \sin\frac{\pi\lambda}{2}\right]$, where $\tilde{F}_s$ and $\tilde{F}_a$ (namely the transforms of symmetric and antisymmetric combinations F_s, F_a of the s-cut and u-cut discontinuities of F) are carlsonian interpolations in Π_N of (respectively) the even and odd "off-shell partial waves" $f_\ell\left(\underline{\zeta}, \underline{\zeta}', t\right)$ of F [Bros, Viano, 1990].

(9) Analytic continuation of $\tilde{F}$ with respect to the mass variables $\underline{\zeta}$, $\underline{\zeta}'$ at t fixed ($t < 0$) can be performed in the domain of validity of dispersion relation, thus including the mass shell (up to the occurrence of a kinematical cut).

(10) Analytic continuation with respect to t in the region $t > 0$ can also be performed by making use ("à la Martin" [N.C. 1966]) of positivity properties of the absorptive parts of F carried by the s-cut and u-cut. This program of analytic continuation of $\tilde{F}$ remains to be explored.

Non-linear program: a review of various properties

(11) In the Bethe-Salpeter type structure of F, namely $F = G + F \circ G = G + \cdots + G \circ \cdots \circ G + \cdots$, the irreducibility of G (i.e. the vanishing of its discontinuity ΔG for $t < (3m)^2$) is the expression of two-particle asymptotic completeness and implies the meromorphic continuation of F locally around $t = 4m^2$ (across the cut $4m^2 < t < 9m^2$) [Bros 1970; Bros, Lassalle, CMP 1977]. In this axiomatic framework, $\circ$ denotes a two-line Feynman-type convolution whose propagators are possibly damped in euclidean directions by an analytic cutoff. The corresponding kernel G (defined in terms of F by $F = G + F \circ G$) is then called a *regularized* Bethe-Salpeter kernel.

(12) Passage to the off-shell partial waves f_ℓ (resp. g_ℓ) of F (resp. G) yields corresponding Bethe-Salpeter equations $f_\ell = g_\ell + f_\ell \hat{\circ} g_\ell$ ($\hat{\circ}$ denoting integration over the mass variables) in a domain of $(\underline{\zeta}, \underline{\zeta}', t)$-space containing physical points of the t-channel on the mass shell.

(13) In all 2- and 3-dimensional weakly-coupled models of constructive field theory, suitable irreducible kernels are introduced [Iagolnitzer, Magnen, CMP 1987-88] and two-particle asymptotic completeness is proved by model-independent methods.

(14) Decoupled Bethe-Salpeter equations for the absorptive parts F_s and F_a of F (namely $F_s = G_s + F_s \diamond G_s$ and $F_a = G_a + F_a \diamond G_a$, where $\diamond$ denotes integration in a double-cone) follow from the analytic framework of (A) [Bros 1970].

(15) In view of (8), the previous Bethe-Salpeter equations are then (partially) diagonalized by the Laplace-Froissart-Gribov transformation, namely one has: $\tilde{F}_{s,a}(\cdot; t, \lambda) = \tilde{G}_{s,a}(\cdot; t, \lambda) + \tilde{F}_{s,a}(\cdot; t, \lambda) \hat{\circ} \tilde{G}_{s,a}(\cdot; t, \lambda)$ where $\cdot$ stands for the mass variables $(\underline{\zeta}, \underline{\zeta}')$ [Faraut, Viano, JMP 1986; Bros, Viano, 1990]; these equations interpolate in λ the corresponding equations for the (even and odd) partial waves (8,12), at least when the point $(\underline{\zeta}, \underline{\zeta}', t)$ lies in the restricted domain specified in (7).

Results on the composite spectrum

(16) Fredholm theory applied to the Bethe-Salpeter equations for the f_ℓ implies that the meromorphic continuation of F may admit poles $t = t_\ell^{(\alpha)}$ with factorized residue appearing in definite (off-shell) partial waves f_ℓ.

(17) In field theories with small coupling g, the composite spectrum can be studied in a fixed neighbourhood of the region $4m^2 \leq t < 9m^2 - \varepsilon(g)$; while no pole occurs in (hypothetical) models of dimension 4, a kinematical threshold effect at $t = 4m^2$ always produces a bound or antibound state in 2-dimensional models, and possibly an $\ell = 0$ bound state in 3-dimensional models whose mass m_B ($m_B < 2m$) tends to $2m$ when $g \longrightarrow 0$ (see [Bros, Iagolnitzer CMP 1988] for a complete and model-independent analysis).

(18) Fredholm theory applied to the Bethe-Salpeter equations for the transforms $\tilde{F}_{s,a}(\cdot; t, \lambda)$ of $F_{s,a}$ (see (15)) implies a meromorphic continuation of $\tilde{F}_{s,a}$ from Π_N into a larger half-plane $\Pi_{\hat{N}}(\hat{N} < N)$, if the Bethe Salpeter kernel G may be shown to have a better behaviour than F in the cut-plane of $\cos\theta$ (i.e. in $|\cos\theta|^{\hat{N}}$).

(19) The transform $\hat{F}$ of F can then possibly present "Regge poles" in the strip $\Pi_{\hat{N}} \backslash \Pi_N$ of the λ-plane; with each such pole $\lambda = \lambda_0(t)$, there corresponds a contribution to F of the form $(2\lambda_0(t)+1)\, P_{\lambda_0(t)}(-\cos\theta) \times \frac{\varphi(t,\zeta)\times\psi(t,\zeta')}{\sin\pi\lambda_0(t)}$ which, on the one hand, may interpolate poles t_ℓ (i.e. $\lambda_0(t_\ell) = \ell$) in the region of analytic continuation $t > 0$, and on the other hand provides the dominant behaviour $|\cos\theta|^{\mathrm{Re}\,\lambda_0(t)}$ of F in the $\cos\theta$-plane (in particular for all $t < 0$).

Asymptotic properties at large euclidean momenta

(20) The exploration of the asymptotic properties of F for $\underline{\zeta}, \underline{\zeta}' \longrightarrow \infty$ in the euclidean region, in a semi-axiomatic approach of field theories of φ_4^4-type (i.e. non-superrenormalizable theories with only two and four-point renormalization parts), can be performed by introducing "renormalized" instead of "regularized" Bethe-Salpeter kernels. A more sophisticated system of Fredholm equations (see the Appendix of [Symanzik, CMP 1973] and [Bros, Ducomet, Ann. IHP 1986]) allows one again to reconstruct F in terms of G and to exhibit the leading (factorized) term of the Wilson expansion of F; moreover a discussion of the "existential problem" of such theories is suggested by the occurrence of a Fredholm denominator whose possible zeros are of the same nature as the "Landau ghost" (and reduce to the latter in a drastic perturbative approximation). This treatment relies in a crucial way on "asymptotic conditions" stating that appropriate increments of G (instead of G itself) decrease as an inverse momentum in euclidean directions. Under these conditions, the convergence of the "renormalized Neumann expansion" $F = G + \Sigma(G \circ G \circ \cdots \circ G)_{\mathrm{ren}}$ is also proved [Iagolnitzer, Ann. IHP 1989].

(21) A complete illustration of these asymptotic properties in the 2-dimensional massive Gross-Neveu model at small coupling, has been obtained in [Iagolnitzer, Magnen CMP 1988]. In particular, the relevant asymptotic conditions on the increments of G have been checked and the leading term of F comes out from the Bethe-Salpeter structure as outlined above (the full Wilson expansion is also derived directly in [Iagolnitzer, Magnen, CMP 1988]). These results of constructive field theory follow from a refined phase-space analysis incorporated to the cluster expansion method.

Recent Developments and Results Concerning the n-Point Functions

(22) Variations (23-28) will be devoted to reviewing the following topics: local analyticity properties in the linear program and their macrocausal interpretation, present status of the crossing property, use of generalized Neumann expansions in a semi-axiomatic formulation of the non-linear program, implementation of this viewpoint in constructive field theory.

Local analyticity

(23) In the linear program, local analyticity properties of $m \longrightarrow n-m$ particle collision amplitudes (obtained by applying "local tube theorems" to the primitive domains of the n-point functions implied by (A)) hold for every n but only near points $P = (P_1, ..., P_n)$ that belong to certain parts of the physical regions. Specific domains have been obtained for the $2 \longrightarrow 3$ amplitudes in [Bros, Epstein, Glaser HPA 1972] (in which case only a small region below five-particle threshold, is excluded) and for the $3 \longrightarrow 3$ forward amplitudes in [Muzafarov, Pavlov TMF 1978].

(24) The method of "localized Fourier transforms" ($\mathcal{F}_{\mathrm{loc}}$), or "microlocal analysis", yields a more general characterization of the local momentum-space properties of n-point (time-ordered) functions τ_n in terms of their "essential support" (or "singular spectrum"); the latter, which describes the local decompositions of τ_n as sums of boundary values of analytic functions (see [Bros, Epstein, Glaser HPA 1972]), can be directly interpreted in phase-space as a set of "causal configurations" (P, u) ($u = (u_1 ... u_n) = |u| \times \omega_u$ denoting space-time translations) outside which exponential decrease properties of $\mathcal{F}_{\mathrm{loc}}(\tau_n)(P, u)$ (of the form $\exp(-\alpha(P, \omega_u)|u|)$) hold. A precise form of such "macrocausality properties" implied by (A) has been given recently in [Iagolnitzer 1991].

(25) In the non-linear program (based on (A)+(B)), extended local analyticity properties including Landau singularity structure can be obtained through structural equations adapted to given energy-momentum regions (see e.g. [Bros, M∩ϕ 1983] which presents substantial results for $2 \longrightarrow 3$ and $3 \longrightarrow 3$ processes but only partial results for $m \longrightarrow m$ processes): difficulties concern the need for a complete recurrent system of Bethe-Salpeter-type equations.

Crossing

(26) The crossing property is the existence of an analyticity domain for the n-point function *on the complex mass shell* which connects the various $m \longrightarrow n-m$ particle regions (namely the local domains obtained in (23,25)) together. It is only for $n = 4$ (i.e. for $2 \longrightarrow 2$ processes) that this property has been shown to follow purely from (A) [Bros, Epstein, Glaser, CMP 1965]. In the framework of ((A) + (B)), a general method (applicable to arbitrary n) has been indicated, which yields a complete proof for $2 \longrightarrow 3$ processes [Bros, Phys. Rep. 1986]; one thus obtains, as in the $2 \longrightarrow 2$ case, "asymptotic crossing domains" (i.e. cut-neighbhourhoods of infinity in suitable energy variables); another procedure applicable to $3 \longrightarrow 3$ forward kinematics yields bounded crossing domains [Bros, M∩ϕ 1979] for the corresponding $3 \longrightarrow 3$ processes (A treatment of the latter based purely on (A) has been presented by Muzafarov, TMF 1978, quoted by us in M∩ϕ, 1979; we now feel obliged to say that this treatment does not make sense).

The approach of generalized Neumann expansions

(27) Considering the structural equations mentioned above in (25) in the sense of "Neumann-type expansions in terms of irreducible kernels" suggests a semi-axiomatic approach [Iagolnitzer, Fizika 1985] supported by perturbation theory, in which n-point functions are expressed by series of Feynman-type amplitudes adapted to arbitrary energy-momentum regions; these expansions should exhibit the complete Landau singularity structure and imply various discontinuity formulae of interest for the n-point functions.

(28) This viewpoint has been successfully implemented [Iagolnitzer, Magnen, CMP 1987-88 and Iagolnitzer, M$\cap\phi$ 1986] in all models of constructive field theory (thanks to the introduction of diversified irreducible kernels via "cluster expansions of higher order"). At sufficiently small couplings, the series of Feynman-type amplitudes mentioned in (27) are shown to be *convergent* and the method leads to a promising program for proving asymptotic completeness in all these models.

Haag-Swieca Property and Asymptotic Completeness

(29) The Haag-Swieca property [CMP, 1965] was proposed as a reasonable "local manifestation" of field theories enjoying a pure particle interpretation. It claims that for all energies E and bounded space-time regions $\mathcal{O}$, a certain set of states $S_E(\mathcal{O})$, defined by means of observables localized in $\mathcal{O}$, should be a compact subset of the Hilbert space, with exponentially decreasing "thickness" in terms of an "approximate dimension" parameter N.

(30) Although this property has been considered recently as a too restrictive criterion for the general case including "infra-particles" [Buchholz, 1991], it keeps its value for theories with purely positive mass spectrum. As a matter of fact, a general method has been given [Bros, 1991] for establishing the Haag-Swieca property from Asymptotic Completeness; this method, based on the local analyticity and uniform boundedness properties of the wave packets representing the states of $S_E(\mathcal{O})$ (expressed in terms of appropriate n-point functions) yields a complete proof for values of E of the two-particle region (the treatment of higher energy ranges calls for further technical results of the non-linear program).

Quantum Field Theory in Curved Spacetime

Bernard S. Kay

Department of Applied Mathematics and Theoretical Physics,
University of Cambridge, 3 Silver Street, Cambridge CB3 9EW, UK

In this talk, I discuss the theory of the covariant Klein Gordon equation, $(\Box_g + m^2)\phi = 0$, on the class of globally hyperbolic spacetimes (M, g). (Here globally hyperbolic means there exists a diffeomorphism $\delta : M \to \mathbb{R} \times \mathcal{C}$ where $\mathcal{C}$ is a three manifold and each of the surfaces $\delta^{-1}(\{t\} \times \mathcal{C})$ is a Cauchy surface i.e. an achronal 3-surface which is cut precisely once by every inextendible causal curve in M.) My aim is to give an impression of what has been achieved in clarifying the conceptual and mathematical foundations of quantum field theory in curved spacetime and to raise some open issues of importance for the further development of the subject. Lack of space enforces the omission of many important topics (especially, the back-reaction problem and the definition of a quantum energy momentum tensor $T_{\mu\nu}$, higher spin fields, interacting fields, and non-globally hyperbolic spacetimes) and there will be space neither for an adequate bibliography nor to always provide adequate motivation and details. The remedy for some of these deficiencies will be found in the references (see especially Section 3 in [1]).

1 Introduction

The basic piece of mathematical structure – familiar from flat spacetime quantum field theory – which survives in a generic globally hyperbolic spacetime consists of a net $\mathcal{A} = \cup_{\mathcal{O}} \mathcal{A}(\mathcal{O})$ ($\mathcal{O}$ will always mean *bounded* open region) of local $*$-algebras whose elements correspond to polynomials in smeared fields and which incorporates the relations (for $f, f_1, f_2 \in C_0^\infty(M)$): $\phi(f)^* = \phi(f)$ (for a Hermitean field) together with

$$\text{(i)} \quad \phi((\Box_g + m^2)f) = 0, \qquad \text{(ii)} \quad [\phi(f_1), \phi(f_2)] = i\Delta(f_1, f_2)\mathbb{1}$$

where Δ denotes the advanced minus retarded fundamental solution (which exists on globally hyperbolic spacetimes). (Really, we take the (C*-) Weyl algebra over the quotient of $C_0^\infty(M)$ by $(\Box_g + m^2)C_0^\infty(M)$ with the symplectic form $\sigma([f_1],[f_2]) = \Delta(f_1,f_2)$).The $\mathcal{A}(\mathcal{O})$ are generated by $\phi(f)$, supp$f \in \mathcal{O}$, and satisfy (see [2]) a number of general properties (axioms) the most important of which are

(a) $\mathcal{O} \subset \mathcal{O}'$ and also, $\mathcal{O}$ causally dependent on $\mathcal{O}' \Rightarrow \mathcal{A}(\mathcal{O}) \subset \mathcal{A}(\mathcal{O}')$
(b) $\mathcal{O}$ and $\mathcal{O}'$ spacelike separated $\Rightarrow [\mathcal{A}(\mathcal{O}), \mathcal{A}(\mathcal{O}')] = 0$.

Quantum field theory in curved spacetime differs from quantum field theory in flat spacetime in that, for a generic (globally hyperbolic) curved spacetime, there is *no preferred state* (and consequently the notions "vacuum" and "particles" lose meaning). For this reason, it becomes essential to think of observables (A etc.) as elements of an abstract $*$-algebra $\mathcal{A}$, and states ω as positive (i.e. $\omega(A^*A) \geq 0$) linear functionals on $\mathcal{A}$. In what follows, we shall make use of the following two elementary notions: The *GNS construction* associates to any state ω a triple $(\rho_\omega, \mathcal{H}_\omega, \Omega_\omega)$ (ρ_ω a representation on the Hilbert space $\mathcal{H}_\omega$ with cyclic vector Ω_ω) such that $\omega(A) = \langle \Omega_\omega | \rho_\omega(A) \Omega_\omega \rangle_{\mathcal{H}_\omega}$. The *folium* of a state ω consists of the set $\{\omega_\delta : \delta > 0,\ \mathrm{tr}\delta = 1,\ \omega_\delta(A) = \mathrm{tr}(\delta \rho_\omega(A))_{\mathcal{H}_\omega}\}$ of density matrix states in the GNS representation belonging to ω.

Two important features of the subject are the "Unruh effect" and the "Hawking effect" [3]. The Unruh effect is the mathematical fact that the usual ground state ω_0 for the usual time-translations of a quantum field in flat spacetime is a thermal (i.e. KMS) state (at "temperature" $1/2\pi$) when restricted to a "Rindler wedge" with respect to the corresponding family of Lorentz boosts. As was first pointed out by Sewell [4], this result may be regarded as a corollary to the Bisognano-Wichmann and Reeh-Schlieder theorems. The Hawking effect is the result that if one quantizes say $(\Box_g + m^2)\phi = 0$ on a (say, spherically symmetric) spacetime representing the collapse of a star to a black hole, then the in vacuum state will look like a state of thermal radiation in the "out" region ("far future and late times") at the Hawking temperature $T_{\text{Hawking}} = 1/8\pi M$. Ultimately, this result may be traced back to the special properties of classical waves which pass through the centre of the star just before the final critical ingoing null ray which enters (and remains in) the horizon. The behaviour of such waves is treated in [3] using a geometrical optics approximation.

2 Further Development of the Mathematical Framework

So far, our conceptual framework consists of the $*$-algebra $\mathcal{A}$ together with the remark that there is no preferred state. To leave things at this would be unsatisfactory for two (related) reasons: (1) The algebra $\mathcal{A}$ is too small. It contains (smeared versions of) products of fields at different points such as $\phi(x)\phi(y)$ but nothing that could correspond to products of fields at the same point such as $\phi(x)^2$ (or $T_{\mu\nu}(x)$). (2) The set of all states on $\mathcal{A}$ is too large. It contains e.g. states with "infinite energy density". The remedy (cf. [5]) is to introduce a preferred class of states with the "correct" short distance behaviour, namely the quasi free "Hadamard" states.

Here, "quasi-free" means all truncated n-point functions vanish unless $n = 2$ and "Hadamard" means (see [1] for details) that the two-point function has the appropriate curved spacetime generalization of the short distance singularity of physically admissible states in flat spacetime (and no other "non-local spacelike singularities"):

$$\omega(\phi(x)\phi(y) + \phi(y)\phi(x)) = -(1/2\pi^2)(u/\sigma + v \ln \sigma + w)$$

where σ denotes the square of the interval between x and y, u and v are smooth two-point functions determined by the local geometry by the demand that such a formula (with u, v, w all smooth) should be a bisolution, and w is a smooth two-point function which depends on and contains all the information about the state.

The rôle this preferred class of states plays in completing the conceptual framework is still in a somewhat tentative state. We recall some old, and present some new, conjectures, questions and definitions:

Conjecture [5] *All quasi-free Hadamard states are locally quasi-equivalent i.e. on any bounded $\mathcal{O}$, the folia of any pair ω_1, ω_2 of states coincide on $\mathcal{A}(\mathcal{O})$.*

A closely related result has now been proven for the special class of Robertson-Walker spacetimes by Lüders and Roberts [6]. Assuming the truth of this conjecture, the (constructive) remedy outlined here would make contact with the proposal made around the same time in [7] which proposed, in an axiomatic spirit, that the net of local $*$-algebras $\mathcal{A}$ should be enlarged to a net of local W*-algebras (actually factors), and that the allowed states should be locally normal. Clearly, the above conjecture would imply the existence of a preferred such net. (For more discussion, see [6] and footnote 2 on page 70 in [1].)

Next, a number of questions concerning the set of physically admissible states. Consider the following three plausible alternative sets of necessary conditions:

(a) States which, when restricted to $\mathcal{A}(\mathcal{O})$ for each element $\mathcal{O}$ of some open cover (by bounded $\mathcal{O}$) of M (no matter how fine) belong to the folium of the quasi-free Hadamard states.
(b) States which, when restricted to $\mathcal{A}(\mathcal{O})$ for any bounded $\mathcal{O}$ (no matter how large) belong to the folium of the quasi-free Hadamard states.
(c) States whose two-point functions are Hadamard and all of whose truncated n-point functions for n other than 2 are smooth.

It is tempting to conjecture that (a) and (b) are actually equivalent. If this were true, it could be interpreted, using notions introduced by Fredenhagen and Haag [8] in their discussion of a possible framework for quantum gravity, as saying that the laws of quantum field theory in curved spacetime are "as local as possible" given the existence of a fixed background metric. Note also that this conjecture is closely related to the conjecture that locally Hadamard states automatically fail to have non-local spacelike singularities. See [9],[10]. It also seems of importance to answer the question: How close is (c) to (a) and (b)?

3 Rigorous Results on the Hawking Effect

Concerning the collapsing star problem (see Section II): the main mathematical problem was to justify the geometrical optics approximation in Hawking's original derivation. There is now a rigorous result on this in [11] which proves a thermal radiation formula corresponding to the result in [3].

For further insight, it is interesting to focus on the question: In what sense do black holes "want" to be hot? One may interpret this question mathematically by studying, say, $(\Box_g + m^2)\phi = 0$ on the Kruskal spacetime and asking about the set of physically admissible states (for this problem, it seems convenient to, provisionally, take alternative (c) above) which are stationary (i.e. invariant under the Schwarzschild isometries). For this setting, it is now possible to prove the following theorem. (For the fully precise statement, we must refer to [1] and [12].)

Theorem *(On a "large" subalgebra of $\mathcal{A}$) there is a unique stationary, physically admissible state – which, moreover, when restricted to the right-wedge subalgebra is a KMS state at the Hawking temperature.*

This theorem contains (modulo some technicalities) some earlier partial results. In [7] a similar result was proved amongst the set of states which are KMS at some temperature. In [1] it was proved amongst the set of quasi-free stationary states. Very recently [12] I have shown how to drop the quasi-free assumption.

References

1. B.S. Kay, R.M. Wald: Physics Reports **207** No. 2, 49-136 (1991)
2. J. Dimock: Commun. Math. Phys. **77**, 219-228 (1980)
3. S.W. Hawking: Commun. Math. Phys. **43**, 199-220 (1975)
4. G.L. Sewell: Ann. Phys. NY **141**, 201-224 (1982)

5. B.S. Kay, in: Proc. 10th Inter. Conf. on General Relativity and Gravitation (Padova, 1983) eds. B. Bertotti et al. (Reidel, Dordrecht, 1984); talk (see Workshop Chairman's Report by A. Ashtekar, pp. 453-456)
6. C. Lüders, J.E. Roberts: Commun. Math. Phys. **134**, 29-63 (1990)
7. R. Haag, H. Narnhofer, U. Stein: Commun. Math. Phys. **94**, 219-238 (1984)
8. K. Fredenhagen, R. Haag: Commun. Math. Phys. **108**, 91-115 (1987)
9. B.S. Kay, in: Differential Geometric Methods in Theoretical Physics, eds. K. Bleuler and M. Werner (Reidel, Dordrecht, 1988)
10. G. Gonnella, B.S. Kay: Class. Quantum Grav. **6**, 1445-1454 (1989)
11. K. Fredenhagen, R. Haag: Commun. Math. Phys. **127**, 273-284 (1990)
12. B.S. Kay: Theorems on the uniqueness and thermal properties of stationary, nonsingular, quantum states on spacetimes with a bifurcate Killing horizon: dropping the quasi-free assumption (In preparation)

Charges in Quantum Field Theory

Karl-Henning Rehren

II. Inst. Theor. Physik, Univ. Hamburg, D-2000 Hamburg 50

In quantum field theory one has to deal with *superselection sectors*, i.e. inequivalent representations π_α of the algebra of observables $\mathcal{A}$. It is then desirable to have a *unified description*, i.e. a Hilbert space which is a direct sum of all sectors together with an algebra of charged fields interpolating among them, and a *gauge principle*, i.e. a transformation law for the fields to single out the observables as the gauge-invariant quantities. Typically

$$\mathcal{H}_{\text{uni}} = \bigoplus_\alpha \mathcal{H}_\alpha \otimes V^{\bar{\alpha}} , \tag{1}$$

where the representations of $\mathcal{A}$ and the gauge symmetry G on $\mathcal{H}_{\text{uni}}$ are

$$\pi = \bigoplus_\alpha \pi_\alpha \otimes \mathbb{1}^{\bar{\alpha}}, \qquad \text{resp.} \qquad \mathcal{U} = \bigoplus_\alpha 1_\alpha \otimes \tau^{\bar{\alpha}} . \tag{2}$$

Equation (1) gives rise to a 1:1 correspondence between two *a priori* different notions of "charge": namely the equivalence classes of representations of the algebra of *local* observables ("superselection charge") and of the *global* symmetry ("gauge charge"). The "Hilbert space vehicle" of this correspondence are the unobservable fields, which on the one hand interpolate the sectors and in particular create charged sectors from the vacuum sector, and on the other hand obey a specific gauge transformation rule; that is:

$$\psi \cdot \pi(A) = (\pi \times \pi_\beta)(A) \cdot \psi , \tag{3}$$

$$\mathcal{U}(g) \cdot \psi = \psi \cdot (\tau^\beta \times \mathcal{U})(g) , \tag{4}$$

where the product of representations of $\mathcal{A}$ in (3) will be discussed below, and (4) is a generalized transformation law: for groups with the natural composition law for their matrix representations τ

$$(\tau^\beta \times \tau^\alpha)(g) := \tau^\beta(g) \otimes \tau^\alpha(g) , \tag{5}$$

it is just a sophisticated notation of the conventional transformation law

$$\mathcal{U}(g)\psi_i\mathcal{U}(g)^{-1} = \sum_j \psi_j \tau_{ji}(g) . \tag{6}$$

We shall encounter "quantum symmetry" generalizations below.

From the physical viewpoint, we consider the superselection structure reflected by the product $\pi_\alpha \times \pi_\beta$ as given (observable) data, while the dual symmetry structure as well as the charged fields implementing *on Hilbert space* the functorial correspondence between superselection and gauge charges by algebraic properties have to be found. The data are provided by the DHR theory [1] of superselection sectors: the relevant information about the decomposition of products of sectors into their irreducible subsectors (superselection rules), the dependence of a product on the order of the factors (statistics), and charge conjugation, is cast into the mathematical form of a *strict monoidal braided C^* category with direct sums, subobjects, and conjugates.* The representation theory of the gauge symmetry should exhibit the same structures of "fusion", "braiding", and "conjugation" as given by the data, and finally the charged fields ψ to be constructed should solve (3) and (4) simultaneously, with a propagation of transformation laws to products of field operators and the Clebsch-Gordon decomposition of the latter, commutation relations at space-like distance (causal independence), and behaviour under operator (Hermitean) conjugation.

In physical space-time dimension 3+1, the superselection category is a *symmetric* one. In that case, there is a reconstruction theorem for the symmetry which is a compact group, and for charged fields with all the desired properties. This is the DR theorem [2]. However, for *braided* categories arising in low-dimensional QFT, a substitute for the DR theorem is missing.

Let us sketch the DHR theory of superselection sectors (see also [3]; and [4],[5] for recent studies of abelian and non-abelian models of chiral current algebras). The observable algebra $\mathcal{A}$ has inequivalent classes of physically relevant (particle-like) positive energy representations $\pi : \mathcal{A} \to \mathcal{B}(\mathcal{H})$. These are equivalently described by endomorphisms $\rho : \mathcal{A} \to \mathcal{A}$ such that π is unitarily equivalent to $\pi_\rho = \pi_0 \circ \rho$, where π_0 is a faithful reference (vacuum) representation. The endomorphisms of interest are *localized*, i.e. they act trivially on observables representing measurements at space-like distance. Together with Haag duality, this *local* set-up makes the DHR theory a powerful tool to understand charges in QFT.

The DHR representations have a natural associative composition law:

$$\pi_\alpha \times \pi_\beta := \pi_0 \circ (\rho_\alpha \circ \rho_\beta) \equiv \pi_\alpha \circ \rho_\beta \,, \tag{7}$$

respecting classes of unitary equivalence. The theory controls the reducibility of the product, and its commutativity up to unitary equivalence is described by the statistics. The distinction between high- and low-dimensional physics is due to the different topology of causal complements, i.e. the global causal structure of space-time: the statistics in 3+1 dimensions defines a representation of the permutation group (acting on $\pi_1 \times \ldots \times \pi_n$), while in less dimensions representations of the braid group may and will occur [6],[7].

It is a crucial fact that intertwining operators (e.g., between $\pi_1 \times \pi_2$ and a subrepresentation π_3, or between $\pi_1 \times \pi_2$ and $\pi_2 \times \pi_1$), are elements of $\mathcal{A}$

itself, and the algebraic relations among them express observable information. E.g., associated with every sector is a positive state on $\mathbb{C}B_\infty$ (a strong Markov trace), which is a class-invariant for the localized endomorphisms describing the sector [6]. A central role plays the value λ_π of the trace state on a generator σ_i of the braid group. The main condition for the existence of a conjugate sector is the non-vanishing of λ_π. In that case, its complex phase is related to the "spin" of a space-time covariant sector such that the violation of spin additivity can be traced back to the statistics monodromy [8]. The significance of its inverse modulus $d_\pi \geq 1$, the statistical dimension, can be seen from the following remarks. An endomorphism ρ is invertible iff $d_\pi = 1$. For permutation group statistics, d_π is the dimension of the corresponding representation of the symmetry group on the charged fields [2]. Although for braid group statistics d_π typically takes non-integer values, there still holds a "sum rule" for the product of sectors and its subsectors, reminiscent of a dimension formula for tensor products of vector spaces:

$$d_1 \cdot d_2 = \sum\nolimits_{r=1}^{n} d^{(r)} \qquad \text{if} \qquad \pi_1 \times \pi_2 \simeq \bigoplus\nolimits_{r=1}^{n} \pi^{(r)} \,. \tag{8}$$

The DHR theory suggests to study various type III_1 and II_1 inclusions of von Neumann factors related to the endomorphisms, and to apply classification methods for these objects. E.g., it predicts that d_π can take values only in the Jones set $\{2\cos(\pi/l)|l = 3,4,5,\dots\} \cup [2,\infty)$. Longo [9],[10] has established the connection with the index for III_1 subfactors, namely

$$\mathrm{Ind}[\mathcal{A}(\mathcal{O}) : \rho(\mathcal{A}(\mathcal{O}))] = d_\pi^2 \tag{9}$$

for ρ localized in $\mathcal{O}$. He gets restrictions on the possible values of $d_\pi > 2$ from the sum rule (8) and the fact that for $\pi \times \pi$ decomposing into few ($n \leq 3$) subrepresentations, the DHR theory predicts additional relations among the braid group representatives (Hecke resp. Birman-Wenzl algebra) for which a classification of positive Markov traces exists. He shows the range of d_π to have a gap between 2 and $\simeq 2.05$ and at most countably many accumulation points below $\simeq 2.25$. Optimizing his methods we can show the gap to extend until $\sqrt{5}$, with only four more values until $\sqrt{6}$. We expect further classification progress from this direction.

Let us turn to quantum symmetries, the required fusion, braiding, and conjugation structures of which we shall exemplify with quantum groups [11] as familiar prototypes. These due to their apparent relation to rational conformal QFT models suggest themselves to play the role of generalized gauge symmetries. The composition law of representations of quantum groups $\mathcal{X}$ as well as the transformation law for covariant fields are induced from the co-product homomorphism $\Delta : \mathcal{X} \to \mathcal{X} \otimes \mathcal{X}$, generalizing (5), (6):

$$(\tau^\alpha \times \tau^\beta)(x) := (\tau^\alpha \otimes \tau^\beta)(\Delta(x)) \,, \tag{10}$$

$$\mathcal{U}(x) \cdot \psi_i = \sum\nolimits_j \psi_j \cdot (\tau_{ji} \otimes \mathcal{U})(\Delta(x)) \,. \tag{11}$$

The invertible element R which implements the non-co-commutativity of the co-product being in $\mathcal{X} \otimes \mathcal{X}$ (quasi-triangularity) is the counterpart of the innerness of the statistics operators in the DHR theory, and is expected to translate into commutation relations of covariant fields. Finally, the antipode map S [11] induces a conjugation of representations $\tau \mapsto \bar{\tau} := {}^t\tau \circ S$.

Problems arise when quantum group covariant fields are supposed to act on a *Hilbert space* with positive metric, for the following reason. Quantum groups are of interest to QFT for their "cut-off" representation theory at the deformation parameter $q = e^{2\pi i/m}$. But omitting the corresponding higher-spin products of field operators solving (3) turns out inconsistent with quantum covariance (4). This was overlooked in the first attempts (e.g., [12]). Moreover, since the operator adjoint * changing the order of factors is in conflict with non-co-commutativity of the co-product, the transformation law does not propagate to products of ψ with ψ^*. Thus, in most proposals for quantum covariant fields, the question of Hilbert space representations is either not addressed, or indefinite metrics are admitted.

Hadjiivanov, Paunov, and Todorov [13] exploit the coincidence of $(6j)$-symbols to construct quantum group covariant chiral fields with numerical commutation relations by coupling field theoretic conformal blocks (chiral correlation functions) to the corresponding symmetry blocks of the quantum group. The degenerate representations have, however, to be included.

Mack and Schomerus [14] find that in order to comply with charged Hilbert space fields postulated to intertwine the superselection sectors of specific models (Ising, minimal), one has to admit a non-co-associative (quasi-Hopf) co-product. Then the operator product of covariant fields (11) is no longer covariant, but a covariant product can be defined with the help of coefficients in the symmetry algebra. Similarly, the covariant conjugate differs from the operator conjugate. With respect to the covariant product, fields have space-like commutation relations with numerical coefficients, while the commutation relations for the operator product involve symmetry algebra valued R-matrices. The quantum symmetry is not a quantum group but rather the direct sum of its "good" representations, i.e. a finite-dimensional algebra like the group ring of a finite group.

A model-independent explicit Hilbert space construction of charged field multiplets in terms of the superselection data [15] envisages a further generalization. Gauge covariance arises as a bi-module action of a hyperfinite type II_1 factor $\mathcal{M}$ which by construction is the commutant of the observables $\mathcal{A}$ on $\mathcal{H}_{\text{uni}}$. Representations of $\mathcal{M}$ are given as homomorphisms $(\rho_{ij}) = \rho_{\text{mat}} : \mathcal{M} \to \text{Mat}_N(\mathcal{M})$, inherited from the endomorphisms $\rho : \mathcal{A} \to \mathcal{A}$ of the DHR theory. Substituting ρ_{mat} for $\tau \times id$ in (4) yields the composition and transformation laws generalizing (10) and (11)

$$(\rho \times \rho)_{ik,jl} := \rho_{ij} \circ \rho_{kl}\,, \tag{12}$$

$$m \cdot \psi_i = \sum_j \psi_j \cdot \rho_{ji}(m)\,. \tag{13}$$

The conjugate representation coincides with the conjugate in the sense of "correspondences" [9]. The fields ψ_i, intertwining the superselection charges as in (3), are partial isometries with source projections $p_i \in \mathcal{M}$, and the statistical dimension arises as the "dimension of the representation" of $\mathcal{M}$

$$d_\pi = \sum_i \mathrm{tr}(\rho_{ii}(1_\mathcal{M})) \qquad \text{where} \qquad \rho_{ij}(1_\mathcal{M}) = \delta_{ij} \cdot p_i \,. \tag{14}$$

This field algebra has $\mathcal{M}$-valued coefficients in commutation relations and product expansions of charged fields, a feature it shares with the Mack-Schomerus algebra and which was already implicitly present in the construction of localized intertwining fields carrying the anyonic charges of chiral $U(1)$ current algebra [4]. These BMT models have a $\mathbb{Z}_{2N}$ gauge *group* symmetry in accordance with their superselection structure, but it is implemented on the unified Hilbert space with a true quasi Hopf (not Hopf!) structure visible from its non-scalar Clebsch-Gordan coefficients [15].

References

1. Doplicher, S., Haag, R., Roberts, J.E.: Comm.Math.Phys. **23**, 199 (1971); Doplicher, S., Haag, R., Roberts, J.E.: Comm.Math.Phys. **35**, 49 (1974)
2. Doplicher, S., Roberts, J.E.: Comm.Math.Phys. **131**, 51 (1990)
3. Fredenhagen, K.: Contribution to these proceedings
4. Buchholz, D., Mack, G., Todorov, I.: In: Proc. Annecy 1988; Nuclear Phys. B (Proc. Suppl.) **5B**, 20 (1988)
5. Wassermann, A.: Preliminary notes (1991) on positive energy representations of loop groups and *Diff*(S^1), Univ. Cambridge (UK)
6. Fredenhagen, K., Rehren, K.-H., Schroer, B.: Comm.Math.Phys. **125**, 201 (1989)
7. Fröhlich, J.: in: Proc. Cargèse 1987, eds. G.'tHooft et al., p. 71 (Plenum 1988)
8. Fredenhagen, K.: Lecture Notes in Phys. **370**, p. 340 (Springer 1990)
9. Longo, R.: Comm.Math.Phys. **126**, 217 (1989); Longo, R.: Comm.Mat.Phys. **130**, 285 (1990)
10. Longo, R.: *Minimal index and braided subfactors.* Univ. Rome preprint (1991)
11. Drinfel'd, V.G.: in: Proceedings Int. Congr. Math. Berkeley 1986, p. 799, ed. A. Gleason (Berkeley 1987) Woronowicz, S.L.: Comm.Math.Phys. **111**, 613 (1987)
12. Moore, G., Reshetikhin, N.Yu.: Nuclear Phys. B **328**, 557 (1989)
13. Hadjiivanov, L., Paunov, R., Todorov, I.: Quantum group extended chiral p-models. Trieste preprint SISSA 160/90/FM, to appear in Nuclear Phys. B
14. Mack, G., Schomerus, V.: Quasi Hopf quantum symmetry in quantum theory. Univ. Hamburg preprint DESY 91-037 (1991), to appear in Nuclear Phys. B
15. Rehren, K.-H.: *Field operators for anyons and plektons.* Univ. Hamburg preprint DESY 91-043 (1991), to appear in Comm.Math.Phys.

Three Exactly Soluble Quantum Field Theory Models in 2-,3- and 4-Dimensional Space Time and Some General Questions They Suggest

Arthur S. Wightman

Princeton University, Department of Physics
NJ 08544, Princeton

1 Introduction

As a result of three decades of hard work on the local algebra formalism, we now have a general theory of quantized fields that provides a satisfactory framework for field theory. On the other hand, constructive quantum field theory has made rather limited progress toward the objective of characterizing and constructing all field theories that satisfy the axioms of the general theory. We do have the nontrivial examples $P(\phi)_2$,Y_2,ϕ^4_3,Y_3,Higgs_2,Higgs_3 and fragments of YM_3 and YM_4. However, these examples do not provide enough information to suggest reasonable guesses for the answers to general questions. For example, how do the perturbatively non-renormalizable theories fit into the general picture? The recent results of da Veiga and coworkers establishing the existence of tempered solutions of the Gross-Neveu model in three dimensional Euclidean space-time, show that a non-perturbative treatment of a perturbatively non-renormalizable theory is possible using rigorous renormalization group methods. What distinguishes such theories? Is the applicability of renormalization group methods to be regarded as always a reliable guide to the existence of non-trivial solutions of theories?

A particular puzzle which may or may not be related to these questions about non-renormalizable theories is the status of solutions of ϕ^4_4. On the one hand, there are no-go theorems of Aizenman and Fröhlich and coworkers that assert that any solution of ϕ^4_4 obtained as a limit of ferromagnetic lattice approximations is trivial. On the other hand, there are the results of Gawedzki and Kupiainen which assert the existence of tempered solutions for negative values of the coupling constant. These results suggest that a detailed non-perturbative study of the solutions of ϕ^4_ν might be enlightening.

More specifically, it would seem worthwile to study the super renormalizable phase of ϕ_ν^4 as a function of the mass gap, m, the space-time dimension ν and the dimensionless scale-invariant coupling constant, g:

$$g = -m^\nu \frac{\overline{S_4^T}}{(\overline{S_2^T})^2}$$

(S_4^T and S_2^T are respectively, the truncated 4 and 2 point Schwinger functions). In particular, one would like to know how the dimensional poles at $g = g_n$

$$g_n = 8\left(1 - \frac{1}{2n}\right), \quad n = 1, 2, \ldots$$

affect the upper bound on g.

As a preliminary to this program, I undertook the construction of some examples of exactly soluble theories for which one can reduce the calculation of g to relatively simple quadratures.

2 The Free Massive Dirac Current, Its Dual And Potential

The free massive Dirac spinor field is defined as

$$\psi(x) = (2\pi)^{(\frac{\nu-1}{2})} \int d\Omega_m(p)$$

$$\sum_\zeta [a(p,\zeta)u(p,\zeta)\exp -ip\cdot x + b^*(p,\zeta)u^c(p,\zeta)\exp ip\cdot x]$$

where

$$p = \{p^0, \boldsymbol{p}\} \in \mathbb{R}^n, \quad p^0 = \sqrt{m^2 + \boldsymbol{p}^2}, \quad d\Omega_m(p) = \frac{d^{\nu-1}p}{\sqrt{m^2 + \boldsymbol{p}^2}}$$

and the annihilation operators $a(p,\zeta)$ and $b(p,\zeta)$ for particles and antiparticles satisfy the CAR:

$$[a(p,\zeta), a(p^1,\zeta^1)]_+ = 0 = [b(p,\zeta), b(p^1,\zeta^1)]_+$$
$$[a(p,\zeta), b(p^1,\zeta^1)]_+ = 0 = [a(p,\zeta), b^*(p^1,\zeta^1)]_+$$
$$[a(p,\zeta), a^*(p^1,\zeta^1)]_+ = \delta_{\zeta\zeta^1} p^0 \delta(\boldsymbol{p} - \boldsymbol{p}^1) = [b(p,\zeta), b^*(p^1,\zeta^1)]_+ \quad .$$

The $u(p,\zeta)$ are standard solutions of the Dirac equation

$$(-p\!\!\!/+m)u(p,\zeta)=0, \qquad p\!\!\!/=\gamma^\mu p_\mu$$

$$\gamma^\rho\gamma^\sigma+\gamma^\sigma\gamma^\rho=2g^{\rho\sigma}\mathbb{1}, \quad g^{\rho\sigma}=\begin{cases}1 & \rho=\sigma=0\\ -1 & \rho=\sigma=1,\dots,\nu-1\\ 0 & \text{otherwise}\end{cases}$$

$$u(p,\zeta)^+\gamma^\mu u(p,\zeta^1)=\delta_{\zeta\zeta^1}p^\mu, \qquad (u(p,\zeta)^+)_\alpha=\sum_\beta \overline{u(p,\zeta)_\beta}(\gamma^0)_{\beta\alpha}$$

$$u^c(p,\zeta)=C^{-1}\overline{u(p,\zeta)}, \qquad -\overline{\gamma^\mu}=C\gamma^\mu C^{-1}$$

$$(p\!\!\!/+m)u^c(p,\zeta)=0 \quad .$$

The Dirac current, j, is defined by

$$j^\mu(x)=:\psi^+\gamma^\mu\psi:(x)=\frac{1}{(2\pi)^{\nu-1}}\iint d\Omega_m(p)d\Omega_m(q)$$

$$\sum_{\rho,\sigma}[a^*(p,\rho)a(q,\sigma)u(p,\rho)^+\gamma^\mu u(q,\sigma)\exp i(p-q)\cdot x$$
$$-b^*(q,\sigma)b(p,\sigma)u^c(p,\rho)^+\gamma^\mu u^c(q,\sigma)\exp -i(p-q)\cdot x$$
$$+a^*(p,\rho)b^*(q,\sigma)u(p,\rho)^+\gamma^\mu u(q,\sigma)\exp i(p+q)\cdot x$$
$$-a(q,\sigma)b(p,\sigma)u^c(p,\rho)^+\gamma^\mu u(q,\sigma)\exp -i(p+q)\cdot x]$$

and satisfies the conservation law

$$\partial_\mu j^\mu(x)=0$$

by virtue of such identities as

$$(p-q)^\mu u(p,\rho)^+\gamma^\mu u(q,\sigma)=0$$
$$(p+q)^\mu u(p,\rho)^+\gamma^\mu u^c(q,\sigma)=0.$$

The Dirac current, j, being a 1-form, has a dual, J, which is a $\nu-1$ form. If we introduce the totally anti-symmetric tensor, $\varepsilon^{\mu_1\dots\mu_\nu}$, with $\varepsilon^{01\dots\nu-1}=1$, we have

$$j^\mu=\frac{1}{(\nu-1)!}\varepsilon^{\mu\mu_1\dots\mu_{\nu-1}}J_{\mu_1\dots\mu_{\nu-1}}; \quad J_{\mu_1\dots\mu_{\nu-1}}=\varepsilon_{\mu_1\dots\mu_{\nu-1}\mu}j^\mu.$$

Now the current conservation law expressed in terms of the dual current, J,

$$\partial_\mu\varepsilon^{\mu\mu_1\dots\mu_{\nu-1}}J_{\mu_1\mu_2\dots\mu_{\nu-1}}=0,$$

says, in the language of differential forms

$$dJ=0.$$

Thus, if some analogue of the Poincaré Lemma holds, there should exist a $\nu - 2$ form K such that

$$J = dK\,,$$

that is, J closed should imply J exact. Explicitely

$$J_{\mu_1 \ldots \mu_{\nu-1}} = \sum_{j=1}^{\nu-1} \left\{ \begin{array}{ll} (-1)^{j+1} & \nu \,\text{odd} \\ 1 & \nu \,\text{even} \end{array} \right\} \partial_{\mu_j} K_{\mu_{j+1} \ldots \mu_{\nu-1} \mu_1 \ldots \mu_{j-1}} \quad .$$

For example,

$$\begin{array}{lll} \nu = 2 & j^\mu = \varepsilon^{\mu\mu_1} J_{\mu_1} & J_{\mu_1} = \partial_{\mu_1} K \\ \nu = 3 & j^\mu = \dfrac{1}{2!} \varepsilon^{\mu\mu_1\mu_2} J_{\mu_1\mu_2} & J_{\mu_1\mu_2} = \partial_{\mu_1} K_{\mu_2} - \partial_{\mu_2} K_{\mu_1} \\ \nu = 4 & j^\mu = \dfrac{1}{3!} \varepsilon^{\mu\mu_1\mu_2\mu_3} J_{\mu_1\mu_2\mu_3} & J_{\mu_1\mu_2\mu_3} = \partial_{\mu_1} K_{\mu_2\mu_3} \\ & & + \partial_{\mu_2} K_{\mu_3\mu_1} + \partial_{\mu_3} K_{\mu_1\mu_2} \quad . \end{array}$$

Now in the Clifford algebras

$$\begin{array}{ll} \nu = 2 & \gamma^\mu = \varepsilon^{\mu\mu_1} \tilde{\gamma}^5 \gamma_{\mu_1} \\ \nu = 3 & \gamma^\mu = \dfrac{i}{2!} \varepsilon^{\mu\mu_1\mu_2} \tilde{\gamma}^5 j_{\mu_1} j_{\mu_2} = \dfrac{1}{2!} \varepsilon^{\mu\mu_1\mu_2} \tilde{\gamma}^5 \sigma_{\mu_1\mu_2} \\ \nu = 4 & \gamma^\mu = \dfrac{i}{3!} \varepsilon^{\mu\mu_1\mu_2\mu_3} \tilde{\gamma}^5 \gamma_{\mu_1} \gamma_{\mu_2} \gamma_{\mu_3} \quad , \end{array}$$

where

$$\tilde{\gamma}^5 = (-i)^{\frac{(\nu-1)(\nu-2)}{2}} \gamma^0 \gamma^1 \cdots \gamma^{\nu-1}.$$

(Recall that for ν odd there are two inequivalent representations of the Clifford algebra and $\tilde{\gamma}^5$ commutes with the γ^μ. In that case, $\tilde{\gamma}^5 = \pm 1$ for the irreducible representations.) Thus,

$$\nu = 2 \quad \partial^\mu K =: \psi^+ \tilde{\gamma}^5 \gamma^\mu \psi : \tag{2.1}$$

$$\nu = 3 \quad \partial_{\mu_1} K_{\mu_2} - \partial_{\mu_2} K_{\mu_1} =: \psi^+ \tilde{\gamma}^5 \sigma_{\mu_1\mu_2} \psi : \tag{2.2}$$

$$\begin{aligned} \nu = 4 \quad \partial_{\mu_1} K_{\mu_2\mu_3} + \partial_{\mu_2} K_{\mu_3\mu_1} + \partial_{\mu_3} K_{\mu_1\mu_2} &= 0 \quad \text{if any two of } \mu_1, \mu_2, \mu_3 \text{ are equal} \\ &=: \psi^+ i \tilde{\gamma}^5 \gamma_{\mu_1} \gamma_{\mu_2} \gamma_{\mu_3} \psi : \text{ if } \mu_1 \neq \mu_2 \neq \mu_3 \neq \mu_1 \,. \end{aligned} \tag{2.3}$$

3 Study of the Dual Potentials

The dual potential for space-time dimension $\nu = 2$ was introduced by the author in 1963 [1]. Here it is in a slightly different form. (In the old version a standard spinor basis was chosen and expressions like $u(p)^+\gamma^5 u(q)$ were expressed as functions of p and q.)

$$\begin{aligned} K(x) = \frac{im}{2\pi} \int\int d\Omega_m(p) d\Omega_m(q) \\ [a^*(p)a(q)\frac{2mu(p)^+\gamma^5 u(q)}{(p-q)\cdot(p-q)} \exp i(p-q)\cdot x \\ + b^*(q)b(p)\frac{2mu^c(p)^+\gamma^5 u^c(q)}{(p-q)\cdot(p-q)} \exp -i(p-q)\cdot x \\ + a^*(p)b^*(q)\frac{2mu(p)^+\gamma^5 u^c(q)}{(p+q)\cdot(p+q)} \exp i(p+q)\cdot x \\ + a(q)b(p)\frac{2mu^c(p)^+\gamma^5 u(q)}{(p+q)\cdot(p+q)} \exp -i(p+q)\cdot x] . \end{aligned} \tag{3.1}$$

K was shown to be local:

$$[K(x), K(y)] = 0, \text{ for } (x-y)^2 < 0 ,$$

but K and ψ not relatively local:

$$[K(x), \psi(y)] \neq 0, \ [K(x), \psi(y)^+] \neq 0, \text{ for } (x-y)^2 < 0 .$$

The equation of motion for K was recognized as PCAC for the Dirac axial current

$$j_5^\mu =: \psi^+\tilde{\gamma}^5\gamma^\mu\psi :, \ \partial_\mu j_5^\mu = 2mi : \psi^+\tilde{\gamma}^5\psi : \text{ or } \Box K = 2mi : \psi^+\tilde{\gamma}^5\psi : . \tag{3.2}$$

However, it was not until the work of Coleman and Lehmann and Stehr that it was recognized that the right hand side of (3.2) can be written as a function of K. The latter authors find it convenient to work with $\tilde{K} = \sqrt{\pi}(-K + Q|_2)$ instead of K and derive

$$\Box\tilde{K} = 2m^2 \vdots\tilde{K}(\cos(2\sqrt{\pi}\tilde{K} - 1))\vdots ,$$

where

$$\vdots \exp i2\pi\lambda\tilde{K} \vdots = \exp 2\pi i\lambda\tilde{K}[\langle \exp 2\pi i\lambda\tilde{K}\rangle_0]^{-1} .$$

For $\nu = 3$, an analogous analysis of (2.2) yields for the dual potential

$$K_\mu(x) = \frac{1}{(2\pi)^2} \int\int d\Omega_m(p) d\Omega_m(q)$$

$$\left\{ \frac{a^*(p)a(q)\exp i(p-q)\cdot x}{-2(p\cdot q+m^2)} \times \right.$$

$$\times \left[(p+q)_\mu u(p)^+ \tilde{\gamma}^5 u(q) + \frac{imu(p)^+ u(q) \varepsilon_\mu^{\rho\sigma} (p-q)_\rho (p+q)_\sigma}{(p\cdot q - m^2)} \right]$$

$$+ \frac{b^*(q)b(p)\exp -i(p-q)\cdot x}{-2(p\cdot q+m^2)} \times$$

$$\times \left[(p+q)_\mu u^c(p)^+ \tilde{\gamma}^5 u^c(q) - \frac{imu^c(p)^+ u^c(q) \varepsilon_\mu^{\rho\sigma} (p-q)_\rho (p+q)_\sigma}{(p\cdot q - m^2)} \right]$$

$$+ \frac{a^*(p)b^*(q)\exp i(p+q)\cdot x}{-2(p\cdot q-m^2)} \times$$

$$\times \left[(p-q)_\mu u(p)^+ \tilde{\gamma}^5 u^c(q) + \frac{imu(p)^+ u^c(q) \varepsilon_\mu^{\rho\sigma} (p-q)_\rho (p+q)_\sigma}{(p\cdot q + m^2)} \right]$$

$$+ \frac{a(q)b(p)\exp -i(p+q)\cdot x}{-2(p\cdot q-m^2)} \times$$

$$\left. \times \left[(p-q)_\mu u^c(p)^+ \tilde{\gamma}^5 u(q) - \frac{imu^c(p)^+ u(q) \varepsilon_\mu^{\rho\sigma} (p-q)_\rho (p+q)_\sigma}{(p\cdot q + m^2)} \right] \right\}$$

This dual potential is hermitian and divergence-free:

$$K_\mu(x)^* = K_\mu(x) \qquad \partial_\mu K^\mu = 0 \,.$$

Its equation of motion is the Gordon decomposition of the current:

$$\Box K^\lambda - 2m\varepsilon^{\lambda\rho\sigma}\partial_\rho K_\sigma = i(\psi^+ \partial^\lambda \psi - (\partial^\lambda \psi^- \psi)) \,.$$

As of this writing, it is not known whether, and if so how the right hand side can be expressed, as a local function of K_μ.

For lack of space, the formula for the dual potential $K_{\mu\nu}$ for space-time dimension $\nu = 4$ will have to appear elsewhere.

References

1. Wightman, A.S.: Introduction to some Aspects of the Relativistic Dynamics of Quantized Fields. Cargése Lectures in Theoretical Physics 1964, Ed. M. Levy, Gordon and Breach 1967, esp. 233-238.
For important further developments:
Pohlmeyer, K.: The Equation Curl $W_\mu(x) = 0$ in Quantum Field Theory. Comm.Math.Phys. **25**, 73-86 (1972)
2. Lehmann, H. and Stehr, J.: The Bose Field Structure Associated with a Free Massive Dirac Current in One Space Dimension. DESY 76/29, June 1976

CPT- and Lorentz-Transformations in Two-Dimensional Q.F.T.

H.J. Borchers

Institut für Theoretische Physik Universität Göttingen
Bunsenstrasse 9, D 3400 Göttingen, Fed. Rep. of Germany

The C.P.T.-theorem plays an important role in relativistic quantum field theory. In 1957 R.Jost [3] gave a proof of the C.P.T.-theorem in the frame of Wightman-field theory in which he revealed the connection of the C.P.T.-symmetry with the assumptions of positivity of the energy, Lorentz-invariance, and the standard locality assumptions. In this proof the existence of a vacuum state was essential. But up to now there is no proof of the C.P.T.-theorem in the theory of local observables in the sense of Araki, Haag, and Kastler.

A different aspect of the C.P.T.-operator has been realized in connection with the Tomita-Takesaki-theory [5],[4] of modular Hilbert-algebras. Bisognano and Wichmann [1],[2] observed that the C.P.T.-operator and the Lorentz-boosts are related to the modular conjugation and to the modular group of the wedge-algebra, provided the theory of local observables is generated from a Wightman-field which is covariant under Lorentz-transformations. Here again one is dealing with the vacuum-sector.

In this representation we want to show that the modular group for the algebra of the wedge domain can always be interpreted as the group of Lorentz-boosts, provided the cyclic and separating vector is the vacuum-vector. In order to obtain this result one must show that the modular group induces outer automorphisms of the translation group. To prove this two properties are essential. One is the spectrum condition for the translations, the other is the structure of the wedge domain, which is mapped by a semi-group of translations into itself. If the wedge is fixed then one is dealing with a two-dimensional problem. Therefore, in all of the investigations we shall only deal with field theories in two space-time dimensions.

Notation: The symbol $\{\mathcal{M}, \mathcal{H}, \Omega, \Delta, J\}$ means an algebra $\mathcal{M}$ acting on a Hilbert space $\mathcal{H}$ such that Ω is cyclic and separating for $\mathcal{M}$. The modular operator and modular conjugation given by this situation is denoted by Δ and J.

First we are looking at one endomorphism of $\mathcal{M}$ induced by unitary operators which leave Ω fixed.

Theorem 1 *Let $\{\mathcal{M}, \mathcal{H}, \Omega, \Delta, J\}$ be a von Neumann algebra and let U be unitary with $U\Omega = \Omega$ and assume U induces an endomorphism of $\mathcal{M}$ i.e. $U\mathcal{M}U^* \subset \mathcal{M}$ then we have:*

(1) *With $S = J\Delta^{\frac{1}{2}}$ the relation*

$$US = SU.$$

(2) *The operator-function*

$$\Delta^{it} U \Delta^{-it}$$

has an analytic extension into the strip,

$$S(-\frac{1}{2}, 0) = \{z; -\frac{1}{2} < \Im m\, z < 0\},$$

and it is continuous on the boundary, and it fulfills the estimate

$$\|\Delta^{iz} U \Delta^{-iz}\| \leq 1.$$

If U induces an automorphism then U commutes as well with Δ as with J and there is no interplay between U and the modular group. Next we are turning to the case where we have a semi-group $U(a)$ of unitaries with positive generator and which induce a semi-group of endomorphisms. Since it induces endomorphisms of $\mathcal{M}$ for positive a, one maps the upper half-plane by $a = e^{\xi}$ onto the strip $(0, \pi)$. For real ξ we have an analytic continuation into $t \in S(-\frac{1}{2}, 0)$ and for real t an analytic continuation into $\xi \in S(0, \pi)$. This gives analyticity in a tube with triangular base. Looking now at the set $\Im m\, t = -\frac{1}{2}$ one finds that one can analytically continue in the variable ξ into the strip $S(-\pi, 0)$. Altogether this gives the econd result.

Theorem 2 *Let $\{\mathcal{M}, \mathcal{H}, \Omega, \Delta, J\}$ be a von Neumann algebra and let $U(a)$ be a one-parametric group of unitaries with $U\Omega = \Omega$ and assume $U(a)$ induce for $a \geq 0$ endomorphisms of $\mathcal{M}$ i.e. $U(a)\mathcal{M}U^*(a) \subset \mathcal{M}$, $a \geq 0$ then we have:*

1. *The expression*

$$\Delta^{it} U(e^{\xi}) \Delta^{-it} =: V(t, \xi)$$

has an analytic extension into the tube T and is continuous on the closure, where T is a tube with quadrangular base

$$T = \{(z, \zeta); (\Im m\, z, \Im m\, \zeta) \in B)\},$$
$$B = \text{int conv}\, \{(0,0), (-\frac{1}{2}, -\pi), (-\frac{1}{2}, 0), (0, \pi)\}.$$

2. *The operator* $V(z,\zeta)$ *is bounded*

$$\|V(z,\zeta)\| \leq 1, \quad \text{for} \quad (z,\zeta) \in T.$$

3. $V(t-\frac{\imath}{2},\xi-\imath t)$ *has the value*

$$V(t-\frac{\imath}{2},\xi-\imath t) = \Delta^{\imath t} J U(-e^{\xi}) J \Delta^{-\imath t}.$$

In order to find further analytic continuations one has to make use of the cyclicity of Ω for $\mathcal{M}$ and for $\mathcal{M}'$. To this end one introduces two functions.

Notation: Let $A \in \mathcal{M}$ and $B' \in \mathcal{M}'$ then we define:

$$F^{+}(t,\xi) = (\Omega, B' \Delta^{\imath t} U(e^{\xi}) \Delta^{-\imath t} A \Omega)$$

and

$$F^{-}(t,\xi) = (\Omega, A \Delta^{\imath t} U(-e^{\xi}) \Delta^{-\imath t} B' \Omega).$$

As one sees from the construction of the two functions, they coincide for real values of t and of ξ. However the third statement of the second result implies much more analyticity.

Theorem 3 *The two functions* $F^{+}(t,\xi)$ *and* $F^{-}(t,\xi)$, *which are holomorphic in the tubes* T *and* $-T$ *respectively, are different representations of one holomorphic function* $H(z,\zeta)$, *which is holomorphic in the domain*

$$\{\Re e\, z, \Re e\, \zeta \in \mathbb{R}^2\} \bigcap \{-\pi < 2\pi \Im m\, z - \Im m\, \zeta < \pi\}. \qquad (*)$$

In this domain $H(z,\zeta)$ *is bounded. One has*

$$|H(z,\zeta)| \leq \max\{\|A\Omega\|\|B'^{*}\Omega\|, \|A^{*}\Omega\|\|B'\Omega\|\}.$$

This latter statement implies that $H(z,\zeta)$ *is constant along the direction given by the strip* (*) *i.e.*

$$H(z,\zeta) = H(z+\frac{w}{2},\zeta+w\pi) \quad \text{for} \quad w \in \mathbb{C}.$$

Using now the fact that $\mathcal{M}\,\Omega$ and $\mathcal{M}'\Omega$ are both dense in $\mathcal{H}$ we obtain:

Theorem 4 *Let* $\{\mathcal{M}, \mathcal{H}, \Omega, \Delta, J\}$ *be as before. Let* $U(a)$ *be a one-parametric group of unitaries with* $U\Omega = \Omega$ *and assume* $U(a)$ *induces for* $a \geq 0$ *endomorphisms of* $\mathcal{M}$ *i.e.* $U(a)\mathcal{M}\, U^{*}(a) \subset \mathcal{M}$, $a \geq 0$ *then we have for* $(t,a) \in \mathbb{R}^2$

(a)
$$\Delta^{\imath t} U(a) \Delta^{-\imath t} = U(e^{-2\pi t} a)$$

and

(b)
$$J U(a) J = U(-a).$$

If we assume, instead, that $U(a)$ induces endomorphisms for $a \leq 0$ then (b) remains unchanged but (a) is replaced by

$$\Delta^{it} U(a) \Delta^{-it} = U(e^{2\pi t} a).$$

Two-dimensional Q.F.T.

Let $\{\mathcal{M}, \mathcal{H}, \Omega, \Delta, J\}$ be as before. Now we assume that we have a continuous representation $U(a), a \in \mathbb{R}^2$ of the vector group of $\mathbb{R}^2$. Furthermore, we assume

(a) $U(a)\Omega = \Omega$

(b) The spectrum of $U(a)$ is contained in the forward light cone $V^+ := \{a \in \mathbb{R}^2; a_0 \geq |a_1|\}$

(c) Let W be the set

$$W := \{\ a \in \mathbb{R}^2; a_1 \geq |a_0|\}$$

then we assume

$$U(a)\mathcal{M}\ U^*(a) \subset \mathcal{M} \quad \text{for} \quad a \in W.$$

Now we introduce the lightcone coordinates

$$a^+ = \frac{a_0 + a_1}{2}, a^- = \frac{a_0 - a_1}{2} \quad \text{or} \quad a_0 = a^+ + a^-, a_1 = a^+ - a^-.$$

Since $U(a)$ fulfills the spectrum condition it follows that $U(a^+)$ and $U(a^-)$ both have positive generators and, moreover, we have

$$U(a^+)\mathcal{M}\ U^*(a^+) \subset \mathcal{M} \quad \text{for} \quad a^+ \geq 0 \quad \text{and}$$
$$U(a^-)\mathcal{M}\ U^*(a^-) \subset \mathcal{M} \quad \text{for} \quad a^- \leq 0.$$

Therefore, we apply the main result and obtain:

Theorem 5 *With the assumptions described above and with*

$$U(\Lambda(t)) = \Delta^{it} \quad \text{and} \Lambda(t) = \begin{pmatrix} \cosh 2\pi t & -\sinh 2\pi t \\ -\sinh 2\pi t & \cosh 2\pi t \end{pmatrix}$$

then $\{U(\Lambda), U(a)\}$ induce a representation of the two-dimensional Poincaré-group which fulfills the spectrum condition and which have Ω as invariant vector. Furthermore, one has

$$JU(a)J = U(-a)$$

All these transformations act local on the net constructed from the algebra of the wedge and its commutant.

References:

1. Bisognano, J., Wichmann, E.H.: J.Math.Phys. **16**, 985-1007 (1975)
2. Bisognano, J., Wichmann, E.H.: J.Math.Phys. **17**, 303-321 (1975)
3. Jost, R.: Helv.Phys.Acta **30**, 409-416 (1957)
4. Takesaki, M.: Tomita's theory of modular Hilbert algebras and its applications. In: Lecture Notes in Math. **128**, Springer-Verlag, Heidelberg (1970)
5. Tomita, M.: Standard forms of von Neumann algebras. In: Fifth functional analysis symposium of Math. Soc. of Japan, Sendai (1967)

The Concept of Spontaneously Broken Symmetry and a New Approach to Goldstone's Theorem

John Roberts

Fachbereich Physik, Universität Osnabrück
D-4500 Osnabrück, Fed. Rep. of Germany

The work reported on here forms part of investigations on spontaneously broken symmetries done in collaboration with D. Buchholz, S. Doplicher and R. Longo.

As is well known, the notion of spontaneously broken symmetry, unlike that of broken symmetry, can be formalized without too much difficulty. The basic idea is as follows: A symmetry of a physical system is described by an automorphism β. If ω is a state then $\omega \circ \beta$ will, in general, be a different state of the same system. However, if ω is a state of special physical significance such as a vacuum state or an equilibrium state at some particular temperature, we might expect ω to be typically unique in which case $\omega \circ \beta = \omega$. Whenever for one of these special states $\omega \circ \beta \neq \omega$, we say that the symmetry β is spontaneously broken in the state in question. Here we shall be interested solely in the case that $\omega = \omega_0$, a vacuum state of a relativistic quantum field theory. Furthermore, we have in mind the case where β is an internal symmetry.

Now it is an empirical fact that internal symmetries turn out to be gauge symmetries. It is not understood why this is so. The problem may be that in a model the observables are specified by fiat but we are not able to check consistency: In principle, which quantities are observable should be determined by the interactions present in the model.

Once we have accepted that we are dealing with spontaneously broken *gauge* symmetries, the above picture must be modified. The relation $\omega_0 \circ \beta \neq \omega_0$ refers to states of an algebra of fields and restricting to the subalgebra of observables we have equality since β is a *gauge* symmetry. Thus the vacuum state of the observables can be unique even in the presence of spontaneously broken symmetries.

To be intrinsic, the notion of spontaneously broken symmetry should be expressed, at least indirectly, in terms of the algebra of observables. Now it has been shown that, for conventional field theories with localizable charges, the observable algebra $\mathfrak{A}$ intrinsically determines a field algebra $\mathfrak{f} \supset \mathfrak{A}$ with normal Bose-Fermi commutation relations on its vacuum Hilbert space (for details and extensions to charges localizable in cones see [1]). For these theories, the group $\mathcal{G}$ of gauge symmetries can be reasonably defined as the group of automorphisms β of $\mathfrak{F}$ which leave $\mathfrak{A}$ pointwise invariant. The construction in [1] determines an intrinsic compact subgroup $G \subset \mathcal{G}$ whose representation theory describes the superselection sectors of $\mathfrak{A}$.

Lemma *If $\beta \in \mathcal{G}$ then*

a) $\beta \in G$ if and only if $\omega_0 \circ \beta = \omega_0$,
b) β is a local automorphism of $\mathfrak{F}$, i.e. if $\mathcal{O}$ is any double cone then $\beta(\mathfrak{F}(\mathcal{O})) = \mathfrak{F}(\mathcal{O})$.

We would expect b) to hold for any internal symmetry β but there seems to be no way of proving that β commutes with translations without making further assumptions. However, we show that if the energy–momentum density is an observable in a technical sense appropriate to algebraic quantum field theory then $\mathcal{G}$ commutes with translations and so consists of internal symmetries.

Turning to Goldstone's Theorem, we consider a continuous 1-parameter group $\lambda \to \beta_\lambda$ of internal symmetries and local automorphisms of $\mathfrak{F}$. Its infinitesimal generator is the derivation δ with domain $D \subset \mathfrak{F}$

$$D = \{F \in \mathfrak{F} : \|\frac{1}{\lambda}(\beta_\lambda(F) - F) - \delta(F)\| \to 0 \quad \text{as} \quad \lambda \to 0\}\,.$$

$\omega_0 \circ \beta_\lambda = \omega_0$ for $\lambda \in \mathbb{R}$ if and only if $\omega_0 \circ \delta = 0$.

For our results on Goldstone's Theorem it is not essential for $\mathfrak{F}$ to be the canonical field algebra constructed above and we exploit this freedom to construct counter-examples which, although they exhibit certain unphysical features, help to round off our results from a mathematical point of view.

Let $\mathcal{O}_R$ denote the double cone with base $\{\boldsymbol{x} : |\boldsymbol{x}| < R\}$ then one can show that there is a skew-adjoint operator J_R inducing δ on $\mathfrak{f}(\mathcal{O}_R)$, i.e. such that

$$(\Phi, \delta(F)\Psi) = (\Phi, [F, J_R]\Psi)\ ,\ \Phi, \Psi \in D(J_R)\ ,\ F \in D \cap \mathfrak{F}(\mathcal{O}_R)\,.$$

If the vacuum vector $\Omega \in D(J_R)$, we have an estimate

$$|\omega_0 \circ \delta(F)| \leq C_R(\|F\Omega\| + \|F^*\Omega\|)\ , \quad F \in D \cap \mathfrak{F}(\mathcal{O}_R),$$

where $C_R \leq \|J_R \Omega\|$.

The traditional approach to Goldstone's Theorem (see e.g. [2]) uses the additional hypothesis that J_R can be derived from a conserved current density j_μ, i.e. that we can take

$$J_R = i \int \mathrm{d}^{s+1}x f(x^0) g\left(\frac{\boldsymbol{x}}{R}\right) j_0(x),$$

where s is the number of space dimensions, $f \in \mathcal{D}(\mathbb{R})$ with $\int \mathrm{d}x^0 f(x^0) = 1$ and $g \in \mathcal{D}(\mathbb{R}^s)$, with $g(\boldsymbol{x}) = 1$ for $|\boldsymbol{x}| \leq 1$. This then implies that

$$C_R \leq \|J_R \Omega\| \leq \text{const.} \cdot R^{\frac{1}{2}(s-1)}.$$

Our point is that even without assuming the existence of a suitable conserved current density j_μ, the above bounds on C_R already allow us to conclude that the symmetry can be broken only in the presence of Bosons of mass zero.

Furthermore, there are examples of spontaneously broken symmetries where the constants C_R grow more rapidly as a function of R.

To be sure that our new approach to Goldstone's Theorem is sufficiently general, we must deal with the case where $\Omega \notin D(J_R)$. In fact, in this case, the following estimate holds. Given $\eta > 0$ there are constants $C_{R,\eta}$ such that

$$|\omega_\circ \circ \delta(F)| \leq C_{R,\eta}(\|F\Omega\| + \|F^*\Omega\|) + \eta\|\delta(F)\|, \quad F \in D \cap \mathfrak{F}(\mathcal{O}_R).$$

Our results are now the following:

a) if $C_{R,\eta} \leq C_\eta R^k$, $k < \frac{s-1}{2}$ then necessarily $\omega_\circ \circ \delta = 0$;
b) if $C_{R,\eta} \leq C_\eta R^{\frac{1}{2}(s-1)}$ then $\omega_\circ \circ \delta = 0$ unless there are Bosons of mass zero in the theory;
c) if $C_{R,\eta} \leq C_\eta R^k$, $k > \frac{s-1}{2}$, then $\omega_0 \circ \delta \neq 0$ is possible even when there are no particles of mass zero in the theory. In fact we give an example with $k = \frac{s}{2}$;
d) In a theory with a mass gap $m > 0$, if $C_{R,\eta} \leq C_\eta e^{\mu R}$, with $\mu < \frac{2}{e\pi}m$ then $\omega_0 \circ \delta = 0$. We give an example to show that if $\mu > m$ then the symmetry can be spontaneously broken.

References

1. Doplicher, S., Roberts, J.E.: Why there is a Field Algebra with a Compact Gauge Group Describing the Superselection Structure in Particle Physics. Comm.Math.Phys. **131**, 51-107 (1990)
2. Kastler, D., Robinson, D.W., Swieca, A.: Conserved Currents and Associated Symmetries: Goldstone Theorem. Comm.Math.Phys. **2**, 108-120 (1966)

Section 9

Nonrelativistic Quantum Mechanics (Stationary)

Organizer: I. Herbst

On the Atomic Energy Asymptotics

C. L. Fefferman [1], L. A. Seco [2]

[1]Princeton University
[2]California Institute of Technology

1 Introduction

Consider an atom consisting of N quantized electrons at positions x_i, and a nucleus fixed at the origin. The Schrödinger Hamiltonian of such a system is given by

$$H_{Z,N} = \sum_{i=1}^{N} \left(-\Delta_{x_i} - \frac{Z}{|x_i|} \right) + \frac{1}{2} \sum_{i \neq j} \frac{1}{|x_i - x_j|}$$

acting on $\mathcal{H} = \wedge_{i=1}^{N} L^2 \left(\mathbb{R}^3 \right)$ (in this exposition, in order to simplify notation, we neglect spin.) Define the ground state of an atom of charge Z by

$$E(Z) = \inf_{N} \inf_{\substack{\|\psi\|=1 \\ \psi \in \mathcal{H}}} \langle H_{Z,N} \psi, \psi \rangle \quad .$$

It is a remarkable fact that $E(Z)$ behaves in a very simple way when $Z \to \infty$, as follows:

Theorem 1.1

$$E(Z) = -c_0 Z^{7/3} + \tfrac{1}{8} Z^2 - c_1 Z^{5/3} + O\left(Z^{\frac{5}{3} - a} \right) \qquad a > 0 \quad .$$

The first term above was conjectured by Thomas ([22]) and Fermi ([11]), and proved rigorously by Lieb-Simon ([16]). Their proof also holds for molecules. The second term was conjectured by Scott ([18]) and proved rigorously by Hughes-Siedentop-Weikard ([12], [19], [20] and [21]); the molecular case was recently proved by Ivrii–Sigal ([13]). The third term was conjectured by Schwinger ([17]), based on the fundamental work of Dirac ([2]); its proof was announced in [3], and the complete details of it appear in [4]-[10].

The purpose of this paper is to give a brief expository presentation of the proof of Theorem 1.1. We refer the reader to [4] – [10] for accurate statements and detailed proofs of what follows.

Before we preceed to the proof, we recall a few basic facts about Thomas–Fermi theory. We refer the reader to [14] for a complete, accurate exposition.

We denote the Thomas–Fermi density by $\rho^Z_{\mathrm{TF}}(x)$. It satisfies the perfect scalling property

$$\rho^Z_{\mathrm{TF}}(x) = Z^2 \rho^1_{\mathrm{TF}}(Z^{1/3}x) \quad .$$

The Thomas–Fermi potential

$$V^Z_{\mathrm{TF}}(x) = \int \frac{\rho_{\mathrm{TF}}(y)}{|x-y|}\, dy$$

is radially symmetric, smooth away from the origin, and vanishing at infinity. The screened Thomas–Fermi potential

$$W^Z_{\mathrm{TF}}(x) = \frac{Z}{|x|} - V^Z_{\mathrm{TF}}(x) \geq 0$$

satisfies the Thomas–Fermi equations

$$6\pi^2 \rho^Z_{\mathrm{TF}}(x) = |W(x)|^{3/2} \qquad \Delta W(x) = 4\pi\rho(x).$$

W^1_{TF} can be written as $y(\gamma r)/r$, with an appropiate constant γ, and y satisfying the Thomas–Fermi Ordinary Differential Equation (ODE)

$$\left.\begin{aligned} y''(r) &= \frac{y^{3/2}(r)}{r^{1/2}} \\ y(0) &= 1 \\ y(\infty) &= 0 \end{aligned}\right\} \quad .$$

2 Reduction to a One–Electron Problem

The first step in understanding $E(Z)$ is to simplify $H_{Z,N}$. We do this in the spirit of [12] and [20], but here we need to improve Lieb's inequality (see [14]) to capture the full $Z^{5/3}$ contribution.

Theorem 2.1

$$H_{Z,N} \geq \sum_{i=1}^{N} \left(-\Delta_{x_i} - \frac{Z}{|x_i|} + V^Z_{\mathrm{TF}}(|x_i|) \right)$$
$$-\tfrac{1}{2} \iint \frac{\rho_{\mathrm{TF}}(x)\rho_{\mathrm{TF}}(y)}{|x-y|}\, dx\, dy - c_D \int \rho^{4/3}_{\mathrm{TF}}(x)\, dx + O\left(Z^{\frac{5}{3}-a}\right) \qquad a > 0 \quad .$$

The complete proof of this theorem appears in [4].

In preparation for the proof of Theorem 2.1, we split the electron–electron repulsion into a short and long range part, as follows

$$\tfrac{1}{2}\sum_{i\neq j}\frac{1}{|x_i-x_j|}=V_L+V_S=\sum_{i<j}K_L(x_i,x_j)+\sum_{i<j}K_S(x_i,x_j) \quad ,$$

where

$$K_L(x,y)=\frac{1}{|x-y|}\chi_{|x-y|>d}$$
$$K_S(x,y)=\frac{1}{|x-y|}\chi_{|x-y|\leq d} \quad ,$$

where d is a variable length–scale, approximately equal to $Z^{-\frac{2}{3}+a_0}$, $a_0>0$, throughout most of the atom.

Now we point out that, if we apply Lieb's inequality to V_L, the error we expect to obtain will behave like

$$\frac{Z}{\text{distance between electrons}}\leq\frac{Z}{Z^{-2/3+a_0}}=Z^{5/3-a_0} \quad ,$$

which is good enough for our purposes. Thus, we only need to understand the short range part V_S of the repulsion. This amounts to understanding our wave functions ψ only at short distances. Motivated by this, we put ourselves in the following set–up:

Let $\psi(x_1,\ldots,x_K)$ be an antisymmetric wave function *supported* on $x_i\in Q$, for $i=1,\ldots,K$, Q being a cube of side $L=Z^{-\frac{2}{3}+a_1}$. For simplicity, we impose on ψ periodic boundary conditions at ∂Q. This condition on ψ will have to be de–localized, in order to allow ψ's of arbitrary support, but this will be entirely omitted in the present exposition.

Furthermore, we assume that ψ satisfies the crucial estimate

$$\|\nabla\psi\|^2\leq c_{\mathrm{TF}}L^{-2}K^{5/3}\left(1+Z^{-a_2}\right) \quad , \tag{1}$$

where c_{TF} is the semiclassically correct constant, and a_2 is a positive constant to be carefully picked during the proof. It follows, as a consequence of Lieb–Simon's theorem, that if ψ is a ground state, then assumption (1) (or some version of it) must be satisfied for *most* cubes Q, or else, the first term in the energy asymptotics would be wrong. Therefore, we are justified in assuming (1). Our immediate goal is to conclude that if ψ satisfies (1), then it is very nearly equal to an antisymmetric product of "plane waves", in the following sense:

Consider Q_0, a subcube of Q, of side much smaller than L, but still larger than $Z^{-\frac{2}{3}+a_3}$, $a_3>0$; next, construct an orthonormal set $\{\varphi_\alpha\}_{\alpha=1}^{N_{\mathrm{TF}}}$ of functions supported on Q_0, which are essentially of the form $e^{i\xi_\alpha\cdot x}$, with $\xi_\alpha\in\Lambda(L)$ away from ∂Q_0. Here, N_{TF} is the number of ξ_α inside the Fermi ball $B(0,2\pi(4\pi/3)^{-1/3}N^{1/3}L^{-1})$. We extend this orthonormal set to a basis for $L^2(Q)$, thus extending α to all integers. Consider also the translates φ_α^z of φ_α to $z \in Q$. Then, in this situation, the following holds:

Lemma 2.2 *Let ψ satisfy (1). Then, for most $z \in Q$, we can write*

$$\psi = \phi^z_{\text{basic}} \wedge \phi^z_{\text{extra}} + \phi^z_{\text{error}} \tag{2}$$

with $\phi^z_{\text{basic}} = \wedge_{\alpha=1}^{N_{\text{TF}}} \varphi^z_\alpha$, ϕ^z_{extra} strongly orthogonal to ϕ^z_{basic}, i.e.,

$$\int \phi^z_{\text{extra}}(y_1, \ldots, y_{K-N_{\text{TF}}}) \phi^z_{\text{basic}}(x_1, \ldots, x_{N_{\text{TF}}}) \delta_{y_j = x_i}\, dx_i\, dy_j = 0$$

for all i, j, and all $y_l \neq y_j$, all $x_n \neq x_i$; ϕ^z_{error} is very small in appropriate norms.

Sketch of Proof: We introduce the positive numbers

$$\mu(z, \alpha) = 1 - \sum_{j=1}^{K} \int |\langle \psi, \varphi^z_\alpha \rangle|^2 \, dx_1 \cdots \hat{dx_j} \cdots dx_K \quad ,$$

where the scalar product is in the x_j variable of ψ, and in the integration we remove the dx_j factor.

These numbers are important mainly for two reasons: first, (1) will imply they are very small. Second, since μ measures how much ψ differs from being an antisymmetric product of the ψ_α, errors of interest in what follows will be bounded by $\mu(z, \alpha)$. We make this *a little* more precise:

If we Fourier–expand ψ, with Fourier coefficients $\xi_i \in \Lambda(L)$, then

$$\mathrm{Av}_{z \in Q}\, \mu(z, \alpha) = \sum_{\xi_1, \ldots, \xi_K} \left| \hat{\psi}(\xi_1, \ldots, \xi_K) \right|^2 \left(1 - \sum_{j=1}^{K} |\hat{\varphi}^z_\alpha(\xi_j)|^2 \right) \quad .$$

Now, (1) implies that $\hat{\psi}(\xi_1, \ldots, \xi_K)$ is negligible unless most of the ξ_j are inside the Fermi ball; on the other hand, since the φ_α are almost equal to $e^{i\xi_j \cdot x}$, stationary phase implies the term in brackets above is close to zero for all ξ_j in the Fermi ball, when $1 \leq \alpha \leq N_{\text{TF}}$. Therefore,

$$\mathrm{Av}_{z \in Q}\, \mu(z, \alpha) = O\big(Z^{-p}\big) \qquad p > 0 \quad .$$

This implies that for most z, $\mu(z, \alpha)$ is very small. Then, we can obtain (2) by – essentially – setting

$$\phi^z_{\text{error}} = \sum_{\{1, \ldots, N_{\text{TF}}\} \not\subset \{1, \ldots, K\}} \hat{\psi}(\xi_1, \ldots, \xi_K)\, \varphi^z{}_1 \wedge \cdots \wedge \varphi^z_K \quad .$$

Thus,

$$\begin{aligned}\psi - \phi^z_{\text{error}} &= \sum_{\{1,\dots,N_{\text{TF}}\}\subset\{1,\dots,K\}} \hat{\psi}(\xi_1,\dots,\xi_K)\varphi^z_1 \wedge \cdots \wedge \varphi^z_K \\ &= \phi^z_{\text{basic}} \wedge \phi^z_{\text{extra}}\end{aligned}$$

as stated. The size of ψ^z_{error} is bounded by $\sum_\alpha \mu(z,\alpha)$, and is therefore small. □

This, with some more work that we will omit here, allows us to simply ignore ϕ^z_{error} from now on. However, ϕ^z_{extra} cannot be ignored in such a way. Fortunately, since $K_S \geq 0$, creating electrons should only increase the energy, which is fine with Theorem 2.1. Indeed,

$$\langle V_S(\phi^z_{\text{basic}} \wedge \phi^z_{\text{extra}}), (\phi^z_{\text{basic}} \wedge \phi^z_{\text{extra}})\rangle \geq \langle V_S\phi^z_{\text{basic}}, \phi^z_{\text{basic}}\rangle$$

and this is trivial to prove using the strong orthogonality between ϕ^z_{basic} and ϕ^z_{extra}.

In orther words, in order to prove Theorem 2.1, we can assume that ψ correlates at short distances like the Hartree–Fock wave function ϕ^z_{basic}. Thus, imposing $a_3 < a_0$ we are justified in writing

$$\langle V_S\psi, \psi\rangle \geq \tfrac{1}{2}\sum_{i\neq j} \langle K_S(x_i, x_j)\phi^z_{\text{basic}}, \phi^z_{\text{basic}}\rangle - \text{junk}$$

and the right hand side can be calculated, since we know explicitely the density and correlation function of ϕ_{basic}. This yields

$$\langle V_S\psi, \psi\rangle \geq \tfrac{1}{2}\iint K_S(x,y)\left(\rho^z(x)^2 - |S^z(x,y)|^2\right) dx\, dy - \text{junk}$$

for

$$\rho^z(x) = \sum_{\alpha=1}^{N_{\text{TF}}} |\varphi^z_\alpha(x)|^2 \qquad S^z(x,y) = \sum_{\alpha=1}^{N_{\text{TF}}} \varphi^z_\alpha(x)\overline{\varphi^z_\alpha(y)} \quad .$$

Recalling the construction of the φ^z_α, note that

$$\begin{aligned}S^z(x,y) &= \sum_{\alpha=1}^{N_{\text{TF}}} e^{i\xi_\alpha(x-y)}(1 + \text{tiny errors}) \\ &= \int_{\text{Fermi Ball}} e^{iz(x-y)}\frac{dz}{(2\pi)^3}(1 + \text{tiny errors})\end{aligned}$$

and similarly, ρ^z is very nearly equal to N_{TF}. Therefore,

$$\langle V_S\psi, \psi\rangle \geq \tfrac{1}{2}\iint K(x,y)\left(N^2_{\text{TF}} - |\widehat{\chi_{\text{Fermi Ball}}}(x-y)|\right) dx\, dy - \text{junk} \quad .$$

This last inequality becomes Theorem 2.1 in passing from ψ supported on cubes Q to arbitrary ψ.

This finishes the exposition of the proof of Theorem 2.1

3 The One–Electron Problem

The goal of this Section is to understand the spectrum of a certain class of operators in $L^2(\mathbb{R}^3)$ given by

$$H = -\Delta - W \quad .$$

Again, we omit a detailed description of the conditions on W, and we simply say that W is a large, slowly varying *radial* potential, which is allowed a Coulomb singularity, and that our screened Thomas–Fermi potential belongs to this class. We start by exhibiting the well–known fact that estimates for the spectrum of H imply bounds for $E(Z)$. Indeed, letting W be the screened Thomas–Fermi potential, if we denote by E_k the negative eigenvalues of H, and by ψ_k the corresponding normalized eigenfunctions, we obtain from the previous section that

$$E(Z) \geq \sum_k E_k - \tfrac{1}{2} \iint \frac{\rho_{\mathrm{TF}}(x)\rho_{\mathrm{TF}}(y)}{|x-y|}\, dx\, dy - c_D \int \rho_{\mathrm{TF}}^{4/3}(x)\, dx + O\left(Z^{\frac{5}{3}-a}\right) . \tag{3}$$

Note that the last two constants in the right hand side scale perfectly in Z.

Knowledge of the eigenfunctions ψ_k, on the other hand, provides us with upper bounds by simply setting $\psi = \wedge_k \psi_k$; then, clearly

$$E(Z) \leq \langle H_{Z,N}\psi, \psi\rangle \quad . \tag{4}$$

A simple computation shows that

$$\langle H_{Z,N}\psi,\psi\rangle = \sum_k E_k - \tfrac{1}{2}\iint \frac{\rho_{\mathrm{TF}}(x)\rho_{\mathrm{TF}}(y)}{|x-y|}\, dx\, dy - \iint \frac{|S(x,y)|^2}{|x-y|}\, dx\, dy + \iint \frac{(\rho_{\mathrm{TF}}(x)-\rho(x))\,(\rho_{\mathrm{TF}}(y)-\rho(y))}{|x-y|}\, dx\, dy \quad ,$$

where

$$\rho(x) = \sum_k |\psi_k(x)|^2 \qquad S(x,y) = \sum_k \psi_k(x)\overline{\psi_k(y)} \quad .$$

Now, it is a very different question whether these bounds (3) and (4) will be any good. It is a remarkable property of the Thomas–Fermi potential that, in fact, these upper and lower bounds agree up to an error $O\left(Z^{\frac{5}{3}-a}\right)$ with $a > 0$. Indeed, the following Lemma holds:

Lemma 3.1

$$\iint \frac{|S(x,y)|^2}{|x-y|}\, dx\, dy \geq c_D \int \rho_{\mathrm{TF}}^{4/3}(x)\, dx + O\left(Z^{\frac{5}{3}-a}\right) \tag{5a}$$

$$\iint \frac{(\rho_{\mathrm{TF}}(x)-\rho(x))\,(\rho_{\mathrm{TF}}(y)-\rho(y))}{|x-y|}\,dx\,dy = O\left(Z^{\frac{5}{3}-a}\right) \quad . \tag{5b}$$

Note that this Lemma is also a consequence of an accurate enough knowledge of the spectrum of H. We remark that in the case of a smooth potential, the formula for the exchange (5a) was known ([1]).

Combining this Lemma with (3), (4) and (5), we note that Theorem 1.1 is reduced to obtaining the asymptotic formula for the sum of the eigenvalues E_k that implies the expression for $E(Z)$. Note also that our result also implies that the Hartree–Fock energy is accurate through the $Z^{5/3}$ term.

We start now our analysis of H. In the present exposition we will take care of the eigenvalue sum only. The main theorem we try to proof now is the following:

Theorem 3.2

$$\sum_k E_k = -\frac{1}{15\pi^2}\int (W(x))_+^{5/2}\,dx + \frac{1}{8}Z^2 + \frac{1}{48\pi^2}\int (W(x))_+^{1/2}\,\Delta W(x)\,dx$$

$$+\textit{lower order terms}$$

assuming that classical periodic orbits of the Hamiltonian given by $p^2 - W(x)$ *have measure 0.*

A crucial property of H in what follows, is that it is radially symmetric; thus, separation of variables into spherical harmonics yields that, if we denote by $E_{k,l}$ the k'th eigenvalue of

$$H^{\mathrm{ODE}} = -\frac{d^2}{dr^2} - W(r) + \frac{l(l+1)}{r^2} \quad ,$$

then the eigenvalues of H come in a strightforward way from the $E_{k,l}$, each taken with multiplicity $2l+1$. Here, l is a non–negative integer.

The analysis of H is thus reduced to the analysis of ODE's

$$\frac{d^2}{dr^2} + p_E(r,l) = 0 \tag{6}$$

$$p_E(r,l) = W(r) - \frac{l(l+1)}{r^2} + E \quad E<0 \quad l=0,1,\ldots$$

in the sense that if we can find a solution of (6) which vanishes at 0, at infinity, and at $k-1$ other points, then $E = E_{k,l}$. The main ingredient in the proof of Theorem 3.2 is the following eigenvalue formula.

Theorem 3.3 *If* $l \geq Z^{a_4}$, a_4 *small,*

$$\phi(E_{k,l}) + \tfrac{1}{48}\psi(E_{k,l}) = \pi(k+\tfrac{1}{2}) + O\left(\|p\|^{-1+\epsilon_2}\right) \quad ,$$

where

$$\phi(E) = \int (p_E(x;l))_+^{1/2}\, dx$$

$$\psi(E) = \lim_{\delta\to 0} \left(\int (p_E(x;l))_+^{-3/2}\, p''(x;l)\, dx - G(E,l)\delta^{-1/2} \right) \quad ,$$

where $G(E,l)$ is uniquely determined by requiring the finiteness of the limit. When $l \le Z^{a_4}$, then

$$\phi(E_{k,l}) = \pi(k+\tfrac{1}{2}) + O(1) \quad .$$

The proof of this Theorem is the main content of [5]. We also refer the reader to [5] for the exact meaning of $\|p\|$ above, and for the exact formulation of the Theorem. The strategy of the proof is as follows:

Note first the following change of variables formula for ODE's. Consider two ODE's

$$F''(x) + p(x)F(x) = 0 \tag{7a}$$
$$A''(y) + V(y)A(y) = 0 \quad . \tag{7b}$$

Then, the change of variables $y = y(x)$ given by

$$(y')^2\, V(y) + \{y,x\} = p(x) \qquad \{y,x\} = -\tfrac{3}{4}\left(\frac{y''(x)}{y'(x)}\right)^2 + \tfrac{1}{2}\frac{y'''(x)}{y'(x)} \tag{8}$$

makes (7a) and (7b) equivalent via the transformation law

$$F(x) = (y'(x))^{-1/2}\, A(y(x)) \quad .$$

What we will do then is, first, to take V a simple potential for which (7b) admits an explicit solution; then, a virtue of (8) is that we can understand the change of variables as an explicit formal power series expansion in $\|p\|^{-1}$; then, we combine both $A(y)$ and $y(x)$ to understand the solution of (7a). The choices of V will be as follows:

a. Near a zero of p, $V(x) = \lambda^2 x$ for λ large. Then, (7b) becomes the well known Airey equation.
b. When $p > 0$, we take $V = \lambda^2$. When $V < 0$, we take $V = -\lambda^2$, λ large.

The previous two choices apply for generic values of l and E. For special values, we need the two additional choices:

c. When l is small, say 1, our two previous choices will not work for x near the Coulomb singularity. The correct choice will be, of course, $V(x) = \frac{\lambda^2}{x}$; then (7b) becomes the well known Hydrogen equation. This is very important, since it gives rise to the non–semiclassical term Z^2

in the energy asymptotics in Theorems 1.1 and 3.2. Smooth potentials don't have it.

d. When E is very near $\min p_0$, then p_E and its first derivative are almost zero when p_E is negative; as a consequence, the previous choices will not work nicely, and we compare in this case with $V(x) = \lambda^2 x^2$, the equation of the Harmonic Oscillator.

Note that this method gives control over the eigenfunctions, too.

The next step is to lift Theorem 3.3 in order to say something about *radial* potentials in $\mathbb{R}^3$. We do this by summing Theorem 3.3 in k and l, recalling the multiplicity of each eigenvalue, obtaining

$$\sum_k E_k = -\frac{1}{15\pi^2}\int W(x)_+^{5/2}\,dx + \frac{1}{8}Z^2 + \frac{1}{48\pi^2}\int W(x)_+^{1/2}\Delta W(x)\,dx + G(W) + \text{junk} \quad ,$$

where junk is negligible, and $G(W)$ is a term that comes from a careful replacement of sums by integrals: it has the form

$$\sum_l c_l \beta\left(\tfrac{1}{\pi}F_W(l) - \tfrac{1}{2}\right)$$

for $\beta(x) = \{x\}^2 - \frac{1}{12}$, $\{\cdot\}$ meaning the distance to the closest half-integer, and

$$F_W(l) = \int \left(W(r) - \frac{l^2}{r^2}\right)_+^{1/2} dr \quad .$$

However, G is far from being just a technical term: it captures the main contribution of classical closed orbits to semiclassical asymptotics.

It is rather easy to see ([Ar]) that a potential W such that $F(t)$ is a linear function of t, with F' a rational multiple of π, then all classical orbits are periodic, our term $G(W)$ is oscillating, and has the same size as the third term in the formula in Theorem 3.2, which would then be wrong in this case; for the atom, this term G would then be of size $O\left(Z^{5/3}\right)$, and therefore Theorem 1.1 would also be wrong if the screened Thomas–Fermi potential were of this type. Potentials of this type are, for example, those of the Hydrogen atom and the Harmonic Oscillator.

The opposite case is a potential W for which $|F''(t)| \geq c > 0$. In this case, classical periodic orbits have measure 0, and $G(W)$ can be seen to be of size $O\left(Z^{\frac{5}{3}-b}\right)$, $b > 0$. An intermediate case of potentials are those for which F'' vanishes to finite order. In this case, closed orbits also have measure 0, and the size of $G(W)$ continues to be $O\left(Z^{5/3-b}\right)$, but the value of b depends on the order of the zeros of F''.

It is therefore very important to determine to which of these classes of potentials the screened Thomas–Fermi potentials belongs to. It is rather easy to see that F'' vanishes only to finite order; therefore, Theorem 1.1 holds, possibly with a bad error term if F'' vanishes very much. However, more is true, and it can be seen that $|F''| \geq c > 0$; this is a very delicate fact since arbitrary small perturbations of V_{TF} in a natural topology allow F'' to vanish as many times as we want. The analysis of F'' is taken care of in [10].

We end by pointing out the similarity between problems in Analytic Number Theory like the Lattice Point Problem, and the effect of the zeros of F'' on the semiclassics for $E(Z)$. In retrospect, this is not so surprising, since the Lattice Point Problem also comes from the semiclassics of a Laplacian on the Torus.

References

1. Conlon, J.: Semiclassical Limit Theorems for Hartree–Fock Theory. Comm. Math.Phys. **88** no 1 (1983)
2. Dirac, P.: Note on Exchange Fenomena in the Thomas–Fermi Atom. Proc. Cambridge Philos.Soc. **26**, 376 – 385 (1930)
3. Fefferman, C., Seco, L.: The Ground State of a Large Atom. Bull. AMS (1990)
4. Fefferman, C., Seco, L.: On the Dirac and Schwinger Corrections to the Ground–State Energy of an Atom. Adv. in Math. (To appear)
5. Fefferman, C., Seco L.: Eigenvalues and Eigenfunctions of Ordinary Differential Operators. Adv. in Math. (To appear)
6. Fefferman, C., Seco L.: The Density in a One–Dimensional Potential. Adv. in Math. (To appear)
7. Fefferman, C., Seco L.: The Eigenvalue Sum for a One–Dimensional Potential. Adv. in Math. (To appear)
8. Fefferman, C., Seco L.: The Density in a Three–Dimensional Radial Potential. Adv. in Math. (To appear)
9. Fefferman, C., Seco L.: The Eigenvalue Sum in a Three–Dimensional Potential. Adv. in Math. (To appear)
10. Fefferman, C., Seco L.: Aperiodicity of the Hamiltonian Flow in the Thomas-Fermi Potential. Adv. in Math. (To appear)
11. Fermi, E.: Un Metodo Statistico per la Determinazione di alcune Priorieta dell'Atome. Rend.Accad.Naz.Lincei **6**, 602 – 607 (1927)
12. Hughes, W.: An Atomic Energy Lower Bound that Agrees with Scott's Correction. Adv. in Math. **79**, 213 – 270 (1990).
13. Ivrii V., Sigal I.M.: To appear in Ann. of Math.
14. Lieb, E.: Thomas–Fermi and Related Theories of Atoms and Molecules. Rev.Modern Phys. **53** no. 4 (1981)
15. Lieb, E. H. A Lower Bound for Coulomb Energies. Phys. Lett. **70A**, 444 – 446 (1979)
16. Lieb E., Simon B.: Thomas–Fermi Theory of Atoms, Molecules and Solids. Adv. in Math. **23**, 22 – 116 (1977)

17. Schwinger, J.: Thomas–Fermi Model: The Second Correction. Phys.Rev. A 24, **5**, 2353 – 2361 (1981)
18. Scott, J. M. C.: The Binding Energy of the Thomas–Fermi Atom. Phil.Mag. **43**, 859 – 867 (1952)
19. Siedentop, H., Weikard, R.: On the Leading Energy Correction for the Statistical Model of the Atom: Interacting Case. Comm.Math.Phys. **112**, 471-490 (1987)
20. Siedentop, H., Weikard, R.: On the Leading Correction of the Thomas–Fermi Model: Lower Bound. Invent.Math., **97**, 159 – 193 (1989)
21. Siedentop, H., Weikard, R.: A Lower Bound of Scott Type by a New Microlocalization Technique. To appear (1990)
22. Thomas, L. H.: The Calculation of Atomic Fields. Proc.Cambridge Philos.Soc. **23**, 542 – 548 (1927)

Section 10

Special Systems in Quantum Field Theory

Organizer: E. Seiler

Existence of the Schwinger Functions of the Three-Dimensional Gross-Neveu Model

P. A. Faria da Veiga

Department of Mathematics, Harvard University,
Cambridge, MA 02138, USA

One of the conclusions of previous rigorous studies on renormalization theory is that the concept of perturbative renormalizability shall not be considered as a fundamental requirement for a quantum field model to exist. Indeed, to solve the question of the ultraviolet (UV) limit, $\rho \longrightarrow \infty$, of a model presenting a set of parameters Ω_ρ [ρ labelling an UV cutoff], in most cases one must drop out completely the traditional intermediate step involving a Taylor expansion in the coupling constant and properly ask whether or not one can prescribe *functions* for the parameters appearing in Ω_ρ, in terms of the variable ρ and the set Ω_{ren} of (finite and often positive) renormalized parameters, such that the n-point Schwinger (correlation) functions $S_{n,\rho}$ exist when $\rho \longrightarrow \infty$ and describe a non-trivial (interacting) system.

In general, there is a tautology between showing the existence of such a limit and our capability of controlling the model around an UV-stable fixed-point solution of the renormalization group (RG).

In collaboration with C. de Calan, J. Magnen and R. Sénéor [1,2], we have constructed the UV and the thermodynamic limits for the normalized Schwinger functions of the three-dimensional Euclidean and massive four-Fermi interaction with a *large* (finite) number of flavor components N.

For the time being, we have not yet checked whether or not the constructed functions satisfy the Osterwalder-Schrader axioms of quantum field theory. This will be the subject of further investigation and there is no a priori belief for this not to be true, despite of any technical difficulty that this analysis may involve. Besides, we believe our method can be adapted to treat the massless case.

The model is formally defined by the Lagragian density

$$L = \bar{\psi}(x)[i\zeta\partial\!\!\!/ + m]\psi(x) + \lambda[:\bar{\psi}(x)\psi(x):]^2/2N$$

Work supported by the National Science Foundation under DMS-PHY-8816214.

and the usual perturbation series in λ is non-renormalizable. Above, for ψ being an N-component fermion field and x a point in $\mathbb{R}^3$, m is the fermion mass, ζ is the field-strength normalization, λ is the coupling constant and :: denotes the Wick ordering.

Our program has been motivated by a former observation (see [3]) that we can render the model perturbatively just-renormalizable if we expand around the theory given by the leading order approximation of the $1/N$ expansion (here denoted by $N = \infty$), instead of the free field. In other words, the good expansion parameter when doing perturbation is the effective coupling constant comprehending the geometric sum of *all* chains with the *bubble* graph (i.e. the four-point diagram with two internal propagators making a loop) and not the coupling constant λ.

It is easy to see that, when $N = \infty$, the theory can be solved exactly and the Schwinger functions can be written as products of free fermion propagators and (convergent) series of bubble chains. Also, the corresponding RG equations for the dimensionless coupling present a non-trivial (non-zero), positive, finite and UV stable fixed-point solution β^*/N.

Such a solution persists when one computes the first corrections in $1/N$ and it is natural to conjecture this behaviour will not change when one includes the remaining contributions if N is taken sufficiently large.

One suitable way to exploit the aforementioned properties from a point of view of constructive quantum field theory is to start from unnormalized Schwinger functions defined by using a Matthews-Salam formalism. This includes an intermediate boson field σ and integrates the fermions out. For fixed ρ, one can easily perform the summation of bubble chains, include it in the boson covariance, introduce spatial cutoffs, and apply the usual (non-perturbative) machinery to decompose the phase space into cells, defined with suitable scales, and then perform multislice cluster and Mayer expansions (see [4] for a review on this technique) to the resulting expression.

When doing this, the most difficult problem one faces is the control of the large fields: the price we pay for having introduced bosons in the model. This is solved, in the one hand, by adopting a technique that exploits the effective potential to dominate badly localized boson fields and, on the other hand by performing a sort of large-field expansion, before any other expansion. The latter is aimed at allowing us to show the integrability of the interaction part appearing in the aforementioned Matthews-Salam formula (essentially, a Fredholm determinant).

Finally, by requiring N to be large, we can inductively check the existence of the fixed-point solution as well as prove convergence of the above expansions.

Since we do not have enough space to go into the deep details here, we refer the reader to [1] and [2].

References

1. de Calan, C., Faria da Veiga, P.A., Magnen, J., Sénéor, R.: Phys.Rev.Lett. **66**, 3233 (1991) and Preprint from the Centre de Physique Théorique - Ecole Polytechnique / Palaiseau (extended version to appear soon)
2. Faria da Veiga, P.A.: PhD Thesis, University of Paris, Orsay, 1991
3. Parisi, G.: Nuclear Phys. **100**, 368 (1975);
   Gross, D.: In: Methods in Field Theory, Les Houches 1975, R. Balian and J. Zinn-Justin eds., North-Holland, Amsterdam, 1976;
   Araf'eva, Ya., Nissimov E.R., Pacheva, S.: Comm.Math.Phys. **71**, 213 (1980);
   Rosenstein, B., Warr, B., Park, S.: Phys.Rev.Lett. **62**, 143 (1989);
   Rosenstein, B., Warr, B., Park, S.: Phys.Lett. **B219**, 469 (1989)
4. Rivasseau, V.: *From Perturbative to Constructive Renormalization.* Princeton University Press, Princeton, 1990

Progress in Completely Integrable Models of Quantum Field Theory

H. Grosse

Institut für Theoretische Physik
Universität Wien

1 Introduction – History

Recently a number of subjects became interrelated. For a long time field theory and statistical mechanics developed separately. From the first subject we mention as a cornerstone the Thirring model. Classical field theoretic models like KdV, NSE, KP, Sine-Gordon and Toda models were seen to be integrable and led to the Lax pair, the AKNS and Zhakarov-Shabat scheme.

As for the second subject Onsagers solution of the two-dimensional Ising model represents the biggest step forward. After work of Yang and Lieb, Baxter "solved" the XYZ-model. More recently IRF models (with interaction round a face) have been studied. We shall return to the ABF and RSOS case later. The Potts models shall be mentioned too.

One unification of both subjects was obtained through quantization of the inverse scattering transform method. At least at four different places this step has been undertaken. For me, V. Glaser has suggested it. The most prominent contributions have been done from the Leningrad group around Faddeev [1],[2].

The structure of quantum groups have been extracted from the QIST method and the central role of the Yang-Baxter relation has been recognized. The S-matrix factorization of Zamolodchikov, being identical to the latter, led to a number of proposed on shell quantities.

Recently, big insight into the algebraic structure behind integrable models has been obtained. The key words are Temperley-Lieb algebra, Hecke algebra and Jone index theory. At a certain stage it appeared that the models were of bigger help for analyzing braid group representations and for obtaining knot invariants than for obtaining physical insight.

But a fourth subject joined: Conformal quantum field theory in two dimensions. In models, the Virasoro algebra with central extension is realized.

In addition, BPZ [3] used the operator product expansion to relate correlation functions. Exchange algebras have been described by Fröhlich and by Fredenhagen, Rehren and Schroer.

We shall introduce lattice models in order to establish quantum integrability and discuss the QIST method. The algebraic schemes are mentioned in part five. Consequences from combining integrability and conformal symmetry are reviewed afterwards. A new proposal to use nonlocal currents for restricting correlation functions is mentioned at the end.

For a finite $2N$-dimensional classical dynamical system, integrability is easily described. N globally defined conserved quantities in involution are needed. There exists then a canonical map to action angle variables and the initial value problem can be solved. The inverse scattering transform method allows to solve certain partial differential equations by first applying the direct step to action angle variables which have a free time evolution and then using the inverse step to solve for the full time evolution. Classical integrability is connected to the classical Yang-Baxter relation. The early attempts to quantize the classical scheme were attempts to find q-deformations of the classical relations.

2 Lattice Model $d = 2$

Consider a $N \times M$ lattice and certain boundary conditions. Assign a quantity to bonds. Allow (for simplicity only) two assignments. For a vertex we have then 16 possibilities and we may represent it by a 4×4 matrix $L^{\alpha_1 \alpha_2}_{\beta_1 \beta_2}$ acting in $\mathbb{C}^2 \otimes \mathbb{C}^2$. The monodromy matrix

$$M_N = L_1 \ldots L_N = \begin{pmatrix} A & B \\ C & D \end{pmatrix}, \qquad V_0 \otimes (V)^N \tag{2.1}$$

describes the row-to-row interaction with open ends. The transfer matrix $T_N = \mathrm{tr}_{V_0} M_N$ allows to write the partition function as $Z = \mathrm{Tr}\ T_N^M$.

Symmetry allows to reduce to the 8 vertex models [4], where

$$L = \left(\begin{array}{cc|cc} a & 0 & 0 & d \\ 0 & b & c & 0 \\ \hline 0 & c & b & 0 \\ d & 0 & 0 & a \end{array}\right) \tag{2.2}$$

$$a = \mathrm{sn}\,(\gamma - \lambda), \qquad b = \mathrm{sn}\,\lambda, \qquad c = \mathrm{sn}\,\gamma, \qquad d = k\ \mathrm{sn}\,\gamma\ \mathrm{sn}\,\lambda\ \mathrm{sn}\,(\gamma - \lambda).$$

For modulus $k = 0$, and periods $K \to \pi/2$, $K' \to \infty$ of the Jacobi elliptic function the trigonometric parametrization of the 6-vertex (or ice) model results. The latter is known to be critical for $|a^2 + b^2 - c^2| < |2ab|$. There is an interesting connection to quantum spin models defined on $\otimes(V)^N$.

The logarithmic derivative of $T_N(\lambda)$ at $\lambda = 0$ equals the XXZ Hamiltonian, where the asymmetry is determined through γ.

SOS models may be defined by assigning heights to plaquettes. Neighbouring values should differ by one. A dual picture defines an IRF-model. Values assigned to four sites of a plaquette determine a weight. An elliptic parametrization allows to relate the models to the 8-vertex one. For $\gamma = 2K/\nu$, with $\nu = 4, 5, \ldots$ it is possible to restrict the heights to integer values $1 \leq \ell \leq \nu - 1$. Four regions with different ground state structure have been found. Huse obtained exponents which suggest that the multicritical behaviour in one region may fit to the minimal conformal series classification.

3 Quantum Integrability – Yang-Baxter Equation

For integrability we need at least an infinite number of conserved quantities. A commuting family of transfer matrices $[T(\lambda), T(\mu)] = 0$ will imply this. The latter holds since

$$R_{12}(\lambda - \mu)L_{13}(\lambda)L_{23}(\mu) = L_{13}(\mu)L_{23}(\lambda)R_{12}(\lambda - \mu) \tag{3.1}$$

holds on $V_1 \otimes V_2 \otimes (V)^N$. Indices denote the vector space in which the matrix acts. If we put $R(\lambda) = PL(\lambda)$ and P denotes the permutation operator the Yang-Baxter relation results

$$R_{23}(\lambda - \mu)R_{12}(\lambda)R_{23}(\mu) = R_{12}(\mu)R_{23}(\lambda)R_{12}(\lambda - \mu)\ . \tag{3.2}$$

There exist rational, trigonometric, elliptic solutions and solutions with parameters lying on a curve of higher genus. There are generalizations to the graded case, higher spin solutions and multiparameter deformations. (3.2) is similar to the S-matrix factorization $S_{12}S_{13}S_{23} = S_{23}S_{13}S_{12}$, where we suppressed the rapidity dependence.

The above structure led to the definition of exchange algebras for quantities T such that $S_{12}T_1T_2 = T_2T_1S_{12}$. Quantum groups are obtained in the limit $\lambda \to i\infty$ after suitable gauging a, b, c of the 6-vertex model. They are Hopf algebras H with coproduct, coinverse and counit, for which an element $\mathcal{R} \in H \otimes H$ exists, which transform from the coproduct to the primed coproduct [5].

4 QIST-Method, Algebraic Bethe-Ansatz

Equation (3.1) implies an algebra for A, B, C, D introduced in (2.1), for example,

$$
\begin{aligned}
A_\lambda B_\mu &= \frac{a(\mu-\lambda)}{b(\mu-\lambda)} B_\mu A_\lambda - \frac{c(\mu-\lambda)}{b(\mu-\lambda)} B_\lambda A_\mu \\
D_\lambda B_\mu &= \frac{a(\lambda-\mu)}{b(\lambda-\mu)} B_\mu D_\lambda - \frac{c(\lambda-\mu)}{b(\lambda-\mu)} B_\lambda D_\mu \,.
\end{aligned}
\tag{4.1}
$$

The aim is to diagonalize $T_N(\lambda) = A_\lambda + D_\lambda$. For the 6-vertex model a pseudovacuum $|\Omega\rangle_N = \bigotimes_{j=1}^N |\uparrow\rangle_j$ is easily found.

Since $C_\lambda|\Omega\rangle_N = 0$ the eigenvalue $a^N(\lambda) + b^N(\lambda)$ for $T_N(\lambda)$ results. An attempt to obtain excited states like $B_{\lambda_1} \dots B_{\lambda_n}|\Omega\rangle_N$ fails in general. The second terms of (4.1) yield vanishing contributions iff the Bethe conditions for λ_ℓ, $\ell = 1, \dots, n$ hold:

$$
\frac{a^N(\lambda_\ell)}{b^N(\lambda_\ell)} = \prod_{j\neq\ell}^{n} a(\lambda_\ell - \lambda_j) b(\lambda_j - \lambda_\ell) b(\lambda_\ell - \lambda_j) a(\lambda_j - \lambda_\ell) \tag{4.2}
$$

(if $c(\lambda)/b(\lambda) = c(-\lambda)/b(-\lambda)$). Solutions are available for small N or large N. For the latter case one may rewrite (4.2) as

$$
\exp[ip_\ell N] = \prod_{j\neq\ell}^{n} \exp[i\phi(\lambda_j - \lambda_\ell)] \rightarrow N \cdot p_\ell = 2\pi I_\ell + \sum_{j\neq\ell}^{n} \phi(\lambda_j - \lambda_\ell) \,. \tag{4.3}
$$

Set of roots $\{p_1, \dots, p_n\}$ are given by set of numbers $\{I_1, \dots, I_n\}$ with $I_1 < I_2 < \dots < I_n$. In the limit $N \rightarrow \infty$ an integral equation results. The ground state = maximal eigenvalue of T_N is obtained by filling real roots. For excited states holes ($I_{j+1} - I_j > 1$) or complex roots (bound states of pseudoparticles) show up.

The 8-vertex model is more complicated. There exists a generalized pseudovacuum and a generalized Bethe ansatz which involve an angle Θ. All eigenvectors of the transfer matrix for this model are also eigenvectors of the transfer matrix for the SOS model. For the special values $\gamma = 2K/\nu$, $\nu = 4, 5, \dots$. Θ is restricted to $m\pi/\nu$ with $m = 1, \dots, \nu - 1$ and the RSOS model is obtained [6].

5 Algebraic Structures

The Yang-Baxter relation without spectral parameter $R_{12}R_{23}R_{12} = R_{23}R_{12}R_{23}$ resembles the braiding relation and an homomorphism $b_i \to \mathrm{End}\,(\bigotimes^N V)$ with $b_i \to \mathbf{1} \otimes \ldots \otimes R_{ii+1} \otimes \mathbf{1} \ldots$ fulfilling

$$b_i b_j = b_j b_i \text{ if } |i-j| \geq 2, \qquad b_i b_{i+1} b_i = b_{i+1} b_i b_{i+1} \tag{5.1}$$

is obtained. Invariants are constructed from solutions of (3.2) by finding a Markov trace. Jones polynomials have been generalized by Akutsu, Deguchi and Wadati [7] and others.

From the gauged 6-vertex model we obtain for $\lambda \to i\infty$

$$R = \left(\begin{array}{cc|cc} 1 & 0 & 0 & 0 \\ 0 & 0 & t & 0 \\ \hline 0 & t & 1-t^2 & 0 \\ 0 & 0 & 0 & 0 \end{array}\right), \qquad t = e^{i\gamma}, \qquad (R-1)(R+t^2) = 0\,. \tag{5.2}$$

Since R has only two different eigenvalues a shift from b_i to $\sigma_i : b_i = (-1-t^2)\sigma_i + 1$ maps to the Hecke algebra

$$\sigma_i^2 = \sigma_i, \qquad \sigma_i\sigma_{i+1}\sigma_i - \left(\frac{t}{1+t^2}\right)^2 \sigma_i = \sigma_{i+1}\sigma_i\sigma_{i+1} - \left(\frac{t}{1+t^2}\right)^2 \sigma_{i+1}\,. \tag{5.3}$$

If both sides of (5.3) are put additionally to zero, the Temperley-Lieb algebra results

$$e_i e_j = e_j e_i, \qquad |i-j| \geq 2, \qquad e_i^2 = e_i = e_i^\dagger, \qquad e_i e_{i\pm 1} e_i = \frac{1}{Q} e_i\,. \tag{5.4}$$

Jones deduced quantization of $Q \in \{4\cos^2 \pi/m\} \cup [4,\infty)$, $m = 4, 5, \ldots$.

The algebra (5.4) showed up in the Q-state Potts model where $T_N = V^{1/2} W V^{1/2}$ with $V = \prod_j^{N-1}(1 + \xi_h e_{2j})$ and $W = \prod_j^N (\xi_v + e_{2j-1})$. The e's may be expressed by $(E_{ij})_{k\ell} = \delta_{ki}\delta_{\ell j}$ as

$$e_{2j} = \mathbf{1} \otimes \ldots \otimes \sum_{k=1}^{Q} E_{kk} \otimes E_{kk} \otimes \mathbf{1} \ldots, \qquad e_{2j-1} = \frac{1}{Q} \mathbf{1} \otimes \ldots \otimes \sum_{k,\ell=1}^{Q} E_{k\ell} \otimes \mathbf{1} \ldots \tag{5.5}$$

It follows that the partition function is determined from the above algebraic structure.

6 Conformal Quantum Field Theory $d = 2$

The lattice distance ε serves as a cut-off. A field theory is obtained in the scaling limit, $T \to T_c$, $\varepsilon \to 0$, $\varepsilon \cdot \xi = const$, where ξ is the correlation length. Full control has been obtained only for the Ising model. A conformal invariant Majorana model results.

Unitary highest weight representations of the Virasoro algebra (the Lie algebra of the group of diffeomorphism of the S^1) are characterized by the value of the central extension c and by the conformal weight h. Irreducible representations follow only if $c \geq 1$, $h \geq 0$ or

$$c = 1 - \frac{6}{m(m+1)}, \qquad h_{p,q(c)} = \frac{((m+1)p - mq)^2 - 1}{4m(m+1)} . \tag{6.1}$$

Again a quantization is obtained. $m = 3$ corresponds to the Ising, $m = 4$ to the tricritical Ising model, $m \geq 5$ to the ABF models. $m = 5$ to the 3 state Potts model.

The operator content is restricted by the allowed values for the conformal weights h, $\bar{h}$. Fusion rules for conformal blocks restrict correlation functions.

There has been great interest in establishing links between the algebraic structure of lattice models and possible continuum limits. We shall mention a few attempts.

Connes and Evans [8] start from the TL algebra, construct currents and use the Sugawara construction to obtain the energy momentum tensor. In this way they obtain a representation of the Virasoro algebra. This construction has been done up to now only for one special value for c.

A number of papers dealt with finite size corrections. The maximal eigenvalue of the transfer matrix $\lambda_{\max}$ scales as $\exp(-Nf + \pi 6Nc)$ as $N \to \infty$. In addition energy and momentum for excited states are determined through the scaling dimensions h, $\bar{h}$ of an operator A which connects the ground state to the appropriate state $\langle 0|A|\psi_i\rangle \neq 0$:

$$E_i - E_0 \overset{N\to\infty}{\simeq} \frac{2\pi}{N}(h + \bar{h}), \qquad P_i - P_0 \overset{N\to\infty}{\simeq} \frac{2\pi}{N}(h - \bar{h}) . \tag{6.2}$$

One may now use the Bethe equations for real and complex roots in order to deduce possible values for c, h and $\bar{h}$ [9].

For the 6-vertex model $c = 1$ results, the RSOS models correspond to $c = 1 - 6\nu(\nu - 1)$, and for the Potts model $c = 1 - 6\nu(\nu - 1)$ with $\nu = 4, 6, \infty$ for $Q = 2, 3, 4$ is obtained [6].

These results suggest that the scaling limit of these models will correspond to a conformal field theory with central extension c. An expansion of the vertex model around $\lambda = 0$ yields an XXZ model with special boundary conditions. After introducing lattice fermions one may apply a Jordan Wigner transformation and one obtains a Thirring model with coupling g

related to ν through $\beta^2 8\pi = 1 - 1\nu$, $4\pi\beta^2 = 1 + g\pi$ [6]. At these values of β dipole condensation is supposed to occur.

7 Common Structures

It is tempting to look for spin systems which respect the quantum group symmetry and study their continuum limit. At first we remark how to obtain a quantum group representation from the L matrix of the 6-vertex model. Through a gauge transformation asymmetric weights are introduced. After taking a limit M_N involves J^3, $J^\pm$ of the form

$$J^3 = \sum_{j=1}^{N} s_j^3, \qquad J^\pm = \sum_{j=1}^{N} q^{s_1^3+\ldots+s_{j-1}^3} s_j^\pm \, q^{-s_{j+1}^3-\ldots-s_N^3} \tag{7.1}$$

which obey the algebra of $U_q(sl(2))$

$$[J^+, J^-] = (q^{2J^3} - q^{-2J^3})/(q - q^{-1}), \qquad [J^3, J^\pm] = \pm J^\pm\,. \tag{7.2}$$

Representations are similar to the $sl(2)$ case if q is not a root of unity; otherwise indecomposable representations occur.

In order to obtain $U_q(sl(2))$ as a symmetry of a lattice model the diagonal to diagonal transfer matrix is introduced $[T_D, U_q(sl(2))] = 0$. By the standard procedure a Hamiltonian with special boundary terms is connected to T_D:

$$\begin{aligned} H &= \sum_{j=1}^{L-1} \left(\frac{q+q^{-1}}{2} - 2e_j \right) \\ &= \sum_{j=1}^{L-1} (\sigma_i^1 \sigma_{i+1}^1 + \sigma_i^2 \sigma_{i+1}^2 + \frac{q+q^{-1}}{2} \sigma_i^3 \sigma_{i+1}^3) + \frac{q-q^{-1}}{2} (\sigma_1^3 - \sigma_L^3)\,. \end{aligned} \tag{7.3}$$

This Hamiltonian has a hidden symmetry. From the Bethe ansatz one deduces for $q = e^{i\pi/\nu}$ the value of $c = 1 - 6\nu(\nu - 1)$. From $E_j - E_0$ the conformal weight $h_{1,2j+1}$ is obtained. In addition the ADE classification of modular invariant partition functions can be used to obtain information on the spectrum of operators contributing to partition functions [10].

For an open chain H_{XXZ} is expressed in terms of TL generators and corresponds to $L_0 \in$ Virasoro algebra. The commutant $U_q(sl(2))$ corresponds to $Q^\pm$, Q^3 in the Feigin-Fuchs construction. For the closed chain irreducible representations are characterized by two integers corresponding to L_0, $\overline{L_0}$.

8 Nonlocal Currents in 2d QFT

From an S-matrix solution of the Yang-Baxter relation it is easy to obtain nonlocal conserved charges. Let V be the vectorspace of asymptotic states. The two body S-operator S_{12} maps $V \otimes V$ into $V \otimes V$. Lorentz invariance demands that S_{12} depends only on the rapidity difference $\Theta = \Theta_1 - \Theta_2$. From the assumed factorization property an exchange algebra follows

$$S_{12}(\Theta)\mathcal{T}_1^\lambda(\Theta_1)\mathcal{T}_2^\lambda(\Theta_2) = \mathcal{T}_2^\lambda(\Theta_2)\mathcal{T}_1^\lambda(\Theta_1)S_{12}(\Theta) \tag{8.1}$$

where $T(\lambda - \Theta) = \mathcal{T}^\lambda(\Theta)$. The logarithmic derivative of (8.1) gives

$$S_{12}(\Theta)(Q^\lambda_{1,ij} \otimes \mathbf{1} + \Theta^{\lambda,k\ell}_{1,ij} \otimes Q^\lambda_{2,k\ell}) = (Q^\lambda_{2,ij} \otimes \mathbf{1} + \Theta^{\lambda,k\ell}_{2,ij} \otimes Q^\lambda_{1,k\ell})S_{12}(\Theta) \tag{8.2}$$

where $Q^\lambda = \partial\partial\lambda \ln T^\lambda$ and $\Theta^{\lambda,k\ell}_{ij} = (\mathcal{T}^\lambda)_{ik}(\mathcal{T}^\lambda)^{-1}_{\ell j}$. (8.2) yields conservation laws for on shell conserved charges Tr Q^λ. But formal expansion in λ gives an infinite number of nonlocal ones too. The idea is to reverse the above arguments. One tries to construct nonlocal charges and currents via some regularization [11]: $J^a_\mu(t,x)$ are obtained by attaching a string from infinity going to the point (t,x). One tries to establish braiding relations for currents

$$J^a_\mu(t,x)J^b_\nu(t,y) = S^{ab}_{cd}J^c_\nu(t,y)J^d_\mu(t,x), \qquad x > y \tag{8.3}$$

and deduces how charge operators act on fields. This yields a coproduct for charge operators. The idea then is to use comultiplication and Lorentz transformation property to restrict the S-matrix and correlation functions.

For the Sine-Gordon model single soliton states carry a topological charge. There exist four nonlocal charges. They restrict the S-matrix so that after using crossing and unitarity it is fixed uniquely.

9 Final Remarks

In this short review we did not mention the closely related developments like: perturbations of integrable models dealt with extensively by Zamolodchikov and others; the principal chiral models and the WZNW model treated by Faddeev at this conference; generalized height models connected to simply laced Lie algebras; hidden Yangians in two-dimensional massive current algebras and restricted Sine-Gordon models of Bernard and Leclair; mappings to the Coulomb gas picture, to mention only a few of them.

At this conference quantum groups have been dealt with by a number of speakers. They were abstracted from the structure of integrable models. It is likely that they will become more useful as symmetries of field theoretic models.

Acknowledgement: I would like to thank E. Seiler for inviting me to give this report and the organizers of the Leipzig conference for their efforts.

References

1. Faddeev, L., Takhtajan, L.: *Hamiltonian Methods in the Theory of Solitons.* Springer, 1987
2. Grosse, H.: *An Introduction to Integrable Models and Conformal Field Theory.* Schladming 1991
3. Belavin, A.A., Polyakov, A.M., Zamolodchikov, A.B.: Nuclear Phys. B **241**, 330 (1980)
4. Baxter, R.J.: *Exactly solved models in statistical mechanics.* Academic Press, 1982
5. Reshetikin, N.Yu., Takhtadzhyan, L.A., Faddeev, L.D.: Leningrad Math.J. **1**, 193 (1990)
6. Karowski, M.: In: Proceedings of the Vth Jorge A. Swieca Summer School, São Paulo, 1989
7. New Developments in the Theory of Knots. Ed. T. Kohns, Advanced Series in Math. Physics **11**, World Scientific, 1989
8. Connes, A., Evans, D.E.: Comm.Math.Phys. **121**, 507 (1989)
9. de Vega, H.: Adv.Stud.Pure Math. **19**, 567 (1989)
10. Pasquier, V., Saleur, H.: Nuclear Phys.B **330**, 523 (1990)
11. Bernard; D., Leclair, A.: Nonlocal Currents in 2*d* QFT: An Alternative to the Quantum Inverse Scattering Method. Preprint CLNS-90/1036 and SPhT/90/173

Section 11

Scattering Theory, Inverse Problems

Organizer: V. Enss

Inverse Scattering at Fixed Energy

Adrian I. Nachman

Department of Mathematics, University of Rochester, Rochester, NY 14627, U.S.A.

1 Introduction

Let $-\Delta + V$ be a quantum mechanical two-body Hamiltonian in $L^2(\mathbb{R}^n)$, $n \geq 3$, and let $S(k)$ be the corresponding scattering matrix at energy k^2. We consider the classical problem of recovering V from knowledge of $S(k)$ at one energy. The potential $V(x)$ is not assumed to have any spherical symmetry. (The spherically symmetric case, including the non-uniqueness which arises if one allows potentials with reasonably mild decay at infinity, has been extensively studied—see [3] and references given there.) We show (Theorem 3.1) that if V has compact support and is in $L^{n/2}$ then it is uniquely determined by $S(k)$; the proof gives a method to reconstruct the potential from the scattering matrix.

To solve this inverse scattering problem we first settle the analogous question for near-field data. The motivation for the latter comes mainly from acoustics. Let $\mathcal{G}(x,y)$ be the scattered pressure field generated by a point source at y oscillating harmonically with frequency ω:

$$\Delta_x \mathcal{G}(x,y) + \frac{\omega^2}{c^2(x)} \mathcal{G}(x,y) = -\delta(x-y)\,. \tag{1.1}$$

Let Ω be a bounded region outside of which the speed of sound $c(x)$ is assumed equal to a known constant c_0 (for instance, the speed of sound in water); transducers (which can act both as emitters and receivers) are placed on the boundary $\partial\Omega$, surrounding the region one would like to image. The problem we consider is to reconstruct $c(x)$ inside Ω from measurements of $\mathcal{G}(x,y)$ at all receivers x on $\partial\Omega$ and sources y on $\partial\Omega$, at *one* fixed frequency ω. To relate the two problems, let

---

This research was supported in part by ONR through grant #N0001491J1107.

$$V(x) = \omega^2 \left(\frac{1}{c_0^2} - \frac{1}{c^2(x)} \right) \quad \text{and} \quad k^2 = \frac{\omega^2}{c_0^2} . \tag{1.2}$$

Then $V(x)$ has compact support in Ω and $\mathcal{G}(x, y, k)$ can be thought of as the kernel of the resolvent $(-\Delta + V - k^2 - i0)^{-1}$. We are given $\mathcal{G}(\cdot, \cdot, k)$ on $\partial\Omega \times \partial\Omega$ or, equivalently, the single layer operator defined on $\partial\Omega$ by

$$\mathcal{S}(k)f(x) = \int_{\partial\Omega} \mathcal{G}(x, y, k) f(y) d\sigma(y) . \tag{1.3}$$

It was shown in [2] that one can determine $\mathcal{S}(k)$ (at least when Ω is a ball) from knowledge of the scattering matrix $S(k)$, and conversely. Thus, it had been known for a long time that an inversion theorem from the near-field data $\mathcal{S}(k)$ would imply one from the far-field $S(k)$, and conversely. The fact that $S(k)$ uniquely determines a compactly supported V was shown by Novikov [15] (see also [16]); uniqueness from $\mathcal{S}(k)$ was independently proved in [11]. The extension of the results of [11] to potentials which are locally in $L^{n/2}$ and domains with Lipschitz boundaries is joint work with Richard Lavine [9].

A crucial ingredient in the solution of both reconstruction problems is the construction of a non-physical "scattering transform" T of V (not to be confused with the standard T-matrix), first introduced by Faddeev [5] and later rediscovered in the treatment of multidimensional inverse scattering by the $\bar{\partial}$ method [12], [1]. The overdeterminacy issue in the multidimensional inverse scattering problem can be understood in terms of T (see [6], [12], [1], [8] and especially [14], [19]); furthermore, once T is known, the recovery of V is straightforward (this was first observed in [12]). A linear integral equation for determining (certain boundary values of) T from $S(k)$ was first given in [5]. Unique solvability of this equation breaks down at certain values of k (corresponding to the so called real exceptional points — see [8], [14]). This difficulty can be avoided by taking sufficiently small V; the first general uniqueness result from $S(k)$ at fixed k was thus obtained in [14], for potentials which are both exponentially decaying and sufficiently small. The global results presented here were made possible by the discovery, following the proof of uniqueness for an inverse boundary value problem in [18], of a new method for obtaining T from the physical data, when V has compact support [13], [11], [15].

In section 2 we motivate the introduction of T from the point of view of fixed energy inversion, give the main analytic result needed in its construction for the class of potentials considered here, and state our theorem on the near-field $\mathcal{S}(k)$. In section 3 we obtain the corresponding result on the far-field $S(k)$; the derivation we outline establishes optimal bounds on appropriate operator norms of $S(k)$ and $\mathcal{S}(k)$ (the estimates in [2] are pointwise). Formula (3.10) appears to be new.

2 The Scattering Amplitude and the Scattering Transform

The scattering matrix can be written as

$$S(k) = I - \frac{i}{2}\frac{k^{n-2}}{(2\pi)^{n-1}} A(k), \tag{2.1}$$

where $A(k)$ is the operator on $L^2(S^{n-1})$ whose kernel is the scattering amplitude $A(k,\omega,\omega')$. Stationary scattering theory gives an expression of the latter as

$$A(k,\omega,\omega') = \int_{\mathbb{R}^n} e^{-ik\omega\cdot x} V(x)\psi^+(x,k,\omega')dx \,, \tag{2.2}$$

where $\psi^+(x,k,\omega)$ is the solution of the Lippmann-Schwinger equation

$$\psi^+(x,k,\omega) = e^{ik\omega\cdot x} - \int_{\mathbb{R}^n} \mathcal{G}^+(x-y,k)V(y)\psi^+(y,k,\omega)dy \,, \tag{2.3}$$

with G^+ the standard outgoing Green's function for the Helmholtz operator

$$G^+(x,k) = \frac{1}{(2\pi)^n}\int_{\mathbb{R}^n} \frac{e^{ix\cdot\xi}}{\xi^2 - k^2 - i0} d\xi \left(= \frac{e^{ik|x|}}{4\pi|x|} \quad \text{when} \quad n=3 \right) . \tag{2.4}$$

To motivate the introduction of the scattering transform T, recall that the solutions $\psi^+(x,k,\omega)$ of the Schrödinger equation obtained from (2.3) turn out to behave like $\exp(ik\omega\cdot x)$ for large k. It follows that if the momentum transfer vector $\xi = k\omega - k\omega'$ is kept fixed, then for large k

$$\lim_{\substack{k(\omega-\omega')=\xi \\ k\to\infty}} A(k,\omega,\omega') = \hat{V}(\xi) \,, \tag{2.5}$$

the Fourier transform of V. One thus obtains a classical uniqueness result: knowledge of the scattering matrix at all large energies uniquely determines the potential.

What if the energy is fixed? Instead of the incident plane waves occurring in (2.3) we will use $\exp(i\zeta\cdot x)$, with ζ appropriate vectors in $\mathbb{C}^n$, (sometimes called nonhomogeneous plane waves). This choice is motivated by the simple observation that, given k, it is possible to have

$$(\Delta + k^2)e^{ix\cdot\zeta} = 0 \tag{2.6}$$

with ζ of arbitrarily large magnitude, while $\zeta^2 = k^2$ is kept fixed. Our first result shows that one can construct a family $\psi(x,\zeta)$ of (non-physical) solutions of the Schrödinger equation which behave like $\exp(i\zeta\cdot x)$ for large ζ, by solving

$$\psi(x,\zeta) = e^{ix\cdot\zeta} - \int_{\mathbb{R}^n} G(x-y,\zeta)V(y)\psi(y,\zeta)dy \,, \tag{2.7}$$

where G is Faddeev's Green function

$$G(x,\zeta) = \frac{1}{(2\pi)^n} \int \frac{e^{ix\cdot(\xi+\zeta)}}{\zeta^2 + 2\zeta\cdot\xi} d\xi \,. \tag{2.8}$$

Theorem 2.1 [8] *Suppose $V \in L^{n/2}(\mathbb{R}^n) \cap L^1(\mathbb{R}^n)$, $n \geq 3$. For $\zeta \in \mathbb{C}^n \setminus \mathbb{R}^n$ sufficiently large, there is a unique solution $\psi(\cdot,\zeta)$ of Equation (2.7) with $\exp(-ix\cdot\zeta)\psi(x,\zeta) - 1$ in $L^{2n/(n-2)}$.*
Furthermore, $\| \exp(-ix\cdot\zeta)\psi(x,\zeta) - 1 \|_{L^{2n/(n-2)}}$ is uniformly bounded in ζ and, for $\delta > 1$,

$$\lim_{|\zeta|\to\infty} \| e^{-ix\cdot\zeta}\psi(x,\zeta) - 1 \|_{L^2_{-\delta}} = 0 \tag{2.9}$$

in the norm of the weighted space

$$L^2_{-\delta} = \left\{ f : \int |f(x)|^2 (1+x^2)^{-\delta} dx < \infty \right\} \,. \tag{2.10}$$

The solutions $\psi(x,\zeta)$ were introduced in [4] and rediscovered three times [12], [1], [18] by different routes. Sylvester and Uhlmann were the first to prove the crucial large ζ estimates needed for global uniqueness. Once $\psi(\cdot,\zeta)$ are constructed, we define (for $\zeta^2 = k^2$, $\xi \in \mathbb{R}^n$, $(\xi+\zeta)^2 = k^2$ and ζ sufficiently large)

$$T(\xi,\zeta) = \int e^{-ix\cdot\xi} V(x)(e^{-ix\cdot\zeta}\psi(x,\zeta))dx \,, \tag{2.11}$$

and use Theorem 2.1 to show that T approaches $\hat{V}(\xi)$ for large $|\zeta|$.

Given the surface data $\mathcal{S}$, we now follow the procedure in [11] to determine T (via the Dirichlet-Neumann map). We thus prove the following

Theorem 2.2 [8] *Let Ω be a bounded Lipschitz domain in $\mathbb{R}^n$, $n \geq 3$, with $\mathbb{R}^n \setminus \overline{\Omega}$ connected. Assume $V \in L^{n/2}(\Omega)$ and V vanishes outside $\overline{\Omega}$.*

i) The operator $\mathcal{S}(k)$ given by (1.3) is bounded: $H^{-1/2}(\partial\Omega) \to H^{1/2}(\partial\Omega)$ and invertible if and only if k^2 is not a Dirichlet eigenvalue of $-\Delta + V$ in Ω.

ii) V can be reconstructed from $\mathcal{S}(k)$.

To use the Dirichlet-Neumann map, as in [11], we first prove part ii) of the above theorem assuming k^2 is not in the Dirichlet spectrum of $-\Delta + q$ or $-\Delta$. That is the only case we need in the proof of Theorem 3.1 (see the comments following the statement of that theorem in Section 3). The proof of Theorem 2.2 for general k is then completed, for instance, by invoking Theorem 3.1.

3 From Near-field to Far-field and Back

Let Ω be a bounded Lipschitz domain in $\mathbb{R}^n$, $n \geq 2$, with $\mathbb{R}^n \setminus \overline{\Omega}$ connected. We define the far- field pattern $\mathcal{F}f$ of any f in $H^{1/2}(\partial\Omega)$ as follows. Let u be the outgoing solution of the exterior Dirichlet problem

$$(\Delta + k^2)u = 0 \quad \text{in} \quad \mathbb{R}^n \setminus \overline{\Omega}, \quad u\big|_{\partial\Omega} = f\,; \tag{3.1}$$

then the behavior of $u(x)$ for large $|x|$ can be shown to be

$$u(r\omega) = \frac{e^{ikr}}{r^{\frac{n-1}{2}}} u_\infty(\omega) + o(r^{(1-n)/2})\,. \tag{3.2}$$

We define the operator $\mathcal{F}(k) : H^{1/2}(\partial\Omega) \to L^2(S^{n-1})$ by

$$(\mathcal{F}(k)f)(\omega) = 4\pi(2\pi i/k)^{\frac{n-3}{2}} u_\infty(\omega)\,. \tag{3.3}$$

A useful expression of $\mathcal{F}(k)$ as an integral operator can be obtained using the solution $\varphi^e(x,k,\omega)$ of the exterior Dirichlet problem

$$(\Delta + k^2)\varphi^e = 0 \quad \text{in} \quad \mathbb{R}^n \setminus \overline{\Omega},\ \varphi^e\big|_{\partial\Omega} = 0 \tag{3.4}$$

with $\varphi^e(x,k,\omega) - \exp(ikx\cdot\omega)$ outgoing. Using Green's formula on the functions u in (3.1) and $\varphi^e(x,k,\omega) - exp(ikx\cdot\omega)$ yields

$$\mathcal{F}(k)f(\omega) = \int_{\partial\Omega} \frac{\partial\varphi^e}{\partial\nu(x)}(x,k,-\omega) f(x) d\sigma(x)\,. \tag{3.5}$$

Note that the scattering amplitude $A_\Omega(k)$ of the domain Ω corresponding to the exterior Dirichlet problem can be written as

$$A_\Omega(k,\omega,\omega') = \mathcal{F}(k)(\exp(ik\langle\cdot\,,\omega'\rangle))(\omega)\,. \tag{3.6}$$

Note also that if $\mathcal{F}_-$ denotes the far-field operator corresponding to an incoming condition at infinity, then its adjoint $\mathcal{F}_-^* : L^2(S^{n-1}) \to H^{-1/2}(\partial\Omega)$ turns out to be

$$\mathcal{F}_-^*(k)g(x) = \int_{S^{n-1}} \frac{\partial\varphi^e}{\partial\nu(x)}(s,k,\omega) g(\omega) d\sigma(\omega)\,. \tag{3.7}$$

The asymptotics of the solutions $\psi^+(x,k,\omega)$ of (2.3) yield

$$\mathcal{F}(k)(\psi^+(\cdot\,,k,\omega') - \exp(ik\langle\cdot,\omega'\rangle))(\omega) = -A(k,\omega,\omega')\,, \tag{3.8}$$

while those of $\mathcal{G}(x,y,k)$, obtained from the resolvent equation, show

$$\mathcal{F}(k)(\mathcal{G}(\cdot\,,y,k))(\omega) = \psi^+(y,k,-\omega)\,. \tag{3.9}$$

Using the equations (3.9), (3.7), (3.5), (3.6), (3.8) and the reciprocity property of the scattering amplitudes, it is now straightforward to verify the identity

$$\mathcal{F}(k)\mathcal{S}(k)\mathcal{F}_-^*(k) = A_\Omega(k) - A(k)\,. \tag{3.10}$$

From Rellich's uniqueness theorem we know that $\mathcal{F}(k)$ and $\mathcal{F}_-(k)$ are injective (hence also that the range of $\mathcal{F}_-^*(k)$ is dense); it follows that knowledge of $A(k)$ determines $\mathcal{S}(k)$ on $\partial\Omega$. Theorem 2.2 now yields the following uniqueness result.

Theorem 3.1 *Let $V_i, i = 1, 2$, be compactly supported, real- valued, potentials in $L^{n/2}(\mathrm{I\!R}^n), n \geq 3$, and let $S_i(k)$ be their respective scattering matrices. If $S_1(k) = S_2(k)$ for some $k > 0$, then $V_1 = V_2$ (a.e.).*

Part of the proof above consists of showing that there is a bounded open Ω containing the supports of V_i such that k^2 is not a Dirichlet eigenvalue of $-\Delta + V_i$ in Ω. This follows from the fact that the eigenvalues are *strictly* decreasing functions of the domain, a result of [10] which we extend to potentials in $L^{n/2}$ using the unique continuation theorem in [7].

For reconstruction purposes we henceforth take Ω to be the ball B_ρ of radius ρ. Expanding the solution $u(r\omega)$ of (3.1), for $r \geq \rho$, in an orthonormal basis of spherical harmonics $Y_\ell^m(\omega)$ we obtain

$$\mathcal{F}(k)f(\omega) = 4(2\pi\rho/k)^{\frac{n-2}{2}} \sum_{\ell,m} (i^{\ell+1} H^{(1)}_{\frac{n-2}{2}+\ell}(k\rho))^{-1} (Y_\ell^m, f(\rho\cdot)) Y_\ell^m(\omega)\,. \tag{3.11}$$

We note that the operator $\mathcal{F}(k)$ is simply a multiplier on the coefficients of the spherical harmonics expansion. This observation allows us to describe explicitly the range of $\mathcal{F}(k)$ and the inverse.

Lemma 3.2 *Let $\mathcal{A}_t^s \subset L^2(S^{n-1})$ be the Hilbert space with norm*

$$\|g\|^2_{\mathcal{A}_t^s} := \sum_{\ell,m} (1 + \ell(\ell + n - 2))^s \left|H^{(1)}_{\frac{n-2}{2}+\ell}(t)\right|^2 |(Y_\ell^m, g)|^2\,. \tag{3.12}$$

The operator $\mathcal{F}(k)$ is, up to a constant, an isometry from $H^s(\partial B_\rho)$ <u>onto</u> $\mathcal{A}^s_{k\rho}$; its inverse is $\mathcal{N}(k)$, the near field map

$$\mathcal{N}(k)g(\rho\omega) = \frac{1}{4}\left(\frac{k}{2\pi\rho}\right)^{\frac{n-2}{2}} \sum_{\ell} i^{\ell+1} H^{(1)}_{\frac{n-2}{2}+\ell}(k\rho)(Y_\ell^m, g) Y_\ell^m(\omega)\,. \tag{3.13}$$

Note that the large ℓ behaviour of the Hankel functions gives the equivalent description

$$\mathcal{A}_t^s = \left\{ g \in L^2(S^{n-1}) : \sum_{\ell,m} \ell^{n-3+2s}(2\ell/et)^{2\ell}|(Y_\ell^m, g)|^2 < \infty \right\} . \qquad (3.14)$$

The identity (3.10) now allows us to deduce properties of $A(k)$ from those of $\mathcal{S}(k)$.

Theorem 3.3

i) If $V \in L^{n/2}(\mathbb{R}^n)$ has support in the ball B_ρ of radius ρ, then its scattering amplitude $A(k)$ extends to a bounded operator from $(\mathcal{A}_{k\rho}^{1/2})^*$ to $\mathcal{A}_{k\rho}^{1/2}$.

ii) The number k^2 is a Dirichlet eigenvalue of $-\Delta+V$ in B_ρ if and only if there is a sequence $f_m \in L^2(S^{n-1})$, Cauchy in the $(\mathcal{A}_{k\rho}^{1/2})^*$ norm, such that $\|(A_{B_\rho}(k) - A(k))f_m\|_{L^2}$ tends to zero.

iii) If V_i, $i = 1,2$, are two potentials as in i), the operator norm of $\mathcal{S}_1(k) - \mathcal{S}_2(k)$ from $H^{-1/2}(\partial B_\rho)$ to $H^{1/2}(\partial B_\rho)$ is equivalent to that of $A_1(k) - A_2(k)$ from $(\mathcal{A}_{k\rho}^{1/2})^*$ to $\mathcal{A}_{k\rho}^{1/2}$.

Part iii) can be used to sharpen somewhat the stability result of [17] (where the operator $A_1(k) - A_2(k)$ is estimated by a norm on its kernel). Part ii) shows that, knowing $A(k)$ on $L^2(S^{n-1})$, one can in principle detect those balls $B_r, r \geq \rho$, for which k^2 is not an interior Dirichlet eigenvalue of $-\Delta + V$. Given $S(k)$, the scattering matrix of a potential V in $L^{n/2}$ with support in B_ρ we first choose such a ball B_r, (taking care to avoid eigenvalues of $-\Delta$ as well) then apply the formula

$$\mathcal{S}(k) = \mathcal{N}(k)(A_{B_r}(k) - A(k))\mathcal{N}_-^*(k) \qquad (3.15)$$

only on the dense subspace of functions $f(r\cdot)$ in $\mathcal{A}_{kr}^0$ and extend by continuity. (This requires only knowledge of $A(k)$ on $L^2(S^{n-1})$, rather than having to deal with the very large space $(\mathcal{A}_{kr}^{1/2})^*$). The reconstruction of V from $S(k)$ is thus reduced to that from the near field data $\mathcal{S}(k)$.

References

1. R. Beals, R.R. Coifman: Proc. Symp. Pure Math., **43**, 45– 70 (1985)
2. Yu. M. Berezanskii: Trudy Moskov. Mat. Obshch. **7**, 3–62 (1958)
3. K. Chadan, P.C. Sabatier: *Inverse Problems in Quantum Scattering Theory*, 2nd Ed., Springer-Verlag (1989)
4. L.D. Faddeev: Soviet Phys. Dokl. **10**, 1033–1035 (1966)
5. L.D. Faddeev: Soviet Phys. Dokl. **11**, 209–211 (1966)
6. L.D. Faddeev: J. Soviet Math. **5**, 334–396 (1976)
7. D. Jerison, C.E. Kenig: Ann. Math. **121**, 463–494 (1985)

8. R.B. Lavine, A.I. Nachman: in *Topics in Soliton Theory and Exactly Solvable Nonlinear Equations*, ed. by M. Ablowitz, B. Fuchssteiner, M. Kruskal, World Scientific Co. (1986)
9. R.B. Lavine, A.I. Nachman: Global uniqueness in inverse problems with singular potentials (in preparation)
10. R. Leis: Math. Z. **96**, 26–32 (1967)
11. A.I. Nachman: Ann. Math. **128**, 531–576 (1988)
12. A.I. Nachman, M.J. Ablowitz: Studies Appl. Math. **71**, 243–250 (1984)
13. A.I. Nachman, J. Sylvester, G. Uhlmann: Comm. Math. Phys. **115**, 595–605 (1988)
14. R.G. Novikov, G.M. Henkin: Russian Math. Surveys **42** (3), 109–180 (1987)
15. R.G. Novikov: Funct. Anal. and Appl. **22** (4), 263–272 (1988)
16. A.G. Ramm: Inverse Prob. **4**, 877–886 (1988)
17. P. Stefanov: Stability of the inverse problem in potential scattering at fixed energy (preprint)
18. J. Sylvester, G. Uhlmann: Ann. Math. **125**, 153–169 (1987)
19. R. Weder: Inverse Prob. **7**, 461–489 (1991)

Semi-Classical and High Energy Asymptotics of the Scattering Phase for Perturbations of Elliptic Operators

Didier Robert

Département de Mathématiques, Université de Nantes,
2, Rue de la Houssinière, 44072-Nantes-Cédex 03, France

1 Introduction

For usual Schrödinger operators the scattering phase can be defined as follows:
Let us consider the free Hamiltonian $H_0 = -\frac{\hbar^2}{2m}\Delta$ and a smooth perturbation $H = H_0 + V$ of H_0 such that $V(x) = O(|x|^{-\rho})$ as $|x| \to +\infty$, $x \in \mathbb{R}^n$, the configuration space. If $\rho > n$ then it is well known that the scattering matrix $S(\lambda)$, $\lambda > 0$, for the system (H, H_0) is a trace-class perturbation of the identity on $L^2(\mathbb{S}^{n-1})$. The scattering phase: $s(\lambda)$, is then defined by the equality:

$$det S(\lambda) = \exp(-2i\pi s(\lambda)) \qquad (1)$$

$s(\lambda)$ and its derivatives have many interesting properties as well as from physical and mathematical point of view. This notion was first introduced for central potentials. In that case $S(\lambda)$ is diagonalized on the spherical harmonics: $S(\lambda)f = (\exp(2i\delta_\ell(\lambda))f \quad \forall f \in \mathcal{H}_\ell$(spherical harmonics of degree ℓ) [11]. We have clearly:

$$s(\lambda) = -\frac{1}{\pi}\sum_{\ell\geq 0}(2\ell+1)\delta_\ell(\lambda) .$$

The energy derivative: $\frac{\partial\delta_\ell}{\partial\lambda}(\lambda)$ can be interpreted as a *Time Delay* for a radial wave packet (Eisenbud-Wigner~1950) [17]. Later on, Jauch-Misra-Sinha gave a natural and rigorous definition of the *Time Delay* for a general quantum process and proved a so-called Eisenbud-Wigner formula in particular cases. The mean value of the *Time Delay* τ in a state ψ is defined as the limiting value in large balls of the difference of sojourn times between the free particle and the asymptotically free one at time -∞ [10]. τ can be extended to a self-adjoint, unbounded operator in $L^2(\mathbb{R}^n)$, commuting with H_0. Furthermore, in the spectral representation of H_0 we have:

$$\tau(\lambda) = -iS(\lambda)^* \frac{dS}{d\lambda}(\lambda) . \qquad \text{(Eisenbud – Wigner)}$$

Several people contributed to state this result: W. Amrein-M. Cibils, Ph. Martin, A. Jensen, S. Nakamura, H. Narnhofer, X.P. Wang (see References). From the Eisenbud-Wigner formula we see that the average *Time Delay*: $\overline{\tau}(\lambda) = tr[\tau(\lambda)]$, where $'tr'$ denotes the usual functional trace defined on the obvious Hilbert space, is related with the energy derivative of the scattering phase: $\overline{\tau}(\lambda) = -2\pi \cdot' (\lambda)$.

M.S Birman and M.G Krein [2], discovered that the scattering phase can also be interpreted as a spectral shift function in the following sense:

$$\forall f \in C_0^\infty(]0,+\infty[) \quad \text{we have: } tr(f(H) - f(H_0)) = -\int f'(\lambda) \cdot s(\lambda) d\lambda . \tag{2}$$

Using (2) and the dilation property of the Laplace operator H_0, we can obtain a nice and useful formula connecting $s'(\lambda)$ and the local spectral density of the perturbed Hamiltonian H : $\frac{\partial e_H}{\partial \lambda}(\lambda, x, x)$ ($e_H(\lambda, x, y)$ denotes the integral kernel of the spectral projector $E_H(\lambda)$) of H on $]-\infty\lambda[$:

$$s'(\lambda) = \frac{1}{2\lambda} \cdot \int_{\mathbb{R}^n} (2V(x) + x \cdot \nabla V(x)) \cdot \frac{\partial e_H}{\partial \lambda}(\lambda, x, x) dx . \tag{3}$$

We are interested here in asymptotics for $\frac{d^k s}{d\lambda^k}$ in $(\hbar, \lambda)$ as $\hbar \searrow 0$ (semi-classical regime) or $\lambda \nearrow +\infty$ (high energy regime).

We divide the rest of this lecture into two parts: in **Part 2** we present very recent results obtained for perturbations of the Laplace operator; in **Part 3** we present new results on the scattering phase for pairs of long range perturbations of Laplace operators.

2 The Scattering Phase for Perturbations of Laplace Operator

Perturbations under consideration are defined by an electric potential V, a magnetic potential $A = (A_1, \ldots, A_n)$ and a Riemaniann metric $g = \{g_{j,k}\}$. We use the usual notations: $G = det g$, $\{g^{j,k}\} = \{g_{j,k}\}^{-1}$. The natural quantum Hamiltonian to compare with H_0 in $L^2(\mathbb{R}^n)$ (with Lebesgue measure) is:

$$H(g, A, V) = -G^{-1/4} \cdot \sum_{\substack{1 \leq j \leq n \\ 1 \leq k \leq n}} (\hbar\partial_j + iA_j) G^{1/2} g^{j,k} (\hbar\partial_k + iA_k) G^{-1/4} + V .$$

We assume that the data are smooth and satisfy the following decreasing assumption: $\exists \rho > 0$ such that :

$$\forall \alpha, \text{multiindex}, \exists C_\alpha \quad \text{such that} \quad \forall x \in \mathbb{R}^n,$$

$$|\partial^\alpha (g(x) - (\delta_{j,k}))| + |\partial^\alpha A(x)| + |\partial^\alpha V(x)| \leq C_\alpha (1 + |x|)^{-\rho - |\alpha|} \quad ((DC)_\rho)$$

Assume the condition $(DC)_\rho$ with $\rho > n$ and let us denote the scattering phase for the scattering pair $(H = H(g, A, V), H_0)$ by $s(\lambda)$. We have the following result:

Theorem 1 *Suppose $\hbar$ fixed ($\hbar = 1$) and the metric g has no trapped geodesics(i.e. each geodesic goes to infinity). Then we have the following asymptotics, as $\lambda \nearrow +\infty$,*

$$s'(\lambda) \asymp \lambda^{\frac{n}{2}-1} \left(\sum_{j \geq 0} \alpha_j \cdot \lambda^{-j} \right) . \tag{4}$$

The coefficient can be expressed with the data (g, A, V). For example we have:
$\alpha_0 = c_0 \cdot \int_{\mathbb{R}^n} (\sqrt{G(x)} - 1) dx$,
$\alpha_1 = c_1 \cdot \int_{\mathbb{R}^n} (K(x) + 6V(x)) \sqrt{G(x)} dx$,
where c_0 and c_1 are universal constants depending only on n, and $K(x)$ is the scalar curvature of g at x.

Remark. This Theorem was proved first, for $g(x)$ equal to the Euclidean metric outside a bounded set, by Majda and Ralston ([6]). It was extended by the author ([13]) to this more general class of perturbations, in particular to answer a question of Schrader [15]. Theorem 1 is also valid for Yang-Mills potentials. If g has trapped geodesics one can't expect a polynomial behaviour for $s'(\lambda)$ (see [4] about the $\ll$ *Breit – Wignerform* $\gg$). Besides, we still have a Weyl type formula for $s(\lambda)$ (see [13]):
$s(\lambda) = d_0 \lambda^{\frac{n}{2}} + O(\lambda^{\frac{n}{2}-1})$, d_0 is a volume constant. Even if g has trapped geodesic we still get complete asymptotics for some average of the scattering phase. That holds for example with $\vartheta(\beta) = \int s'(\lambda) e^{-\beta\lambda} d\lambda = tr(\exp(-\beta H) - \exp(-\beta H_0))$ as $\beta \searrow 0$. This can be used to compute the coefficients α_j (see [13]).

Proof. (For Theorem 1 – a sketch) First we replace the pair (H, H_0) by (P, P_0) with $P = \sqrt{E + H}$ and $P_0 = \sqrt{E + H_0}$, E being fixed, large enough. Let us denote the scattering phase for (P, P_0) by $\sigma(\lambda)$. We use a scaling argument putting $\lambda = \frac{1}{h}$ and $P - \lambda = \frac{P(h) - 1}{h}$; $P(h)$ is a h-admissible pseudodifferential operator ($h \searrow 0$) (see [12]). If σ_h denotes the scattering phase for $(P(h), P_0(h))$ we have: $\sigma'(\lambda) = h\sigma'_h(1)$.

Now it is useful to remark that we can obtain a representation formula similar to (3):

$$\sigma_h'(1) = tr\left[W(h)\cdot\frac{\partial E_P}{\partial\lambda}\right], \tag{5}$$

where $W(h)$ is a h -admissible pseudodifferential operator. With (5) we can prove that a complete asymptotics holds for $\sigma_h'(1)$ as $h \searrow 0$ using time-dependent methods and semi-classical, microlocal propagation estimates. Theorem 1 follows easily. (See [13] for details.) □

Remark. Our proof of validity of (5) depends on the fact that the unperturbed Hamiltonian does not vary with the position. It was for us an interesting challenge to look at what happens if we start from an Hamiltonian depending on position and momentum. The answer is in the next paragraph. Our method for proving Theorem 1 clearly gives asymptotics for $s'(\lambda,\hbar)$ as λ is fixed and $\hbar \searrow 0$, assuming a non trapping condition [14].

3 Comparaison of Long Range Perturbations of Laplace Operator

Let us start with two Hamiltonians $H_k{=}H(g_{(k)},A_{(k)},V_{(k)})$, k=1,2, where the data $(g_{(k)},A_{(k)},V_{(k)})$ satisfy the condition $(DC)_\rho$, $\rho>0$. Assume also that $(g_{(2)}-g_{(1)}+\mathbb{1}_n, A_{(2)}-A_{(1)}, V_{(2)}-V_{(1)})$ satisfies the short range condition $(DC)_\delta$ for some $\delta>n$. Let us denote the scattering phase for the pair (H_2,H_1) by $s_{(2,1)}(\lambda)$. Under the previous assumptions we have:

Theorem 2 *The scattering phase $s_{(2,1)}(\lambda)$ is C^∞ on $]0,+\infty[$. If the metrics $g_{(k)}, k = 1,2$, have no trapped geodesics then $\forall k\in\mathbb{N}$, $\frac{d^k s_{(2,1)}}{d\lambda^k}(\lambda)$ has a complete asymptotics for $\lambda \nearrow +\infty$*

$$\frac{d^k s_{(2,1)}}{d\lambda^k}(\lambda) \asymp \lambda^{\frac{n}{2}-k}\left(\sum_{j\geq 0}\alpha_j{}^{(k)}\cdot\lambda^{-j}\right). \tag{6}$$

In particular we have: $\alpha_0{}^{(1)} = c_0\cdot\int_{\mathbb{R}^n}\left(\sqrt{G_{(2)}(x)}-\sqrt{G_{(1)}(x)}\right)dx$.

Proof. (A sketch) We start as in the proof of Theorem 1 by reducing the problem to a semi-classical problem for a pair: $(P_2(h),P_1(h))$ of h-admissible pseudodifferential operators.Let us denote by $\sigma_{(2,1)}$ the corresponding scattering phase. The new technical point here is the following representation formula:

$\exists\theta \in C_0^\infty(\mathbb{R}^n)$ and two h-admissible pseudodifferential operators $K(h)$ and $M(h)$, with symbols compactly supported in momentum variables, $O(|x|^{-\rho})$ $(\rho > n)$ in position variables, such that:

$$\begin{aligned}\frac{\partial\sigma_{(2,1)}}{\partial\lambda}(\lambda) = tr\left[\theta\cdot\left(\frac{\partial E_{P_2}}{\partial\lambda} - \frac{\partial E_{P_1}}{\partial\lambda}\right)\right] + tr\left(K(h)\frac{\partial E_{P_0}}{\partial\lambda}\right) \\ + tr\left((M(h)\mathcal{PV}\left(\frac{1}{P_0-\lambda}\right)\right) + O(h^{+\infty})\,.\end{aligned} \tag{7}$$

Where we have denote: $\mathcal{PV}(\frac{1}{P_0-\lambda}) =: \frac{1}{2}\left(P_0 - \lambda - i0)^{-1} + (P_0 - \lambda + i0)^{-1}\right)$ This decomposition is locally uniform in $\lambda \in]0, +\infty[$. From this formula it is not very difficult to deduce a complete asymptotics in $h \searrow 0$ for $\sigma_{(2,1)}(h, \lambda)$ and hence we get Theorem 2 for $k = 1$. For $k \geq 2$ the proof is similar. □

Remark. The proof of (7) does not use a nontrapping condition and reduces essentially the problem in a bounded set of the configuration space. The above semi-classical expansions are valid if we replace H_0 by more general Hamiltonians $\omega(hD)$ and $H_j = \omega(hD)+q_j(h, x, hD)$ where q_j are long range h-admissible symbols and $q_2 - q_1$ is short range .

References

1. Amrein, W.O., Cibils, M.B.: Helv.Phys.Acta **60**, 481 (1987)
2. Birman, M.S., Krein, M.G.: Dokl.Akad.Nauk SSSR5.**5**, 475 (1962)
3. Buslaev, V.S.: Soviet Math.Dokl. **12**, 591 (1971)
4. Gérard, C., Martinez, A., Robert, D.: Comm.Math.Phys **121**, 323 (1989)
5. Jensen, A.: Comm.Math.Phys. **82**, 435 (1981)
6. Majda, A., Ralston, J.: Duke Math.J. ; **45**, 183 (1978); **45**, 513 (1978); **46**, 725 (1979)
7. Martin, Ph.: Acta Phys. Austriaca Suppl. **23**, 157 (1981)
8. Nakamura, S.: Comm.Math.Phys. **109**, 397 (1987)
9. Narnhofer, H.: J.Math.Phys. **25**, 987 (1984)
10. Rauch, J.M., Sinha, K.B., Misra, B.N.: Helv.Phys.Acta **45**, 398 (1972)
11. Reed, M., Simon, B.: *Scattering Theory.* Academic Press (1979)
12. Robert, D.: *Autour de l'Approximation Semi-Classique.* **PM 68** Birkhäuser (1987)
13. Robert, D.: Asymptotic Anal. **3**, 301 (1991);
    Robert, D.: Ann. ENS. Paris, to appear;
    Robert, D.: Conf. in Honour of S. Agmon, to appear in J.d'Anal. Math.
14. Robert, D., Tamura, H.: J.Funct.Anal. **80:1**, 124 (1988)
15. Schrader, R.: Z.Phys.C. Part. and Fields **4**, 27 (1980)
16. Wang, X.P.: Helv.Phys.Acta **60**, 501 (1987)
17. Wigner, E.P.: Phys.Rev. **98:1**, 145 (1955)

On the Mapping Properties of the Wave Operators in Scattering Theory

Arne Jensen [1], Shu Nakamura [2]

[1]Department of Mathematics and Computer Science, Institute for Electronic Systems, Aalborg University, Fredrik Bajers Vej 7, DK-9220 Aalborg Ø, Denmark
[2]Department of Pure and Applied Sciences, University of Tokyo, 3-8-1 Komaba, Meguro-ku, Tokyo 153, Japan

1 Statement of Results

In this note we describe some recent results on the mapping properties of the wave operators for two-body Schrödinger operators with long range potentials. We show that the wave operators and adjoint wave operators, localized in energy, map a weighted L^2-space into a slightly larger weighted L^2-space. We also give some preliminary results on the mapping properties between the L^p-spaces. The precise statement of the results requires some preparation.

We denote by $L^p(\mathbb{R}^n)$ the usual L^p-space. The weighted L^2-space is given for any $s \in \mathbb{R}$ by

$$L^{2,s}(\mathbb{R}^n) = \{f \in L^2_{\text{loc}}(\mathbb{R}^n) \mid \|f\|_{L^{2,s}} = \|(1+x^2)^{s/2} f\|_{L^2} < \infty\}\,.$$

Inclusion gives a continuous embedding of $L^{2,s}(\mathbb{R}^n)$ into $L^{2,s'}(\mathbb{R}^n)$ for any $s > s'$. The dual of $L^{2,s}(\mathbb{R}^n)$ is identified with $L^{2,-s}(\mathbb{R}^n)$ via the duality given by the inner product on $L^2(\mathbb{R}^n)$. We introduce the notation $\langle x\rangle = (1+x^2)^{1/2}$.

Let $H_0 = -\Delta$ denote the free Hamiltonian on $L^2(\mathbb{R}^n)$. We introduce the following class of potentials.

Definition 1 A smooth long range potential is a real-valued function $V \in C^\infty(\mathbb{R}^n)$ such that for some $\varepsilon > 0$ and all multi-indices α we have

$$|\partial_x^\alpha V(x)| \leq C_\alpha \langle x\rangle^{-\varepsilon-|\alpha|}\,.$$

The full Hamiltonian is given by $H = H_0 + V$. It is well known that for $0 < \varepsilon \leq 1$ in Definition 1 the usual wave operators between H and H_0 do not exist. Instead, we have to introduce a modified free evolution. Here we choose the stationary modifier introduced by Isozaki and Kitada, see [4]. The modifier J is a bounded operator on $L^2(\mathbb{R}^n)$, constructed as a Fourier integral operator. The modified wave operators are given by

$$W_J^{\pm} = \operatorname*{s-lim}_{t \to \pm\infty} e^{itH} J e^{-itH_0} .$$

Our main result on the maping properties between the weighted L^2-spaces can be stated as follows: (Note that $(\cdot)^*$ denotes the L^2-adjoint of an operator)

Theorem 1 *Let V be a smooth long range potential. Let $f \in C_0^\infty((0, \infty))$ and let $s, s' \in \mathbb{R}$ with $s > s'$. Then $W_J^{\pm} f(H_0)$ and $(W_J^{\pm} f(H_0))^*$ determine bounded operators from $L^{2,s}(\mathbb{R}^n)$ to $L^{2,s'}(\mathbb{R}^n)$.*

This result states that the localization in x-space is preserved in some sense during the scattering process, when looking at corresponding initial states. The result has applications in the study of time-delay, see [1],[5].

Our result on the mapping properties in L^p-spaces is much weaker and of a preliminary nature. We denote by $[s]$ the integer part of a real number s. We consider only very short range potentials, so no modifier is needed. Hence we write $W^{\pm} = \text{s-lim}_{t\to\pm\infty} \exp(itH)\exp(-itH_0)$ for the ordinary wave operators.

Theorem 2 *Assume $n \geq 3$. Let $V \in C^{[n/2]+1}(\mathbb{R}^n)$ be a realvalued function such that for all multi-indices α with $|\alpha| \leq [n/2] + 1$ we have $|\partial_x^\alpha V(x)| \leq C_\alpha \langle x \rangle^{-n}$. Let $f \in C_0^\infty((0, \infty))$. Let p, q with $1 \leq p \leq q \leq \infty$ satisfy (i) if $1 \leq p < 2$: $1/q < (n-2)/(np) + 1/n$ or (ii) if $2 < p < n$: $1/q < n/((n-2)p) - 1/(n-2)$. Then $W^{\pm} f(H_0)$ and $(W^{\pm} f(H_0))^*$ determine bounded operators from $L^p(\mathbb{R}^n)$ to $L^q(\mathbb{R}^n)$.*

We expect that it should be possible to improve this result by allowing a larger range of p and q values. We will not discuss the proof of Theorem 2, since it is quite involved, see [6]. It seems that there are no previous results of the type obtained in Theorem 2.

The results in Theorem 1 for $(W_J^{\pm})^*$ have been obtained by Herbst and Skibsted for $s' \geq 0$, see [2]. For $s > 2$, $s' = 2$, and for a restricted class of potentials with decay rate $V(x) = O(|x|^{-2-\delta})$ as $|x| \to \infty$, the results in Theorem 1 were obtained in [1]. There are also results by Isozaki [3] on the properties of the generalized eigenfunction expansion operators which imply some of the results in Theorem 1.

2 Sketch of Proof of Theorem 1

We briefly sketch the proof of Theorem 1. The idea is to use the properties of the modifier constructed by Isozaki and Kitada, together with some propagation estimates, in a Cook-type argument.

Using duality arguments and complex interpolation, we can reduce the proof of Theorem 1 to showing that the operators

$$\langle x\rangle^{2k} W_J^{\pm} f(H_0)\langle x\rangle^{-2k-2} \quad \text{and} \quad \langle x\rangle^{2k}(W_J^{\pm} f(H_0))^* \langle x\rangle^{-2k-2}$$

extend to bounded operators on $L^2(\mathbb{R}^n)$. Here $k \geq 0$ is an integer. We will consider only W_J^+. The other cases are treated analogously.

Let $f \in C_0^\infty((0,\infty))$ and choose $f_1 \in C_0^\infty((0,\infty))$ such that $ff_1 = f$. By the usual Cook-type argument, and the intertwining property of the wave operators, we can write

$$\begin{aligned} W_J^+ f(H_0) &= f_1(H) J f(H_0) \\ &\quad + i\int_0^\infty f_1(H) e^{itH}(HJ - JH_0)e^{-itH_0} f(H_0)\,dt\,. \end{aligned} \tag{1}$$

Using the results in [4] it is not difficult to show that the operator $\langle x\rangle^{2k} f_1(H) J f(H_0)\langle x\rangle^{-2k-2}$ is bounded on L^2. To treat the integrand we need two lemmas:

Lemma 1 *Let $f \in C_0^\infty((0,\infty))$. Then for any integer $k \geq 0$ we have on the Schwartz space $\mathcal{S}(\mathbb{R}^n)$*

$$\langle x\rangle^{2k} e^{itH} f(H) = \sum_{\ell=0}^{2k} B_\ell(t)\langle t\rangle^\ell \langle x\rangle^{2k-\ell}\,,$$

where $B_\ell(t)$, $\ell = 0,\ldots,2k$, extend to bounded operators on $L^2(\mathbb{R}^n)$ with operator norm uniformly bounded in $t \in \mathbb{R}$.

This lemma can be proved by direct commutator computations, see [5] for the details.

The main property of the modifier used here is the propagation estimate given in the next lemms. The proof can be extracted from the computations in [4]. Note that we are combining two terms into the one we call P_- below. One part has phase space support in the non-propagation set and the other part is a rapidly decaying term.

Lemma 2 *We have $(HJ - JH_0)f(H_0) = P_-$, where P_- is a bounded operator on $L^2(\mathbb{R}^n)$ with the following property: For any $s_2 > s_1 \geq 0$, and $\delta > 0$ we have*

$$\|\langle x\rangle^{s_1} P_- e^{-itH} f(H)\langle x\rangle^{-s_2}\| \leq C\langle t\rangle^{-(s_2-s_1)+\delta}$$

for all $t \geq 0$. The same result holds for H_0.

To complete the proof, we estimate the integrand from (1) with the x-weights. We write $(HJ - JH_0)f(H_0) = P_-$ as in Lemma 2 and use the lemmas to get

$$\langle x\rangle^{2k} f_1(H)e^{itH}(HJ - JH_0)e^{-itH_0} f(H_0)\langle x\rangle^{-2k-2}$$
$$= \sum_{\ell=0}^{2k} B_\ell(t)\langle t\rangle^\ell \left(\langle x\rangle^{2k-\ell} P_- e^{-itH_0} f_1(H_0)\langle x\rangle^{-2k-2}\right) .$$

From Lemma 1 we then get the estimate for each term

$$O(\langle t\rangle^\ell \langle t\rangle^{-(2k+2)+(2k-\ell)+\delta}) = O(\langle t\rangle^{-2+\delta}) .$$

Since $\delta > 0$ can be arbitrarily small, the term is integrable, and the boundedness of $\langle x\rangle^{2k} W_J^+ f(H_0)\langle x\rangle^{-2k-2}$ follows. This completes the sketch of the proof of Theorem 1.

References

1. Amrein, W., Cibils, M., Sinha, K.B.: Ann.Inst.H.Poincaré, Phys.Théor. **47**, 367 (1987)
2. Herbst, I., Skibsted, E.: Preprint, Aarhus University, 1990
3. Isozaki, H.: J.Math. Kyoto Univ. **25**, 789 (1985)
4. Isozaki, H., Kitada, H.: J.Fac.Sci.Univ. Tokyo Sect.IA Math. **32**, 77 (1985); Isozaki, H., Kitada, H.: Sci.Papers College Arts Sci. Univ Tokyo **35**, 81 (1985)
5. Jensen, A., Nakamura, S.: Preprint, Aalborg University, June 1991
6. Jensen, A., Nakamura, S.: In preparation

Section 12

Conformal and Topological Field Theory

Organizer: K. Gawedzki

Reduction of Kac-Moody Systems

Andreas W. Wipf

Institut für Theoretische Physik, Eidgenössische Hochschule,
Hönggerberg, Zürich CH-8093, Switzerland

Abstract: The general structure of reductions of the Wess-Zumino-Novikov-Witten theory by first class Kac-Moody constraints is analysed. Algebraic conditions are derived which ensure exact integrability, conformal invariance and $\mathcal{W}$-symmetry for the reduced effective theories. Solutions to the general algebraic conditions are given leading to new generalized conformal Toda theories. A Lagrangean implementation of the reduction is established in general and is used for setting up the framework for quantizing the effective theories.

1 Introduction

One of the most urgent problems in conformal field theory (CFT) is the classification of the possible chiral symmetry algebras or (bosonic) extensions of the conformal symmetry. Initially this were the Kac-Moody (KM) algebras and Zamolodchikovs chiral $\mathcal{W}$-algebras [1]. $\mathcal{W}$-algebras are extensions of the Virasoro algebra by primary fields. The primary fields close non-linearly onto themselves under commutation (thus form not Lie algebras).

Now it has been realized that certain first class constraints reduce the simple current algebras to a variety of new chiral symmetry algebras containing Zamolodchikovs $\mathcal{W}$-algebras as very particular examples. This led and certainly will further lead to new insight into the hidden symmetries of CFT and their relation with integrable systems. The general reductions described below [2] are based on earlier results on the reduction of Wess-Zumino-Novikov-Witten (WZNW) to Toda theories [3],[4].

2 General Structure of KM Reductions

Let $J(x)$ be a $\mathcal{G}$-valued current forming a KM algebra where $\mathcal{G}$ is an arbitrary semi-simple Lie algebra. To describe the possible linear *first class constraints* on J we shift it as

$$J(x) = M + j(x) \tag{1}$$

with constant $M \in \mathcal{G}$ so that such constraints can be written as

$$\big(\Gamma, j(x)\big) = 0, \qquad \Gamma \subset \mathcal{G}, \tag{2}$$

where $(.,.)$ is the Cartan-Killing form. Now one can prove the following

Lemma 1 *The constraints (1), (2) are first class iff*

$$[\Gamma,\Gamma] \subset \Gamma, \qquad \omega(\Gamma,\Gamma) = 0 \quad \textit{and} \quad (\Gamma,\Gamma) = 0, \tag{3}$$

where ω is the Kostant-Kirillov 2-form evaluated at the identity: $\omega(a,b) = (M,[a,b])$.

The first condition shows that Γ is a subalgebra, the second that $[M,\Gamma] \subset \Gamma^{\perp}$ and the last one that Γ is contained in its orthogonal complement. The reduced theory is unaffected by a shift of M by an element from $\Gamma^{\perp}$ (this follows from the Lagrangean reduction (7)) so that the condition on M and the ambiguity in choosing it can be summarized as

$$M \in [\Gamma,\Gamma]^{\perp}/\Gamma^{\perp}. \tag{4}$$

Now one can prove from (3) that Γ is solvable, so that the coset is non-empty implying the existence of at least one M for any admissable Γ. Γ is in general not nilpotent. One can find solutions to (3) with non-nilpotent Γ's.

The constraints (1), (2) generate the KM transformations corresponding to the subalgebra Γ, and the first class conditions (3) are equivalent to these gauge transformations respecting the constraints (2). The reduced phase space can be identified as the space of gauge orbits in the constraint surface.

Let us now consider the conditions for *conformal invariance*. With respect to the Sugawara Virasoro density L_{Sug} the currents have weight one and the constraints (1), (2) would break the corresponding conformal symmetry. Thus one improves it by a local term to

$$L_H(x) = L_{\text{Sug}} - \partial_x\big(H, J(x)\big), \qquad H \in \mathcal{G}. \tag{5}$$

We must demand that L_H respects the constraints (is gauge invariant). One can prove

Lemma 2 *The reduced theories are conformally invariant iff*

$$H \in \Gamma^{\perp}, \qquad [H,\Gamma^{\perp}] \subset \Gamma^{\perp} \quad \textit{and} \quad \big([H,M]+M\big) \in \Gamma^{\perp}. \qquad (6)$$

The second condition is equivalent to $[H,\Gamma] \subset \Gamma$ so that H leaves both Γ and its orthogonal complement invariant.

Below we assume $\mathcal{G}$ to be the maximally non-compact real form (although most of our result apply to the complex case as well) and that $H \in \mathcal{G}$ is diagonalizable and thus is contained in a splitting Cartan subalgebra. Then H grades $\mathcal{G}$, Γ and $\Gamma^{\perp}$. Under these assumptions the last condition in (6) can be replaced by $[H,M] = -M$ and one sees that the current components corresponding to H-grade h have weights $h+1$, with the exception of the H-component which transforms as spin one connection.

3 Effective Field Theories

The constraints (2) together with similar constraints for the rightmoving current $\tilde{J}$ reduce the WZNW theories to effective CFT's. In principle Γ and $\tilde{\Gamma}$ can be chosen independently. If one wants the reduction to preserve the parity invariance of the WZNW theory one is lead to choose them dual to each other. We choose dual bases $\gamma_i \in \Gamma$ and $\tilde{\gamma}_j \in \tilde{\Gamma}$. Next one observes that the reduced theories can be identified as the gauge invariant content of a corresponding gauged WZNW theory with action

$$I(g,A,\tilde{A}) = S_{\mathrm{WZ}}(g) + \int d^2x\Big((A,J-M)-(\tilde{A},\tilde{J}-\tilde{M})+(A,g\tilde{A}g^{-1})\Big), \quad (7)$$

where $A(x) \in \Gamma$ and $\tilde{A}(x) \in \tilde{\Gamma}$ [2],[3]. Note that I only depends on the projection of M to Γ. It is invariant under the non-chiral gauge transformations

$$g \to \alpha g \tilde{\alpha}^{-1}; \quad A \to \alpha A \alpha^{-1} + \alpha\partial_{-}\alpha^{-1}; \quad \tilde{A} \to \tilde{\alpha}\tilde{A}\tilde{\alpha}^{-1} + \partial_{+}\tilde{\alpha}\tilde{\alpha}^{-1}, \quad (8)$$

where e.g. $\alpha \in e^{\Gamma}$. To derive the effective theories one decomposes $\mathcal{G} = \Gamma + \mathcal{B} + \tilde{\Gamma}$ and accordingly Gauss-decomposes $g = abc$. Upon eliminating the (non-dynamical) gauge fields one obtains the effective actions

$$I_{\mathrm{eff}} = S_{\mathrm{WZ}}(b) - \kappa \int d^2x (\tilde{T}(b), bT(b)b^{-1}), \qquad (9)$$

where

$$T = \tilde{\gamma}_i V_{ij}^{-1}(\gamma_j, M - \partial_{+}bb^{-1}) \qquad \tilde{T} = (\tilde{\gamma}_i, \tilde{M} - b^{-1}\partial_{-}b)V_{ij}^{-1}\gamma_j \qquad (10)$$

and

$$V_{ij}(b) = (\gamma_i, b\tilde{\gamma}_j b^{-1}). \qquad (11)$$

4 Solutions to the Conformal First Class Constraints

We decompose $\mathcal{G}$ as $\mathcal{G}$=Ker$M + T$, that is into the kernel of M and the tangential space to the group orbit of M at M. On T the Kostant-Kirillov form ω is symplectic (i.e. non-degenerate) and polarizes T:

$$T = P + Q; \qquad \omega(P,P) = \omega(Q,Q) = 0. \tag{12}$$

The (say) P-space fulfils the second condition of (3) and is a candidate for Γ. A diagonalizable H grades Ker$M$ and thus (a properly chosen) T. This grading allows for simple solutions of the conditions on the triple (Γ, M, H).

4.1 Toda Theories

Here H is the unique element in the Cartan subalgebra for which $[H, E_\alpha] = E_\alpha$ for all simple roots α, Γ is generated by all positive root vectors and $M = \sum E_{-\alpha}$, α simple. By grading-arguments one easily proves that this triple solves the constraints. This reductions lead to Toda theories [3].

4.2 Interacting WZNW Theories

Let H be an integral grading operator whose spectrum in the adjoint contains ± 1 and take

$$\Gamma = \mathcal{G}_{\geq 1} \quad , \quad \tilde{\Gamma} = \mathcal{G}_{\leq -1} \tag{13}$$

and any non-zero M, $\tilde{M}$ in the grade -1 and $+1$ subspaces. In this case b in (9) is in the little group of H and the effective theory simplifies to

$$I_{\text{eff}}(b) = S_{\text{WZ}}(b) - \int d^2x \big(b\tilde{M}b^{-1}, M\big). \tag{14}$$

It describes WZNW fields in interaction with themselves and gravity [4].

4.3 Sl(2)-Based Generalized Toda Theories

Lets assume that Γ consists of nilpotent elements. Since M may be chosen from Γ^t it is nilpotent as well and can always be imbedded into a $sl(2)$ subalgebra $(M_- = M, M_0 = H, M_+)$. This H grades $\mathcal{G}$ in half-integers and one takes $\Gamma = \mathcal{G}_{\geq 1} + P_{1/2}$, where $\mathcal{G}_{1/2} = Q_{1/2} + P_{1/2}$ is a polarization of the grade 1/2 subspace. The effective theories contain primary fields with half-integer weights [2],[5]. No auxiliary fields are needed to implement these constraints as has been claimed.

One can show that (only) for $\mathcal{G}$=$sl(n)$ the last two reductions lead to isomorphic theories, but with different implementation of the conformal symmetry.

5 $\mathcal{W}$-Symmetries, Lax-Pairs, BRST and Solutions

The chiral symmetry algebras are generated by the gauge invariant functions of the constrained currents. For a diagonalizable grading H one can prove

Lemma 3 $\Gamma \cap \mathrm{Ker(ad)}_M = \{0\}$ *and* $\Gamma^\perp \subset \mathcal{G}_{>-1}$ *imply that there is a complete set of gauge invariant polynomials of the constrained currents.*

These conditions guarantee that the constrained currents can be gauged into a generalized Drinfeld-Sokolov form by an iterative algorithm and allow for an explicit construction of the $\mathcal{W}$-generators. One can also derive algebraic conditions on the triple $\{\Gamma, M, H\}$ which imply that the algebraic relations of the $\mathcal{W}$-algebras are polynomial [2],[3],[4].

Starting from the gauged theories (7) one can prove that the *quantum reductions* lead to the same effective theories (9) as the classical reduction. This way one also derives the BRST-action, BRST-energy momentum tensor and BRST-charge

$$Q_{\mathrm{BRST}} = \int dx^+ \operatorname{tr} c\Big(\partial_+ g g^{-1} - M - c\bar{c}_+\Big) \tag{15}$$

for the reduced theories with Lie-algebra valued ghost fields, $c \in \Gamma$ and $\bar{c}_+ \in \tilde{\Gamma}$. This BRST charge simplifies to the one derived in [6] for the simplest reductions. The central charges of the reduced theories contain KM-contributions from (5) and ghost contributions and explicit formulae for the above reductions can be derived.

For further results like the *Lax-pair formulation* and *explicit solutions* for the reduced effective theories we refer to [2].

References

1. Zamolodchikov, A.B.: Theoret.and Math.Phys. **65**, 1205 (1986)
2. Feher, L. O'Raifeartaigh, L., Ruelle, P., Tsutsui, I., Wipf, A.: ETH-TH/91-17; ETH-TH/91 (to appear)
3. Balog, J., Feher, L., O'Raifeartaigh, L., Wipf, A.: Phys.Lett. **244B**, 435 (1990); Ann.Phys. **203** 76 (1990) and references therein
4. O'Raifeartaigh, L., Wipf, A.: Phys.Lett. **251 B**, 361 (1990); O'Raifeartaigh, L., Ruelle, P., Tsutsui, I., Wipf, A.: DIAS-STP-91-01
5. Bais, F.A., Thin, T., van Driel, P.: Nucl.Phys. **B357**, 632 (1991); Fuchs, J.: Phys.Lett. **262 B**, 249 (1991)
6. Bershadski, M., Ooguri, H.: Comm.Math.Phys. **126**, 49 (1989)

Quantum Symmetry of Rational Conformal Models

Ludmil K.Hadjiivanov, Yassen S.Stanev, Ivan T.Todorov

Institute for Nuclear Research and Nuclear Energy
Bulgarian Academy of Sciences
Boul. Trakia 72, BG-1784 Sofia, Bulgaria

Abstract: Recent progress in understanding the $U_q(sl_2)$ symmetry of $s\hat{u}(2)_k$ current algebra and of "thermal" minimal conformal models is previewed. New features include the introduction of a pair of regular bases of 4-point invariants in the space of conformal blocks and in the quantum group space as well as a derivation of the results for minimal models from the current algebra results for a fractional level k.

1. Single valued two dimensional (2D) local fields $\phi(\tau,\xi)$ on a cylinder (satisfying the periodicity condition $\phi(\tau,\xi+2\pi)=\phi(\tau,\xi)$) in a rational conformal field theory can be split into a finite sum of products

$$\phi(\tau,\xi)=\sum_i \varphi_i(\tau-\xi)\overline{\varphi}_i(\tau+\xi) \tag{1}$$

of multivalued "chiral fields" depending on a single light cone variable. The periodicity of ϕ is restored if φ and $\overline{\varphi}$ have inverse monodromies:

$$\varphi(x_R-2\pi)=\varphi(x_R)M \ \ (\text{i.e.}\ \varphi_i(x_R-2\pi)=\sum_j \varphi(x_R)M_i^j\,) \tag{2.a}$$

$$\overline{\varphi}(x_L+2\pi)=M^{-1}\overline{\varphi}(x_L)\,. \tag{2.b}$$

The chiral components φ_i and $\overline{\varphi}_i$ are not determined by the observable 2D field ϕ. It leaves room for a (first kind) gauge freedom: $\varphi\to\varphi V\,,\,\overline{\varphi}\to V^{-1}\overline{\varphi}$. Correlation functions of φ carry a representation of the braid group with a nontrivial monodromy ($B^2=M$). Assuming that braid transformations commute with the statistics operator (that can be indentified with the braid generator B) we deduce that the tensor product of two gauge transformations V acting on a product of fields should be substituted by a (twisted) coproduct $\Delta(V)$, a characteristic feature of a quantum symmetry. In a large family of cases it corresponds to a quantum universal enveloping (QUE) algebra or quantum group.

The hidden quantum symmetry of (the chiral constituent of) a 2D conformal models is being gradually understood at both the classical [1] and the quantum level [2]-[6]. We outline here recent progress in the axiomatic approach to the problem (started in a joint effort by D.Buchholz, G.Mack and I.Todorov back in 1989 and reflected in [3] and [6]). The physical state space of a chiral model admits a non-associative "weak quasi-Hopf" symmetry algebra [6]. Here we follow [3] in considering instead an extended indefinite metric chiral state space that gives room to an associative (Hopf) QUE symmetry algebra. The correlation functions of physical 2D fields are recovered by using a semidefinite sesquilinear form that contracts the quantum group's degree of freedom of the left and right sectors, the non-unitary contributions being absorbed by its kernel.

We consider the familiar $\hat{su}(2)_k$ current algebra model with QUE symmetry algebra

$$U_q = U_q(sl_2) \text{ for } q = e^{\frac{i\pi}{k+2}} . \quad (3)$$

(A representation in the tensor product of two irreducible U_q modules of) Drinfeld's universal R-matrix - followed by a permutation - gives rise to a QUE algebra representation of the braid group (on top of the one obtained by exchanging neighbouring factors in a correlation function through analytic continuation). Our chief objective is to establish the invariance of (U_q extended) chiral correlation functions under the diagonal action of the braid group. A new feature in the argument summarized here is the introduction of a regular basis of n-point U_q invariants [7] and a corresponding basis of conformal blocks (Sec.3) which allows to work with a deformation parameter q at a root of unity (like in (3)) without having to resort to a regularization. We only outline the study of diagonal ("A-series") models although a simple analysis [3] of 3-j symbols suggests that one can deal as well with the non-diagonal models of the ADE classification [8]. We point out (Sec.4) a way to prove the braid invariance of thermal minimal models by extending the current algebra results to fractional levels k.

2. The irreducible finite dimensional U_q modules $\mathfrak{V}_I$, $2I \in \mathfrak{N}_{k+2} = \{0, 1, \ldots, k+1\}$ can be realized as polynomials in a formal variable u of degree $2I$. (For a q-deformation of a simple Lie algebra of rank r the variable u is to be replaced by a set of r non-commuting elements parametrizing a nilpotent quantum group matrix.) The result of [7] is summarized in the following proposition. For each $I = (I_1, \ldots, I_n)$ with $2I_i \in \mathfrak{N}_{k+2}, i = 1, \ldots, n$, and $\sum_{i=1}^n I_i - 2I_j \in \mathbb{Z}_+$ for $j = 1, \ldots, n$, there exists a basis of n-point U_q invariant products

$$J^I_\lambda(u) = \prod_{1 \le 1 < j \le n} W_{i,j} , \quad (4)$$

where $W_{i,j}$ are polynomials of degree $\lambda_{i,j}$ (going into $(u_i - u_j)^{\lambda_{i,j}}$ for $q \to 1$); the non-negative integers $\lambda_{i,j}$ satisfy the conservation laws

$$\sum_j \lambda_{i,j} = 2I_i \tag{5}$$

and the selection rule

$$\lambda_{i,j}\lambda_{k,l} = 0 \text{ for } i < k < j < l \text{ or } k < i < l < j \quad . \tag{6}$$

There is a finite family of solutions to (5) and (6) labeled by $n-3$ integer parameters.

The solution has been worked out explicitly for a general 4-point function. The associated realization of the exchange algebra [9] is characterized by the following properties of its generators: $\mathcal{B}\begin{bmatrix} I_1 & I_2 \\ 0 & I_{34} \end{bmatrix}$ $(I_{ij} = I_i - I_j)$ is a lower triangular matrix, $\mathcal{B}\begin{bmatrix} I_2 & I_3 \\ I_1 & I_4 \end{bmatrix}$ is an upper triangular matrix and the involutive fusion matrix $\mathcal{F}$ appearing in the Yang-Baxter type relation

$$\mathcal{B}\begin{bmatrix} I_2 & I_3 \\ 0 & I_{14} \end{bmatrix}\mathcal{B}\begin{bmatrix} I_1 & I_3 \\ I_2 & I_4 \end{bmatrix}\mathcal{B}\begin{bmatrix} I_1 & I_2 \\ 0 & I_{34} \end{bmatrix} = \mathcal{B}\begin{bmatrix} I_1 & I_2 \\ I_3 & I_4 \end{bmatrix}\mathcal{B}\begin{bmatrix} I_1 & I_3 \\ 0 & I_{24} \end{bmatrix}\mathcal{B}\begin{bmatrix} I_2 & I_3 \\ I_1 & I_4 \end{bmatrix}$$
$$= (-1)^{I_1+I_2+I_{34}} q^{I_1(I_1+1)+I_2(I_2+1)+I_3(I_3+1)-I_4(I_4+1)} \mathcal{F} \tag{7}$$

has unit elements on the second diagonal ($\mathcal{F}^\mu_\lambda = 1$ for $\mu + \lambda = m = \min 2I_i$) and all other elements zero.

3. The Green functions of an extended chiral $\hat{su}(2)_k$ model depends on three types of variables: the world sheet variables $z = (z_1, \ldots, z_n)$ where each z_i is related to the cylinder coordinates in (1) by $z = e^{-t-i\xi}$, $t = i\tau$ being the (real) Euclidean time, the variables $\zeta = (\zeta_1, \ldots, \zeta_n)$, with ζ_i replacing the (undeformed) SU(2) index, and the quantum group variables $u = (u_1, \ldots, u_n)$. In particular, the (diagonal) braid invariant 4-point function can be written in the form

$$G^I(z,\zeta,u) = \eta^{\gamma_{12}}(1-\eta)^{\gamma_{23}} \sum_{\lambda=0}^{m} f^I_\lambda(\eta,\zeta) J^I_\lambda(u) \prod_{1\le i<j\le 4} z_{i,j}^{-\gamma_{i,j}} \quad ; \tag{8}$$

here $m = \min 2I_i$, $z_{i,j} = z_i - z_j$, the Möbius invariant world sheet variable η and the parameters γ are

$$\eta = \frac{z_{12}z_{34}}{z_{13}z_{24}}, \ \gamma_{i,j} = \Delta_i + \Delta_j - \frac{1}{3}\sum_{s=1}^{4}\Delta_s, \ \Delta_i = \frac{I_i(I_i+1)}{k+2}; \tag{9}$$

f^I_λ satisfy the Knizhnik-Zamolodchikov (KZ) equation [10]:

$$(k+2)\frac{\partial}{\partial\eta} f^I_\lambda = (\frac{\Omega_{12}}{\eta} - \frac{\Omega_{23}}{1-\eta}) f^I_\lambda \quad , \tag{10}$$

$$\Omega_{i,j} = 2\vec{I}_i\vec{I}_j = 2I_iI_j + 2\zeta_{i,j}(I_i\frac{\partial}{\partial\zeta_j} - I_j\frac{\partial}{\partial\zeta_i}) - \zeta_{i,j}^2\frac{\partial^2}{\partial\zeta_i\partial\zeta_j} \tag{11}$$

($\vec{I}_i$ being the 3-vector of isospin generators acting on ζ_i). Assuming $|I_{12}| \leq I_{34}$, $2I_4 = m$ ($\leq 2I_i$) and setting

$$f_\lambda^I(\eta,\zeta) = \sum_{\ell=0}^{m} \eta^{\ell-\delta_{1,2;34}}(1-\eta)^{m-\ell-\delta_{2,3;14}} g_\lambda^\ell(\eta) \frac{J^\ell(\zeta)}{\ell!\,(m-\ell)!} \tag{12}$$

($\delta_{1,2;34} = \Delta_1 + \Delta_2 - \Delta_{34}$ etc.), where J_ℓ^I is the $q \to 1$ limit of J_ℓ^I and $g_\lambda^\ell(\eta)$ is an m-fold parametric integral (similar though not identical to previously considered ones [11]),

$$g_\lambda^\ell(\eta) = \int_0^\eta dt_1 \int_0^{t_1} dt_2 \ldots \int_0^{t_{\lambda-1}} dt_\lambda \int_\eta^1 dt_{\lambda+1} \int_{t_\lambda+2}^1 dt_{\lambda+2} \ldots$$
$$\ldots \int_{t_{m-1}}^1 dt_m P_\ell^\lambda(t_i;\alpha,\beta,\gamma) \qquad (0<\eta<1), \tag{13}$$

$$P_\ell^\lambda(t_i;\alpha,\beta,\gamma) = \prod_{i=1}^m t_i^\alpha(1-t_i)^\beta \prod_{i=1}^\lambda (\eta-t_i)^{\gamma-1} \prod_{j=\lambda+1}^m (t_j-\eta)^{\gamma-1} \times$$
$$\times \prod_{i<j} |t_{i,j}|^{2\delta} \sum_\sigma \prod_{s=1}^\ell t_{i_s}^{-1} \prod_{r=\ell+1}^m (1-t_{i_r})^{-1}, \tag{14}$$

$$(k+2)\alpha = I_{34} - I_{12} + 1\,, \quad (k+2)\beta = I_{14} + I_{23} + 1\,,$$
$$(k+2)\gamma - 1 = I_{12} + I_{34} = I_{14} - I_{23}\,, \quad (k+2)\delta = 1\,, \tag{15}$$

and the sum in (14) extends over all $m!$ permutations of t_i. We prove that the product ${}^t B\, \mathcal{B}$ of braid group generators acting on (z,ζ) and on u, respectely, is 1.

4. According to [11] the correlation functions of a thermal minimal model (with Virasoro central charge $c = 1 - 6\frac{(p-p')^2}{pp'}$) are obtained from the corresponding KZ correlation functions for a rational level k,

$$k+2 = \frac{p}{p'}\,, \quad (p,p'\,\text{coprimes}) \tag{16}$$

by equating the formal SU(2) variables ζ with the world sheet variables z. Applying this algorithm to the 2-point functions we recover the thermal conformal weights

$$\Delta_{2I} = I(I+1)\frac{p'}{p} - I \quad . \tag{17}$$

The braid invariance of the resulting U_q-extended chiral Green functions is then a consequence of the braid invariance of the (U_q) KZ model for

$$q = \exp(i\pi\frac{p'}{p}) \quad . \tag{18}$$

5. The single valued 2D (Euclidean) Green functions are recovered by a monodromy invariant contraction in the U_q-variables. In particular, the norm square of the 2-point function is given by the "quantum dimension"

$$\| \prod_{n=0}^{2I-1} (u_1 - q^{2I-2n} u_2) \|^2 = (-1)^{2I} [2I+1] \quad . \tag{19}$$

We note that for the unitary minimal series ($p' = p-1$) and q given by (18) and $2I \leq p-2$ the right hand side of (19) is positive, since

$$[2I+1] := \frac{q^{2I+1} - q^{-2I-1}}{q - q^{-1}} = (-1)^{2I} \frac{\sin(2I+1)\frac{\pi}{p}}{\sin\frac{\pi}{p}} \quad . \tag{20}$$

A more detailed exposition of the new results reported here is presented in [13].

I.T. would like to thank Krzysztof Gawedzki for the invitation to his Session at the Leipzig Congress.

References

1. Alekseev, A., Faddeev, L., Semenov-Tian-Shansky, M., Volkov, A.: The unravelling of the quantum group structure in the WZNW theory. Geneva, preprint CERN-TH-5981/91;
Gawedzki, K.: Classical origin of quantum group symmetries in Wess-Zumino-Witten conformal field theory. Bures-sur-Yvette preprint IHES/P/90/92;
Chu, M., Goddard, P., Halliday, I., Olive, D., Schwimmer, A.: Quantization of the Wess-Zumino-Witten model on a circle. Cambridge (U.K.) and Santa Barbara preprint DAMTP 90-41, NSF-ITP-90-230 (April 1991)
2. Ganchev, A., Petkova, V.: Phys.Lett. **B233**, 374 (1989);
Moore, G., Reshetikhin, N.: Nucl.Phys. **B328**, 557 (1989);
Pasquier, V., SaleurH.: Nucl.Phys. **B330** 523, (1990);
Fröhlich, J., Kerler, T.: *On the role of quantum groups in low dimensional local quantum field theories.* Lecture Notes in Mathematics, Springer (to be published)
3. Todorov, I.T.: Quantum Groups as symmetries of chiral conformal algebras. In: *Quantum Groups, Proc. of the 8th International Workshop on Math. Phys. at ASI, Clausthal 1989,* H.-D. Doebner, J.-D. Hennig (Eds.), Lecture Notes in Physics **370** Springer, Berlin 1990, pp. 231-277;
Hadjiivanov, L.K., Paunov, R.R., Todorov, I.T.: Nucl.Phys. B (Proc. Suppl.) **18B**, 141 (1990); Nucl.Phys. **B356**, 387 (1991)
4. Gómez, C., Sierra, G.: Phys.Lett. **B240**, 149 (1990); Nucl.Phys. B (Proc. Suppl.) **18B**, 107 (1990); Nucl.Phys. **B352**, 791 (1991)
5. Gawedzki, K.: Nucl.Phys. B (Proc. Suppl.) **18B**, 78 (1990)
6. Mack, G., Schomerus, V.: Comm.Math.Phys. **134**, 139 (1990);

Quasi Hopf quantum symmetry in quantum theory. Hamburg preprint DESY 91-037

7. Furlan, P., Stanev, Ya.S., Todorov, I.T.: Coherent state operators and n-point invariants for $U_q(sl(2))$. Bures-sur-Yvette & Orsay preprint IHES/P/91/28, IPNO/TH91-16; Lett.Math.Phys. (to be published)
8. Cappelli, A., Itzykson, C., Zuber, J.-B.: Nucl.Phys. **B280**, 445 (1987)
9. Rehren, K.-H., Schroer, B.: Nucl.Phys. **B312**, 715 (1989)
10. Knizhnik, V.G., Zamolodchikov, A.B.: Nucl.Phys.**B247**, 83 (1984); Zamolodchikov, A.B., Fateev, V.A.: Yad.Fiz **43**, 1031 (1986) (English transl.: Sov.J.Nucl.Phys. **43**, 657 (1986))
11. Christe, P., Flume, R.: Nucl.Phys. **B282**, 466 (1987)
12. Furlan, P., Ganchev, A., Paunov, R., Petkova, V.: Reduction of the rational spin $sl(2,\mathbb{C})$ WZNW conformal theory. SISSA & Clausthal preprint SISSA-67/91/FM, KA-THEP-1991-3
13. Stanev, Ya.S., Todorov, I.T., Hadjiivanov, L.K.: Quantum group extended chiral current algebra models. (To be submitted to Phys.Lett.B.)

Section 13

General Relativity and Classical Field Theory

Organizer: G. Neugebauer

Quasi-Normal Modes of the Schwarzschild Space-Time

Bernd G. Schmidt

Max-Planck-Institut für Astrophysik, Karl-Schwarzschild-Str.1,
8046 Garching bei München, Fed.Rep. of Germany

1 Normal Modes Versus Quasi-Normal Modes

"Loosely speaking, the black hole vibrates around spherical symmetry in a quasi-normal mode, and the mode is slowly damped by gravitational radiation." With this sentence W.Press [1] introduced 1971 the notion of "quasi-normal modes" into the physics of black holes. The purpose of this note is to clarify the mathematical meaning of that notion. [1] This seems to be useful, because in the literature there is some confusion about which properties of normal modes are shared by quasi-normal ones, and which are not. An example of this confusion is the question about the completeness of the quasi-normal modes.

Compact linear oscillating systems like finite strings, membrans or cavities have preferred time harmonic states of motion if dissipation is neglected. There is always a countable number of such solutions, characterized by normal mode frequencies ω_n and by the corresponding eigenfunctions, which describe the spatial pattern of the time harmonic motion. Any motion of the system is a — in general infinite — superposition of harmonic motions, which means that the eigenfunctions form a complete system.

Mathematically, this happens because the equation has the structure

$$\frac{\partial \phi}{\partial t} = \mathbf{A}\phi, \tag{1}$$

where $\mathbf{A}$ is a differential operator, acting only on the spatial variables, which is selfadjoint with pure point spectrum on a Hilbert space because of the boundary conditions dictated by the physical problem. As a prototype one can consider the motion of a string whose both ends are fixed. The energy in the system stays there forever " travelling forwards and backwards ".

[1] This note is based on work done with H.-P. Nollert [2]

Completely different is the behaviour — and consequently the mathematical properties — of infinite systems which loose energy to infinity. Examples are waves propagating on an infinite string, or electromagnetic waves reflected at a metallic body. Let us discuss and compare for clarity the following two problems:

1.) Waves described by the scalar wave equation in a spherical cavity which is perfectly reflecting. ($\Phi = 0$ at the boundary.)
2.) Waves outside a perfectly reflecting sphere.

For simplicity we consider only $l = 1$ spherical harmonic angular behaviour.

The general spherically symmetric solution of the scalar wave equation which is regular (smooth) at the origin is

$$\frac{1}{r}(f(t+r) - f(t-r)). \tag{2}$$

Differentiation with respect to x gives the general $l = 1$ solution regular at the origin with the radial part

$$\Phi(t,r) = \frac{1}{r}(f'(t+r) + f'(t-r)) - \frac{1}{r^2}(f(t+r) - f(t-r)). \tag{3}$$

We try to find solutions with harmonic time behaviour which vanish at $r = 1$. The ansatz $f(z) = e^{i\omega z}$ leads to the condition

$$\omega = \tan(\omega). \tag{5}$$

This equation has an infinite number of solutions ω_n. The corresponding time harmonic solutions of the wave equation are

$$\Phi_n = e^{i\omega_n t}\left(\frac{\omega_n}{r}cos(\omega_n r) - \frac{1}{r^2}sin(\omega_n r)\right). \tag{6}$$

Any solution of the problem is a superposition of these solutions, which is a consequence of the completeness of the eigenfunctions.

In the second case the situation is quite different. Since we want a solution of the wave equation outside the sphere $r = 1$, the regularity at the center is not needed, and the general solution is

$$\phi(t,r) = \frac{1}{r}(A'(t-r) + B'(t+r)) + \frac{1}{r^2}(A(t-r) - B(t+r)). \tag{7}$$

The boundary condition that ϕ vanishes at $r = 1$ implies

$$0 = A'(t-1) + B'(t+1)) + A(t-1) - B(t+1). \tag{8}$$

We see that only one of the two functions can be chosen freely. If we choose B, then A is determined as a solution of the following linear differential equation

$$A'(t) + A(t) = -B'(t+2) + B(t+2) = f(t), \tag{9}$$

whose general solution is

$$A(t) = e^{-t}\left(\int_0^t e^{t'} f(t')dt' + A(0)\right). \tag{10}$$

The function $B(t)$ describes a wave coming in from " past null infinity ". This wave is modified by the reflection at the sphere and goes out to "future null infinity "as a wave described by $A(t)$.

Choose $B(t)$ such that $f(t)$ vanishes outside the intervall $2 < t < c$. This is an incoming pulse of finite duration. Then we have to put $A(0) = 0$ to obtain the solution of finite energy. (Otherwise the solution would grow exponentially in the past.) There is a finite pulse coming in, and a pulse of infinite duration going out. Inserting (10) into (7) we observe that the solution is for large t

$$\phi(t,r) = \alpha e^{-t} e^{r}\left(-\frac{1}{r} + \frac{1}{r^2}\right) \quad , \quad \alpha = \int_0^\infty e^{t'} f(t')dt' \, . \tag{11}$$

Note that the solution (11) is independent of the particular structure of the incoming wave, which is characterised by the function $B(t)$. The constant α however depends on the function $B(t)$.

This solution represents a "quasi-normal mode " with frequency $\omega = -1$. In this particular example there exists just one quasi-normal mode, and furthermore it is real, so that the solution is not oscillatory but just damped. (If we would treat the case $l = 2$, there would be an $-2/3 \pm i\sqrt{3}$ in the exponent; for higher l there are l quasi-normal modes.) In other cases (more complicated obstacles) there may be an infinite number.

The radial part of function in (11) is in a way analogous to the eigenfunctions of the cavity. It is this function which is defined by the usual "quasinormal mode boundary conditions" $r\phi \sim e^{-t+r}$ for $r \to \infty$, which characterizes " outgoing waves ". The physical meaning of the quasi normal modes is that they describe the asymptotic behaviour of the waves as $t \to \infty$ at fixed r.

If we ask for solutions with $e^{i\omega t}$ we will find such a solution for an arbitrary ω. Only for $i\omega = -1$ the radial part will be "outgoing at infinity" as in (11).

What we describe in this example is true in more general cases. The scattering theory by Lax and Phillips [3] demonstrates the following for the reflection of waves at quite general static obstacles in Minkowski space: there is a finite or infinite number of "quasi-normal modes" — this terminology is not used in the mathematical literature — which appear as poles of the scattering matrix and govern the asymptotic behaviour for $t \to \infty$ at fixed r. A theorem by Lax and Phillips shows that these numbers can be calculated as solutions of the Helmholtz equation, which satisfy the quasi- normal mode boundary condition.

The question of completeness of the quasi normal modes is never an adequate one. In the case discussed, the $l = 1$ scattering at the sphere there exists just one quasi normal mode. (Furthermore the analogues of the eigenfunctions grow in general at infinity and are not elements of the Hilbert spaces of finite energy solutions.) It is only the late-time asymptotic expansion which is determined by these solutions. For generic data the mode with the weakest damping dominates this behaviour.

2 Definition of Quasi-Normal Modes

When studying perturbations on the static part of the Schwarzschild space-time, the following master equation has been found to describe scalar as well as electromagnetic or gravitational fields:

$$\frac{\partial^2}{\partial t^2} r\phi_l + \mathbf{A} r\phi_l = 0. \tag{12}$$

Here $\Phi(t, x, \theta, \phi) = \phi_l(t, x) Y_{lm}(\theta, \phi)$ is the perturbing field, where the spherical harmonics Y_{lm} appropriate to the type of perturbation have to be used. The radial coordinate is $x = r + ln\ (r - 1)$, i.e. the horizon is at $x = -\infty$.

A is the following differential operator:

$$\mathbf{A} = -\frac{\partial^2}{\partial x^2} + \left(1 - \frac{2M}{r}\right)\left(\frac{2M\sigma}{r^3} - \frac{l(l+1)}{r^2}\right), \tag{13}$$

where r is a function of x. σ denotes the type of the perturbation, namely scalar (1), electromagnetic (0) and gravitational (-3).

The theory of partial differential equations shows that smooth initial data $\Phi(0, x, \theta, \phi)$ and $\frac{\partial}{\partial t}\Phi(0, x, \theta, \phi)$ determine a unique, smooth solution $\Phi(t, x, \theta, \phi)$ for all t and (x, θ, ϕ) . The treatment as an evolution equation on the Hilbert space of data of finite energy by Dimock and Kay [4] shows that there are no normal modes. The spectrum of **A** is continuous. This is plausible because an eigensolution would define in space-time a solution of finite energy which is periodic in time. Such a solution is identically zero!

To find and define the quasi-normal modes we proceed as follows. All solutions $\phi_l(t, x)$ with data of compact support are bounded in space-time (Kay and Wald [5]). Therefore $\phi_l(t, x)$ has a unique Laplace transform

$$\hat{\phi}_l(s, x) = \int_0^\infty e^{-st}\ \phi_l(t, x) dt, \tag{14}$$

which is an analytic function of $s = s_0 + is_1$ for $s_0 > 0$ and satisfies the differential equation

$$s^2 r\hat{\phi}_l + \mathbf{A} r\hat{\phi}_l = J_l, \tag{15}$$

where $J_l(x)$ is determined by the data at $t = 0$ (in the following we shall omit the subscipt l).

The function $f = r\hat{\phi}$ satisfies the following ordinary differential equation:

$$f''(x) - s^2 f(x) - V(x)f(x) = j(s,x) \tag{16}$$

with the potential

$$V(x) = \left(1 - \frac{2M}{r}\right)\left(\frac{l(l+1)}{r^2} + \frac{2M\sigma}{r^3}\right). \tag{17}$$

Without loss of generality we consider only Cauchy data for which the field vanishes at $t = 0$. Then j depends only on x .

As the solution in space-time is uniquely determined by the data, we have to find a unique solution of the inhomogeneous equation (16). Furthermore we know that the solution we seek is bounded because the space-time solution is bounded. Any two linearly independent solutions $f_+(s,x)$, $f_-(s,x)$ of the *homogeneous* equation define a particular Greens function for the *inhomogeneous* equation.

$$G(s,x,x') = \frac{1}{W(s)}\begin{cases} f_-(s,x')f_+(s,x) & x' < x \\ f_-(s,x)f_+(s,x') & x' > x\ , \end{cases} \tag{18}$$

where $W(s)$ is the Wronskian of f_- and f_+. The solution of the inhomogeneous equation (16) is then

$$f(s,x) = \int_{-\infty}^{\infty} G(s,x,x')j(s,x')dx'. \tag{19}$$

We see that $f(s,x)$ can only be bounded if f_+ is bounded in $x > x_0$ and f_- in $x < x_0$ for some x_0 . The explicitely known asymptotic behaviour of the homogeneous solutions [6] shows that for real s there is a unique f_+ and a unique f_-, which therefore define the correct Greens function. This Greens function has a unique analytic extension onto the halfplane $s_0 > 0$. Further analytic extension to a part of the left halfplane is possible as long as the Wronskian does not vanish. *We define the quasi-normal modes as the isolated zeros of the Wronskian.* For such a value of s, f^+ and f_- are linearly dependent and satisfy the condition of being outgoing at $+\infty$ and ingoing at $-\infty$. (Note that this property cannot be used as a definition, because s is not real, as was demonstrated explicitely in [7].)

The meaning of these solutions is not completely clear. In cases in which the Lax - Phillips - scattering theory is applicable, the quasi-normal mode with the largest real part describes the ultimate decay in time of any solution at a fixed point in space. The corresponding "quasi-normal mode eigenfunction" describes the ultimate form of the damped waves in a fixed finite spatial interval. This form is independent on the particular initial data. Numerical solutions indicate that for some intermediate time the quasi-normal mode also in the Schwarzschild case describes some universal exponential decay before ultimately the polynomial decay takes over.

References

1 Press, W.H.: Ap,J, 170: L105-L108 (1971)

2 Nollert, H.-P., Schmidt, B.G.: To appear

3 Lax, P.D., Phillips, R.S.: *Scattering Theory.* Academic Press, New York (1967)

4 Dimock, J., Kay, B.S.: Ann.Phys. **175** (1987)

5 Kay, B.S., Wald, R.M.: Classical Quantum Gravity **4** (1987)

6 Persides, S.: J.Math.Phys. **14** (1971)

7 Nollert, H.-P.:In: Proceedings of the Fourth Marcsel Grossmann Meeting on General Relativity. Ed. R.Ruffini, Science Publishers B.V.(1986)

The Initial Value Problem for Self-Gravitating Fluid Bodies

Alan D. Rendall

Max-Planck-Institut für Astrophysik, Karl-Schwarzschild-Str. 1,
8046 Garching bei München, Fed. Rep. of Germany

1 Local Existence

Surprisingly little is known concerning the initial value problem for material bodies within the Newtonian and Einsteinian theories of gravity. An understanding of this question has an obvious significance for the study of the problem of motion in these two theories. The following is a summary of some new results concerning the case where the matter model chosen is a perfect fluid. The fact that fluid *bodies* are being described will be encoded by the assumption that the matter density has spatially compact support; it is important to note that replacing this by the requirement that the density tends to zero at infinity at each fixed time does not lead to a significant simplification. In this section results concerning local in time existence will be discussed while in the following section some global statements will be collected.

Recall that the Euler equations (relativistic or non-relativistic) for a perfect fluid can be written as a symmetric hyperbolic system *provided the density is strictly positive*. It follows that the local in time existence of a solution of the Einstein-Euler equations for initial data where the density ρ is bounded below by a positive constant C is easily established. However the study of fluid bodies necessarily involves situations where this does not hold. Apparently the only general result available in the Newtonian case is due to Makino[1]. It depends on the assumptions that the equation of state is polytropic, $p = K\rho^{\frac{n+1}{n}}$, for some positive real numbers K and n and that the quantity $w = \sqrt{4Kn(n+1)}\rho^{\frac{1}{2n}}$ is regular for the initial data. Makino observed that if w is used as a basic variable instead of ρ the Euler equations can be written as a symmetric hyperbolic system which remains regular for $\rho = 0$. In the Newtonian case the system to be solved is the Euler-Poisson system which is a mixture of hyperbolic and elliptic equations. A local existence theorem can be proved by means of an iteration where the Euler and Poisson equations are solved alternately[1]. In solving the Euler equation the regularisation provided by the use of w must be used.

I have recently shown that the work of Makino can be generalised to the relativistic case[2]. The equation of state used was formally identical to the Newtonian polytropic one. Such an equation of state is physically unrealistic at high pressures in the relativistic case since the speed of sound can exceed the speed of light. However this is irrelevant for the local existence theorem since evolving data where the density is not too large for a short time will result in a solution where the density is not too large. If, on the other hand, the aim is to study the formation of singularities as will be done in the next section it is important how the density behaves at high pressures. Thus it is worth mentioning here that a local existence theorem can also be obtained for the equation of state given parametrically by

$$\rho = r + Knr^{\frac{n+1}{n}} , \tag{1}$$

$$p = Kr^{\frac{n+1}{n}} . \tag{2}$$

For this equation of state, which was taken from [3], the speed of sound tends to $\sqrt{1/n}$ as $\rho \to \infty$ in units where the speed of light is 1 and so provided $n \geq 1$ the speed of sound is always less than the speed of light. In this case defining $w = 2nr^{\frac{1}{2n}}$ gives a regularisation and a local existence theorem can be obtained. Note that the relativistic case is in a certain sense simpler than the Newtonian case because all equations are (essentially) hyperbolic.

The above results guarantee the existence of solutions describing relativistic or non-relativistic self-gravitating fluid bodies which are rather general. (In particular they need not have any continuous symmetries.) Unfortunately they all have the property that the boundary of the support of ρ is freely falling. In the relativistic case this means that the flow lines of the fluid contained in the boundary are geodesic. As a consequence the solutions obtained using the regularisation technique probably constitute only a small subset of physically interesting fluid bodies. Nevertheless these solutions themselves seem to be physically reasonable and the relativistic ones can be used to model sources of gravitational radiation. It should also be mentioned that the method of proof does not provide a uniqueness statement for the solution in terms of the physically relevant initial data. Solutions of the regularised system are uniquely determined by certain data but these include quantities which have no physical meaning.

2 Global Non-Existence

In this section cases will be discussed where solutions describing fluid bodies fail to exist globally in time. Note first that the regularisation technique can also be applied to obtain local existence theorems for fluid bodies in the absence of gravitation. The global behaviour of such solutions was discussed in [4]. The conclusion is that for initial data where the boundary is initially at rest the corresponding solution never exists globally in time. The proof is as follows. Define

$$E = \int \left(\frac{1}{2}\rho|v|^2 + np \right) , \tag{3}$$

$$J = \frac{1}{2} \int \rho|x|^2 . \tag{4}$$

Then $dE/dt = 0$ and $d^2J/dt^2 \geq cE$ for a positive constant c. Thus in the case of global existence $J \to \infty$ for $t \to \infty$. On the other hand it can be shown that if the boundary is initially at rest it remains at rest and this implies that J is uniformly bounded. Hence global existence would lead to a contradiction. A singularity must be formed but the proof gives no useful information as to its nature.

It will now be shown that variations on this proof can be used to obtain a number of other results on the formation of singularities in fluids. In fact the same pattern of argument occurs in non-existence proofs in very different contexts. The basic idea is the following. First choose a suitable 'density' for the system under consideration and form from it the quantity J. Now compare the second time derivative of J with the energy. This idea was used a long time ago by Jacobi to prove a result on instability in the n-body problem. It was applied by Glassey[5] to nonlinear Schrödinger equations and in [6] the relativistic Vlasov-Poisson system was treated with its help.

The result of [4] will now be generalised to the special relativistic case. Define

$$E = \int (T^{00} - ru^0) , \tag{5}$$

$$J = \int T^{00}|x|^2 . \tag{6}$$

Here r is a conserved density. In the case of the equation of state defined by (1) and (2) it is identical with the quantity r appearing there. For this equation of state exactly the same argument applies as in the non-relativistic case to give the same conlusion. When trying to adapt this argument to the case with gravitation two difficulties come to light. The one is that it is difficult to control the exact behaviour of the boundary of the support of the density which can no longer be made to stand still. The other is the lack of a useful conserved energy density in the general relativistic case. The first difficulty is much reduced by assuming spherical symmetry. Up to now I have found no way of overcoming the second. Consider now a spherically symmetric solution of the Euler-Poisson equation of Makino type whose boundary is initially at rest. Since the boundary falls freely the body will collapse to a point if the solution exists that long. It turns out, however, that for certain polytropic equations of state there are solutions where a singularity occurs before the collapse can be completed. In the present case the energy is defined by

$$E = \int \left(\frac{1}{2}\rho|v|^2 + np - \frac{1}{2}|\nabla\phi|^2 \right) , \tag{7}$$

where ϕ is the Newtonian gravitational potential. E is independent of time. If $n \leq 3$ then calculating the second time derivative of J, comparing with E and integrating twice with respect to time gives the inequality

$$J(t) \geq J(0) + \dot{J}(0)t + Et^2 \tag{8}$$

for $t \geq 0$. Now choose initial data with the properties that $\dot{J}(0) \geq 0$ and $E \geq 0$. Then $J(t) \geq J(0)$ for $t \geq 0$ as a consequence of (8). On the other hand if the solution collapses to a point at $t = t_c$ then $J(t) \to 0$ as $t \to t_c$. This is only consistent if a singularity develops before $t = t_c$.

As has already been mentioned, the method of proof here is of little help in indicating the nature of the singularities which occur. A plausible explanation is that a shock wave is formed. This fits very well with the usual intuition concerning shock waves. In both the cases with and without gravitation the initial data contain an initial concentration of matter which is not bounded. In the absence of gravitation this is the case because there is no force available to do the binding. It is true for the solutions with gravitation because of the positivity of the energy. Note that in that case the matter is also initially expanding on average. The restriction on the polytropic index n can usefully be compared with the generally accepted result that unstable static solutions exist for $3 < n < 5$. It is therefore to be expected that initial data of this type lead to an expanding wave. Since this wave comes into regions of lower and lower density (and in particular since the density vanishes so fast near the boundary in the Makino solutions) the conditions for the production of shock waves are ideal. However it is probable that at present the only practical way of finding out what really happens is to do a numerical investigation.

One question of particular interest is whether the non-existence results reported here depend essentially on the fact that a fluid body is being investigated or whether similar phenomena could occur for a more or less uniformly distributed fluid. The latter situation has been studied in [7] which is concerned with perturbations of the constant density, zero velocity solution for a non-relativistic fluid without gravity. The paper contains two types of results. The second type, which is concerned with formation of singularities due to initially small perturbations, is not easy to generalise to the relativistic case. The first type, which is concerned with large perturbations, fits well into the present discussion since it uses yet another variation on the technique using J and E. It can be generalised without too much difficulty to the special relativistic case and only the latter will be discussed here. The idea is to consider a perturbation initially supported in a ball of radius R where the fluid is on average outgoing. The quantity J is here infinite and so instead a quantity F is considered which is formally identical to the expression for $\dot{J}$ when the density is compactly supported. It is given by

$$F = \int T^{0a} x_a \tag{9}$$

and satisfies the differential inequality

$$\frac{dF}{dt} \geq F^2 \left[\frac{4\pi(n+1)}{3n} (R + \sigma t)^5 max\ T^{00}(0) \right]^{-1} , \tag{10}$$

where σ denotes the sound speed in the external (unperturbed) region. This is almost identical to what was obtained in [7] and can be used to show that for certain initial data the corresponding solution cannot exist globally in time. More specifically, integration of the inequality (10) gives the estimate

$$F(0)^{-1} \geq \left[\frac{4\pi(n+1)\sigma}{3n} max\ T^{00}(0) \right]^{-1} (R^{-4} - (R + \sigma t)^{-4}) . \tag{11}$$

If the initial data satisfy the condition

$$F(0) \geq \frac{4\pi(n+1)}{3n} \sigma R^4 max\ T^{00}(0) \tag{12}$$

then (11) cannot be satisfied for arbitrarily large values of t.

References

1. Makino, T.: In: Patterns and Waves. Edited by T. Nishida et. al., North Holland, Amsterdam, 1986
2. Rendall, A.D.: Preprint MPA 579, MPI für Astrophysik, Garching (1991)
3. Blanchet, L., Damour, T., Schäfer, G.: Mon.Not.R.Astron.Soc. **242**, 289 (1990)
4. Makino, T., Ukai, S., Kawashima, S.: Japan J.Appl.Math. **3**, 249 (1986)
5. Glassey, R.: J.Math.Phys. **18**, 1794 (1977)
6. Glassey, R., Schaeffer, J.: Comm.Math.Phys. **101**, 459 (1985)
7. Sideris, T.: Comm.Math.Phys. **101**, 475 (1985)

Poster Session

Program

Poster Session

Session 1: Analysis on Manifolds and Classical Mechanics

Invited/Contributed Posters

Dostoglou, S.; **Salamon, D.A.** (Coventry): "Instanton and Symplectic Homology"

Golé, Ch. (Zürich): "Symplectic Twist Map: Conley-Zehnder Theorem for General Cotangent Bundles"

Kröger, P. (Erlangen-Nürnberg): "On the Spectral Gap for Compact Manifolds"

Rogers, A. (London): "Fermionic Paths and the Atiyah-Singer Index Theorem"

General Posters

Popov, V.L.; **Tshertova, N.V.** (Tomsk): "Dissipative Extension of Gauge Theory of Plastically Incompressible Elastic-Plastic Medium"

Stanilov, G. (Sofia): "Curvature Operators in the Riemannian Geometry"

Venables, H.A. (Portsmouth): "Newton-D'Alembert Basis for Quantum Theory"

Session 2: Quantum Groups and Non-Commutative Differential Geometry

Invited/Contributed Posters

Dabrowski, L. (Trieste): "Nonstandard q-Deformation of SL(3)"

Dobrev, V.K. (Sofia): "Singular Vectors of Quantum Groups Representations for Straight Lie Algebra Roots"

Jurčo, B. (Olomouc): "Differential Calculus on Quantized Simple Lie Groups"

Lukierski, J. (Wroclaw); **Ruegg, H.** (Geneva); **Nowicki, A.** (Zielona Gora); **Tolstoy, V.N.** (Moscow): "Q-Deformation of Poincaré Algebra"

Matthes, R. (Leipzig): "A Covariant Differential Calculus on the Quantum Group C_q^n"

Osipov, E.P. (Novosibirsk): "Kac-Moody-Malcev Algebras"

Sánchez-Valenzuela, O.A. (Guanajuato): "Hamilton's Equations on (2,2)-Dimensional Superspace"

Todorov, I. (Sofia): "Simple Formula for the $U_q(sl_{r+1})$ Universal R-Matrix"

Zakrzewski, S. (Warsaw): "Poisson Lie Groups and Pentagonal Transformations"

General posters

Aneva, B. (Sofia): "A Deformed Quantum SU(2) Superalgebra"

Kanatchikov, I.V. (Moscow): "Weyl Spinor Field from the Hamilton-Jacobi Formulation of Null Strings"

Krause, J. (Santiago): "SU(2) Quantum Kinematics"

Oziewicz, Z. (Wroclaw): "An Introduction to Calculus for Associative Rings"

Samoilenko, Yu.S. (Kiev): "Representations of Quadratic and Polynomial $*$-Algebras"

Schmüdgen, K. (Leipzig): "Covariant Differential Calculi on Quantum Spaces"

Woronowicz, S.L. (Warsaw): "Solutions of the Braid Equation Related to a Hopf Algebra"

Session 3: Chaotic Quantum Systems

Invited Posters

Albeverio, S.; **Šeba, P.** (Bochum): "Wave Chaos in Quantum Systems with Point Interaction"

Benatti, F. (Trieste); **Narnhofer, H.** (Vienna); **Sewell, G.L.** (London): "Non-Commutative Arnold Cat Map: Ergodic Properties"

Jauslin, H.R.; **Lebowitz, J.L.** (New Brunswick): "Generalized Floquet Operator for Quasiperiodically Driven Quantum Systems"

Keating, J. (Bristol): "The Trace Formula and Quantum Chaology"

Klein, M.; **Knauf, A.** (Berlin): "Chaotic Motion in Coulombic Potentials"

Leboeuf, P. (Orsay); **Voros, A.** (Gif-sur-Yvette): "Chaos-Revealing Multiplicative Representation of Quantum Mechanics"

Contributed Posters

Cheremshantsev, S.E. (Leningrad): "Delocalization for Slowly Decaying Potentials Randomly Depending on Time"

Duclos, P. (Marseille); **Karner, G.** (Bielefeld): "The Schrödinger Equation on Oscillating Domains - a Solvable Example"

Steiner, F. (Hamburg): "Dynamical Zeta Functions and the Quantization of Chaos"

General Posters

Barashenkov, I.V. (Dubna); **Bogdan, M.M.** (Kharkov); **Korobov, V.I.** (Dubna): "Stability Properties of Exact Soliton Solutions of the Parametrically Driven, Damped Nonlinear Schrödinger Equation"

Session 4: Equilibrium Statistical Mechanics

Invited/Contributed Posters

Fannes, M.; **Nachtergaele, B.** (Leuven); **Werner, R.F.** (Osnabrück): "Finitely Correlated States on Quantum Spin Chains"

Fontes, L. (New York): "Continuum $1/r^2$ Ising Model - New Rigorous Results"

Borgs, Ch. (Berlin); **Kotecký, R.** (Prague); **Miracle-Solé, S.** (Marseille): "Finite-Size Effects at Asymmetric First-Order Phase Transitions"

Maes, Ch. (Leuven): "Constructive Criterion for the Ergodicity of the Interacting Particle Systems"

Mazel, A. (Moscow): "Random Surfaces with Two-Sided Constraints"

Miekisz, J. (Louvain-la-Neuve): "Low Temperature Stability of Nonperiodic Structures - a Quasicrystal Problem"

Newman, Ch. (New York): "The Wetting Transition in a Random Surface Model"

Petrina, D.Ya. (Kiev): "Exactly Solvable Models of Quantum Statistical Mechanics and Spaces of Translation Invariant Functions"

Dorlas, T.C.; **Lewis, J.T.**; **Pulé, J.V.**; **Van den Berg,** M. (Dublin): "Condensation in Models of Systems of Interacting Bosons"

Rybko, A.N. (Moscow): "Phase Transitions for Some Classes of Infinite Queuing Network"

van Enter, A.C.D. (Groningen); **Fernández, R.** (Zürich); **Sokal, A.D.** (New York): "Problems with Block-Spin Transformations"

Zagrebnov, V.A. (Dubna): "Product Formula for Gibbs Semigroups"

General Posters

Brankov, I.G.: "Critical Free Energy of a Möbius Strip"

Jüttner, H.-U.; **Huber, A.** (Kiel): "Temperature Dependence of the Thermodynamic Density Functional for One-Matrices"

Koukiou, F. (Palaiseau);**Petritis, D.** (Rennes);**Zahradnik, M.** (Prague): "Low Temperature Phase Diagram for a Class of Finite Range Interactions on the Penrose Lattice"

Messer, J. (Göttingen): "Equilibrium Statistical Mechanics of Terrestrial Ecosystems"

Nazin, G.I. (Tyumen): "Branches of Solutions of the Vlasov Equation"

Oberaigner, E.R.; **Fischer, F.D.** (Leoben); **Tanaka, K.** (Tokyo): "On Martensitic Phase Transformations in Temperature and Stress-Driven Processes"

Romano, S. (Pavia): "Low-Dimensional Classical Lattice Spin Systems with Long-Range Interactions: Simulation Results and Open Questions"

Session 5: Classical Dynamical Systems and Random Perturbations

Invited/Contributed Posters

Blank, M.L. (Moscow): "Chaotic Mappings and Stochastic Markov Chains"

Eizenberg, A. (Jerusalem): "Divergence of Invariant Measure under Small Random Perturbations"

Hamm, A.; **Graham, R.** (Essen): "The Quasipotential Approach to Small Random Perturbations of S-Unimodal Maps"

General Posters

Fournier, J.-D. (Nice); **Tabor, M.** (New York): "Psi-Series and their Summability for Nonintegrable Dynamical Systems"

Soloveitchik, M.R. (Heidelberg): "Entropy of the Classical Model of Brownian Motion"

Session 6: Disordered Systems

Invited Posters

Cook, J. (Julich): "Lyapunov Exponents of Large, Sparse Random Matrices and the Directed Polymers with Complex Random Weights"

den Hollander, F. (Delft); **Frigessi, A.** (Rome): "A Dynamical Phase Transition in a Caricature of a Spin Glass"

Koukiou, F. (Palaiseau): "Rigorous Results for Spin Systems with Long- and Short-Range Random Interactions"

Tirozzi, B. (Rome): "Some Rigorous Results for Neural Networks"

Wang, W.-M. (Palaiseau): "An Upper Bound on the Critical Exponent for the Diffusion Constant Near the Mobility Edge"

Contributed Posters

Englisch, H. (Leipzig); **Xiao, Y.** (Hubei); **Yao, K.** (Beijing): "Neural Networks as Perpetual Information Generators"

Figotin, A. (Charlotte): "The General Structure of Lines of a Random Steady Flow on a Lattice"

Goldsheid, I.Ya. (Bochum); **Sorets, E.**: "Lyapunov Exponents of the Schrödinger Equation with Quasi-Periodic Potential"

Mirlin, A.D.; **Fyodorov, Y.V.** (Gatchina): "Analytical Solution of the Anderson Model on the Bethe Lattice"

General Posters

Khorushenko, B.; **Shcherbina, M.** (Kharkov): "The Large-n Limit of the Random Anisotropy n-Vector Model"

Session 7: Nonequilibrium Statistical Mechanics

Invited Posters

Boldrighini, C. (Camerino): " 'Navier-Stokes' Equation for a Mechanical Model: the One-Dimensional Hard Rods"

Deuschel, J.-D. (Zürich): "Algebraic L^2 Decay of Critical Processes"

Dittrich, P. (Berlin): "Long-Time Behaviour of the Weakly Asymmetric Simple Exclusion Process"

Maes, Ch. (Leuven): "Long Range Spatial Correlations in Stationary States of Stochastic Dynamics with Conservation Law"

Pilyavsky, A.I.; **Rebenko, A.L.**; **Skrypnik, W.I.** (Kiev): "The Cluster Expansion for Weak Solutions of Bogoliubov Diffusion Hierarchy"

Contributed Posters

Busch, P. (Köln): "Geometric Characterization of Irreversibility"

Duffield, N.G. (Dublin); **Roos, H.** (Göttingen); **Werner, R.F.** (Osnabrück): "Macroscopic Limiting Dynamics of a Class of Inhomogeneous Mean Field Quantum Systems"

Gonzage de Sá, P. (Louvain-la-Neuve); **Maes, Ch.** (Leuven): "The Gac-Kurdyumov-Levin Automaton Revisited"

Quadt, R. (Köln): "Statistical Description of Irreversibility"

Zegarlinski, B. (Bochum): "Gibbsian Description and Description by Stochastic Dynamics in the Statistical Mechanics of Lattice Spin Systems with Finite Range Interactions"

General Posters

Kiessling, M.K.-H. (New York): "Statistical Mechanics of Classical Particles with Logarithmic Interactions"

Session 8: General Theory of Quantized Fields

Invited Posters

Borchers, H.J. (Göttingen): "CPT- and Lorentz-Transformations in Two Dimensional Quantum Field Theories"

Buchholz, D.; **Porrmann, M.**; **Stein, U.** (Hamburg): "Particle Weights: A Generalized Particle Concept in Local Quantum Field Theory"

Kopper, Ch. (Palaiseau): "General Theory of Quantized Fields"

Landsmann, N.P. (Cambridge): "Thermal Quantum Field Theory"

Roberts, J.E. (Osnabrück): "Spontaneously Broken Gauge Symmetries and Goldstone's Theorem"

Summers, S.J. (Gainesville): "Quantum Field Theory and Bell's Inequalities"

Contributed Posters

Brüning, E.A.K. (Cape Town): "How to Construct a Function from its Asymptotic Expansion"

Hofmann, G. (Leipzig): "An Explicit Realization of a GNS-Representation in a Krein-Space"

Pachucki, K. (Warsaw): "Time Evolution in Quantum Electrodynamics"

Ziep, O. (Berlin): "Quantum Transport and Breaking of Space-Time Symmetries"

General Posters

Hübner, M. (Leipzig): "Generalizing the Berry Phase - Do Density Matrices Interfere ?"

Reifler, F. (Moorestown); **Vogt, A.** (Washington): "The Space-Time Support of Free Particle Wave Functions"

Session 9: Nonrelativistic Quantum Mechanics (Stationary)

Invited/Contributed Posters

Ashbaugh, M.S. (Columbia); **Benguria, R.D.** (Santiago): "Proof of the Payne-Polya-Weinberger Conjecture"

Benguria, R.D. (Santiago); **Hoops, S.** (Trondheim); **Siedentop, H.** (Birmingham): "Proof of the Excess Charge Problem for an Atom in the Hellmann-Weizsäcker Model"

Boutet de Monvel-Berthier, A. (Paris); **Georgescu, V.** (Bucharest): "Some Developments of the Abstract Mourre Theory"

Demuth, M. (Berlin); **van Casteren, J.** (Antwerp): "Generalized Schrödinger Operators - Stochastic Spectral Analysis"

Duclos, P. (Marseille); **Exner, P.** (Rez): "On the Critical Thickness of Quantum Waveguides"

Hagedorn, G.A. (Blacksburg): "Electron Energy Level Crossings in the Time-Dependent Born-Oppenheimer Approximation"

Hempel, R. (München): "Spectral Properties of Neumann Laplacians on Some Singular Domains"

Holthaus, M. (Bonn): "Quasiclassical Calculations for Periodic Problems - Floquet Operators and Wavefunctions"

Joye, A.; **Kunz, H.**; **Mileti, G.**; **Pfister, Ch.-Ed.** (Lausanne): "Adiabatic Theorem of Quantum Mechanics"

Vugalter, S.; **Zhislin, G.** (Niznij Novgorod): "On the Discrete Spectrum of the N-Body Problem with Homogeneous Magnetic Field"

General Posters

Beckers, J.; **Debergh, N.** (Liege): "On Generalized Superalgebras in Nonrelativistic Quantum Mechanics"

Bose, S.K. (Gorakhpur): "The Path-Integral Technique and the Large-Quantum-Number Behaviour of Energy Eigenvalues of Coupled Anharmonic Oscillators"

Bose, S.K. (Gorakhpur): "Exact Solution of the Schrödinger Equation for the Central Nonpolynomial Potential $V(r) = r^2 + \lambda r^2/(1+gr^2)$ in N-Dimensions"

Breuer, H.P. (Freiburg); **Holthaus, M.** (Bonn): "A Semiclassical Theory of Quasienergies and Floquet Wave Functions"

Doebner, H.-D. (Clausthal-Zellerfeld); **Goldin, G.A.** (New Brunswick): "On a General Nonlinear Schrödinger Equation Admitting Diffusion Currents"

Grabowski, M. (Torun): "Homodyne Detection and Positive Operator-Valued Measures"

Hoffmann-Ostenhof, M. (Vienna): "Local Properties of Solutions of Schrödinger Equations"

Katriel, J. (Haifa): "Products of Class-Sums of the Symmetric Group: A Status Report"

Session 10: Special Systems in Quantum Field Theory

Invited Posters

Borgs, Ch. (Berlin): "Finite-Size Scaling, Surface Tension and the Vanishing Theorem in SUSY Field Theories"

Keller, G. (München): "Perturbative Renormalization via Flow Equations - Composite Operators"

Narnhofer, H. (Vienna): "Galilei Invariant Quantum Field Theory"

Salmhofer, M.; **Seiler, E.** (München): "Chiral Symmetry Breaking"

Contributed Posters

Antoine, J.-P.; **Moschella, U.** (Louvain-la-Neuve): "Poincaré Coherent States: The Two-Dimensional Massless Case"

Dimock, J. (Buffalo); **Hurd, T.R.** (Hamilton): "A Renormalization Group Analysis of Infrared Models: Dipole Gas, Coulomb Gas, QED"

Gavrilik, A.M. (Kiev): "Stiefel Sigma Model Bifurcation and Fractionalization of Effective Coupling Constants"

Dimock, J. (Buffalo); **Hurd, T.R.** (Hamilton): "The Sine-Gordon Quantum Field Theory for $\beta > 8\pi$ and $\beta < 8\pi$"

Löffelholz, J. (Leipzig): "The Vacuum of $(QED)_2$ on a Cylinder, and OS-Axioms"

Osipov, E.P. (Novosibirsk): "Two-, Four-, Eight-Dimensional Fields and Solutions of Stochastic Partial Differential Equation"

General Posters

Scholz, E.; **Timmermann, W.** (Dresden): "Positive and Completely Positive Maps in Algebras of Unbounded Operators"

Session 11: Scattering Theory, Inverse Problems

Invited/Contributed Posters

Exner, P. (Rez): "A Model of Two-Channel Scattering"

Gesztesy, F. (Columbia); **Race, D.** (Guildford); **Weikard, R.** (Birmingham): "Auto-Bäcklund Transformation for Gelfand-Dikii-Systems of Nonlinear PDE's"

Hempel, R. (München); **Weder, R.** (Mexico City): "On the Completeness of Wave Operators in Obstacle Scattering under Loss of Local Compactness"

Neidhardt, H. (Berlin); **Zagrebnov, V.** (Dubna): "Approximation and Convergence for the Schrödinger Operator"

Stefanov, P. (Sofia): "Generic Uniqueness for Two Inverse Problems in Potential Scattering"

Poerschke, Th.; **Stolz, G.** (Frankfurt/M.): "On Eigenfunction Expansions and Scattering Theory"

Šťovíček, P. (Prague): "Multi-Solenoid Aharonov-Bohm Effect"

Thaller, B. (Graz): "Dollard's Modification and Supersymmetry for the Dirac Equation"

General Posters

Dittrich, J.; **Exner, P.** (Rez): "Spherically Symmetric Dirac Hamiltonian with Interactions Supported by Concentric Spheres"

Session 12: Conformal and Topological Field Theory

Invited Posters

Kraus, E. (Hamburg); **Siebold, K.** (München): "Conformal Transformation Properties of the Energy-Momentum Tensor in Four Dimensions"

Mack, G.; **Schomerus, V.** (Hamburg): "Quantum Symmetry in Quantum Field Theory"

Narasimhan, M.S.; **Ramadas, T.R.** (Bombay): "Factorization of Generalized Theta Functions"

Wieczerkowski, Ch. (Zürich): "Topological Representations of $U_q(sl_2\mathbb{C})$"

Contributed Posters

Bogolubskaya, A.A.; **Bogolubsky,I.L.** (Dubna): "Topological Solitons in Two- and Three-Dimensional Heisenberg Magnet Models"

Nill, F. (Berlin): "A Quantum Field Theoretic Approach to Chern-Simons Theory"

Schlichenmaier, M. (Mannheim): "Algebras of Vector Fields on Riemann Surfaces (Generalized Krichever-Novikov-Algebras)"

Silvestrov, P.G.; **Yelkhovsky, A.S.** (Novosibirsk): "Non-Perturbative Parameter in the Exact Solution of 2-D Gravity via the Matrix Model"

General Posters

Nesterenko, V.V. (Dubna): "Models of Spinning Particles with Higher Derivatives and the Boson-Fermion Transmutations"

Session 13: General Relativity and Classical Field Theory

Invited/Contributed Posters

Chruśiel, P.T. (Canberra): "On Solutions of the Constraint Equations with Hyperboloidal Boundary Conditions"

Demkin, P. (Vilnius): "Solutions of the Equations of Motion and Constraints of the Bosonic-Relativistic Membrane in Orthogonal Parametrization"

Doliwa, A. (Warsaw): "Integrable Geometry of Non-Constant Gaussian Curvature"

Grigore, D.R. (Geneva): "Noetherian Symmetries in Classical Field Theories"

Herdegen, A. (Krakow): "Classical Electrodynamics of a Point Particle from Action Integral"

Jezierski, J.; **Kijowski, J.** (Warsaw): "Quasi-Local Hamiltonian for Gravitational Field"

Kanatchikov, I.V. (Moscow): "On the Caratheodory's Hamilton-Jacobi Formulation of Null Strings and its Relation to the Weyl Spinor Field: Explicit Examples"

McLenaghan, R.G. (Waterloo): "On the Validity of Huygen's Principle for Relativistic Wave Equations on Curved Space-Time"

Menotti, P. (Menotti); **Seminara, D.** (Pisa): "The Radial Gauge in $2+1$ Dimensional Gravity"

Rendall, A. (München): "The Initial Value Problem for Self-Gravitating Fluid Bodies"

Richter, O.; **Rudolph, G.** (Leipzig): "On Dimensional Reduction of Einstein-Cartan Theory"

Schmidt, B. (München): "Quasi-Normal Modes on the Schwarzschild Space-Time"

Shima, K. (Saitama): "Spin 1/2 Dirac Field as a Gauge Field"

Whiting, B. (Gainesville): "Boundary Value Problem in General Relativity"

General Posters

Barashenkov, I.V.; **Getmanov, B.S.** (Dubna); **Kovtun, V.E.** (Kharkov): "The Unified Approach to Integrable Relativistic Equations: Soliton Solutions Over Non-Vanishing Backgrounds"

Doliwa, A.; **Sym, A.** (Warsaw): "H-Constant Surfaces as an Example of Soliton Surfaces"

Levi, D. (Rome); **Sym, A.** (Warsaw): "Integrable Systems Describing Surfaces of Non-Constant Curvature"

Kanatchikov, I.V. (Moscow): "Towards the Causal Interpretation of the Weyl Spinor Field Using Null Strings"

Nikolova, L. (Sofia): "Gauge Coupling Fields in Terms of Generalized Connection Forms"

Neugebauer, G. (Jena): "Phase Transition in Gravitating Systems"

Program

Tuesday, July 30

9:15 Opening Session

Plenary Talks

Chairman: A.Jaffe (Cambridge, Mass.)

9:30-10:45 R. de la Llave (Austin): "Recent Developments in Classical Mechanics"

11:30-12:45 B. Zumino (Berkeley): "Noncommutative Differential Calculus on Quantum Spaces and Quantum Groups"

Afternoon Sessions

Section 1 (Analysis on Manifolds and Classical Mechanics)

Org.: K.D. Elworthy (Coventry)

Topical Session

14:00 C.E. Wayne
15:00 V.V. Trofimov

Poster Session

16:45 C. Golé
17:15 S. Dostoglou
17:45 P. Kröger
18:15 A. Rogers

Section 2 (Quantum Groups and Non-Commutative Differential Geometry)

Org.: S.L. Woronowicz (Warsaw)

Poster Session

14:00 J. Lukierski
14:15 S. Zakrzewski
14:25 V. Dobrew
14:40 E.P. Osipov
14:50 B. Jurčo
15:00 O.A. Sanchez-Valenzuela
15:10 I. Todorov
15:35 L. Dabrowski
15:45 R. Matthes

Topical Session

16:45 A. Jaffe
17:45 Y. Nakagami
18:30 A. Lazar

20:00 Meeting of the IAMP Executive Committee

Wednesday, July 31

Plenary Talks

9:30-10:45 B.V. Chirikov (Novosibirsk): "Chaotic Quantum Systems"
Chairman: G. Emch (Gainesville)
11:30-12:45 D. Ruelle (Bures-sur-Yvette): "Dynamical Zeta Functions: Where Do They Come From and What Are They Good For"
Chairman: D.A. Dubin (Milton Keynes)

Afternoon Sessions

Section 3 (Chaotic Quantum Systems)

Org.: M. Combescure (Paris)

Topical Session

14:00 P. Bleher
14:40 D. Voiculescu
15:20 A. Knauf

Poster Session

16:45 P. Laboeuf
17:00 J. Keating
17:15 H. Jauslin
17:30 F. Benatti
17:45 P. Šeba
18:00 G. Karner
18:15 F. Steiner
18:30 S. Cheremshantsev

Section 4 (Equilibrium Statistical Mechanics)

Org.: S.B. Shlosman (Moscow)

Poster Session

14:00 L. Fontes
14:10 Ch. Maes
14:20 A. Rybko
14:30 D.Ya. Petrina
14:40 J. V. Pule
14:50 V. Zagrebnow
15:00 M. Fannes
15:10 J. Miekisz
15:20 A. Mazel
15:30 A. van Enter
15:40 R. Kotecky

Topical Session

16:45 P. Bleher
17:15 A. Klein
17:45 S.B. Shlosman
18:15 Ch. Newman

20:00 Meeting of the IUPAP Commission C18

Thursday, August 1

Plenary Talks

9:30-10:45 L.A. Bunimovich (Bielefeld/Moscow): "Billiard Type Systems with Chaotic Behavior and Space-Time Chaos"
Chairman: J.R. Klauder (Gainesville)

11:30-12:45 L.A. Pastur (Kharkov): "Statistical Physics and Spectral Theory of Disordered Systems: Some Recent Developments"
Chairman: J. Łopuszanski (Wrocław)

Afternoon Sessions

Section 5 (Classical Dynamical Systems and Random Perturbations)

Org.: Y. Kifer (Jerusalem)

Topical Session

14:00 L.S. Young
15:00 Y. Kifer

Poster Session

16:45 M. Blank
17:25 A. Eizenberg
18:10 A. Hamm and A. Graham

Section 6 (Disordered Systems)

Org.: A. Kupiainen (New Brunswick)

Poster Session

14:00 F. den Hollander
14:20 B. Tirozzi
14:40 H. Englisch
14:50 F. Koukiou
15:00 W.-M. Wang
15:10 I. Goldsheid
15:20 A. Mirlin
15:30 J. Cook
15:40 A. Figotin

Topical Session

16:45 A. Georges
17:25 A. Klein
18:05 J. Wehr

Friday, August 2

Plenary Talks

Chairman: H. Araki (Kyoto)

9:30-10:45 F. Martinelli (Rome): "Low Temperature Stochastic Spin Dynamics: Metastability, Convergence to Equilibrium and Phase Segregation"

11:30-12:45 S.R.S. Varadhan (New York): "Entropy and Entropy Production as Tools in Establishing Hydrodynamic Limits"

Afternoon Sessions

Section 7 (Nonequilibrium Statistical Mechanics)

Org.: J. Fritz (Budapest)

Topical Session

14:00 H.-T. Yau
14:50 M. Bramson
15:30 J. Gärtner

Poster Session

16:45 C. Boldrighini
17:00 J.-D. Deuschel
17:10 P. Dittrich
17:20 Ch. Maes
17:30 A.L Rebenko
17:40 B. Zegarlinski
17:50 P. Gonzaga de Sá
18:00 H. Roos
18:10 P. Busch
18:20 R. Quadt

17:00 Round Table Discussion: "On the Foundations of Quantum Mechanics"
Chairman: A.S. Wightman (Princeton)
with S. Goldstein (New Brunswick), R. Haag (Hamburg), R. Omnés (Paris), H. Primas (Zürich), W.H. Zurek (Los Alamos).

Sunday, August 4

9:30-10:45 General Poster Session

11:30-12:45 General Poster Session

Invited Lectures

14:00-16:00 A. Connes (Bures-sur-Yvette): "About Standard Model"
Chairman: I. Todorov (Sofia)

16:45-18:45 Y.I. Manin (Moscow): "De Rham Complex in Non-Commutative Geometry and Quantum Groups"
Chairman: A. Rogers (London)

20:00-20:30 St. Thomas Church, Organ Concert with W. Thirring for Participants of the IAMP Congress

Monday, August 5

Invited Lecture

Chairman: H. Reeh (Göttingen)

11:00-12:45 L.D. Faddeev (Leningrad): "Hamiltonian Methods in Conformal Field Theory"

Seminars

"Geometric Phase"

Org.: A. Bohm (Austin)

14:00 A. Bohm
14:10 L.J. Boya
14:40 L. Dabrowski
15:00 A. Uhlmann
15:20 G. Rudolph
15:40 open for new developments

"Energy Levels and Moments of the Schrödinger Equation and Relativistic Generalizations"

Org.: A. Martin (Geneva)

14:00 H. Grosse
14:30 A. Martin
14:45 J. Stubbe
15:30 N.N. Khuri

16:45 General IAMP Assembly Meeting

18:00 Scientific Movies
P. Möbius (Dresden): "Propagation and Superposition of Nonlinear Waves (Solitons)"
W. Thirring (Wien): "Dynamics of Thermodynamically Unstable Systems"

20:00 Meeting of the IAMP Executive Committee

Tuesday, August 6

Plenary Talks

Chairman: R. Haag (Hamburg)

9:30-10:45 K. Fredenhagen (Hamburg): "On the General Theory of Quantized Fields"

11:30-12:45 J.P. Solovej (Princeton): "Atomic and Molecular Structure - A Renormalized Picture"

Afternoon Sessions

Section 8 (General Theory of Quantized Fields)

Org.: D. Iagolnitzer (Saclay)

Topical Session

14:00 J. Bros
14:30 B.S. Kay
15:00 K.H. Rehren
15:30 A.S. Wightman

Poster Session

16:45 H.J. Borchers
17:00 D. Buchholz
17:15 C. Kopper
17:30 N.P. Landsmann
17:45 J.E. Roberts
18:00 S.J. Summers
18:15 E.A.K. Brüning
18:22 K. Pachucki
18:30 O. Ziep
18:37 G. Hofmann

Section 9 (Nonrelativistic Quantum Mechanics (stationary))

Org.: I. Herbst (Washington)

Poster Session

14:00 G. Hagedorn
14:10 R. Benguria
14:20 H. Siedentop
14:30 V. Georgescu
14:40 R. Hempel
14:50 A. Joye
15:00 G. Zhislin
15:15 M. Demuth
15:25 M. Holthaus
15:35 P. Exner

Topical Session

16:45 L. Seco
17:45 M.B. Ruskai
18:15 J. Avron

Wednesday, August 7

Plenary Talks

9:30-10:45 J. Fröhlich (Zürich): "$U(1) \times U(2)$-Gauge Invariants in Non-relativistic Quantum Mechanics and the Quantum Hall Effect"
Chairman: H.J. Borchers (Göttingen)

11:30-12:45 G.M. Graf (Pasadena): "Asymptotic Completeness for N-Body Quantum Systems"
Chairman: J. Ginibre (Orsay)

Afternoon Sessions

Section 10 (Special Systems in Quantum Field Theory)

Org.: E. Seiler (München)

Topical Session

14:00 P. Faria da Veiga
15:00 H. Grosse

Poster Session

16:45 H. Narnhofer
17:00 G. Keller
17:15 C. Borgs
17:30 M. Salmhofer
17:45 J. Löffelholz
17:55 E. Osipov
18:05 J.P. Antoine
18:15 A.M. Gavrilik
18:25 J. Dimock
18:35 T.R. Hurd

Section 11 (Scattering Theory, Inverse Problems)

Org.: V. Enss (Aachen)

Poster Session

14:00 P. Stefanov
14:15 R. Weikard
14:30 P. Exner
14:45 H. Neidhardt
15:00 B. Thaller
15:15 G. Stolz
15:30 P. Stovicek
15:45 R. Hempel

Topical Session

16:45 A.I. Nachman
17:30 D. Robert
18:15 A. Jensen

Thursday, August 8

Plenary Talks

9:30-10:45 R. Jackiw (Cambridge, Mass.): "Chern-Simons Dynamics"
Chairman: L.D. Faddeev (Leningrad)

11:30-12:45 S. Klainerman (Princeton): "On the Mathematical Theory of Classical Fields and General Relativity"
Chairman: A.G. McLenaghan (Waterloo)

Afternoon Sessions

Section 12 (Conformal and Topological Field Theory)

Org.: K. Gawedzki (Bures-sur-Yvette)

Topical Session

14:00 E. Verlinde
14:55 A. Wipf
15:35 I. Todorov

Poster Session

16:45 Ch. Wieczerkowski
17:05 V. Schomerus
17:25 T.R. Ramadas
17:45 E. Kraus
18:05 I.L. Bogolubsky
18:15 M. Schlichenmaier
18:25 A. Yelkhovsky
18:35 F. Nill

Section 13 (General Relativity and Classical Field Theory)

Org.: G. Neugebauer (Jena)

Poster Session

14:00 K. Shima
14:10 J. Kanatchikov
14:20 J. Jezierski and J. Kijowski
14:30 A. Doliwa
14:40 D.R. Grigore
14:50 A. Herdegen
15:00 O. Richter and G. Rudolph
15:10 P. Menotti and D. Seminara
15:20 B. Whiting

Topical Session

15:30 P. Demkin

16:45 R.G. McLenaghan
17:15 B. Schmidt
17:45 P.T. Chruściel
18:15 A. Rendall

Friday, August 9

Invited Lecture

Chairman: W. Thirring (Wien)

9:30-11:00 J. Cuntz (Heidelberg): "A Survey of Some Results in Operator Algebras and Non-Commutative Geometry"

11:00 Closing of the Congress